Portland Community College

HANDBOOK OF NUTRITION AND PREGNANCY

NUTRITION AND HEALTH

Adrianne Bendich, PhD, FACN, SERIES EDITOR

Recent Volumes

HANDBOOK OF NUTRITION AND PREGNANCY

Edited by

CAROL J. LAMMI-KEEFE, PhD, RD

Louisiana State University College of Agriculture,
AgCenter, and Pennington Biomedical Research Center,
Baton Rouse, LA

SARAH C. COUCH, PhD, RD

Department of Nutritional Sciences, University of Cincinnati
Cincinnati, OH

ELLIOT H. PHILIPSON, MD

Maternal Fetal Medicine, Cleveland Clinic Lerner College of Medicine
Cleveland, OH

Foreword by
E. ALBERT REECE, MD, PhD, MBA

School of Medicine, University of Maryland
College Park, MD

※ Humana Press

Editors

Carol J. Lammi-Keefe
Louisiana State University
   College of Agriculture
AgCenter, and Pennington
   Biomedical Research Center
Baton Rouse, LA
USA

Sarah C. Couch
   Department of Nutritional Sciences
University of Cincinnati
   Cincinnati, OH
USA

Elliot H. Philipson
Maternal Fetal Medicine
Cleveland Clinic Lerner
   College of Medicine
Cleveland, OH
USA

Series Editor
Adrianne Bendich
GlaxoSmithKline Consumer Healthcare
Parsippany, NJ
USA

ISBN: 978-1-58829-834-8   e-ISBN: 978-1-59745-112-3
DOI: 10.1007/978-1-59745-112-3

Library of Congress Control Number: 2008920315

Cover illustration: Drawing created by David C. Lovelace

Printed on acid-free paper

9 8 7 6 5 4 3 2 1

springer.com

Dedications

To Christopher and Liam—joys of my life—from little boys to young men. Find your passion and make it *your* life.

Carol J. Lammi-Keefe

To my husband Peter and my son Paul. May you too be inspired to follow your dreams.

Sarah C. Couch

To Sandy, Rebecca, and Julia. Thank you for nurturing my life and my work, and for always providing me with food for thought.

Elliot H. Philipson

Series Editor's Introduction

The *Nutrition and Health*™ series of books has an overriding mission to provide health professionals with texts that are considered essential because each includes: (1) a synthesis of the state of the science; (2) timely, in-depth reviews by the leading researchers in their respective fields; (3) extensive, up-to-date, fully annotated reference lists; (4) a detailed index; (5) relevant tables and figures; (6) identification of paradigm shifts and the consequences; (7) virtually no overlap of information between chapters, but targeted, inter-chapter referrals; (8) suggestions of areas for future research; and (9) balanced, data-driven answers to patient–health professionals' questions, which are based on the totality of evidence rather than the findings of any single study.

The series volumes are not the outcome of a symposium. Rather, each editor has the potential to examine a chosen area with a broad perspective, both in subject matter as well as in the choice of chapter authors. The international perspective, especially with regard to public health initiatives, is emphasized where appropriate. The editors, whose trainings are both research and practice oriented, have the opportunity to develop a primary objective for their book, define the scope and focus, and then invite the leading authorities from around the world to be part of their initiative. The authors are encouraged to provide an overview of the field, discuss their own research, and relate the research findings to potential human health consequences. Because each book is developed *de novo*, the chapters are coordinated so that the resulting volume imparts greater knowledge than the sum of the information contained in the individual chapters.

Handbook of Nutrition and Pregnancy, edited by Carol J. Lammi-Keefe, Sarah C. Couch, and Elliot H. Philipson, is a very welcome addition to the *Nutrition and Health* series and fully exemplifies the series' goals. This volume is especially timely since it includes in-depth discussions relevant to the changing health status of women of child-bearing potential around the world. As but one example, there is an extensive chapter on the obesity epidemic that continues to grow even in underdeveloped nations; the chapter includes an analysis of the comorbidities, such as gestational diabetes and related adverse pregnancy outcomes that continue to be seen in increased numbers annually. As indicated by E. Albert Reece, MD, PhD, MBA, in the volume's Foreword, the editors have "...assembled 23 superb chapters on the latest, evidence-based approaches for managing the nutritional requirements of pregnant women in a variety of settings."

This volume has been given the title of handbook because of its inclusive coverage of virtually all of the relevant topics including, but not limited to, the role of nutritional status prepregnancy, during pregnancy, and afterwards; body composition; usual and recommended dietary intakes and intakes in those with eating disorders; dietary components and alternative dietary patterns including vegetarianism and vegan diets; drug–nutrient and drug–supplement interactions; bariatric surgery and pregnancy outcomes; adolescent pregnancy and multifetal pregnancy; pregnancy in HIV-infected women; pregnancy complications including preeclampsia; and the nutritional needs of the lactating woman and her nutritional needs postpartum, whether or not she is breastfeeding. This text is the first to synthesize the knowledge base for the health provider who is counseling both the woman anticipating pregnancy as well as the pregnant woman concerning diet, popular diets and diet supplements, and diet components and their effects on gastrointestinal function. Likewise, this volume contains valuable information for the health provider about the nutritional requirements following pregnancy. In addition to an expected single chapter on specific nutrients such as iron and folate, these essential nutrients are discussed in two chapters from the viewpoints of pregnancy in developed compared to underdeveloped countries, and thus these contrasting chapters will be of great value to the graduate student and academic researcher as well as the practicing nutritionist. Two examples of novel chapters that are unique to this volume include a review of postpartum depression and the nutrients that may be of benefit and a chapter on the role of flavors and fragrances on the fetus and their effects on food preferences later in life. Several chapters contain extensive lists of relevant Internet resources and screening tools that could be implemented in an office setting. Thus, this volume contains valuable information for the practicing health professional as well as those professionals and students who have an interest in the latest, up-to-date information on the full spectrum of data on nutrition and pregnancy and its implications for human health and disease.

The editors of this comprehensive volume are internationally recognized authorities on the role of nutrition in the health of women of childbearing potential and each provides both a practice as well as research perspective. Carol Lammi-Keefe, PhD, is Alma Beth Clark Professor and Division Head, Human Nutrition and Food at the School of Human Ecology, and also has an Adjunct Faculty appointment at the Pennington Biomedical Research Center, Louisiana State University (LSU) in Baton Rouge, LA. Prior to moving to LSU, Dr. Lammi-Keefe served as Professor and Head of the Department of Nutritional Sciences at the University of Connecticut, Storrs, CT. Sarah C. Couch, PhD, is Associate Professor and Chair of Undergraduate Studies in the Department of Nutritional Sciences, College of Allied Health Sciences, University of Cincinnati, Cincinnati, OH. Eliot H. Philipson, MD, currently serves as Vice Chairman, Department of Obstetrics and Gynecology, and Head of the Section of Obstetrics and Maternal–Fetal Medicine, at the Cleveland Clinic Lerner College of Medicine, Cleveland, OH. He is a Diplomate of the American Board of Obstetrics and Gynecology as well as a Diplomate of the American Board in Maternal–Fetal Medicine. The editors are excellent communicators, and they have worked tirelessly to develop a book that is destined to be the benchmark in the field because of its extensive, in-depth chapters covering the most important aspects of the complex interactions between cellular functions, diet and fetal development, and the impact of maternal health and disease states on both optimal pregnancy outcomes and enhanced maternal health and well-being.

The introductory chapters provide readers with the basics so that the more clinically related chapters can be easily understood. The editors have chosen 42 of the most well recognized and respected authors to contribute the 23 informative chapters in the volume. Hallmarks of all of the chapters include complete definitions of terms, with the abbreviations fully defined for the reader and consistent use of terms between chapters.

Key features of this comprehensive volume include the informative abstract and key words that are at the beginning of each chapter; more than 100 detailed tables and informative figures; an extensive, detailed index; and more than 1,500 up-to-date references that provide the reader with excellent sources of worthwhile information about diet, nutrition, and pregnancy.

In conclusion, *Handbook of Nutrition and Pregnancy*, edited by Carol J. Lammi-Keefe, Sarah C. Couch, and Elliot H. Philipson, provides health professionals in many areas of research and practice with the most up-to-date, well-referenced volume on the importance of nutrition in determining the potential for optimal pregnancy outcome. This volume will serve the reader as the benchmark in this complex area of interrelationships between preconception nutritional status; nutritional needs during pregnancy for those at low as well as high risk for adverse outcomes; postpregnancy nutritional recommendations in both lactating and nonlactating mothers; exercise needs for women during their childbearing years; dietary intakes, micronutrient requirements, global issues such as obesity and HIV infections, and the functioning of the human body during these transitions. Moreover, these interactions are clearly delineated so that students as well as practitioners can better understand the complex interactions. The editors are applauded for their efforts to develop the most authoritative resource in the field of nutrition and pregnancy to date, and this excellent text is a very welcome addition to the *Nutrition and Health* series.

Adrianne Bendich, PhD, FACN
Series Editor

Foreword

I have spent the past several decades investigating the consequences of what happens to infants when their mothers have an imbalance of nutrients, such as glucose, during pregnancy. My laboratory and others have discovered a number of biochemical changes that result from chronically high blood glucose levels related to obesity and diabetes during pregnancy, all of which strongly correlate with damage to the developing fetus.

One of the most important lessons we have learned over the years is that nutrition-related malformations are much easier (and less costly) to prevent than to treat after the fact. Inexpensive folate supplementation, for example, has been shown to significantly reduce the risk of neural tube defects, a neurological birth defect that often has devastating consequences for both the infant and mother.

On the other hand, there is now overwhelming evidence that when it comes to providing optimal nutrition during pregnancy, there is no "one-size-fits-all" approach. For example, we now know that women who are overweight and obese have a higher need for folate supplementation than women of normal weight. Similarly, studies have shown that pregnant women in developing countries or women with HIV often have their own unique nutritional challenges.

It is for the above reasons that it is critically important for healthcare professionals who treat pregnant women to be well-versed in the latest information on proper nutritional support for a broad spectrum of pregnancies. In *Handbook of Nutrition and Pregnancy*, co-editors Carol J. Lammi-Keefe, PhD, RD, at Louisiana State University, Sarah C. Couch, PhD, RD, at the University of Cincinnati, and Elliot H. Philipson, MD, at Cleveland Clinic, have significantly simplified that process by assembling 23 superb chapters on the latest, evidence-based approaches for managing the nutritional requirements of pregnant women in a variety of settings.

In addition to the expected sections (1 and 2) on nutritional requirements during normal and high risk pregnancy, section 3 is devoted to special diets, such a vegetarian and vegan diets, and the potential risks/benefits of using selected nutrients and supplements during pregnancy. Section 4 is devoted to special nutritional requirements during the postpartum period, including the role of nutritional factors in post-partum depression, the number one complication of pregnancy.

The last section (5) is devoted to special issues surrounding the developing world, including the consequences of women transitioning from traditional diets to more western diets. This section also includes an important chapter on nutrition and maternal survival in developing countries and discusses the latest science on the consequences of iron and micronutrient deficiencies on birth outcomes in many regions of the world. The final chapter deals with micronutrient status and pregnancy outcomes in HIV-infected women, the fastest growing population of people infected by the AIDS virus.

A good start in life is important, and maternal nutritional status during pregnancy has repeatedly been demonstrated to be associated with pregnancy outcomes for the infant. This Handbook is designed to make that "good start on life" possible for more and more children by giving doctors, nurses, dieticians, and other health care professionals the tools they need to manage the range of pregnancies they are likely to encounter.

E. Albert Reece, MD, PhD, MBA
Vice President for Medical Affairs, University of Maryland
The John Z. and Akiko K. Bowers Distinguished Professor and
Dean, School of Medicine

Preface

Handbook of Nutrition and Pregnancy is written for the clinician and other healthcare professionals who treat and counsel pregnant women and women of childbearing age. Thus, physicians, physicians' assistants, nurses, and dietitians, in particular, as well as dietetic students and graduate and medical students, will find this book a useful resource. In addition to the historical perspective and background to support recommendations that are provided in each chapter, important for the practitioners, recommendations and guidelines have been summarized and provided in tables that are easy to locate and interpret. It is the intent of the editors that *Handbook of Nutrition and Pregnancy* serves as a reliable resource that is shelved at arm's reach by practitioners and researchers around the world. By combining the historical and background information with the easy-to-use practical information of a handbook, the volume is unique among the contemporary books that deal with the topics of nutrition in pregnancy and outcomes both for the mother as well as for the neonate.

At a time when the scientific community is looking to complete the weaving of the threads between genes and function, and to determine to what extent prenatal and perinatal environmental factors are linked to childhood and adult obesity and chronic diseases and metabolomics, *Handbook of Nutrition and Pregnancy* includes relevant chapters that bring contemporary assessments of nutrition knowledge about these cutting-edge areas and their relationships to the pregnant woman and women of childbearing age. The overall objective is to take the most up-to-date information and to translate it into clinically relevant practice recommendations.

A second major goal of this volume is to examine issues that are common to both the developed and the developing worlds and to include chapters that are specific to nutritional and reproductive factors seen mainly in developing countries. These chapters discuss contemporary issues that affect both the woman and the developing infant. For the developed world, contemporary topics for the woman experiencing a normal pregnancy include the Food and Nutrition Board of the National Academy of Science's new dietary reference intakes (DRIs), optimal weight gain, and physical activity/exercise. Part 2 of this book, addressing nutrient needs related to high-risk pregnancies, includes topics on nutrient needs of the pregnant woman who has undergone bariatric surgery, multiple fetuses, eating disorders of women of childbearing age, diabetes, preeclampsia, or HIV/AIDS. The positive and negative effects of specific nutrients and dietary factors are covered in Part 3, including topics on popular diets; vegetarian diets; the need, efficacy, and safety of dietary supplements; folate fortification; iron adequacy; and calcium, vitamin D, and bone health. Part 4 addresses the postpartum period for the mother, with topics on lactation success related to nutrient needs and physical exercise, as well as nutrition and postpartum depression. For the developing world, implications of nutrition transition for the pregnant woman, nutrient adequacy and maternal survival, anemia, and micronutrient status and pregnancy outcomes for HIV-infected women are topics included. Naturally, some topics span both the developed and developing worlds. Our aim has been to include current recommendations and policies, but where these have not been definitively established, to offer guidelines that can be made with reasonable assurances of safety and efficacy, based on current knowledge. For recommendations and policies related to maternal nutrition that have been put in place in the last decade, we review the data, where available, regarding efficacy for the recommendations and policies, e.g., folic acid fortification of the food supply. Students especially will find the coverage of gaps in knowledge useful, while researchers will turn to this volume for information relating to application of the nutrition knowledge.

A third aim of the book, covered in several chapters, is a review of nutritional as well as physiological factors that either increase or decrease the potential for high-risk pregnancies, such as gestational diabetes mellitus, Types 1 and 2 diabetes mellitus, preeclampsia, anemia, and so forth.

Required nutrients are provided primarily by the food we grow in our gardens, fish from the sea or fresh water sources, the animals we tend, or the food we barter for or purchase from markets or supermarkets. Additionally, in the developed world, the market shelves and media ads are now becoming inundated with products and information about nutrient supplements and functional foods/bioactive foods. What roles do and should these products have in the diets of women of childbearing age? How do we go about assessing the importance of these foods in a healthy pregnancy? What can we recommend? The answers to some of these questions are found herein.

In conclusion, *Handbook of Nutrition and Pregnancy* is a comprehensive volume that includes up-to-date information in 23 chapters written by the leaders in the fields of diet, nutrients, ingredients, environmental factors, and physiological consequences, addressing the needs of women of childbearing potential and pregnant women. The volume contains information that permits the reader to answer confidently practical questions from patients, family members, students and researchers, because the information represents the totality of the data rather than findings of a single study. There is not another book in the marketplace that duplicates the breadth of information found herein. Thus, this volume can serve as the benchmark in this field.

Carol J. Lammi-Keefe
Sarah C. Couch
Elliot H. Philipson

Contents

Contributors

RAUL ARTAL, MD • *Department of Obstetrics, Gynecology, and Women's Health, Saint Louis University, School of Medicine, St. Louis, MO*

LYNN B. BAILEY, PHD • *Food Science and Human Nutrition, University of Florida, Gainesville, FL*

JOHN BEARD, PHD • *Department of Nutritional Sciences, The Pennsylvania State University, University Park, PA*

CHERYL TATANO BECK, DNSC, CNM, FAAN • *School of Nursing, University of Connecticut, Storrs, CT*

LINDA BLOOM, CNM, ND • *Department of OB/GYN, Cleveland Clinic, Macedonia, OH*

ROSE CATANZARO, MS, RD, LD, CDE • *Department of Obstetrics, Gynecology, and Women's Health, Saint Louis University, St. Louis, MO*

PARUL CHRISTIAN, PHD • *Program in Human Nutrition, Department of International Health, Bloomberg School of Public Health, Johns Hopkins University, Baltimore, MD*

KRISTIN H. COPPAGE, MD • *Tristate Maternal Fetal Medicine, Seton Center, Cincinnati, OH*

SARAH C. COUCH, PHD, RD • *Department of Nutritional Sciences, University of Cincinnati, Cincinnati, OH*

RICHARD J. DECKELBAUM, MD • *Institute of Human Nutrition, Department of Pediatrics, Department of Epidemiology, Columbia University, New York, NY*

ARLENE ESCURO, MS, RD, LD, CNSD • *Department of OB/GYN, Cleveland Clinic, Macedonia, OH*

GRACE A. FALCIGLIA, PHD, RD • *Department of Nutritional Sciences, University of Cincinnati Medical Center, Cincinnati, OH*

BETH THOMAS FALLS, PHD, RD • *Indian River Community College, Ft. Pierce, FL*

WAFAIE W. FAWZI, MBBS, DRPH • *Departments of Nutrition and Epidemiology, Harvard School of Public Health, Boston, MA*

JULIA L. FINKELSTEIN, MPH • *Department of Nutrition, Harvard School of Public Health, Boston, MA*

AMY FLEISHMAN, MS, RD, CDN • *Department of Surgery, Mount Sinai School of Medicine, New York, NY*

CATHERINE A. FORESTELL, PHD • *Research Associate, Philadelphia, PA, Department of Psychology, College of William and Mary, Williamsburg, VA*

SARAH GOPMAN, MD • *Department of Family and Community Medicine, University of New Mexico Health Sciences Center, Albuquerque, NM*

DANIEL M. HERRON, MD, FACS • *Department of Surgery, Mount Sinai School of Medicine, New York, NY*

LISA A. HOUGHTON, PHD, RD • *Department of Clinical Dietetics, The Hospital for Sick Children, University of Toronto, Toronto, ON, Canada*

BETH IMHOFF-KUNSCH, MPH • *Nutrition and Health Sciences Program, Emory University, Atlanta, GA*

MICHELLE PRICE JUDGE, PHD, RD • *School of Nursing, University of Connecticut, Storrs, CT*

JANET C. KING, PHD • *Children's Hospital Oakland Research Institute, Oakland, CA*

KATHERINE KUNSTEL, RD, CNSD • *Community Nutrition Resource Manager, The Partnership for the Homeless, New York, NY*

CAROL J. LAMMI-KEEFE, PHD, RD • *Louisiana State University College of Agriculture, AgCenter, and Pennington Biomedical Research Center, Baton Rouse, LA*

LARRY LEEMAN, MD, MPH • *Department of Family and Community Medicine, University of New Mexico Health Sciences Center, Albuquerque, NM*

ANN REED MANGELS, PHD, RD • *The Vegetarian Resource Group, Baltimore, MD*

MICHELLE K. MCGUIRE, PHD • *Department of Food Science and Human Nutrition, Washington State University, Pullman, WA*

SAURABH MEHTA, MBBS, MS • *Departments of Nutrition and Epidemiology, Harvard School of Public Health, Boston, MA*

JULIE A. MENNELLA, PHD • *Monell Chemical Senses Center, Philadelphia, PA*

SHARON M. NICKOLS-RICHARDSON, PHD, RD • *Department of Nutritional Sciences, College of Health and Human Development, The Pennsylvania State University, University Park, PA*

DEBORAH L. O'CONNOR, PHD • *Department of Clinical Dietetics, The Hospital for Sick Children, University of Toronto, Toronto, ON, Canada*

CARMEN GLORIA PARODI, MSC • *Department of Nutrition, Diabetes and Metabolic Diseases, School of Medicine, Pontificia Universidad Catolica de Chile, Santiago de Chile, Chile*

ELLIOT H. PHILIPSON, MD • *Maternal Fetal Medicine, Cleveland Clinic Lerner College of Medicine, Cleveland, OH*

MARY FRANCES PICCIANO, PHD • *Office of Dietary Supplements, National Institute of Health, Bethesda, MD*

USHA RAMAKRISHNAN, PHD • *Hubert Department of Global Health, Rollins School of Public Health, Emory University, Atlanta, GA*

LORRENE D. RITCHIE, PHD • *Center for Weight and Health, Department of Nutritional Sciences and Toxicology, University of California at Berkeley, Oakland, CA*

NANCY RODRIGUEZ, PHD • *Department of Nutritional Sciences, University of Connecticut, Storrs, CT*

JAIME ROZOWSKI, PHD • *Department of Nutrition, Diabetes and Metabolic Diseases, School of Medicine, Pontificia Universidad Catolica de Chile, Santiago de Chile, Chile*

KELLY L. SHERWOOD, MSC, RD • *Department of Clinical Dietetics, The Hospital for Sick Children, University of Toronto, Toronto, ON, Canada*

ALYCE M. THOMAS, RD • *Department of Obstetrics and Gynecology, St. Joseph's Regional Medical Center, Paterson, NJ*

LANA K. WAGNER, MD • *Department of Family and Community Medicine, University of New Mexico Health Sciences Center, Albuquerque, NM*

I

NUTRIENT AND HEALTH NEEDS
DURING NORMAL PREGNANCY

1 Nutrient Recommendations and Dietary Guidelines for Pregnant Women

Lorrene D. Ritchie and Janet C. King

Summary The requirements for selected nutrients increase appreciably during pregnancy. The recommended intakes for the following nutrients are >25% higher than are the amounts recommended for nonpregnant women: protein, α-linolenic acid, iodine, iron, zinc, folate, niacin, riboflavin, thiamin, and vitamin B_6. The needs for protein, iron, folate, and vitamin B_6 are about 50% higher. Good food sources of these nutrients are grains, dark green or orange vegetables, and the meat, beans, and nuts groups. Additional energy is also required to meet the needs for moving a heavier body, the rise in metabolic rate, and tissue deposition. Approximately 340–450 kcal are needed in the second and third trimesters, respectively. Although these increased nutrient requirements are significant, the same food pattern recommended for nonpregnant women can be recommended to pregnant women because that food pattern meets pregnancy nutrient Recommended Daily Allowances (RDAs) for all nutrients except iron and vitamin E. The shortfall in iron and vitamin E can be provided by any vitamin–mineral supplement supplying at least 10 mg iron and 9 mg vitamin E. Use of a common food pattern for women at all stages in the reproductive cycle enables dietitians and other health care providers to teach pregnant women the elements of a quality diet that will better ensure good health for a life time.

Keywords: Pregnancy, Nutrient requirements, Dietary guidelines, Food patterns, Nutrient intakes, Diet counseling

1.1 INTRODUCTION

Nutrient needs typically increase more during pregnancy than during any other stage in a woman's adult life. Additional nutrients are required during gestation for development of the fetus as well as for growth of maternal tissues that support fetal development. The materials required for this rapid growth and development depend on supply from the maternal diet. However, because of the differing roles nutrients play in tissue development and growth as well as nutrient-specific changes in maternal homeostasis during pregnancy, nutrient requirements do not increase uniformly. Changes in the efficiency

From: *Nutrition and Health: Handbook of Nutrition and Pregnancy*
Edited by: C.J. Lammi-Keefe, S.C. Couch, E.H. Philipson © Humana Press, Totowa, NJ

of absorption from the gastrointestinal tract and excretion by the renal system, as well as changes in maternal storage or tissue reserve, are examples of homeostatic mechanisms that must be considered in establishing nutrient requirements during gestation. Because the demand for some nutrients is great relative to others, care must be taken in selecting the optimal diet during pregnancy.

The purpose of the first section of this chapter is to describe the Dietary Reference Intakes (DRIs) for pregnancy, outline how they compare to the DRIs in the nonpregnant state, and explain the physiological reasons for adjusting nutrient requirements during pregnancy. Emphasis is on nutrients with relatively high increases in demand relative to prepregnancy. This does not imply that other nutrients are not critical for a healthy pregnancy outcome, but that if increased intake for nutrients with the largest relative demand is achieved, and a mixed diet is consumed, then it is likely that the needs for other nutrients will be met as well. In the subsequent section of the chapter, nutrient requirements are translated into foods according to the most recent dietary guidelines.

1.2 NUTRIENT RECOMMENDATIONS FOR PREGNANCY

The DRIs, released from 1997 to 2005 by the Institute of Medicine of the National Academies (IOM), differ from previous recommendations [1]. The recommendations continue to be based on scientifically valid experiments with emphasis on in vivo studies in humans (rather than in vitro or animal experiments), reliable intake data, and whenever possible, measurements of relevant biomarkers. In the most recent recommendations, however, the role of nutrients in promoting and protecting health is emphasized. Prevention of nutrient deficiencies was not the only criterion used. Further, differences in the strength of the scientific evidence available for establishing nutrient requirements were delineated. For nutrients with sufficient available evidence, an RDA was established equivalent to the amount needed to meet the nutrient requirements of nearly all ($\approx$97.5%) healthy individuals for a given gender and stage of life. When insufficient evidence was available to formulate an RDA, an Adequate Intake (AI) was provided. An AI is typically based on the amount that people normally consume, and, because it involves more expert discretion, must be applied with greater caution than does an RDA. Despite these differences, both RDAs and AIs are reference values for normal, healthy individuals eating a typical mixed North American diet. A given individual may have physiological, health, or lifestyle characteristics that require tailoring of specific nutrient values.

Table 1.1 outlines the most recent DRIs for women 19–30 years old. The changes from nonpregnancy differ slightly between younger and older women for several nutrients (as noted in the table footnotes), but in general these differences are small and for the majority of nutrients recommendations do not vary by maternal age.

During the first trimester of pregnancy, nutrient needs generally do not increase above the nonpregnant state. Although the fetus is undergoing rapid developmental change early in gestation, most of the nutrients for growth in maternal and fetal tissues are required later in pregnancy. For this reason, the DRIs were generally based on needs during the last half of pregnancy. To allow for optimal storage and accumulation of functional reserve in early pregnancy, however, recommendations were not varied by trimester, with the exception of dietary energy (see discussion below).

Table 1.1
Dietary Reference Intakes for Women 19–30 Years of Age. (Adapted from [4, 14, 19, 29, 30, 31, 36, 39])

Nutrient	Nonpregnancy	Pregnancy	Increase (%)	Criterion for increase	Comment
Energy and macronutrients					
Energy (kcal/day)	2,403	2,855	19	Maternal and fetal deposition	a
Carbohydrate (g/day)	130	175	35	Fetal brain glucose utilization	b
Total fiber (g/day)	25	28	12	Extrapolation based on increased energy intake	
Protein (g/day)	46	71	54	Maternal and fetal deposition	c
n-6 PUFA (g/day)	12	13	8	Median linoleic acid intake from CSF II	
n-3 PUFA (g/day)	1.1	1.4	27	Median α-linolenic acid intake from CSF II	
Elements					
Calcium (mg/day)	1,100	1,100	0	Adequate adjustments in maternal homeostasis in pregnancy	—
Fluoride (mg/day)	3	3	0	Limited data available to suggest increased need in pregnancy	—
Phosphorus (mg/day)	700	700	0	Adequate adjustments in maternal homeostasis in pregnancy	—
Chromium (mcg/day)	25	30	20	Extrapolation based on average maternal weight gain	d
Copper (mcg/day)	900	1,000	11	Maternal and fetal deposition	d
Iodine (mcg/day)	150	220	47	Maternal and fetal deposition and for maternal iodine balance and prevention of goiter during pregnancy	
Iron (mg/day)	18	27	50	Maternal and fetal deposition, basal losses, and expansion of hemoglobin	d
Magnesium (mg/day)	310	350	13	Maternal and fetal deposition of lean body mass	b
Manganese (mg/day)	1.8	2	11	Extrapolation based on average maternal weight gain	d
Molybdenum (mcg/day)	45	50	11	Extrapolation based on average maternal weight gain	d
Selenium (mcg/day)	55	60	9	Fetal deposition	
Zinc (mg/day)	8	11	38	Maternal and fetal deposition	b

(continued)

5

Table 1.1

Dietary Reference Intakes for Women 19–30 Years of Age. (Adapted from [4, 14, 19, 29, 30, 31, 36, 39])

Vitamins					
Choline (mg/day)	425	450	6	Median intake from CSF II	d
Folate (mcg/day)	400	**600**	50	Maintain normal folate status	e
Niacin (mg/day)	14	**18**	29	Maternal and fetal deposition plus increased energy utilization	
Pantothenic acid (mg/day)	5	6	20	Maternal and fetal deposition	
Riboflavin (mg/day)	**1.1**	**1.4**	27	Maternal and fetal deposition plus increased energy utilization	d
Thiamin (mg/day)	**1.1**	**1.4**	27	Maternal and fetal deposition plus increased energy utilization	d
Vitamin A (mcg/day)	**700**	770	10	Fetal liver vitamin A deposition	b
Vitamin B$_{12}$ (mcg/day)	**2.4**	**2.6**	8	Fetal deposition and changes in maternal absorption	
Vitamin B$_6$ (mg/day)	**1.3**	**1.9**	46	Maternal and fetal deposition	d
Vitamin C (mg/day)	**75**	**85**	13	Amount needed to prevent scurvy in infant and estimated fetal transfer	d
Biotin (mcg/day)	30	30	0	Limited data available to suggest increased need in pregnancy	d
Vitamin D (mcg/day)	5	5	0	Daily accretion in pregnancy is small	
Vitamin E (mg/day)	**15**	**15**	0	Circulating concentrations normally increase in pregnancy; lack of clinical deficiency	
Vitamin K (mcg/day)	90	90	0	Comparable concentrations in pregnancy; lack of clinical deficiency	
Water and Electrolytes					
Water (l/day)	2.7	3	11	Median intake from NHANES III	
Chloride (g/day)	2.3	2.3	0	Limited data available to suggest increased need in pregnancy	b

Potassium (g/day)	**4.7**	4.7	0	Daily accretion in pregnancy is small
Sodium (g/day)	**1.5**	1.5	0	Daily accretion in pregnancy is small

RDAs are in *boldface type*, AIs are in *Roman*.

PUFA polyunsaturated fatty acids, *CSF II* Continuing Survey of Food Intakes by Individuals, 1989–1991, US Dept. of Agriculture, Agricultural Research Service, *NHANES III* Third National Health and Nutrition Examination Survey, 1988–1994, Center for Disease Control, National Center for Health Statistics.

[a]For healthy moderately active individuals, third trimester; requirements for first trimester are not increased above nonpregnancy, and requirements for second trimester are 2,708 kcal/day. Subtract 7 kcal/day for females for each year of age above 19 years.

[b]Percent increase for pregnant women 14–18 y is slightly *lower* than for age 19–30 years.

[c]Percent increase for pregnant women 31–50 y is slightly *higher* than for age 19–30 years.

[d]Percent increase for pregnant women 14–18 y is slightly *higher* than for age 19–30 years.

[e]Low maternal folate status in very early pregnancy (before women typically know they have conceived) has been associated with the birth of offspring with a neural tube defect (NTD). Therefore, the nonpregnant DRI for women in their childbearing years was formulated for preventing NTDs. In view of evidence linking the use of supplements containing folic acid before conception and during early pregnancy with reduced risk of NTDs in the fetus, it is recommended that all women capable of becoming pregnant take a supplement containing 400 mcg of folic acid every day in addition to the amount of folate consumed in a healthy diet.

For most nutrients, the criterion for the increase in the nutrient recommendation was based on deposition in the fetus and maternal tissues (e.g., placenta, amniotic fluid, breast tissue, fat storage), with adjustments made for changes in nutrient absorption, urinary excretion, and/or storage during pregnancy, when these were relevant and well characterized. For most nutrients, limited data were available to adjust the DRIs on the basis of changes in maternal homeostasis. Surprisingly, nutrient balance studies or supplementation trials during pregnancy have been conducted infrequently. When no direct data were available for determining the additional daily requirement for a nutrient during pregnancy, increased nutrient needs were extrapolated based on a median weight gain of 16 kg reported for women with good pregnancy outcomes [2] and the estimated nutrient need per unit of weight gain. In other cases, the median intake of pregnant women from national diet surveys was used. A brief description of the needs for select nutrients during pregnancy and the rationale for increased nutrient requirements follows. Focus was placed on those nutrients with relatively high increases in requirements above the nonpregnant state.

1.2.1 Macronutrients

1.2.1.1 Energy

Energy needs during pregnancy vary according to a woman's basal metabolic rate, prepregnancy weight, amount and composition of weight gain, stage of pregnancy, and physical activity level. It is estimated that on average a pregnant woman requires a total of 85,000 additional calories over the course of 40 weeks of pregnancy, which extrapolates to approximately 300 extra calories per day [3]. For most women, however, energy needs in the first trimester of pregnancy are minimal. While the first trimester is characterized by rapid development of fetal organs and tissues, these processes are not very energy intensive. Maternal basal metabolic rate, for example, does not measurably increase until the fourth month of pregnancy when notable increases in growth of the uterus, mammary glands, placenta and fetus, and increases in blood volume and the work of the heart and respiratory system begin. As a woman's weight increases, she also requires more energy to accomplish the same amount of physical work such that even if physical activity levels remain unchanged from prior to pregnancy, the energy costs of these activities increase.

When the new DRIs for macronutrients were released in 2005 [4], a new approach was used to estimate the energy requirements of pregnancy. Since total energy expenditure (TEE) had been measured using doubly labeled water in several hundred pregnant women, those data were used as the basis for the recommendation. The Estimated Energy Requirement (EER) for pregnancy is derived, therefore, from the sum of the TEE in nonpregnant women plus a median change in TEE of 8 kcal/week plus 180 kcal/day in the second and third trimesters to account for the energy deposition in tissue gained. At 20 weeks' gestation, the additional energy required totals 340 kcal/day; at 34 weeks gestation the additional energy need is 450 kcal/day. Table 1.2 illustrates how energy needs for pregnant women vary with body mass index (BMI) and physical activity level.

Because energy needs are influenced by a variety of factors, they can vary dramatically between individuals. For this reason, monitoring weight gain during pregnancy is the best way to ensure adequacy of energy intake [5]. The IOM [3] released recommendations for

Table 1.2
Estimated Energy Requirements (*EER*) for Women 30 Years of Age During Pregnancy.
(Adapted from [4])

Height m (in)	PAL	EER Prepregnancy BMI of 18.5 kg/m²	EER Prepregnancy BMI of 24.99 kg/m²
First trimester			
1.50 (59)	Sedentary	1,625	1,762
	Low active	1,803	1,956
	Active	2,025	2,198
	Very active	2,291	2,489
1.65 (65)	Sedentary	1,816	1,982
	Low active	2,016	2,202
	Active	2,267	2,477
	Very active	2,567	2,807
1.80 (71)	Sedentary	2,015	2,211
	Low active	2,239	2,459
	Active	2,519	2,769
	Very active	2,855	3,141
Second trimester			
1.50 (59)	Sedentary	1,965	2,102
	Low active	2,143	2,296
	Active	2,365	2,538
	Very active	2,631	2,829
1.65 (65)	Sedentary	2,156	2,322
	Low active	2,356	2,542
	Active	2,607	2,817
	Very active	2,907	3,147
1.80 (71)	Sedentary	2,355	2,551
	Low active	2,579	2,799
	Active	2,859	3,109
	Very active	3,195	3,481
Third trimester			
1.50 (59)	Sedentary	2,075	2,212
	Low active	2,253	2,406
	Active	2,475	2,648
	Very active	2,741	2,939
1.65 (65)	Sedentary	2,266	2,432
	Low active	2,466	2,652
	Active	2,717	2,927
	Very active	3,017	3,257
1.80 (71)	Sedentary	2,465	2,661
	Low active	2,689	2,909
	Active	2,969	3,219
	Very active	3,305	3,591

For each year below 30, add 7 kcal/day; for each year above 30 subtract 7 kcal/day

PAL physical activity level

[a]Derived from the following regression equation based on doubly labeled water data:

EER = 354 − 6.91 × age (years) + PA × (9.36 × wt [kg] + 726 × ht [m]),

where PA refers to coefficient for PAL

PAL = (total energy expenditure) / (basal energy expenditure)

PA = 1 if PAL ≥ 1 < 1.4 (sedentary)

PA = 1.12 if PAL ≥ 1.4 < 1.6 (low active)

PA = 1.27 if PAL ≥ 1.6 < 1.9 (active)

PA = 1.45 if PAL ≥ 1.9 < 2.5 (very active)

weight gain during pregnancy based on prepregnancy weight status. For women classified as being within a normal weight range prior to pregnancy (BMI 19.8–26 kg/m²), the recommended weight gain is 11.3–15.9 kg (25–35 lbs.), and the recommended rate of weight gain is 0.9–1.8 kg (2–4 lbs.) in the first trimester and about 0.5 kg (1 lb.)/week thereafter. A slightly higher total gain of 12.7–18.1 kg (28–40 lbs.) is recommended for underweight women (BMI < 19.8). A slightly lower total gain of 6.8–11.3 kg (15–25 lbs.) and at least 6.8 kg (15 lbs.) is recommended for women who, prior to pregnancy, are overweight (BMI > 26–29) and obese (BMI > 29), respectively. Even among obese women, inadequate weight gain during pregnancy can lead to increased risk of preterm delivery [6].

The additional energy requirements of pregnancy are small relative to the needs for many other nutrients. While an extra 340–450 kcal could be consumed by simply adding a glass of 2% milk and a small sandwich, this would not meet increased nutrient needs for pregnancy. The fact that the relative increase for many other nutrients is more dramatic than for energy indicates the importance of emphasizing nutrient-dense foods during pregnancy. Following the dictum of "eating for two" may result in excessive maternal weight gain. Further, for obese women, sedentary women, and women whose activity levels decline during pregnancy (e.g., bed rest) the recommendations of 340–450 kcal/day may be too high. On the other hand, underweight women, young adolescent mothers who are still growing (<14 years), and women carrying multiple fetuses may need 500 kcal/day or more [7].

The goal is to avoid both ends of the spectrum, both excessive energy intake as well as inadequate energy intake. Overnutrition and excess weight gain in pregnancy impart risk of gestational diabetes, macrosomia, delivery complications such as shoulder dystocia, cesarean delivery and post operative problems, difficulty initiating breastfeeding, and risk of subsequent maternal and child obesity [8–10]. Conversely, undernutrition and inadequate weight gain during pregnancy can lead to impaired intrauterine growth and consequent low birth weight of the newborn. In addition to complications at birth, intrauterine growth retardation has been associated with metabolic abnormalities in adulthood, such as hyperlipidemia, hypertension, cardiovascular disease, glucose intolerance, and type 2 diabetes [10, 11].

1.2.1.2 PROTEIN

During pregnancy, additional dietary protein is used for fetal growth, placental development, production of amniotic fluid, increased maternal blood volume, and gain of other maternal tissues. Increases in protein needs mirror maternal and fetal growth rates; early in pregnancy, the requirements for extra protein are relatively small, but increase progressively as pregnancy proceeds. Approximately 82% of the total demand for the 925 g of protein required for maternal and fetal needs is accumulated over the last half of gestation [12]. Inadequate maternal protein intake incurs risk of low birth weight.

A factorial approach was used to calculate the protein DRI during pregnancy. The summation of the additional lean tissue accumulated in pregnancy and the additional protein required to maintain an increased body weight were estimated as outlined in Table 1.3. By the second half of pregnancy, this translates into a 25-g/day increased requirement for a total of approximately 70–75 g/day (or 1.1 g protein per kg of body weight per day) [4].

Table 1.3
Derivation of Protein Requirements During Pregnancy. (Adapted from [4])

Maintenance of increased body weight		Fetal and maternal tissue deposition			
Average weight gain (kg)[a]	Estimated average requirement (g/kg/day)[b]	Additional protein required (g/day)	Protein deposition (g/day)	Utilization efficiency of dietary protein	Additional protein required (g/day)[c]	Total increase in protein required for pregnancy (g/day)[d]
12.8	0.66	8.5	5.4	0.43	12.6	25

[a]Average for trimesters 2 and 3; protein requirements in trimester 1 are estimated to be minimal
[b]Estimated average requirement for maintenance of protein in adults
[c]Based on slope of regression line of protein intake versus nitrogen balance (recalculated from [13])
[d]Adjusted for normal variation to meet the needs of 97.5% of pregnant women

The relative increased need above nonpregnancy is greater for protein (54% increase) than for any other nutrient. However, because protein intakes tend to be high relative to needs in the nonpregnancy state, averaging approximately 60 g/day for nonpregnant women [4], inadequate protein intake is not common in the United States, even among pregnant women. However, vegans and women carrying multiple fetuses may need to pay close attention to their protein intakes.

1.2.2 Water-Soluble Vitamins

Requirements for most water-soluble vitamins increase during pregnancy. Folate and vitamin B_6 will be emphasized in the following discussion because increases in demand associated with pregnancy are relatively high (50% for folate, 46% for vitamin B_6), and average intakes of these water-soluble vitamins relative to requirements are generally lower than for other water-soluble vitamins.

1.2.2.1 FOLATE

Folate is involved in single-carbon transfer reactions, notably important for the synthesis of nucleic acids and certain amino acids for new cell and tissue production. Erythrocyte folate is considered the best marker of long-term folate status in pregnancy; serum folate can also be used but reflects more recent changes in dietary intake. With inadequate folate intake, serum and erythrocyte folate concentrations decline, and megaloblastic anemia can develop. Impaired folate status during pregnancy may be involved with adverse outcomes such as pregnancy complications, spontaneous abortion, preterm delivery, and low birth weight [14]. Results from supplementation trials suggest that an additional 200 mcg of dietary folate equivalent* is required to maintain optimal folate status during pregnancy [15].

Neural tube defects (NTDs), a group of heterogeneous malformations involving neural tissue in the brain and/or spinal cord, occur in less than 1 per 1,000 births in the United States [16]. The etiology of NTDs is an ongoing area of research; however, inadequate maternal folate status prior to and in the first few weeks after conception appears to play a role in at least some cases of neural tube defects. According to 1999–2000 National Health and Nutrition Examination Survey (NHANES) data, the average folate intake of 20- to 39-year-old women in the United States is 327 mcg/day [17]. Results from supplementation studies suggest that women capable of becoming pregnant should consume an additional 400 mcg/day of folic acid from supplements and/or fortified foods in addition to consuming food folate from a varied diet.**

It is recommended that women consume 400 mcg/day of synthetic folate at least 1 month prior to conception to optimize folate status at the time of neural tube closure [5]. Based on evidence from randomized controlled trials, it has been estimated that this level of folate supplementation could prevent up to half of NTD cases [18]. A 19%

---

*Dietary folate equivalents are used to account for the differences in bioavailability between food folate ($\approx$50% bioavailable) and folic acid used in supplements and food fortification (~85% available)[15].

**Available evidence suggests that synthetic folic acid (found in supplements and fortified foods) is more effective at preventing neural tube defects than is food folate [15].

reduction in NTD prevalence occurred after the mandatory fortification with folate of enriched breads, cereals, flours, and other grain products in 1998 [16]. It should be noted that the additional 400 mcg/day folic acid supplementation is not included in the recommendations for pregnant women because by the time a woman is normally aware that she is pregnant, the window of opportunity for the effective prevention by folate of NTD (due to the embryological timing of the initial development and closure of the neural tube) has passed.

1.2.2.2 VITAMIN B$_6$

Vitamin B$_6$, in the form of pyridoxal phosphate, is a coenzyme involved in over 100 metabolic reactions, most of which involve amino acid and protein metabolism. During pregnancy, vitamin B$_6$ plays important roles in the synthesis of nonessential amino acids, heme, erythrocytes, immune proteins, and hormones. In observational studies, vitamin B$_6$ has been positively associated with improved pregnancy outcomes such as reduced incidence of preeclampsia and higher Apgar scores and neonatal behavior [3]. Results from randomized, controlled supplementation trials suggest limited clinical benefit of vitamin B$_6$ supplementation [19]. However, few such trials have been done and few pregnancy outcomes have been investigated.

Maternal and fetal accumulation during pregnancy totals approximately 25 mg, which translates into an increase in the daily requirement of about 0.25 mg after accounting for an average 75% bioavailability of food B$_6$ and allowing for increased weight of the mother [15]. Because the needs for vitamin B$_6$ predominate in the last half of pregnancy and because vitamin B$_6$ is not stored in the body to any appreciable extent, increased intake in early pregnancy is not likely to be adequate to meet needs later in pregnancy. Therefore, the DRI was set at an additional 0.6 mg/day [15].

1.2.3 Minerals

1.2.3.1 IODINE

Iodine needs increase during pregnancy for the synthesis of thyroid hormones. Maternal iodine deficiency during pregnancy can result in the enlargement of a woman's thyroid gland, development of goiter, and hypothyroidism. Maternal hypothyroidism increases the risk of a variety of poor fetal outcomes including stillbirth, spontaneous abortion, congenital anomalies, mental retardation, deafness, spastic dysplegia, and cretinism [3]. To avoid risk of harm to the fetus, maternal iodine deficiency should be corrected prior to conception.

During gestation, fetal iodine deposition is approximately 75 mcg/day. Results from iodine balance studies as well as iodine supplementation trials to prevent thyroid enlargement and goiter during pregnancy corroborate that an additional 70 mcg/day is required to cover the pregnancy needs of 97–98% of the population during pregnancy [20].

1.2.3.2 IRON

If iron is not readily available from the diet, then iron from maternal liver stores is mobilized. Thus, the production of fetal hemoglobin is usually adequate even if the mother is severely iron deficient, and anemia in the newborn due to iron deficiency is relatively rare. However, maternal iron deficiency is relatively common, and anemia is the most common nutrition-related complication of pregnancy. Although the prevalence of anemia in pregnancy is difficult to quantify, it has been estimated that 2–4% of pregnant women in the United States suffer from anemia [21]. In the majority of cases (≈90%), it is due to a deficiency of dietary iron. Maternal iron deficiency increases the risk of premature delivery and consequent low birth weight and may reduce a mother's risk of tolerating hemorrhage during delivery and postpartum iron deficiency [22].

The total requirement for iron during pregnancy is approximately 1,070 mg, most of which is accumulated over the last half of pregnancy [20, 23]. A large part of iron needs (≈500 mg) are used by the bone marrow for blood hemoglobin synthesis [9]. Red blood cell mass increases by approximately 33%, and blood volume increases by about 50% over the course of a healthy pregnancy [24]. An augmented blood supply is required for extra blood flow to the uterus and placenta, the extra metabolic needs of the growing fetus, and increased perfusion of other maternal organs, especially the kidneys for removal of the additional generation of metabolic waste products during pregnancy. Fetal iron storage also occurs, primarily during the last trimester. It has been estimated that 250–300 mg are accumulated in fetal and placental tissues [9].

Although the efficiency of absorption of dietary iron may increase during pregnancy,*** the daily increased requirement of 9 mg/day is not easy to achieve by diet alone. Further, women rarely enter pregnancy with optimal iron stores [9]. The 50% increase in the iron requirement during pregnancy compared to prior to conception is larger than for any other nutrient except protein. However, while average intakes of pregnant women are generally sufficient to meet pregnancy needs for protein, dietary intakes of iron tend to be low relative to requirements. The average intake of women in the United States is approximately 13 mg/day [17], below the nonpregnant DRI of 18 mg/day. The typical US diet contains about 6 mg iron per 1,000 calories. A pregnant woman consuming an additional 400 kcal/day is therefore likely to consume only 2.5 mg/day additional iron, less than a third of the 9 mg/day increase recommended.

Because of the inherent difficulties in meeting the DRI for iron in pregnancy, the Centers for Disease Control and Prevention recommends an iron supplement of 30 mg/day for all pregnant women, beginning at the first prenatal visit [25]. When hemoglobin levels are low, a 60–120 mg/day iron supplement may be prescribed. Because large amounts of iron can interfere with the absorption of other trace minerals important during pregnancy, pregnant women taking over 30 mg/day of iron should also take 15 mg of supplemental zinc and 2 mg of supplemental copper [20].

1.2.3.3 ZINC

Zinc is another nutrient with a large (38%) increase in demand during pregnancy relative to the nonpregnancy state. Zinc is involved in the synthesis of deoxyribonucleic acid, ribonucleic acid, and ribosomes and is therefore required for gene expression, cell differentiation, and cell replication. In rare cases of maternal zinc deficiency due to acrodermititis enteropathica, a genetic inability to absorb dietary zinc properly, increased risk of congenital malformations in the newborn occurs [26]. Supplementation trials involving populations with habitually low zinc intakes suggest that increased zinc is also important for preventing premature delivery and promoting proper neurological development in the fetus [20].

***Absorption of non-heme iron increases; the efficiency of heme iron, which is normally very high, does not notably increase during pregnancy.

The total requirement for zinc has been estimated as 100 mg for synthesis of maternal and fetal tissues, most of which is accumulated during the last half of pregnancy [27]. The efficiency of absorption of zinc during pregnancy does not appear to change sufficiently to meet zinc needs in the absence of an increased dietary intake [28]. The increased recommendation for zinc of 3 mg/day in pregnancy is based on the accumulation of fetal and maternal zinc of 0.73 mg/day during the last quarter of pregnancy, accounting for a 27% efficiency of absorption [20].

1.2.4 Nutrients without Increased Requirements during Pregnancy

The fact that requirements for some nutrients do not increase during pregnancy does not imply that these nutrients are not critical to maternal and fetal health. Calcium is a case in point. The needs of the fetus for calcium are substantial, averaging 300 mg/ day. However, due to homeostatic adjustments, the dietary requirements for calcium do not change during pregnancy. An integrated system of hormones, namely parathyroid hormone and 1,25-dihydroxyvitamin D, regulate intestinal absorption, urinary excretion, and bone flux of calcium. During pregnancy, the efficiency of calcium absorption increases by nearly 50%, such that fetal needs appear to be met without increasing calcium intake or net losses of maternal bone mineral [11, 29].

Even though the DRI for calcium does not increase during pregnancy, it should be noted that many women fail to meet calcium requirements. According to data from the 1999–2000 NHANES, the average calcium intake of women of childbearing age is 797 mg/day, far below recommended levels [16]. Phosphorus absorption is also increased in pregnancy by changes in calcitropic hormone concentrations. Therefore, as with calcium, the DRI in pregnancy for phosphorus remains the same as for nonpregnant women [30].

For some of the other nutrients, the available evidence is generally not sufficient to warrant recommending an increased intake during pregnancy (e.g., biotin, vitamin K, vitamin E, chloride, fluoride). For yet other nutrients, the intake of nonpregnant women already appears ample to meet the small increased demands during gestation (e.g., sodium, potassium, vitamin D) [15, 20, 30–32].

1.3 DIETARY GUIDELINES

The *Dietary Guidelines for Americans* translates scientific information on nutrient requirements and dietary characteristics that promote good health into recommendations and advice for the food intake by the general public. Thus, the *Dietary Guidelines* is the backbone of nutrition education efforts throughout the country. They also reflect nutrition policy in the United States because it provides the basis for the all federal food and nutrition programs, i.e., food stamps; Women, Infants, and Children (WIC); school meal programs; and emergency feeding efforts.

The first edition of the *Dietary Guidelines* was released in 1980, and then it has been revised every 5 years. The sixth, and latest, edition was released in 2005 [33]. The first five editions of the *Guidelines* consisted of 7 or 10 statements providing guidance on how to adopt a pattern of eating that supports good health. The statements were remarkably consistent from one edition to the next [34]. Common themes in all five editions included eating a variety of foods, maintaining body weight, and limiting dietary fat, sugar, sodium, and alcohol intakes. A recommendation to eat foods with adequate starch and fiber in the first two editions evolved into recommendations to choose a diet with plenty of fruits, vegetables, and grains in 1990 and thereafter.

A very different approach was taken by the 2005 Dietary Guidelines Advisory Committee [35]. Specifically, the Committee was charged with conducting an evidence-based review of the scientific literature on diet and health rather than writing a more general document. To address that charge, the Committee initially posed over 40 specific questions related to dietary guidance, thoroughly reviewed the scientific literature pertaining to those questions, and deliberated on the results. Some of the questions were dropped because of incomplete or inconclusive data. Consequently, the Committee wrote conclusive statements and a comprehensive rationale supporting those statements for 34 of the original questions. After some minor revisions and modifications, the 2005 *Dietary Guidelines for Americans* was drafted from the conclusions put forth by the Advisory Committee. The 2005 *Dietary Guidelines* included 23 key recommendations and 18 recommendations for specific population groups, for a total of 41 recommendations [33].

These 41 recommendations are intended to be the primary source of dietary information for policymakers, nutrition educators, and health providers in the United States. However, it is impossible for the public to assimilate and apply so many different recommendations to their own food choices. Therefore, additional documents were developed specifically for the public. One is a bulletin entitled, "Finding your way to a healthier you," written jointly by the Departments of Agriculture and Health and Human Services. The bulletin synthesizes the 41 recommendations from the "policy document" into three primary messages:

- Make smart choices from every food group.
- Find your balance between food and physical activity.
- Get the most nutrition out of your calories.

The bulletin emphasizes the kinds of foods, appropriate amounts, and how often to eat certain foods or food groups. A basic underlying premise of the *Dietary Guidelines* is that nutrient requirements should be met primarily from foods, and that this is best accomplished by choosing a diverse, nutrient-dense diet within one's dietary energy needs. Support for this premise is provided by the research of Foote and colleagues who found that the diets of American adults who met the current DRIs selected a variety of foods daily from each of the five food groups (grains, fruits, vegetables, dairy, and meat/protein) [36].

1.4 DEVELOPMENT OF A FOOD PATTERN MEETING NUTRIENT RECOMMENDATIONS

The key recommendations in the 2005 *Dietary Guidelines* regarding nutrient adequacy emphasize the importance of consuming a variety of nutrient-dense foods and beverages within and among the five basic food groups. The 2005 Dietary Guidelines Advisory Committee recognized, however, that the general public would benefit from guidance on specific food patterns that meet the DRIs. The *Dietary Guidelines'* key recommendations emphasize the types of foods to select, but lack the specificity needed by an individual to make selections that meet his/her requirements within dietary energy needs. Thus, the Advisory Committee collaborated with staff from the US Department of Agriculture's (USDA's) Center for Nutrition Policy and Promotion (CNPP) on their update of the food pattern from the original Food Guide Pyramid to meet the nutrient recommendations from the Institute of Medicine *Dietary Reference Intake* reports [30, 37–41]. This new food pattern is the basis for the revised USDA Food Guidance System, MyPyramid [42].

The method for developing the food pattern, based on the model used to develop the Food Guide Pyramid, involves a five-step process.

1. *Establish energy levels.* Appropriate energy levels for various population groups based on age, gender, and activity level were established using IOM's EER equations [4]. Based on these results, modifications of the food pattern were developed for caloric levels from 1,000 to 3,200 kcal/day in 200-calorie increments.

2. *Establish nutrition goals for the food pattern.* Nutrient goals were the RDAs for vitamins, minerals, electrolytes, and macronutrients published by the IOM between 1997 and 2004 [4–9].

3. *Assign nutrient goals to each specific energy level.* The nutrition goals assigned to each energy level were the goals for age/gender groups that most closely matched the specific energy level. For example, the 1,800 kcal/day level included the goals of females aged 31–50 years, males/females aged 9–13 years, and females aged 14–18 years.

4. *Assign nutrient values for each food group and subgroup.* The nutrient values assigned to each food group (i.e., fruits, milk, meat and beans, whole grains, enriched grains, dark green vegetables, orange vegetables, legumes, starchy vegetables, and other vegetables) were weighted averages of the nutritional value of foods consumed by Americans within that group based on results of the nationwide food consumption surveys (i.e., the National Health and Nutrition Examination Survey 1999–2000). For example, broccoli makes up 44% of the dark green vegetables consumed, spinach is 21%, and the remaining 35% is composed of other dark green vegetables [43]. Therefore, the nutritional values assigned to dark green vegetables were based on 0.44 for broccoli, 0.21 for spinach, and 0.35 for others. Nutrient values for each group were a weighted average of nutrient-dense forms—low-fat and no-added-sugar forms—of the various foods in each group. An exception is that fat-free milk was the single food item used for the milk group.

5. *Determine the daily intake amounts for each food group or subgroup.* Starting from the original Pyramid food pattern at three calorie levels, the amounts of each food group or subgroup were increased or decreased in an iterative manner until the pattern for each of the twelve energy levels met its nutrition goals or came within a reasonable range.

There are advantages and disadvantages to this approach for developing food patterns that meet the RDAs for all Americans. The fact that it builds on the model used for the previous Pyramid provides continuity in food guidance over time. Also, it integrates the entire gamut of IOM recommendations into a lone food intake pattern. Limitations include basing the nutrient profile for each food group on the food consumption patterns of Americans within that group. Americans may not choose foods rich in certain nutrients within that group. For example, Americans eat very few nuts relative to other choices in the meat, poultry, fish, dry beans, eggs, and nuts group. Consequently, the vitamin E content of that food group tends to be low. Thus, the intake pattern, like typical American diets, is low in vitamin E, and it is difficult for most individuals to meet the vitamin E DRI. In the future, separation of this diverse food group into animal and plant sources may improve the capacity to meet the DRI for vitamin E as well as for other nutrients, such as potassium. However, the complexity of the food pattern increases with each subgroup added to the pattern. For example, vegetables were broken down into "dark green," "orange," "legumes," "starchy," and "other"; grains were subdivided into "whole" and "enriched." This was done in recognition of their different nutrient contents and to encourage increased consumption of some subgroups to meet the AIs for several nutrients. Finally, only the lowest fat forms of milk products (i.e., fat-free milk) and lean meats are used in the patterns. If higher fat forms are consumed, that energy needs to be considered to assure that the total intake of energy and saturated fat does not exceed the recommendations. It is important to remember that foods with added sugar will also contribute to the total energy intake.

1.5   RECOMMENDED FOOD PATTERNS FOR PREGNANCY

Specific food patterns are available for pregnant women at the MyPyramid.gov website. The woman is asked to specify her age, due date, height, weight, and physical activity level. She will then receive a menu plan for the first, second, and third trimesters. The amounts of food increase slightly with advancing pregnancy to meet increased energy and nutrient needs in late gestation. We compared the nutrients provided in each of the food intake patterns between 2,000 and 3,000 kcal/d for non-pregnant women to the recommended nutrient intakes for pregnancy and found that the amount of macronutrients, vitamins, and minerals in the patterns for non-pregnant women meet pregnancy standards except for three nutrients—iron, vitamin E, and, to a lesser extent, potassium (Table 1.4). Thus, one could continue to base dietary guidance for pregnant women on the MyPyramid food patterns recommended for non-pregnant women. As mentioned above, the iron DRI during pregnancy, 27 mg/day, is higher than the amount that can be met from foods. At the first prenatal visit, all women are advised, therefore, to take a 30 mg iron supplement daily [44]. The food patterns only provide about 60–80% of the pregnancy vitamin E recommendation. The patterns are also insufficient in vitamin E for nonpregnant adults, providing only 50–70% of the requirement. The Dietary Guidelines Advisory Committee found that it was very difficult to develop food patterns meeting the vitamin E RDAs that also remained within the guidelines for dietary fat since vegetable oils are a primary source of vitamin E. Nuts are also a good source of vitamin E, and the Committee considered making nuts a subgroup of the meat group in order to emphasize their importance in the diet. But, since evidence of health problems among Americans due to insufficient intakes of vitamin E was lacking, the Committee decided to allow the vitamin E intakes to fall short of the RDAs. The vitamin E DRI can be met, however, by selecting vitamin E fortified ready-to-eat cereals, almonds, sunflower seeds, avocados, and certain oils (i.e., sunflower and cottonseed). Potassium intakes in the 2,000-, 2,200-, and 2,400-kcal patterns range from 85 to 95%, the pregnancy DRI. This potassium intake is likely within an acceptable range, but the amount in the diet can be enhanced by increasing milk, white potato, tomato, or orange juice intakes.

The amounts of protein, carbohydrate, fat, and types of fat in the six food patterns are shown in Table 1.5. The percent of energy as protein is about 18%, as carbohydrate about 55%, and as fat about 27%. Saturated fat makes up about 7.4% of the energy, monounsaturated fat about 10%, and polyunsaturated fat about 8%. The recommended

Table 1.4
Nutrient content of 2,000-, 2,200-, 2,400-, 2,600-, 2,800-, and 3,000-kcal food patterns expressed as a percentage of the pregnancy DRI

Nutrient	Pregnancy DRI	%DRI					
		2,000 kcal	2,200 kcal	2,400 kcal	2,600 kcal	2,800 kcal	3,000 kcal
Protein (g/day)	71	128	139	148	154	163	172
Carbohydrate (g/day)	175	155	175	191	211	231	247
Fiber (g/day)	28	111	125	132	146	157	164
Vitamins							
A (mcg/day)	770	137	141	146	161	166	172
E (mg/day)	15	63	67	71	77	82	87
C (mg/day)	85	182	192	192	200	231	267
Thiamin (mg/day)	1.4	143	164	171	186	207	214
Riboflavin (mg/day)	1.4	200	207	221	229	243	243
Niacin (mg/day)	18	122	139	152	163	177	187
B_6 (mg/day)	1.9	126	147	153	163	179	184
Folate (mcg/day)	600	116	128	137	151	164	142
B_{12} (mcg/day)	2.6	319	338	354	362	381	362
Minerals							
Calcium (mg/day)	1,100	120	123	126	131	135	138
Phosphorus (mg/day)	700	249	267	280	295	310	316
Magnesium (mg/day)	350	109	119	126	135	145	152
Iron (mg/day)	27	65	73	80	88	96	95
Zinc (mg/day)	11	130	142	152	160	171	169
Copper (mg/day)	1	150	170	180	200	210	220
Sodium (g/day)	1.5	119	133	140	153	167	180
Potassium (g/day)	4.7	86	94	96	102	109	117

Content of the food patterns is described in Table 1.6. Nutrient values of the food patterns from [42]

Table 1.5
Amounts of Protein, Carbohydrate, Total Fat, and Types of Fat in the Six Food Patterns for Pregnant Women [35]

Calorie levels	2,000	2,200	2,400	2,600	2,800	3,000
Protein (% kcal)	19	19	18	18	17	16
Carbohydrate (% kcal)	56	56	57	57	58	58
Fat (% kcal)	27	28	27	27	27	28
Saturated fat (% kcal)	7.4	7.4	7.4	7.3	7.1	7.4
Monounsaturated fat (% kcal)	10	10	10	10	10	10
Polyunsaturated fat (% kcal)	8	8	8	8	8	9
Linoleic acid						
(g)	16.2	18.2	19.4	21.2	22.6	25.9
(% kcal)	7.3	7.4	7.2	7.4	7.3	7.8
α-Linolenic acid						
(g)	1.6	1.8	1.9	2.1	2.2	2.5
(% kcal)	0.7	0.7	0.7	0.7	0.7	0.8

AI for linoleic acid is 13 g/day during pregnancy and the AI for α-linolenic acid is 1.4 g/day [4]. All six of these food patterns exceed the AI for the polyunsaturated fatty acids, with the linoleic acid intakes ranging from 125 to 199% of the AI, and the α-linolenic acid intakes ranging from 114 to 179%.

The six specific food patterns that can be recommended for individual pregnant women are shown in Table 1.6. The energy levels of the food patterns range from 2,000 to 3,000 kcal/day in 200-kcal increments. This range of energy intakes should cover the energy needs of most pregnant women with BMIs between 18.5 and 25 who have sedentary, moderate, or active lifestyles (see Table 1.2). The 2,200-kcal pattern would be appropriate for a sedentary woman weighing about 60 kg and 15.5 m tall. That pattern includes 2 cups of fruits, 3 cups of vegetables, 200-g (7 oz) equivalents of grains with at least half as whole grains, 171-g (6 oz) equivalents of meat or beans, 3 cups of milk, and 27 g (or about 75 ml) of oils. It is important to remember that the nutrient and energy contributions from each food group are calculated using the most nutrient-dense forms (e.g., lean meats and fat-free milk). Selection of foods with higher fat content will increase the intakes of energy and saturated fat. A small allowance for selecting some foods higher in fat and/or with added sugars is included at each calorie level. Examples of these are a choice of 2% milk, 80% lean ground beef, or cereal with added sugars. In Table 1.6, this allowance is noted as the "discretionary calorie allowance." Selection of foods or beverages with added sugars and with more fat forms should be limited to this allowance.

The standard of care for pregnant women generally involves recommending a prenatal vitamin-mineral supplement. The nutrient analysis of these food patterns shows that there is only a short-fall between the DRIs and nutrient levels in the food patterns for two nutrients—iron and vitamin E. Although a 30 mg/day iron supplement is recommended for all pregnant women, a multivitamin–mineral supplement that provides at least 10 mg iron daily is probably sufficient for women who are adhering to these dietary patterns.

Table 1.6
Amounts of food from each food group in food patterns, with energy levels ranging from 2,000 to 3,000 kcal/day

Food group[a]	Energy level (kcal/day)					
	2,000	2,200	2,400	2,600	2,800	3,000
Fruits[a]	2 cups[a,b]	2 cups	2 cups	2 cups	2.5 cups	2.5 cups
Vegetables	2.5 cups	3 cups	3 cups	3.5 cups	3.5 cups	4 cups
Dark green	3 cups/week	3 cups/week	3 cups/week	3 cups/week	3 cups/week	3 cups/week
Orange	2 cups/week	2 cups/week	2 cups/week	2.5 cups/week	2.5 cups/week	2.5 cups/week
Legumes	3 cups/week	3 cups/week	3 cups/week	3.5 cups/week	3.5 cups/week	3.5 cups/week
Starchy	3 cups/week	6 cups/week	6 cups/week	7 cups/week	7 cups/week	9 cups/week
Other	6.5 cups/week	7 cups/week	7 cups/week	8.5 cups/week	8.5 cups/week	10 cups/week
Grains	6-oz. eq.	7-oz. eq.	8-oz. eq.	9-oz. eq.	10-oz. eq.	10-oz. eq.
Whole	3	3.5	4	4.5	5	5
Enriched	3	3.5	4	4.5	5	5
Meat and Beans	5.5-oz. eq.	6-oz. eq.	6.5-oz. eq.	6.5-oz. eq.	7-oz. eq.	7-oz. eq.
Milk	3 cups	3 cups	3 cups	3 cups	3 cups	3 cups
Oils	27 g	29 g	31 g	34 g	36 g	44 g
Discretionary calorie allowance[c]	267 kcal	290 kcal	362 kcal	410 kcal	426 kcal	512 kcal

For details regarding the types of food items included in each group and subgroup, see [42]

[a]Amounts are per day unless otherwise indicated

[b]The following each count as one cup or oz. equivalent in their respective food groups:

Grains (oz eq): 1/2 cup cooked rice, pasta, or cooked cereal; 1 oz. dry pasta or rice, 1 slice bread; 1 small muffin; 1 cup ready-to-eat cereal flakes

Fruits and vegetables (cup eq.): 1 cup cut-up raw or cooked fruit or vegetable, 1 cup fruit or vegetable juice, 2 cups leafy salad greens

Meat and beans (oz eq.): 1 oz. lean meat, poultry, or fish; 1 egg; 1/4 cup cooked dry beans or tofu; 1 tbsp. peanut butter; 1/2 oz. nuts or seeds

Milk (cup eq): 1 cup milk or yogurt, 1 1/2 oz. natural cheese such as Cheddar cheese or 2 oz. process cheese. Note that discretionary calories must be counted for all choices, except fat-free milk

[c]The discretionary calorie allowance is the amount of calories remaining in each food pattern after selecting the specified number of nutrient-dense forms of foods in each food group. The number of discretionary calories assumes that food items in each food group are selected in nutrient-dense forms (e.g., forms that are fat free or low fat and that contain no added sugars). Solid fat and sugar calories always need to be counted as discretionary calories, as in the following examples:

- The fat in low-fat, reduced-fat, or whole milk, or milk products or cheese and the sugar and fat in chocolate milk, ice cream, pudding, etc.
- The fat in higher fat meats (e.g., ground beef with more than 5% fat by weight, poultry with skin, higher-fat luncheon meats, sausages)
- The sugars added to fruits and fruit juices with added sugars or fruits canned in syrup
- The added fat and/or sugars in vegetables prepared with added fat or sugars
- The added fats and/or sugars in grain products containing higher levels of fats and/or sugars (e.g., sweetened cereals, higher-fat crackers, pies and other pastries, cakes, cookies)

Also, a supplement providing at least 9 mg of vitamin E will bring the total intake up to the DRI recommendation for pregnancy. When prenatal vitamin–mineral supplements are given, it should be emphasized that those supplements do not replace a healthy diet composed of a variety of nutrient-dense foods because an array of other compounds, such as phytochemicals and antioxidants that may benefit health, are present in foods.

1.6 LIFE-CYCLE APPROACH TO NUTRITION

Good nutrition begins *before* conception. A woman's nutritional status at conception can have positive or negative impacts on her pregnancy outcome. For example, inadequate folic acid intake before pregnancy increases the risk of NTDs [45]. Also, iron insufficiency at conception increases the risk for developing anemia during late pregnancy, when the iron demands are high. As previously mentioned, maternal iron deficiency increases preterm births and coincident low birth weights as well as the mother's ability to tolerate hemorrhage during delivery [22]. Entering pregnancy with excessive amounts of body fat stores also increases the risk for metabolic complications during pregnancy such as glucose intolerance or preeclampsia [46]. Implementing a healthy food pattern prior to conception may reduce the prevalence of these complications during pregnancy.

This analysis of the food patterns recommended for nonpregnant women of reproductive age in the United States shows that the same general food patterns can be followed throughout pregnancy, and that the recommended intake of all but two nutrients (iron and vitamin E) will be met. The only change necessary in the second or third trimester is to increase total energy intake by about 200 or 400 kcal to cover the additional energy needed for tissue energy deposition and the metabolic costs of pregnancy. Thus, the food pattern for nonpregnant women only needs minor adjustments for pregnancy. This continuity makes it easier to provide guidance to women planning pregnancies. Furthermore, the general food pattern for pregnant women is appropriate for all family members as well as the mother after pregnancy. This means that dietary counseling provided to pregnant women is a great opportunity to promote good nutrition for everyone in the household. Pregnant women generally tend to have a heightened interest in food and nutrition making this period a very "teachable" time. Teaching a couple of concepts to the pregnant women should enable her to modify the food pattern for all family members. Those concepts are (1) choose diverse, nutrient-dense foods within and among the five food groups every day; (2) make fruits and vegetables part of every meal or snack, and (3) make at least half of grains consumed whole grains.

Surveys show that about only about 3–4% of all Americans follow all of the *Dietary Guidelines* [35]. Although information about the principal sources of foods contributing to the nutrient intakes of pregnant women is scarce, one prospective study showed that low nutrient-dense foods were the major contributors of energy, fat, and carbohydrate whereas fortified foods were the primary sources of iron, folate, and vitamin C [47]. This study was done in a population of black and white women living in North Carolina; over 50% of the women were <185% of the poverty level. Biscuits, muffins, French fries, whole milk, white bread, and soft drinks were the top five food sources of energy. Mayonnaise and salad dressings, cheese and cheese spreads, along with whole milk, French fries, and biscuits/muffins were the top five sources of fat. Whole milk, hamburgers, cheese and cheese spreads, beef steak and roasts, and fried chicken were the top protein sources. Soft drinks and fruit juices were the major sources of carbohydrates. These data suggest that the diets were not only high in total fat, but also high in saturated fat as illustrated by the frequent consumption of high-fat animal products. Soft drinks were often used in place of more nutrient-dense foods or beverages. These food patterns are similar to those reported for African-American adults [48], suggesting that lower income women do not change their food habits appreciably during pregnancy. This population could benefit from prenatal dietary guidance that emphasizes the importance of modifying the food pattern for the entire household.

The 2005 *Dietary Guidelines* makes two specific recommendations for women of reproductive age [33]. The first focuses on the high prevalence of iron deficiency in this population. About 9–11% of adolescent girls and women of childbearing age have laboratory evidence of iron depletion [49]. Consequently, it was recommended that women of childbearing age who may become pregnant should eat foods high in heme iron and/or consume iron-rich plant foods or iron-fortified foods, with an enhancer of iron absorption, such as vitamin C–rich foods. The second recommendation focused on reducing the risk of NTDs. Specifically, women of childbearing age who may become pregnant or those in the first trimester of pregnancy should consume adequate synthetic folic acid daily (from fortified foods or supplements) in addition to food forms of folate from a varied diet. The new folic acid fortification program may influence the need for obtaining folic acid from supplements. However, until further data are available confirming this trend, it seemed prudent to continue to recommend folic acid supplements for women of reproductive age.

ACKNOWLEDGMENTS

The authors acknowledge the contributions of Patricia Britten from the Center for Nutrition Policy and Promotion with the US Department of Agriculture for her advice regarding uses of the MyPyramid food intake patterns for pregnant women, and her sharing the nutrient contents of various MyPyramid patterns with the authors.

REFERENCES

1. Institute of Medicine of the National Academies (2003) Dietary Reference Intakes: applications in dietary planning. National Academy Press, Washington, D.C.
2. Carmichael S, Abrams B, Selvin S (1997) The pattern of maternal weight gain in women with good pregnancy outcomes. Am J Public Health 87:1984–1988
3. Institute of Medicine of the National Academies (1990) Nutrition during pregnancy: weight gain, nutrient supplements. National Academy Press, Washington, D.C.
4. Institute of Medicine of the National Academies (2005) Dietary Reference Intakes for energy, carbohydrate, fiber, fat, fatty acids, cholesterol, protein, and amino acids (macronutrients). National Academy Press, Washington, D.C.
5. Kaiser LL, Allen L (2002) Position of the American Dietetic Association: nutrition and lifestyle for a healthy pregnancy outcome. J Am Diet Assoc 102:1479–1490
6. Schieve LA, Cogswell ME, Scanlon KS et al (2000) Prepregnancy body mass index and pregnancy weight gain: associations with preterm delivery. The NMIHS Collaborative Study Group. Obstet Gynecol 96:194–200
7. Gutierrez Y, King JC (2000) Nutrition during teenage pregnancy. Pediatr Ann 22:99–108
8. Galtier-Dereure F, Boegner C, Bringer J (2000) Obesity and pregnancy: complications and cost. Am J Clin Nutr 71(Suppl):S1242—S1248
9. Vause T, Martz P, Richard F, Gramlich L (2006) Nutrition for healthy pregnancy outcomes. Appl Physiol Nutr Metab 31:12–20

10. Whitaker RC, Dietz WH (1998) Role of the prenatal environment in the development of obesity. J Pediatr 132:768–776

11. Ritchie LD, Ganapathy S, Woodward-Lopez G, Gerstein DE, Fleming SE (2003) Prevention of type 2 diabetes in youth: etiology, promising interventions and recommendations. Pediatr Diabetes 4:174–209

12. King JC (2000) Physiology of pregnancy and nutrient metabolism. Am J Clin Nutr 71(Suppl): S1218–S1225

13. King JC, Calloway, DH, Margen S (1973) Nitrogen retention, total body 40 K and weight gain in teenage pregnant girls. J Nutr. 103(5):772–785

14. Scholl TO, Johnson WG (2000) Folic acid: influence on the outcome of pregnancy. Am J Clin Nutr 71(Suppl):S1295–S1303S

15. Institute of Medicine of the National Academies (1998a) Dietary Reference Intakes: thiamin, riboflavin, niacin, vitamin B_6, folate, vitamin B_{12}, pantothenic acid, biotin, and choline. National Academy Press, Washington, D.C.

16. Honein MA, Paulozzi LJ, Mathews TJ, Erickson JD, Wong LY (2001) Impact of folic acid fortification of the US food supply on the occurrence of neural tube defects. JAMA 285:2981–2986

17. Briefel RR, Johnson CL (2004) Secular trends in dietary intake in the United States. Annu Rev Nutr 24:401–431

18. Centers for Disease Control and Prevention (2005) Medical Progress in the Prevention of Neural Tube Defects. Available via http://www.cdc.gov/ncbddd/bd/mp.htm

19. Thaver D, Saeed MA, Bhutta ZA (2006) Pyridoxine (vitamin B_6) supplementation in pregnancy. Cochrane Database Syst Rev 2006:CD000179

20. Institute of Medicine of the National Academies (2000a) Dietary Reference Intakes: vitamin A, vitamin K, arsenic, boron, chromium, copper, iodine, iron, manganese, molybdenum, nickel, silicon, vanadium, and zinc. National Academy Press, Washington, D.C.

21. Adebisi OY, Strayhorn G (2005) Anemia in pregnancy and race in the United States: blacks at risk. Fam Med 37:655–662

22. Allen LH (2000) Anemia and iron deficiency: effects on pregnancy outcome. Am J Clin Nutr 71(Suppl): S1280–S1284

23. Hallberg L (1988) Iron balance in pregnancy. In: Berger H (ed) Vitamins and minerals in pregnancy and lactation. Raven, New York, pp 115–127

24. Hytten F (1985) Blood volume changes in normal pregnancy. Clin Haematol 14:601–612

25. Centers for Disease Control and Prevention (1998) Recommendations to prevent and control iron deficiency in the United States. MMWR Recommend Report 47(RR-3):1–29

26. Hambidge KM, Neldner KH, Walravens PA (1975) Zinc, acrodermatitis enteropathica, and congenital malformations. Lancet 1:477–578

27. Swanson CA, King JC (1987) Zinc and pregnancy outcome. Am J Clin Nutr 46:763–771

28. Fung EB, Ritchie LD, Woodhouse LR, Roehl R, King JC (1997) Zinc absorption in women during pregnancy and lactation: a longitudinal study. Am J Clin Nutr 66:80–88

29. Ritchie LD, Fung EB, Halloran BP et al (1998) A longitudinal study of calcium homeostasis during human pregnancy and lactation and after resumption of menses. Am J Clin Nutr 67:693–701

30. Institute of Medicine of the National Academies (1997) Dietary Reference Intakes for calcium, phosphorus, magnesium, vitamin D, and fluoride. National Academy Press, Washington, D.C.

31. Institute of Medicine of the National Academies (2000b) Dietary Reference Intakes for vitamin C, vitamin E, selenium, and carotenoids. National Academy Press, Washington, D.C.

32. Institute of Medicine of the National Academies (2004a) Dietary Reference Intake for water, potassium, sodium, chloride, and sulfate. National Academy Press, Washington, D.C.

33. US Department of Health and Human Services, US Department of Agriculture (2005) Dietary Guidelines for Americans. US Department of Health and Human Services, US Department of Agriculture, Washington, D.C.

34. Ballard-Barbash R (2001) Designing surveillance systems to address emerging issues in diet and health. J Nutr 131:437–439

35. Dietary Guidelines Advisory Committee (2005) Report of the Dietary Guidelines Advisory Committee on the Dietary Guidelines for Americans, 2005. National Technical Information Service, Springfield, Va.

36. Foote JA, Murphy SP, Wilkens LR, Basiotis PP, Carlson A (2004) Dietary variety increases the probability of nutrient adequacy among adults. J Nutr 134:1779–1784
37. Institute of Medicine of the National Academies (1998b) Dietary Reference Intakes for thiamin, riboflavin, niacin, vitamin B_6, folate, vitamin B_{12}, pantothenic acid, biotin, and choline. National Academy Press, Washington D.C.
38. Panel on Micronutrients Institute of Medicine (2002) Dietary Reference Intakes: vitamin A, vitamin K, arsenic, boron, chromium, copper, iodine, iron, manganese, molybdenum, nickel, silicon, vanadium, and zinc. National Academy Press, Washington, D.C.
39. Institute of Medicine of the National Academies (2002b) Dietary Reference Intakes for energy, carbohydrate, fiber, fat, fatty acids, cholesterol, protein, and amino acids. Part I. The National Academies Press, Washington, D.C.
40. Institute of Medicine of the National Academies (2004b) Dietary Reference Intakes: water, potassium, sodium, chloride, and sulfate. National Academies Press, Washington, D.C.
41. Food and Nutrition Board Institute of Medicine of the National Academies (2000) Dietary Reference Intakes of vitamin C, vitamin E, selenium, and carotenoids. National Academy Press, Washington, D.C.
42. Britten P, Marcoe K, Yamini S, Davis C (2006) Development of food intake patterns for the MyPyramid Food Guidance System. J Nutr Educ Behav 38(Suppl):S78–S92
43. Marcoe K, Juan W, Yamini S, Carlson A, Britten P (2006) Development of food group composites and nutrient profiles for the MyPyramid Food Guidance System. J Nutr Educ Behav 38(Suppl):S93–S107
44. Centers for Disease Control and Prevention (1998) Recommendations to prevent and control iron deficiency in the United States. MMWR 47:1–36
45. Rush D (1994) Periconceptional folate and neural tube defect. Am J Clin Nutr 59:511–516
46. King JC (2006) Maternal obesity, metabolism, and pregnancy outcomes. Annu Rev Nutr 26:271–291
47. Siega-Riz AM, Bodnar LM, Savitz DA (2002) What are pregnant women eating? Nutrient and food group differences by race. Am J Obstet Gynecol 186:480–486
48. Gary TL, Baptiste-Roberts K, Gregg EW et al (2004) Fruit, vegetable and fat intake in a population-based sample of African Americans. J Natl Med Assoc 96:1599–1605
49. Looker AC, Dallman PR, Carroll MD, Gunter EW, Johnson CL (1997) Prevalence of iron deficiency in the United States. J Am Diet Assoc 277:973–976

2 Optimal Weight Gain

Grace A. Falciglia and Kristin H. Coppage

Summary Optimal birth weight and outcome are influenced by maternal weight gain. Low gestational weight gain is associated with poor fetal growth and risk of preterm delivery. Excessive weight gain affects infant growth, body fatness in childhood, and the potential for postpartum weight retention and future obesity. Guidelines from the Institute of Medicine recommend that a woman with a normal body mass index (BMI) of 19.8 to 26 should gain 11.5–16 kg (25 to 35 lb). Women with a lower-than-normal BMI should gain slightly more, and those with a BMI greater than 26 should gain 5.9–11.5 kg (13 to 25 lb). Ideally, weight gain recommendations should be individualized to promote the best outcomes while reducing risk for excessive postpartum weight retention and reducing the risk of later chronic disease for the child and adult.

Keywords: Gestational weight gain, Energy cost of pregnancy, Body mass index (BMI), Postpartum weight retention, Fetal growth

2.1 INTRODUCTION

Optimal weight gain in pregnancy has distinct implications for both the mother and fetus. It is associated with a favorable outcome for the mother in terms of maternal mortality, complications of pregnancy, labor and delivery, postpartum weight retention, and the ability to lactate. For the newborn, it is defined in terms of fetal growth (birth weight, length, head circumference), gestational age, mortality and morbidity [1, 4]. In considering the relationship between gestational weight gain and pregnancy outcome, attention has centered on birth weight [1, 4]. One reason for this is that birth weight is always recorded and thus is the pregnancy outcome most frequently examined in epidemiological studies. A more fundamental reason for the emphasis on birth weight is its widely recognized association with infant mortality and morbidity. Therefore, optimal gestational weight gain represents a balance between the benefits of appropriate birth weight (>2.5 kg) on the one hand and the possible risks to the mother and infant, including complicated labor and delivery with increased birth weight (>4 kg) and metabolic complications including insulin resistance, on the other.

Factors associated with gestational weight gain include maternal prepregnancy weight for height, ethnic background, age, parity, cigarette smoking, socioeconomic status, and

From: *Nutrition and Health: Handbook of Nutrition and Pregnancy*
Edited by: C.J. Lammi-Keefe, S.C. Couch, E.H. Philipson © Humana Press, Totowa, NJ

energy intake [1, 4]. Accordingly, gestational weight gain should be individualized rather than generalized due to the differences in body size and lifestyles, e.g., underweight pregnant women have different weight gain needs than do overweight and obese women. Superimposed on this is high variability in activity patterns and energy intake.

The *components of weight gain* include the products of conception and maternal tissue accretion. The products of conception comprise the fetus, placenta, and amniotic fluid. The fetus represents 25% of the total gain, the placenta approximately 5%, and the amniotic fluid approximately 6% [5]. Expansion of maternal tissues accounts for approximately two thirds of the total gain. The expansion includes maternal blood volume and extracellular fluid along with an increase in uterine and mammary tissues, and fat stores. Expansion of the blood volume and extracellular fluid accounts for 10 and 13% of the total weight gain, respectively [5]. Fat and protein accretion accounts for the remainder.

In terms of the *composition of the weight gain*, Hytten and colleagues [5] estimated that on average, water contributes approximately 62% of the total gain at term, fat contributes 30%, and protein contributes 8%. Most of the total fat gain is deposited in maternal stores. Protein is deposited predominantly in the fetus, but also in the uterus, blood, placenta, and breasts.

2.2   GESTATIONAL WEIGHT GAIN RECOMMENDATIONS

Although the need for appropriate weight gain during pregnancy has long been recognized, recommendations for weight gain have changed over the years as new data have become available. The changes in recommended ranges for gestational weight gain are summarized in Table 2.1.

Prior to 1970, it was standard obstetric practice to restrict gestational weight gain to between 18 and 20 lb (8–9 kg) [2]. Overeating was believed to cause large babies and, as a consequence, more difficult deliveries.

In 1970, the Food and Nutrition Board's Committee on Maternal Nutrition [6] recommended a higher gestational weight gain, 20–25 lb (9–11.5 kg). The increase was based on new evidence that low weight gain was related to increased risk of delivering low-birth-weight infants, with those infants at increased risk of mortality and developmental problems. This recommendation was followed by heightened interest in helping pregnant women achieve appropriate weight gain and nutrient intake. For example, the US Department of Agriculture (USDA) established the Special Supplemental Food Program for Women, Infants, and Children (WIC) to provide both food and nutrition education for nutritionally vulnerable women. This change in gestational weight gain recommendation, among other factors such as participation in the WIC program, contributed to an increase in gestational weight gain and fetal growth, as evidenced by an increase in mean birth weight, and a reduction in low birth weights [2].

In 1990, the Food and Nutrition Board of the Institute of Medicine (IOM) [1] recommended gestational weight ranges for women on the basis of their prepregnancy BMI. BMI is calculated by the weight in kilograms divided by the square of height in meters (kg/m^2). The recommended total weight gain ranges were 28–40 lb (12.5–18 kg) for women with low BMI (<19.8), 25–35 lb (11.5–16 kg) for women with normal BMI (19.8–26), and 15–25 lb (7–11.5 kg) for women with high BMI (>26–29). It was recommended that young adolescents and African American women should strive for gains at the upper end of the recommended range. Short women (<157 cm)

Table 2.1

Historical Perspective: Total Gestational Weight Gain Recommendations

Reference	Year	Recommended total gain (kg)	(lb)
Standard obstetric practice[a]	Prior to 1970	8–9	18–20
National Research Council, Food & Nutrition Board Committee on Maternal Nutrition	1970	9–11.5	20–25
Institute of Medicine[b] Food & Nutrition Board	1990	11.5–16	25–35
World Health Organization	1995	10–14	22–31

[a] From [2]

[b] Recommended total weight gain for pregnant women with normal body mass index (19.8–26)

should strive for gains at the lower end of the range. The recommended target weight gain for obese women (BMI > 29) was at least 15 lb (6 kg). These ranges were derived from the 1980 US National Natality Survey and based on the observed weight gains of women delivering full-term (39–41 weeks), normal-growth (3–4 kg) infants without complications. Furthermore, the rate of weight gain recommended by the IOM was approximately 1 lb (0.4 kg) per week in the second and third trimesters of pregnancy for women with a normal prepregnancy BMI, slightly more than 1 lb (0.5 kg) per week for underweight women, and 0.66 lb (0.3 kg) per week for overweight women.

A comprehensive review of studies conducted by Abrams [7] showed that gestational weight gains within the IOM's recommended ranges were associated with the best outcome for both infants, in terms of birth weight, and for mothers, in terms of delivery complications and postpartum weight retention.

In 1995, the World Health Organization (WHO) Collaborative Study on Maternal Anthropometry and Pregnancy Outcomes [4] reviewed information on 110,000 births from 20 different countries to define desirable maternal weight gain. The range of gestational weight gain associated with birth weights greater than 3 kg was 22–31 lb (10–14 kg). Comparing the WHO weight ranges with the IOM's recommended weight ranges for women with low and normal prepregnancy BMI, the WHO's ranges are slightly lower than the IOM's ranges (10–14 kg versus 12.5–18 kg [low BMI] and 11.5–16 kg [normal BMI]).

In 2000, the US Department of Health and Human Services released the *Healthy People 2010* document, with specific objectives for maternal and infant health [8]. One of these objectives (developmental) refers to increasing the proportion of mothers who achieve a recommended weight gain during their pregnancies. Unfortunately, approximately 50% of women receive no prenatal advice or inappropriate advice regarding gestational weight gain [9]. Cogswell and colleagues [10] found that among those who received advice, 14% were advised to gain less than the recommended weight, while 22% were advised to gain more than recommended. Further, the probability of being advised to gain more than the recommended weight were higher in women with high BMI (>26 kg/m^2). When no advice on gestational weight gain was given, pregnant women tended to gain outside the IOM recommendations. Two groups of women who continue to gain less than the recommended level of weight during pregnancy are adolescents and African American women. As a result, they are at particular high risk for having low-birth-weight infants and preterm delivery. On the other hand, women who are overweight or obese are more likely than are women of normal weight to gain more weight than is recommended [11]. Exceeding the recommendations for weight gain was also found to be more likely in low-income women [12]. Outcomes related to excessive gestational weight gain include postpartum weight retention for the mother, future maternal overweight or obesity, and for the infant, childhood overweight/obesity [13]. Future attention should be directed to the relationship of pregnancy body weight gain to body fat gain for the development of dietary and weight gain recommendations.

2.3   ENERGY COST OF PREGNANCY

Defining the energy cost of pregnancy requires that optimal gestational weight gain be established. Extra dietary energy is required during pregnancy for the energy deposited in maternal/fetal tissues, the rise in energy expenditure attributable to increased basal metabolism, and to changes in the energy cost of physical activity [14, 15].

2.3.1   Protein and Fat Deposition

During pregnancy, protein is mainly deposited in the fetus (42%). The remainder is accounted for by the gain of uterine (17%), blood (14%), placenta (10%), and mammary (8%) tissues. The increment in total body protein in well-nourished pregnant women has been estimated from changes in total body potassium. King et al. [16], Pipe et al. [17], Forsum et al. [18], and Butte et al. [19] estimated that an average of 686 g of protein are deposited unequally throughout pregnancy, mainly in late pregnancy. Butte and colleagues [15] studied total protein deposition in relation to BMI and reported that protein accretion did not differ significantly among low to high BMI groups.

Total fat accretion, the major contributor to energy deposition, was studied in well-nourished pregnant women using multicomponent body composition models based on total body water, body volume, and body mineral content [17, 18, 20–25, 28, 30]. The average estimate was 3.7 kg (range = 2.4–5.9). Mean fat gains in the study of Butte and colleagues [15] were 5.3 kg, 4.6 kg, and 8.4 kg for women in the low-, normal-, and high-BMI groups, respectively. Also, maternal fat retention at 27 weeks postpartum was significantly higher in women who gained weight above the IOM recommendations than in those who gained weight within or below those recommendations.

2.3.2   Basal Metabolism

The energy cost for maintenance rises throughout pregnancy due to increased tissue mass. Several longitudinal studies [18, 20, 26–28, 30, 31] in which changes in basal metabolic rate (BMR) or resting metabolic rate (RMR) were measured have been published. The first, BMR, is measured in the morning after awakening, whereas RMR is measured at any time during the day, after resting for at least 30 min. Both measurements reflect lean body tissue. In these studies, BMR increased over prepregnancy or early pregnancy values by 5, 11, and 24% in the first, second, and third trimesters, respectively. When BMR increase was examined in relation to BMI [15], the values were similar for the women in the low and normal BMI groups, but the increase in BMR was greater for women with high BMI (7, 16, and 38% across trimesters). A common characteristic was the wide variability in metabolic response among women. BMR decreased during the first and second trimesters in some women and increased steadily during pregnancy in others.

2.3.3   *Total Energy Expenditure*

In pregnancy, total energy expenditure (TEE) has been estimated by respiratory calorimetry or by using a doubly labeled water method (DLW) [14, 15]. Whole-room, 24-h respiratory calorimetry demonstrates changes in the components of TEE under standardized protocols, e.g., sedentary conditions and minimal daily energy expenditure for basic survival, but makes no allowance for free-living physical activity. The DLW method, complemented with a measure of BMR, provides a quantitative estimate of the amount of energy expended in physical activity (AEE).

Whole-room respiration calorimetry studies carried out in well-nourished women [28, 29, 37] revealed that EE increases above prepregnancy values on average by 1, 4, and 20% in the first, second, and third trimesters, respectively. This increment was due largely to the increase in BMR.

Free-living TEE has been measured by DLW in well-nourished women [15, 30–33]. In these studies, TEE increased on average by 1, 6, and 19% over baseline values in the first, second, and third trimesters. Furthermore, BMR increased by 2, 9, and 24%, and AEE (TEE – BMR) changed by −2, 3, and 6% relative to baseline. On the basis of the larger increment in BMR, physical activity decreased as pregnancy advanced. These findings support the idea that women may conserve energy by reducing the pace or the intensity with which an activity is performed. Pregnant women may also change their activity patterns and thereby reduce the amount of time spent in activities. However, reduction in physical activity does not compensate for increases in BMR and energy deposited in maternal and fetal tissues. Thus, extra dietary energy is ordinarily required as pregnancy progresses [15].

2.3.4   *Total Energy Cost of Pregnancy*

The total energy cost of pregnancy in well-nourished women has been estimated from the sum of BMR or TEE and energy deposited in laying down maternal and fetal tissues [14]. Gestational weight gain is a major determinant of the incremental energy needs during pregnancy because it reflects not only energy deposition, but also the increase in BMR and TEE resulting from the energy cost of metabolism and moving a larger body mass. Using the gestational weight gain recommended by IOM for pregnant women with normal BMI (mean value = 13.8 kg), the estimated total energy cost of pregnancy was about 88,850 kcal [14]. The incremental needs for energy are not equally distributed over pregnancy. For example, energy needs associated with protein deposition occur primarily in the second and third trimesters, while the distribution of energy deposited as fat, is about 11, 47, and 42% across trimesters. Thus, the increments in BMR and TEE are greater in the second half of pregnancy. Reflecting the deposition of new-weight tissue across pregnancy, the average distribution of energy was estimated at approximately 105, 330, and 535 kcal per day for the first, second, and third trimesters, respectively. Using the gestational weight gain recommended by the WHO Collaborative Study on Maternal Anthropometry and Pregnancy Outcomes (mean value = 12 kg), the total energy cost of pregnancy was estimated at about 77,147 kcal, distributed as 90, 286, and 465 kcal per day for the first, second, and third trimesters, respectively [14].

The 2004 Dietary Reference Intakes (DRI) recommendations for energy intake in pregnant women with a normal prepregnancy BMI are an additional 0, 340, and 452 kcal/day for the first, second, and third trimesters, respectively, over the nonpregnant state [34]. The 2005 USDA Dietary Guidelines for Americans [35] and the 2005 USDA MyPyramid [36] provide dietary guidance to meet the energy needs associated with pregnancy, as described and discussed in Chap. 1 ("Nutrient Recommendations and Dietary Guidelines for Pregnant Women").

2.3.5 Metabolic Adaptations

Under certain physiological conditions, such as undernutrition, adjustments may occur in BMR, efficiency in performing work, and thermogenesis to meet the increased energy requirements of pregnancy. For example, Gambian women showed a pronounced suppression of basal metabolism that persisted into the third trimester of pregnancy, and although there was a later increase in BMR, the overall effect was in fact a slight net saving of energy over the entire gestational period. On the other hand, in well-nourished women, BMR usually begins to rise soon after conception and continues to increase until delivery [37].

King and colleagues [14] reviewed previous studies in which changes in energy efficiency for both weight-bearing (treadmill exercise) and non–weight-bearing (cyclo-ergometer exercise) activities were measured at standard pace and/or intensity [38]. The net energy cost of non–weight-bearing activity did not change until the last month of pregnancy, at which time it increased by approximately 10%. During the first two trimesters, the net energy cost of weight-bearing activity remained stable but then increased in the third trimester by about 15%. As most women had already gained an average of 6 kg by the end of the second trimester, when the net energy cost of weight-bearing activities remained stable, the data suggest that those activities were performed with higher efficiency in late pregnancy.

The thermic effect of feeding refers to the increase in energy expenditure above basal metabolism after the ingestion of food. This increase is related to the energy costs of digestions, absorption, transport, and storage and is usually about 10% of energy intake. In studies of pregnant women, the thermic effect of feeding, as a percent of total energy intake, has been shown to be unchanged [39–41] or lower [42] than those values of nonpregnant women. Although metabolic adaptations represent powerful mechanisms for sustaining pregnancy under marginal nutritional conditions, they should not be viewed as perfect processes that eliminate the need for proper energy intake required to maintain optimal fetal growth.

2.4 BODY WEIGHT CHANGES AFTER PREGNANCY

Both mean gestational weight gain and prevalence of overweight women in the US population have increased over the past two decades [44, 45]. Gunderson and Abrams [43] reviewed the literature to examine whether increased gestational weight gain is responsible in part for the increasing prevalence of overweight women. The majority of the epidemiological studies provided data on average body weight gain among pregnant women, without comparison groups. The estimate for weight gain from these studies ranged from 1.4 kg to 1.5 kg by 6–12 months postpartum [46–48]. Gunderson and Abrams also reviewed data from the 1988 National Maternal and Infant Health Survey (NMIHS) [44], a US representative sample, which revealed a median weight change of 1 kg at 10–18 months postpartum. These relatively small average maternal weight increases suggest that pregnancy may not have a significant influence on body weight for a sizable percentage of women. However, approximately 20% of the women studied experienced a 5 kg or greater weight increase after pregnancy [44, 46–48]. In the NMIHS, almost 16% of women were more than 6.4 kg heavier by 10–18 months postpartum. Therefore, use of the mean value to assess body weight increases after pregnancy fails to adequately reflect the population at risk [44].

In a recent study of over 1,000 mother–child pairs investigators found that mothers with greater gestational weight gain had children with more adiposity at 3 years of age, measured by skin-fold thickness as well as by BMI [51]. This association was independent of parental BMI, maternal glucose tolerance, breastfeeding duration, fetal and infant growth, and child behaviors. Children of mothers who gained more weight also had somewhat higher systolic blood pressure, a cardiovascular risk factor related to adiposity even in young children. Noticeably, mothers with adequate gain, as recommended by the Institute of Medicine [1], had a substantially high risk of having children who were overweight. This new evidence suggests that the current recommendations for gestational weight gain may need to be revised in this era of epidemic obesity.

2.5 APPLICATIONS

With the understanding that total weight gain during pregnancy varies widely among women with similar ages, weights, heights, ethnic backgrounds, and socioeconomic status, recommendations for weight gain should be used only as guides. The following clinical recommendations based on current knowledge [3, 49,50] can aid practitioners in defining the weight gain goals for achieving a desirable body weight for women throughout pregnancy and postpartum. The recommendations relate to anthropometric measurements and counseling and are summarized in Table 2.2.

Table 2.2
General Guidelines for Achieving Optimal Gestational Weight Gain

Preconception

1. Assess weight and height
2. Assess dietary intake
4. Offer guidance regarding healthy eating
5. Provide individualized care to address risk factors such as undernutrition and obesity

During Pregnancy

1. Assess weight and height
2. Determine BMI based on height and prepregnancy weight
3. Recommend overall and incremental weight gain and monitor weight gain throughout pregnancy
4. Assess dietary intake
5. Provide education to all women on weight gain and healthy eating
6. Provide individual assessment and counseling, by a dietitian/Women, Infants and Children (WIC) Food Program, for women with inadequate or excessive weight gain

Postpartum Period

1. Provide advice regarding weight management related to postpartum weight retention
2. Encourage healthy eating

Adapted from [50]

2.5.1 Anthropometric Measures

Prepregnancy weight is an important measurement. Objective data, such as those obtained from a medical record, are preferred over self-reported values. Information provided by the patient should be evaluated for its accuracy.

At the first prenatal visit, the woman's height without shoes should be determined, preferably with a wall stadiometer, the accuracy of which has been verified. Gestational age should be determined from the onset of the woman's last menstruation, supplemented by estimates based on the obstetric clinical examination and by early ultrasound if available. A weight-for-height category derived from the patient's height and prepregnancy weight needs to be established. The resulting BMI should be compared to reference values for BMI. This comparison will provide the bases for the creation of a plan for overall and incremental weight gain and dietary counseling. Other measures that provide information on body composition would add substantially to understanding of the meaning of a given weight gain. Fetal growth may be influenced more by specific maternal tissue changes (accretion of lean tissue, fat or body water), than by total gestational weight gain. For example, skin-fold thickness has potential for clinical use but needs to be standardized for pregnant women; this has yet to be done.

At the beginning of each prenatal visit, the woman's weight should be assessed. Consistency in the method used for obtaining weight is necessary (e.g., without outdoor clothing, purse, and shoes). Values should be recorded on a chart that shows weight gain by gestational age (weight on the vertical axis and week of gestation on the horizontal axis). A slightly lower or higher rate of weight gain than the recommended is not cause for alarm, as long as there is a progressive increase in weight that approximately equals the recommended rate of gain. Reasons for abrupt or persistent deviations from the expected pattern of gain should be investigated. Health care providers should be trained in proper measurement techniques and the equipment used should be calibrated periodically.

2.5.2 Counseling

Ideally, healthy eating patterns should be established before pregnancy. During pregnancy, women may be particularly receptive to guidance regarding behaviors that may influence their health and that of their developing babies. All women should be encouraged to gain enough weight to achieve at least the lower limit of the weight range specified for their weight-for-height category. To help women achieve desirable gestational weight gain, they should be given appropriate dietary information or referred to a dietitian or WIC program to learn how to obtain adequate nutrients within calorie needs. Women should be encouraged to accept a diet rich in a variety of healthy foods consistent with ethnic, cultural, and financial considerations. The USDA Dietary Guidelines for Americans [35] and the USDA MyPyramid [36] provide specific guidance for women of childbearing age who may become pregnant and for pregnant women.

Continuation of dietary counseling after delivery is necessary to regain prepregnancy weight before attempting another pregnancy.

REFERENCES

1. Institute of Medicine and Food and Nutrition Board (1990) Total amount and pattern of weight gain: physiologic and maternal determinants. In: Nutrition during pregnancy. National Academy Press, Washington, D.C., pp 96–120

2. Institute of Medicine and Food and Nutrition Board (1990) Historical trends in clinical practice, maternal nutritional status, and the course and outcome of pregnancy. In: Nutrition During Pregnancy. National Academy Press, Washington, D.C., pp 37–62

3. Institute of Medicine and Food and Nutrition Board (1990) Assessment of gestational weight gain. In: Nutrition during pregnancy. National Academy Press, Washington, D.C., pp 63–95

4. World Health Organization (1995) Maternal anthropometry and pregnancy outcomes—a WHO Collaborative Study. Bulletin of the World Health Organization 73:1–69

5. Hytten, FE (1980) Weight gain in pregnancy. In: Hytten F, Chamberlain G (eds) Clinical Physiology in Obstetrics. Blackwell Scientific, Oxford, pp 193–233

6. National Research Council (1970) Maternal nutrition and the course of pregnancy. National Academy of Sciences, Washington, D.C.

7. Abrams B, Altman SL, Pickett KE (2007) Pregnancy weight gain: still controversial. Am J Clin Nutr 71:1233–1241

8. US Department of Health and Human Services (2000) Maternal, infant, and child health. In: Healthy People 2010: understanding and improving health, 2nd edn. US Government Printing Office, Washington, D.C., 16:3–56

9. Kleinmen RE (ed) (2004) Pediatric nutrition handbook, 5th edn. American Academy of Pediatrics, Elk Grove Village, Ill., 167–190

10. Cogswell ME, Scanlon KS, Fein SB et al (1999) Medically advised, mother's personal target, and actual weight gain during pregnancy. Obstet Gynecol 94:616–622

11. Strychar IM, Chabot C, Champagne F et al (2000) Psychosocial and lifestyle factors associated with insufficient and excessive maternal weight gain during pregnancy. J Am Diet Assoc 100:353–356

12. Schieve LA, Cogswell ME, Scanlon KS (1998) Trends in pregnancy weight gain within and outside ranges recommended by the Institute of Medicine in a WIC population. Matern Child Health J 2:111–116

13. Linne Y, Dye L, Barkeling B et al (2003) Weight development over time in parous women-the SPAWN study-15 years' follow-up. Int J Obes Relat Metab Disord 27:1516–1522

14. Butte NF, King JC (2005) Energy requirements during pregnancy and lactation. Pub Health Nutr 8:1010–1027

15. Butte NF, Wong WW, Treuth MS, Ellis K, Smith EO (2004) Energy requirements during pregnancy based on total energy expenditure and energy deposition. Am J Clin Nutr 79:1078–1087

16. King JC, Calloway DH, Margen S (1973) Nitrogen retention, total body ^{40}K and weight gain in teenage pregnant girls. J Nutr 103:772–785

17. Pipe NGJ, Smith T, Halliday D, Edmonds CY, Williams C, Coltart TM (1979) Changes in fat, fat-free mass and body water in normal human pregnancy. Br J Obstet Gynaec 86:929–940

18. Forsum E, Sadurskis A, Wager J (1988) Resting metabolic rate and body composition of healthy Swedish women during pregnancy. Am J Clin Nutr 47:942–947

19. Butte NF, Hopkinson JM, Ellis K, Wong WW, Treuth MS, Smith EO (2003) Composition of gestational weight gain impacts maternal fat retention and infant birth weight. Am J Obstet Gynec 189:1423–1432

20. Spaaij CJK (1993) The efficiency of energy metabolism during pregnancy and lactation in well–nourished Dutch women. The University of Wageningen, Wageningen, The Netherlands

21. Raaij JMA van, Peek MEM, Vermaat–Miedema SH, Schonk CM, Hautvast JGAJ (1988) New equations for estimating body fat mass in pregnancy from body density or total body water. Am J Clin Nutr 48:24–29

22. Lindsay CA, Huston L, Amini SB, Catalano PM (1997) Longitudinal changes in the relationship between body mass index and percent body fat in pregnancy. Obstet Gynecol 89:377–382

23. Lederman SA, Paxton A, Heymsfiled SB, Want J, Thronton J, Pierson RN Jr (1997) Body fat and water changes during pregnancy in women with different body weight and weight gain. Obstet Gynecol 90:483–488

24. Kopp–Hoolihan LE, Van Loan MD, Wong WW, King JC (1999) Fat mass deposition during pregnancy using a four-component model. J Appl Physiol 87:196–202

25. Sohlström A, Forsum E (1997) Changes in total body fat during the human reproductive cycled as assessed by magnetic resonance imaging, body water dilution, and skinfold thickness: a comparison of methods. Am J Clin Nutr 66:1315–1322

26. Durnin JVGA, McKillop FM, Grant S, Fitzgerald G (1987) Energy requirements of pregnancy in Scotland. Lancet 2:897–900
27. Raaij JMA van, Vermaat–Miedema SH, Schonk CM, Peek MEM, Hautvast JGAJ. Energy requirements of pregnancy in The Netherlands. Lancet 2:953–955
28. de Groot LCPGM, Boekholdt HA, Spaaij CJK, van Raaij JMA, Drijvers JJMM, van der Heihden LJM, Hautvast JGAJ (1994) Energy balances of Dutch women before and during pregnancy: limited scope for metabolic adaptations in pregnancy. Am J Clin Nutr 59:827–832
29. Butte NF, Hopkinson JM, Mehta N, Moon JK, Smith EO (1999) Adjustments in energy expenditure and substrate utilization during late pregnancy and lactation. Am J Clin Nutr 69:299–307
30. Goldberg GR, Prentice AM, Coward WA, Davies HL, Murgatroyd PR, Wensing C, Black AE, Harding M, Sawyer M (1993) Longitudinal assessment of energy expenditure in pregnancy by the doubly labeled water method. Am J Clin Nutr 57:494–505
31. Kopp-Hoolihan LE, Van Loan MD, Wong WW, King JC (1999) Longitudinal assessment of energy balance in well-nourished, pregnant women. Am J Clin Nutr 69:697–704
32. Goldberg GR, Prentice AM, Coward WA, Davies HL, Murgatroyd PR, Sawyer MB, Ashford J, Black AJ (1991) Longitudinal assessment of the components of energy balance in well-nourished lactating women. Am J Clin Nutr 54:788–798
33. Forsum E, Kabir N, Sadurskis A, Westerterp K (1992) Total energy expenditure of healthy Swedish women during pregnancy and lactation. Am J Clin Nutr 56:334–342
34. Institute of Medicine (2002) Energy. In: Dietary Reference Intakes for energy, carbohydrate, fiber, fat, fatty acids, cholesterol, protein, and amino acids. National Academies Press, Washington, D.C. pp 1–114
35. US Department of Health and Human Services & US Department of Agriculture (2005) Dietary guidelines for Americans. Home and Garden bulletin no. 232, US Department of Health and Human Services, Washington, D.C.
36. US Department of Agriculture Center for Nutrition Policy and Promotion (2005) MyPyramid Food Guidance System. Available via http://www.mypyramid.gov.
37. Prentice AM, Goldberg GR, Davies HL, Murgatroyd PR, Scott W (1989) Energy sparing adaptations in human pregnancy assessed by whole-body calorimetry. Br J Nutr 62:5–22
38. Prentice AM, Spaaij CJK, Goldberg GR, Poppitt SD, van Raaij JMA, Totton M, Swann D, Black AE (1996) Energy requirements of pregnant and lactating women. Euro J Clin Nutr 50:82–111
39. Bronstein MN, Mak RP, King HC (1995) The thermic effect of food in normal-weight and overweight pregnant women. Br J Nutr 74:261–275
40. Nagy LE, King JC (1984) Postprandial energy expenditure and respiratory quotient during early and late pregnancy. Am J Clin Nutr 40:1258–1263
41. Spaaij CJK, van Raaij JMA, van der Heihden LJ, Schouten FJ, Drijvers JJ, de Groot LC, Boekholt HA, Hautvast JG (1994) No substantial reduction of the thermic effect of a meal during pregnancy in well-nourished Dutch women. Br J Nutr 71:335–344
42. Schutz Y, Golay A, Jéquier E (1988) 24-h energy expenditure (24-EE) in pregnant women with a standardized activity level. Experientia 44(abstract):A31
43. Gunderson EP, Abrams B (2000) Epidemiology of gestational weight gain and body weight changes after pregnancy. Epidemiol Rev 22:261–274
44. Keppel KG, Taffel SM (1993) Pregnancy-related weight gain and retention: Implications of the 1990 Institute of Medicine guidelines. Am J Public Health 83:1100–1103
45. Kuczmarski RJ, Flegal KM, Campbell SM, Johnson CL (1994) Increasing prevalence of overweight among US adults. The National Health and Nutrition Examination Surveys, 1960 to 1991. JAMA 272:205–211
46. Schauberger CW, Rooney BL, Brimer LM (1992) Factors that influence weight loss in puerperium. Obstet Gynecol 79:424–429
47. Ohlin A, Rossner S (1990) Maternal body weight development after pregnancy. Int J Obes 14:159–173
48. Green GW, Smiciklas-Wright H, Scholl TO, Karp RJ (1988) Postpartum weigh change: how much of the weight gained in pregnancy will be lost after delivery? Obstet Gynecol 71:710–717
49. Suitor CW (1997) Maternal weight gain: a report of an expert work group. National Center for Education in Maternal and Child Health, Arlington, Va.
50. Abrams B, Pickett KE (1999) Maternal nutrition. In: Creasy RK, Resnik R (eds) Maternal–fetal medicine. Saunders, Philadelphia, Pa., pp 122–131
51. Oken E, Taveras E, Kleinman KP, Rich-Edwards JW, Gillman MW (2007) Gestational weight gain and child adiposity at age 3 years. Am J Obstet Gynecol 196:322.e1–322.e8

3 Physical Activity and Exercise in Pregnancy

Rose Catanzaro and Raul Artal

Summary The benefits of exercise in the general population have been well-recognized. There is ample evidence to demonstrate that moderate exercise in a healthy pregnancy results in no adverse effects and provides consequential benefits.

Despite anatomical and physiological changes in pregnancy, women with healthy pregnancies and without contraindications can combine aerobic and resistance elements in their workouts. Clinical evaluation by an obstetrician is recommended before beginning an exercise program. Consideration must be given to the type, intensity, duration, and frequency of exercise when providing the patient with an exercise prescription. Scuba diving and contact sports or exercises with a high risk of falling or abdominal trauma should be avoided. Women who are beginning an exercise program during pregnancy should start slowly and gradually increase to moderate intensity. Women engaging in strenuous physical activities require additional medical supervision.

The nutritional needs of active pregnant women are not clearly defined; however, it should be recognized that there is an additional caloric allowance for increased metabolism and greater energy expenditure both during and after activity. Pregnant women use carbohydrates at a higher rate than do nonpregnant women; this is further increased during exercise, thus adequate carbohydrate intake is essential. Adequate fluid intake helps control the core body temperature and is essential to replace fluid loss during exercise.

Because habits adopted during pregnancy can result in persistent lifestyle improvements, exercise during pregnancy could significantly reduce the lifetime risks of obesity, chronic hypertension and diabetes—not only for pregnant women, but also for their families as well. Overall, a woman whose exercise habits have become firmly entrenched during pregnancy stands a much better chance of maintaining them after her child is born.

Keywords: Physical activity, exercise, nutrition, pregnancy

3.1 INTRODUCTION

Pregnancy is a unique time in a woman's life in which health awareness increases, and she may be more inclined to accept medical advice to either adopt or continue an active lifestyle. Exercise is considered safe for most women during pregnancy as long as

From: *Nutrition and Health: Handbook of Nutrition and Pregnancy*
Edited by: C.J. Lammi-Keefe, S.C. Couch, E.H. Philipson © Humana Press, Totowa, NJ

there are no medical or obstetrical complications [1]. Although physical activity is often considered part of a healthy lifestyle and leisure time activity, some pregnant women may choose to participate in highly competitive sports.

While the health benefits of physical activity are well recognized in the general population [2–6], exercise is still not adequately accepted as a benefit for pregnant women. Healthcare providers remain cautious and often reluctant to encourage exercise during pregnancy, despite the well-recognized benefits. The hesitation is set in conservative ideas that pregnancy is a time of confinement. With abundant evidence to show that moderate exercise in healthy pregnancies results in benefits and without adverse maternal or fetal outcomes, exercise recommendations made by healthcare providers should be a top priority. It is well recognized that healthy lifestyle behaviors adopted in pregnancy can result in persistent lifestyle modifications that could significantly reduce risk factors associated with obesity, chronic hypertension, and diabetes—not only for the pregnant mother, but also for all members of her family.

In this chapter, we review physiological changes that provide the basis for exercise guidelines and nutritional recommendations in pregnancy, as well as maternal and fetal responses to the potential risks and benefits associated with exercise in pregnancy.

3.2   PHYSIOLOGICAL CHANGES IN PREGNANCY

Under the influence of estrogen, progesterone, and elastin, pregnancy is associated with generalized connective tissue laxity, potentially leading to ligament and joint instability [1]. Additional strain on the musculoskeletal system comes from the change in the body's center of gravity, resulting in progressive lordosis (accentuation of the lumbar curvature of the spine) and kyphosis (curvature of the upper spine) [7]. The change in center of gravity requires greater muscular effort with certain movements, such as rising from a squatting or sitting position or changing directions quickly. The progressive lordosis in pregnancy frequently results in lower back pain, which could be prevented by improving posture and muscular strength preferably prior to pregnancy [8]; such preventative measures are also effective during pregnancy [9]. Providing exercise guidelines to increase core strength prior to pregnancy minimizes these injuries.

Special consideration must be given to changes that occur during each trimester of pregnancy that could result in injury from physical activity in pregnancy. Physical activity in pregnancy can be affected by the following progressive anatomical and physiological changes: change in center of gravity, increased connective tissue laxity resulting in joint instability, lordosis and kyphosis, generalized edema possibly resulting in nerve compression syndrome, increase in blood volume, tachycardia, hyperventilation, and reductions in cardiac reserve and residual lung capacity [10]. Figure 3.1 lists the etiology for potential injury that can occur during exercise in pregnancy and the gestational age during which the injury is most likely to occur. The goal of exercise is to maintain physical fitness within the physiological limitations of pregnancy. Exercise prescriptions should be geared towards muscle strengthening to minimize risk of joint injury and towards correcting postural changes thus diminishing lower back pain.

Physical activity may increase uterine activity (contractions). The effect of exercise on uterine activity has little or no change during the last 8 weeks of pregnancy [11]. While there are no studies reporting that strenuous activity results in preterm labor, until the impact is fully studied, women at risk of preterm labor should be advised to reduce

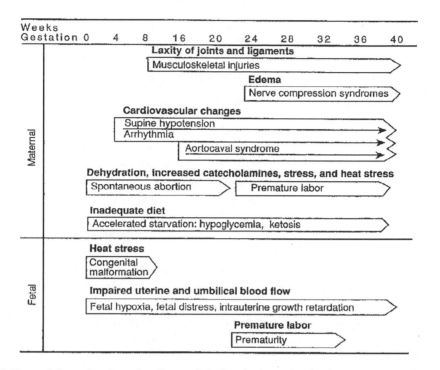

Fig. 3.1. Potential mechanisms leading to injuries during exercise in pregnancy. (From [10])

activity in the second and third trimesters [12]. There is a link between strenuous physical activity and the development of intrauterine growth restriction in the presence of dietary restrictions. Mothers with physically demanding and repetitive jobs were reported in several studies to deliver early and give birth to small-for-gestational-age infants [13–15]; meanwhile, other studies on vigorous exercise found no difference [16] or an increase [17] in infant birth weight. It appears that infant birth weight is not affected by exercise if energy intake is adequate [18], and that fetal weight can be maintained with adequate nutritional intake.

3.3 HYPERTHERMIA

An increase in body temperature during exercise is directly related to the intensity of the activity. During moderate intensity exercise in normal temperature conditions, the body's core temperature tends to rise an average of 1.5°C during the first 30 min of activity, followed by a plateau if the same level of activity continues [19]. Heat is dissipated predominantly through the skin. If heat production exceeds heat dissipation—which can occur if exercise takes place in hot humid conditions, with vigorous exercise, if there is exposure to hot tubs, saunas, or if the woman is running a fever—the core temperature continues to rise. Animal studies have shown that an increase in core temperature greater than 39°C during embryogenesis from 3 to 8 weeks gestation can result in congenital malformations [20–22]. For humans, the association of hyperthermia and congenital malformations is primarily acknowledged in case studies, which suggests a relationship but does not prove causality [23–25]. One prospective study using 165 women exposed to hyperthermia during the first trimester failed to confirm teratogenic effects [26]. As the risk of hypothermia is a concern, pregnant women should be advised to avoid hyperthermic conditions during the entire pregnancy, and particularly the first trimester.

3.4 CARDIOVASCULAR AND RESPIRATORY ADAPTATIONS IN PREGNANCY

During exercise, there is a redistribution of blood flow away from the visceral organs and toward the exercising muscles. The redistribution of blood away from the uterus is related to the intensity and duration of exercise. However, in pregnancy there are corresponding adaptations that are characterized by an increase in blood volume, compensated for by increased venous capacity and decreased peripheral vascular resistance [27]. Although fetal oxygen and substrate availability could be counterbalanced by an increase in the amount of oxygen and substrate taken from the maternal blood supply, the question remains as to whether the redistribution of blood flow during regular or extended physical activity impacts transplacental transport of oxygen, carbon dioxide, and nutrients.

Exercise intensity is usually expressed in terms of demand on the cardiovascular system as percentage of maximal heart rate. The values for volume of oxygen consumed during exercise at maximum capacity ($\%VO_{2max}$), metabolic equivalents (METs), and maximal heart rate ($\%HR_{max}$) for the average nonpregnant woman can be found in Table 3.1 [10]. Maternal heart rate response to strenuous exercise is blunted and does not follow a linear relationship; this is the reason why target heart rates cannot be used for exercise prescriptions in pregnancy. In order to track the level of intensity during physical activity, pregnant women may make use of the following easy-to-use methods. First, the talk test may be used to monitor level of intensity. A subject who is exercising at a moderate intensity (3–4 METs) should be able to comfortably hold a conversation; however, if winded or out of breath during the activity, she may be exercising too vigorously. Another helpful method for measuring intensity is the Borg Scale Rating of Perceived Exertion (RPE) [28]. The RPE is a subjective measure that correlates to a person's physical perception of exercise intensity including heart rate, respiration, perspiration, and muscle fatigue. The Borg RPE scale ranges from a level of 6, which is no exertion at all, to a level of 20, which is maximal exertion. An RPE level of 12–14 would be perceived as "somewhat hard," which corresponds to moderate activity. If exertion were reported as 19 or "extremely hard" on the Borg scale, decreasing to a lower intensity would be beneficial, thus modifying the intensity according to maternal symptoms. To estimate an individual's heart rate during exercise, RPE can be multiplied by 10 (i.e., an RPE of 12–14 × 10 = a heart rate of 120–140 beats per minute). Increased energy expenditure may be estimated using METs, a unit of resting oxygen uptake. One MET is equivalent to 1 kcal per kg of body weight per hour. For example, if a 70-kg woman walks at a brisk pace of 3–4 METs for a half hour, she would increase her caloric requirement by 70–105 kcal(2–3 MET increase over resting × 70 kcal × 0.5 hrs.

Cardiovascular response to body position should be considered. After the first trimester, the supine position results in relative obstruction of the venous return due to the enlarging uterus [29]. Pregnant women may experience a decrease in cardiac output reflective of symptoms associated with this supine hypotensive syndrome.

Table 3.1
**Intensity of Exercise for the Nonpregnant
Woman: $\%VO_{2max}$, METs, and $\%HR_{max}$**

Intensity	$\%VO_{2max}$	METs	$\%HR_{max}$
Light	15–30	1.2–2.7	40–50
Moderate	31–50	2.8–4.3	51–65
Heavy	51–68	4.4–5.9	66–80
Very heavy	69–85	6–7.5	81–90
Unduly heavy	86+	7.6+	90+

From [10] $\% VO_{2max}$ = percentage of aerobic capacity, METs = metabolic equivalent, $\% HR_{max}$ = percentage of maximal heart rate. Adapted from [10]

Pregnancy is associated with profound respiratory changes: minute ventilation (tidal volume × breaths/minute) increases by approximately 50%, primarily as a result of increased tidal volume (volume of gas inhaled and exhaled during one respiratory cycle) [30, 31]. Because of the increased resting oxygen requirements and the increased work of breathing caused by pressure of the enlarged uterus on the diaphragm, there is decreased oxygen availability for performance of aerobic exercise during pregnancy. Thus, both workload and maximum exercise performance are decreased [31, 32].

3.5 FETAL RESPONSE TO MATERNAL EXERCISE

One of the main concerns related to exercise in pregnancy is the effect of maternal activity on the fetus, whereas any maternal benefits may be offset by adverse fetal outcomes. The potential concerns, although theoretical, are related to the selective redistribution of blood flow during exercise and the resultant effects on the transplacental transport of O_2, CO_2, and nutrients. Studies addressing fetal heart response to exercise have focused on fetal heart rate changes before, during, and after exercise. Moderate exercise appears to cause a minimal to moderate increase in fetal heart rate by approximately 10–30 beats/minute over baseline [62].

3.6 EXERCISE GUIDELINES FOR HEALTHY PREGNANCIES

The exercise recommendations from the American College of Obstetricians and Gynecologists (ACOG) mirror those of the Center of Disease Control (CDC), and the American College of Sports Medicine (ACSM). The ACSM recommends moderate intensity exercise for 30 min or more on most days of the week as part of a healthy lifestyle in the nonpregnant population [4]. A moderate level of exertion for 30 min duration has been associated with significant health benefits decreasing risk of chronic diseases including coronary heart disease, hypertension, type 2 diabetes mellitus, and osteoporosis [33]. Women who are sedentary prior to pregnancy should gradually increase their duration of activity to 30 min. Those who are already fit should be advised that pregnancy is not the time to greatly enhance physical performance and that overall activity and fitness tend to decline during pregnancy. Pregnant women should exercise caution in increasing intensity, especially when an exercise session extends beyond 45 min because body core temperature may rise above safe levels, and energy reserves could become depleted. The ACOG guidelines [1] for exercise during pregnancy are established for pregnant women without maternal or obstetrical complications. These recommendations are summarized in Table 3.2. Clinical evaluation by an obstetrician is recommended prior to prescribing an exercise program during pregnancy, with special consideration given to the type and intensity of the exercise—as well as the duration, frequency of the sessions, the level of fitness, and familiarity with the various activities.

In some circumstances, uterine activity has been shown to occur during and after physical activity in pregnancy; however, it could have potential for clinical significance only in those at risk for premature labor. Women who are at risk or have a significant history of premature labor should be advised to refrain from exercise during pregnancy. Table 3.3 lists the absolute and relative contraindications for exercise in pregnancy, and Table 3.4 lists warning signs to terminate exercise while pregnant [1].

Safety is the primary concern for exercise during pregnancy and caution should be implemented. Contact sports and recreational activities with increased risk of falling and abdominal trauma, such as hockey, soccer, baseball, gymnastics, skiing, horseback riding, and racquet sports, should be limited or avoided. Exercise in water has several advantages for the pregnant woman: the safety from the buoyancy of the water, a shift in extracellular fluid back to the vascular system, which can decrease edema, thermoregulation of the core body temperature, less fetal heart rate change as compared to other activities, and increased uterine blood flow [34]. Increases in circulating blood volume are proportional to the depth of immersion, resulting in lower maternal heart rate and

Table 3.2

Excerpts from ACOG Recommendations for Exercise during Pregnancy and the Postpartum Period

1. In the absence of either medical or obstetric complications, 30 min or more of moderate exercise a day on most, if not all, days of the week is recommended for pregnant women.
2. Recreational and competitive athletes with uncomplicated pregnancies can remain active during pregnancy and should modify their usual exercise routines as medically indicated.
3. Generally, participation in a wide range of recreational activities appears to be safe during pregnancy. Each sport should be reviewed individually for its potential risk. Activities with a high risk of falling or risk of abdominal trauma should be avoided. Scuba diving should be avoided.
4. Inactive women and those with medical or obstetric complications should be evaluated before recommendations for physical activity are made. Women engaging in strenuous exercise require close medical supervision.
5. Women with a history of or risk of preterm labor or fetal growth restriction should be advised to reduce activity in the second and third trimesters.
6. Exercise in the supine position, after the first trimester, and prolonged periods of motionless standing should be avoided as much as possible.
7. Exercise during pregnancy may provide additional health benefits to women with gestational diabetes mellitus (GDM) including reducing insulin resistance, postprandial hyperglycemia, and excessive weight gain.
8. Prepregnancy exercise should be gradually resumed postpartum because the physiological changes of pregnancy may persist 4–6 weeks postpartum.

Adapted from [1]

Table 3.3
Contraindications for Exercise during Pregnancy

Absolute contraindications:

- Hemodynamically significant heart disease
- Restrictive lung disease
- Incompetent cervix/cerclage
- Multiple gestation at risk for premature labor
- Persistent second- or third-trimester bleeding
- Placenta previa after 26 weeks of gestation
- Premature labor during the current pregnancy
- Ruptured membranes
- Preeclampsia/pregnancy-induced hypertension

Relative contraindications (patients may be engaged in medically supervised programs):

- Severe anemia
- Unevaluated maternal cardiac arrhythmia
- Chronic bronchitis
- Poorly controlled type 1 diabetes
- Extreme morbid obesity
- Extreme underweight (BMI < 12)
- History of extremely sedentary lifestyle
- Intrauterine growth restriction in current pregnancy
- Poorly controlled hypertension
- Orthopedic limitations
- Poorly controlled seizure disorder
- Poorly controlled hyperthyroidism
- Heavy smoker

From [1]

Table 3.4
Warning Signs to Terminate Exercise While Pregnant

- Vaginal bleeding
- Dyspnea before exertion
- Dizziness
- Headache
- Chest pain
- Muscle weakness
- Calf pain or swelling (need to rule out thrombophlebitis)
- Preterm labor
- Decreased fetal movement
- Amniotic fluid leakage

From [1]

blood pressure in comparison with land exercises. Water aerobics for 30-min sessions have been shown to be as beneficial as static immersion in relieving edema [35]. Scuba diving should be avoided because this puts the fetus at risk for decompression sickness secondary to the inability of the fetal pulmonary circulation to filter bubbles.

In the past, weight lifting in pregnancy was unheard of; however, since women often do not want to give up their prepregnancy routine, they need to learn how to lift weights safely [10]. Weight training is a beneficial way to stay fit during pregnancy, while keeping in mind that the fitness goals should be geared toward maintenance instead of dramatic gains. The use of lighter weights to avoid overloading joints that are loosened by the hormones of pregnancy, along with more repetitions will assist in maintaining muscle mass without stressing the joints. Caution should be executed by avoiding walking lunges that could strain connective tissue. Protecting the abdomen from swinging weights could prevent harm to the fetus. Women should be advised not to lift weights while laying flat on the back, which puts pressure on the vena cava, thus restricting blood flow to the heart. Use of an incline bench would assist with the position change. Proper breathing technique such as exhaling during a lift can lessen the risk on transient hypertension associated with the Valsalva maneuver, seen in inexperienced weight lifters who forcibly exhale air against closed lips and a pinched nose. Any program that works the entire body to promote tone and fitness can be incorporated into a physical activity routine with certain limitations.

Moderate exercise in a low-risk pregnancy does not result in adverse fetal or maternal outcomes but instead helps to maintain fitness and well being in the mother [36]. An exercise prescription for the improvement and maintenance of fitness in nonpregnant women consists of activities recommended to improve cardiorespiratory capacity (aerobic exercise), muscle tone (resistance exercise), and flexibility [37]. These activities include walking, low-impact aerobics, stationary biking, and swimming. In pregnant women, a similar prescription can be made; however, additional consideration should be given to the type and intensity of exercise along with duration and frequency to achieve health benefits minimizing any potential risks to the mother and fetus [18]. Aerobic exercise consists of activities that use large muscle groups in a continuous rhythmic fashion, such as walking, jogging, swimming, stationary biking, or dancing. Indeed many women have familiarity with these activities, and most are able to comply with recommendations to incorporate aerobic activities into their daily schedules. Women should be advised to wear comfortable shoes and avoid activities that may result in falling.

The prevalence of leisure activity among pregnant women in the United States is 66%, compared with 73% in nonpregnant women; however, when examining whether women meet the recommended amounts of physical activity per week, the prevalence was lower at 16% in pregnant women compared with 26% in those not pregnant [38]. The most common leisure time activity reported was walking, followed by swimming, weight lifting, gardening, and aerobics. While women report that exercise positively impacts pregnancy, the greatest influence on whether a woman exercises during her pregnancy was reported to be the encouragement provided by her physician and healthcare provider [39]. Sedentary women, prior to beginning an exercise program, should receive clinical and obstetrical evaluation.

3.7   THE ELITE ATHLETE

The elite athlete experiences limitations and physiological changes in pregnancy similar to the recreational athlete, including ligament relaxation, change in posture, and weight gain. These changes may in turn impact a woman's competitive ability and increase her desire for strenuous training, making her more prone to injury [12]. The elite athlete should be aware that pregnancy is not a time for improving competitive fitness, but instead she should focus on remaining physically active, modifying her exercise routine if medically indicated. Caution and careful evaluation by an obstetrician is essential with high-intensity, prolonged, frequent exercise during pregnancy, since there is evidence of low birth weight and greater risk of thermoregulatory complications associated with this type of exercise regimen.

3.8 NUTRITIONAL REQUIREMENTS FOR THE ACTIVE PREGNANT WOMAN

Although the nutritional needs of active pregnant women are not clearly defined, nutritional needs in pregnancy have been well researched. Energy requirements during the second and third trimesters of pregnancy are an average of 300 kcal a day above prepregnancy requirements [40]. A wide variability in metabolic energy expenditure in pregnancy makes it difficult to set standards for energy requirements [41]. Exercise during pregnancy requires an additional caloric allowance for increased metabolism and greater energy expenditure both during and after the activity. Other factors affecting caloric requirements in pregnancy include prepregnancy body mass index, maternal age, and appetite. Estimation of caloric needs is further complicated by pregnancy changes in maternal extracellular fluid, maternal fat stores, the weight of the fetus and supporting tissue (uterus, placenta, amniotic fluid, and mammary glands), as well as changes in fat-free muscle mass due to variations in activity during pregnancy. Level of activity may either increase or decrease caloric requirements. For example, a competitive athlete who decides to reduce the intensity of the activity may have lower caloric needs in pregnancy compared with prepregnancy needs, while a sedentary person who has started a moderate exercise program may have increased calorie needs above those of normal pregnancy requirements.

Estimation of body composition is more complicated in pregnancy. As gestational age progresses, body water continues to increase, while fat mass stays relatively constant. Changes in hydration, along with errors in measures used to estimate percent body fat, make it more difficult to provide reliable measures of body composition [42].

The Dietary Reference Intakes (DRIs) for macronutrient and micronutrient intakes have not been defined for the active pregnant woman compared with those who are sedentary. Protein requirements in pregnancy have been estimated at 1.1 g/kg/day (71 g/day for someone 163 cm tall, weighing 65 kg.), while in active people there is a slightly higher estimated requirement of 1.2 to 1.4 g/kg body weight per day [43]. The 2005 *Dietary Guidelines for Americans* recommend 20–35% of calories from fat, with most coming from polyunsaturated and monounsaturated fatty acids, while limiting intake of saturated fats to less than 10% of calories and keeping *trans* fatty acids as low as possible. Fat intake should not be restricted to less than 15% of energy requirements because fat is important not only as a source of calories, but also to aid in the absorption of fat-soluble vitamins and provides essential fatty acids [44]. Carbohydrate intake of 40–55% of energy requirements is needed to replace the muscle glycogen stores lost during exercise, minimize maternal hypoglycemia, and limit ketonuria. All pregnant women and athletes should strive to consume foods that provide at least the RDA/DRI for all vitamins and minerals in pregnancy and lactation, as discussed in Chap. 1 ("Nutrient Recommendations and Dietary Guidelines for Pregnant Women") [43].

Women who are diet conscious often do not obtain the necessary nutrients required to maintain a normal pregnancy. Inadequate nutritional intake along with the increased energy requirements for exercise may lead to poor weight gain and fetal growth restriction. Although the data linking low birth weight and maternal exercise are conflicting, for pregnant women who exercise, it is unclear if adequate energy intake can offset a decrease in fetal weight [44]. A meta-analysis of 30 research studies concluded that vigorous exercise during the third trimester of pregnancy has been associated with a 200- to 400-g decrease in fetal weight [45]. When deficient energy intake occurs in combination with chronic strenuous exercise during pregnancy, fetal growth may be adversely affected.

Since pregnancy and exercise place higher demands on oxygen requirements, women who exercise during pregnancy should be monitored for suboptimal iron status and inadequate intake. Many women enter pregnancy with depleted iron stores, as discussed in Chap. 16 ("Iron Requirements and Adverse Outcomes"). This, along with expansion of maternal blood volume and increased fetal demand for oxygen, makes it more of a challenge for many women to achieve adequate iron status. If a woman enters pregnancy with iron deficiency anemia, repletion of iron stores may be difficult. Prenatal vitamin and mineral supplements are routinely prescribed to provide additional iron and folic acid. However, these should not replace a healthy balanced diet containing a variety of foods from all food groups so as to ensure adequate intake of antioxidants, fiber, and the necessary nutrients to support maternal health and growth of the fetus [46].

Oftentimes, active women enter pregnancy underweight, with increased awareness of body image and may resort to caloric intake below recommendations to prevent weight gain in pregnancy. To compensate for nutrient deficiencies, women may over compensate by taking large amount of vitamins or minerals. Although vitamin and mineral supplementation may be beneficial, women should be counseled to avoid excessive micronutrient intake, particularly of the fat-soluble vitamins A and D, which can lead to fetal malformations. Excessive amounts of vitamin D can result in congenital anomalies consisting of supravalvular aortic stenosis, elfin facies, and mental retardation [47]. Women taking high amounts of vitamin A >10,000 IU in supplement form showed higher rates (1 infant in 57) of cranial–neural crest tissue defects [48]. The use of dietary supplements is further discussed in Chap. 14 ("Dietary Supplements during Pregnancy: Need, Efficacy, and Safety").

Athletes may choose to consume nutritional ergogenic aids and dietary supplements to enhance athletic performance with hopes of boosting their competitive edge. Nutritional supplements are a multibillion-dollar industry targeting a wide range of populations, including women of childbearing age. Supplement companies are not required to prove supplement safety, effectiveness, and potency before a product is placed on the market as long as the supplement makes the claim that it has not been evaluated by the US Food and Drug Administration (FDA), and that the product is not intended to diagnose, treat, mitigate, cure or prevent disease [43]. Many may believe that since these products are natural and legal that they are safe; however, there is little scientific evidence demonstrating the safety or effectiveness of these products for the general population. Women of childbearing age should be counseled or warned that supplements and nutritional ergogenic aids have not been shown to be safe and therefore should be avoided prior to and during pregnancy. The reader is also directed to Chap. 13 ("Popular Diets").

Water is a critical yet often forgotten nutrient for healthy pregnancies. Exercise induces significant fluid loss and places the woman at higher risk of dehydration. Weighing before and after exercise can help monitor fluid balance. Weight loss of 2 lb is equivalent to approximately a 1-liter fluid loss. Pregnant women should be encouraged to drink 8 to 12 cups of hydrating fluids per day, with water being the preferential source. Sports drinks help replenish carbohydrate, fluid, and electrolyte losses during exercise sessions lasting 30–45 min. Drinking 1–2 cups of water prior to exercise, replacing fluids every 15–20 min. during activity, and replacing fluids lost after exercise helps maintain hydration and keeps body temperature within normal limits.

Physical activity and diet quality are interconnected behaviors. Individuals following a suboptimal diet tend to be more sedentary, less educated, not married, and non-Caucasians [49]. Hormonal alterations during pregnancy have been shown to cause a 1.5- and threefold increase in maternal cholesterol and triglyceride levels, respectively by the mid-third trimester [50]. One study examined the relationship between recreational physical activity in early pregnancy and found reductions in total cholesterol and triglyceride levels in women who spent a greater amount of time (12.7 h/week) on recreational physical activity [51]. Results of this study, along with others conducted in the nonpregnant population suggest that physical activity in pregnancy may lessen pregnancy-associated dyslipidemia.

3.9   FUEL UTILIZATION IN EXERCISE AND PREGNANCY

Measurements by indirect calorimetry reveal preferential use of carbohydrates during exercise in pregnancy [53]. The respiratory exchange ratio (RER) reflects the ratio between CO_2 output and oxygen uptake (VO_2). The RER provides information on the proportion of substrate derived from various macronutrients. For carbohydrate to be completely oxidized to CO_2 and H_2O, one volume of CO_2 is produced for each volume of O_2 consumed. An RER of 1 indicates carbohydrates are being utilized, while an RER of 0.85 indicates mixed substrate. Assessment of fuel utilization during pregnancy is important because of the possible effect of exercise-induced maternal hypoglycemia [53]. Such events are unlikely to occur during 45 min of moderate exercise, but could occur after 60 min of continuous moderate to strenuous exercise (Fig. 3.2). The tendency for higher respiratory exchange ratios during pregnancy and during exercise in pregnancy suggests a preferential utilization of carbohydrates. Soultanakis et al. [53] found that with exercise greater than 20 min in length, both glucose and glycogen stores were depleted, resulting in higher levels of ketones and free fatty acids being used as fuel sources. Increased carbohydrate metabolism along with lower glycogen stores may further predispose women to hypoglycemia during pregnancy. Protein utilization in pregnancy during exercise does not increase above nonpregnancy levels, and since most people in the United States get more than the required amount of protein in their diets, additional/supplemental dietary protein is unnecessary fuel during moderate bouts of exercise [54].

3.10   CLINICAL APPLICATIONS FOR EXERCISE IN PREGNANCY

Results from the National Health and Nutrition Examination Survey (NHANES) reveal that from 2003–2004 an estimated 66% of adults (over age 20) were either overweight or obese [55]. The obesity rate in women of childbearing age is increasing. In 2003, 19.6% of US women of reproductive age (18–44 years) were classified as obese (BMI > 30) [56]. Whether this trend in weight status is associated with the liberalization of the weight gain

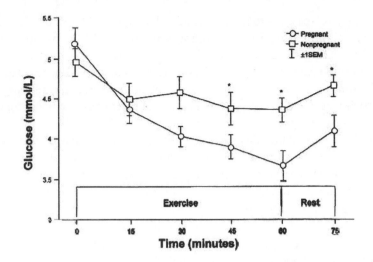

Fig. 3.2. Glucose concentrations during prolonged exercise: pregnant versus nonpregnant. *$p < 0.05$. (From [19])

guidelines in pregnancy is unclear. However, data show that with each subsequent pregnancy, there is a greater risk of postpartum weight retention [57]. A greater focus is needed to prevent excessive weight gain in pregnancy; this may be accomplished in part through exercise. One study revealed that women who gained excessive weight and failed to lose weight by 6 months postpartum were 8.3 kg heavier 10 years later [58]. A 15-year follow up study to determine the effects of weight gain in pregnancy revealed that the 1-year postpartum timeframe was the greatest predictor of long-term weight retention, regardless of weight gain in the pregnancy or prepregnancy BMI [59]. O'Toole et al. reported that a 12-week comprehensive exercise and nutrition intervention program resulted in greater postpartum weight loss than a single 1-hr educational session [60]. Although weight loss is usually not recommended in pregnancy, losing excess weight prior to pregnancy and a gradual weight loss postpartum may be beneficial in overweight and obese women, as described in Chap. 5 ("Obesity and Pregnancy"). Lactating women should not attempt to lose more than 2 kg/month [61].

Gestational diabetes mellitus (GDM) is a condition of glucose intolerance that is first detected during pregnancy. The elevated hormonal response more commonly found in the second and third trimesters of pregnancy further amplifies the reduction in insulin peripheral sensitivity. Through the use of exercise, both insulin sensitivity and the effectiveness of insulin may increase. Exercise has been recognized as an adjunctive and alternative therapy to assist with glycemic control in patients with type 2 diabetes mellitus [62], and this is further discussed in Chap. 10 ("Diabetes and Pregnancy").

The American Diabetes Association endorses exercise as adjunctive therapy for GDM when glycemic control is not achieved with diet alone [63, 64]. Women diagnosed with GDM during pregnancy are at increased risk of developing type 2 diabetes within the first 5 years after delivery [65]. Studies have shown that through exercise and diet therapy, glycemic control can be achieved and may prevent the onset of type 2 diabetes [66, 67]. Epidemiological data [63] suggest that obese women with a BMI > 30 kg/m² can lower the incidence of GDM with exercise during pregnancy compared with obese women who

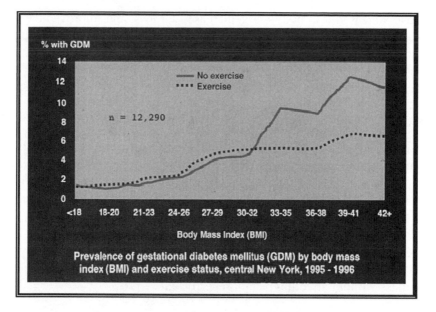

Fig. 3.3. Prevalence of gestational diabetes mellitus (*GDM*) by body mass index (*BMI*) and exercise status, central New York, 1995–1996. (From [63])

do not exercise (Fig. 3.3). These studies demonstrate that exercise as prescribed may be beneficial in the primary prevention of GDM in overweight and obese women. Aerobic exercise plays a role in decreasing the hyperinsulinemia associated with obesity along with decreasing fasting and postprandial blood glucose levels.

Based on the findings of several studies, exercise has been prescribed to improve carbohydrate tolerance and avoid insulin therapy. These studies have been aimed at assessing maternal and fetal safety along with efficacy of the exercise prescription. Artal et al. [68] advised 20 min of bicycle ergometry at 50% VO_{2max} after each meal at least 5 days/week for 6 weeks prior to the expected day of delivery. Jovanovic-Peterson et al. [69] recommended 20 min of arm ergometry at less than 50% VO_{2max} daily for 6 weeks prior to delivery. Bung et al. [70] utilized 45 min of bicycle ergometry at 50% VO_{2max} three to four times a week for 6 weeks prior to delivery. These studies demonstrated exercise, as prescribed above, was sufficient to maintain euglycemia. Therefore, exercise should be viewed as a viable option for women with GDM to improve glycemic control and pregnancy outcomes.

Physical activity offers benefits to those at risk for developing gestational hypertension and preeclampsia, which is characterized by hypertension and proteinuria in pregnancy. Preeclampsia and cardiovascular disease share similar pathways including hypertension, dyslipidemia, insulin resistance, and obesity [71]. Women engaging in physical activity early in pregnancy reduced their risk of preeclampsia by 35% compared with inactive women [72].

3.11   POSTPARTUM

Many of the physiological and morphological changes of pregnancy persist 4–6 weeks postpartum. Therefore, exercise routines should be resumed gradually only when medically and physically safe. Weight loss is often desired during the postpartum period, a time when women are often anxious to resume exercise routines quickly after delivery. Exercise complements the benefits of restricting calories and limiting portion sizes. Exercise helps achieve increased lean body mass, increased fat loss, and improved cardiovascular fitness. Women will be more successful at postpartum exercise if they have a plan and are confident in their ability to carry out the plan [73]. Healthcare providers can help women to identify barriers to exercise, including inclement weather, safety issues, lack of transportation, childcare, time, and cost, and to develop strategies to overcome these obstacles.

3.12   LACTATION

During the postpartum period, many women are eager to lose weight and improve muscle tone. Of concern to many women is whether an energy deficit will affect the quality of breast milk, thus impairing infant growth. Aerobic exercise performed four to six times per week at a moderate intensity of 60–70% maximal heart rate for 45 min per day does not appear to affect breast milk volume and composition [74]. The Institute of Medicine recommends lactating women should lose no more than 2 kg/month [75]. However, one study reveals that short-term weight loss of approximately 1 kg/week through a combination of aerobic exercise and dietary energy restriction helped preserve lean body mass without affecting lactation performance [76].

3.13   CONCLUSION

Women should obtain medical advice from their healthcare providers about the type of activity they should engage in during pregnancy. There is ample evidence that moderate exercise in women with healthy pregnancies is beneficial and has no adverse effects on the mother or the baby. Pregnant women should be encouraged to include 30 min of physical activity into their daily lifestyle on most, if not all days of the week. Despite profound physiological changes in pregnancy, women with healthy pregnancies may engage in a combination of aerobic and resistance training in their workouts. Contact sports, scuba diving, and exercise with a high risk of falling or of abdominal trauma should be avoided. Women who exercise for the first time should start slowly, gradually increasing to moderate-intensity workouts. The elite athlete should be aware that pregnancy is not the time to enhance physical fitness performance. Strenuous exercise with intense workouts or sessions lasting longer than 45 min could raise the body core temperature to levels that could be harmful to the fetus. Energy requirements vary during exercise in pregnancy due to the variability in metabolic energy expenditure and the frequency, intensity, duration, and level of the physical activity.

Exercise habits prior to and during pregnancy may decrease the risk of gestational hypertension and GDM. Because habits adopted during pregnancy can result in persistent lifestyle changes, exercise during pregnancy could significantly reduce lifetime risks for obesity, chronic hypertension, and diabetes. Women whose exercise habits have become firmly engrained before and during pregnancy stand a much better chance of maintaining them after the child is born, and these exercise habits can positively impact the entire family.

REFERENCES

1. American College of Obstetricians and Gynecologists (2002) Exercise during pregnancy and the postpartum period ACOG Committee Opinion No. 267. Obstet Gynecol 99:171–173
2. Blair SN, Kohl HW, Paffernberges RS, Clark DG, Cooper KH, Gibbons LW (1989) Physical fitness and all-cause mortality: a prospective study of healthy men and women. J Am Med Assoc 262:2395–23401

3. Bouchard C, Shephard RL, Stephens RL, Stephens T (eds) (1994) Physical activity fitness and health: international proceeding and consensus statement. Human Kinetics, Champaign, Ill.

4. American College of Sports Medicine (1990) Position stand: the recommended quantity and quality of exercise for developing and maintaining cardiorespiratory and muscular fitness in healthy adults. American College of Sports Medicine, Indianapolis, Ind.

5. Pate RR, Pratt M, Blair SN, Haskell WL, Macera CA, Bouchard C, Buchner D, Ettinger W, Heath GW, King AC, Kriska A, Leon AS, Marcus BH, Morris, J, Paffenbarger RS, Patrick K, Pollock ML, Rippe JM, Sallis J, Wilmore JH (1995) Physical activity and public health: A recommendation from the Centers for Disease Control and Prevention and the American College of Sports Medicine. J Am Med Assoc 273:402–407

6. American College of Sports Medicine (2000) ACSM's guidelines for exercise testing and prescription, 6th edn. Lippincott, Williams and Wilkins, Philadelphia, Pa.

7. Calganeri M, Bird HA, Wright V (1982) Changes in joint laxity occurring during pregnancy. Ann Rheum Dis 41:126–128

8. Gleeson PB, Panol JA (1988) Obstetrical physical therapy. Phys Ther 68:1699–1702

9. Garshasbi A, Faghih ZS (2005) The effect of exercise on the intensity of low back pain in pregnant women. Int J Gynecol Obstet 88:271–275

10. Artal R, Wiswell RA, Drinkwater BL, St. John-Repovich WE (1991) Exercise guidelines for pregacy. In: Artal R, Wiswell RS, Drinkwater B (eds) Exercise in Pregnancy, 2nd edn. Williams & Wilkins, Baltimore, Md.

11. Veille JC, Hohimer AR, Burry K, Speroff L (1985) The effect of exercise on uterine activity in the last eight weeks of pregnancy. Am J Obstet Gynecol 151:727–730

12. Hale RW, Milne L (1996) The elite athlete and exercise in pregnancy. Semin Perinatol 20:277–284

13. Naeye RL, Peters E (1982) Working during pregnancy, effects on the fetus. Pediatr 69:724–727

14. Launer LH, Villar J, Kestler E, de Onis M (1990) The effect of maternal work on fetal growth and duration of pregnancy: a prospective study. Br J Obstet Gynaecol 97:62–70

15. McDonald AD, McDonald JC, Armstrong B, Cherry NM, Nolin AD, Robert D (1988) Prematurity and work in pregnancy. Br J Ind Med 45:56–62

16. Sternfeld B, Quesenberry CP, Eskenazi B, Newman LA (1995) Exercise during pregnancy and pregnancy outcome. Med Sci Sports Exerc 27:634–640

17. Hatch MC, Shu XO, McLeon DE, Levin B, Begg M, Reuss L, Susser M (1993) Maternal exercise during pregnancy, physical fitness, and fetal growth. Am J Epidemiol 137:1105–1114

18. Artal R, O'Toole M (2003) Guidelines of the American College of Obstetricians and Gynecologists for exercise during pregnancy and the postpartum period. Br J Sports Med 37:6–12

19. Soultanakis HN, Artal R, Wiswell RA (1996) Prolonged exercise in pregnancy: Glucose homeostasis, venilatory and cardiovascular responses. Semin Perinatol 20:315–327

20. Sasake J, Yamaguchi A, Nabeshima Y, Shigemitsu S, Mesaki N, Kubo T (1995) Exercise at high temperature causes maternal hyperthermia and fetal anomalies in rats. Teratology 51:223–226

21. Martinez-Frias ML, Mazario MJ, Caldas CF, Conejero Gallego MP, Bermejo E, Rodriguez-Pinilla E (2001) High maternal fever during gestation and severe congenital limb disruptions. Am J Med Genet 98:201–203

22. Milunsky A, Ulcickas M, Rothman M, Willett KJ, Willett W, Jick SS, Jick H (1992) Maternal heat exposure and neural tube defects. J Am Med Assoc 268:882–885

23. Miller P, Smith DW, Shepard TH (1982) Maternal hyperthermia in early pregnancy. Epidemiology in a human embryonic population. Am J Med Genet 12:281–288

24. Shiota K (1978) Neural tube defects and maternal hyperthermia as a possible cause of anencephaly. Lancet 1:519–21

25. Edwards MJ (1986) Hyperthermia as a teratogen. Teratol Carcinog Mutagen 6:563–582

26. Clarren SK, Smith DW, Harvey MAS (1979) Hyperthermia—a prospective evaluation of a possible teratogenic agent in man. J Pediatr 95:81–83

27. Hartman S, Bung P (1999) Physical exercise during pregnancy—physiological considerations and recommendations. J Perinat Med 27:204–215

28. Borg G (1962) Physical performance and perceived exertion. Gleenup, Lund, Sweden, p 63

29. Clark SL, Cotton DB, Pivarnik JM, Lee W, Hankins GD, Benedetti TJ, Phelan JP (1990) Position change and central hemodynamic profile during normal third-trimester pregnancy and post partum. Am J Obstet Gynecol 75:954–959

30. Prowse CM, Gaensler EA (1965) Respiratory and acid-base changes during pregnancy. Anesthesiology 26:381–392

31. Artal R, Wiswell R, Romen Y, Dorey F (1986) Pulmonary responses to exercise in pregnancy. Am J Obstet Gynecol 154:378–383

32. Clapp JF III (1990) Exercise in pregnancy: a brief clinical review. Fetal Med Rev 161:1464–149

33. Pate RR, Pratt M, Blair SN, Haskell WL, Macera CA, Bouchard CA, Buchner D, Ettinger W, Heath GW, King AC, Kriska A, Leon AS, Marcus BH, Morris, J, Paffenbarger RS, Patrick K, Pollock ML, Rippe JM, Sallis J, Wilmore JH (1995) Physical activity and public health: a recommendation from the Centers for Disease Control and Prevention and the American College of Sports Medicine. JAMA 273:402–407

34. Katz VL (1996) Water exercise in pregnancy. Semin Perinatol 20:285–290

35. Kent T, Gregor J, Deardorff L, Katz V (1999) Edema of pregnancy: a comparison of water aerobics and static immersion. Obstet Gynecol 94:726–729

36. Morris SW, Johnson NR (2005) Exercise during pregnancy: a critical appraisal of the literature. J Reprod Med 50(3):181–188

37. Pollack ML, Gaesser GA, Butcher JD (1998) The recommended quantity and quality of exercise for developing and maintaining cardiorespiratory and muscular fitness, and flexibility in healthy adults. Med Sci Sports Exerc 30:975–979

38. Evenson KR, Savitz DA, Huston SL (2004) Leisure-time physical activity among pregnant women in the US. Paediatr Perinat Epidemiol 18:400–407

39. Krans EE, Gearhart JG, Dubbert PM, Klar PM, Miller AL, Replogle WH (2005) Pregnant women's beliefs and influences regarding exercise during pregnancy. J Miss State Med Assoc 46:67–73

40. Position of the American Dietetic Association (2002) Nutrition and lifestyle for a healthy pregnancy outcome. J Am Dict Assoc 102:1479–1490

41. King JC (2000) Physiology of pregnancy and nutrient metabolism. Am J Clin Nutr 71:1218–1225

42. Jacque-Fortunato SV, Khodiguian N, Artal R, Wiswell RA (1996) Body composition in pregnancy. Semin Perinatol 20:340–342

43. Position of the American Dietetic Association, Dietitians of Canada, and the American College of Sports Medicine (2000) Nutrition and athletic performance. J Am Diet Assoc 100:1543–1556

44. Pivarnik JM (1998) Potential effects of maternal physical activity on birth weight: brief review. Med Sci Sports Exerc 30:400–406

45. Leet TL, Flick L (2003) Effect of exercise on birthweight. Clin Obstet Gynecol 46:423–431

46. Rosenbloom CA (ed) (2000) Exercise in pregnancy. In: Sports nutrition: a guide for the professional working with active people, 3rd edn. American Dietetic Association, Chicago, Ill.

47. Garcia RE, Friedman WF, Koback MM (1964) Idiopathic hypercalcemia and supravalvular stenosis: documentation of a new syndrome. N Engl J Med 271:117–120

48. Rothman KJ, Moore LL, Singer MR, Nguyen UD, Mannino S, Milunsky A (1995) Teratogenicity of high vitamin A intake. N Engl J Med 333:1369–1373

49. Gillman MW, Pinto BM, Tennstedt S, Glanz K, Marcus B, Friedman RH (2001) Relationships of physical activity with dietary behaviors among adults. Prev Med 32:295–301

50. Potter JM, Nestel PJ (1979) The hyperlipidemia of pregnancy in normal and complicated pregnancies. Am J Obstet Gynecol 133:165–171

51. Butler CL, Williams MA, Sorensen, TK, Frederick IO, Leisenring WM (2004) Relation between maternal recreational physical activity and plasma lipids in early pregnancy. Am J Epidemiol 160:350–359

52. Artal R, Masaki D, Khodignian N, Romen Y, Rutherford SE, Wiswell RA (1989) Exercise prescription in pregnancy: weightbearing exercise. Am J Obstet Gynecol 161:1464–149

53. Soultanakis H, Artal R, Wiswell R (1996) Prolonged exercise in pregnancy: glucose homeostasis ventilatory and cardiovascular responses. Semin Perinatol 20:315–324

54. Bessinger RC, McMurray RG, Hackney AC (2002) Substrate utilization and hormonal responses to moderate intensity exercise during pregnancy and after delivery. Am J Obstet Gynecol 186:757–764

55. Ogden CL, Carroll MD, Curtin LR, McDoowell MA, Tabak CJ, Flegal KM (2006) Prevalence of overweight and obesity in the United States, 1999–2004. J Am Med Assoc 295:1549–1555

56. March of Dimes Perinatal Data Center (2004) Analysis of data from the Behavioral Risk Factor Surveillance Survey, CDC. March of Dimes, White Plains, N.Y.

57. Beazley JM, Swinhoe JR (1979) Body weight in parous women is there any alteration between successive pregnancies? Acta Obstet Gynecol Scand 58:45–47

58. Rooney BL, Schauberger CW (2002) Excess pregnancy weight gain and long-term obesity: one decade later. Obstet Gynecol 100:245–252

59. Linne Y, Dye L, Barkeling B, Rossner S (2004) Long-term weight development in women: A 15-year follow-up of the effects of pregnancy. Obes Res 12:1166–1178

60. O'Toole ML, Sawicki MA, Artal R (2003) Structured diet and physical activity prevent postpartum weight retention. J Womens Health (Larchmt) 2:991–998

61. Dewey KG, Lovelady CA, Nommsen-Rivers LA, McCrory MA, Lonnerdal B (1994) A randomized study of the effects of aerobic exercise by lactating women on breast-milk volume and composition. N Engl J Med 330:449–453

62. Artal R, Posner M. Fetal responses to maternal exercise (1991) In: Artal R, Wiswell RS, Drinkwater B (eds) Exercise in Pregnancy, 2nd edn. Williams & Wilkins, Baltimore, Md. p. 213–224

63. Dye TD, Knox KL, Artal R, Aubry RH, Wojtowycz MA (1997) Physical activity, obesity, and diabetes in pregnancy. Am J Epidemiol 146:961–965

64. Jovanovic-Peterson, L Peterson CM (1996) Exercise and the nutritional management of diabetes during pregnancy. Obstet Gynecol Clin North Am 23:75–86

65. Kim C, Newton K, Knopp RH (2002) Gestational diabetes and the incidence of type 2 diabetes. Diabetes Care 25:1862–1868

66. Pan XR, Li GW, Hu YH, Wang JX, Yang WY, An ZX, Hu ZX, Lin J, Xiao JZ, Cao HB, Liu PA, Jiang XG, Wang JP, Zhang H, Bennett PH, Howard BV (1997) Effects of diet and exercise in preventing NIDDM in people with impaired glucose tolerance. The Da Qing IGT and diabetes study. Diabetes Care 20:537–544

67. Helmrich SP, Ragland DR, Paffenbarger RS (1994) Prevention of non-insulin-dependent diabetes mellitus with physical activity. Med Sci Sports Exerc 26:824–830

68. Artal R, Masaki D (1989) Exercise in gestational diabetes. Pract Diabetol 8:7–14

69. Jovanovic-Peterson L, Durak EP, Peterson CM (1989) Randomized trial of diet versus diet plus cardiovascular conditioning on glucose levels in gestational diabetes. Am J Obstet Gynecol 161:415–419

70. Bung P, Artal R, Khodiguian N, Kjos S (1991) Exercise in gestational diabetes: an optional therapeutic approach? Diabetes 40(Suppl):182–185

71. Saftlas AF, Logsden-Sackett N, Wang W, Woolson R, Bracken MB (2004) Work, leisure-time activity, and risk of preeclampsia and gestational hypertension. Am J Epidemiol 160:758–765

72. Sorensen TK, Williams MA, Lee I, Dashow EE, Thompson ML, Luthy DA (2003) Recreational physical activity during pregnancy and risk of preeclampsia. Hypertension 41:1273–1280

73. Hinton PS, Olson CM (2001) Postpartum exercise and food intake: The importance of behavior-specific self-efficacy. 101:1430–147

74. Dewey KG, Lovelady CA, Nommsen-Rivers LA, McCrory MA, Lonnerdal B (1994) A randomized study of the effects of aerobic exercise by lactating women on breast-milk volume and composition. N Engl J Med 330:449–453

75. Institute of Medicine (1991) Nutrition during lactation. National Academy Press, Washington, D.C.

76. McCrory M, Nommsen-Rivers AL, Mole PA, Lonnerdal B, Dewey KG (1999) Randomized trial of the short-term effects of dieting compared with dieting plus aerobic exercise on lactation performance. Am J Clin Nutr 69:959–967

4 Food, Folklore, and Flavor Preference Development

Catherine A. Forestell and Julie A. Mennella

Summary Food choices during pregnancy and lactation are influenced by a variety of factors. While internal factors, such as cravings and aversions, play an important role especially during the first trimester of pregnancy, environmental factors such as cultural food practices and beliefs often dictate the types of foods eaten throughout pregnancy and lactation. Such traditional food practices serve to predispose infants to flavors that are characteristic of their mother's culture and geographical region. As discussed herein, amniotic fluid and human milk are composed of flavors that directly reflect the foods, spices, and beverages eaten by or inhaled by (e.g., tobacco) the mother. Because the olfactory and taste systems are functioning by the last two trimesters, these flavors are detected early in life, and early experience can bias behavioral response to these flavors later in life. Although more research is needed to understand the mechanisms involved in early flavor learning, these pre- and early-postnatal flavor exposures likely serve to facilitate the transition from fetal life through the breastfeeding period to the initiation of a varied solid food diet. Such learning is the first, but not the only, way in which children learn to appreciate and prefer the flavors of the foods cherished by their culture.

Keywords: Flavor, food choice, taste, smell, lactation, pregnancy, culture

4.1 INTRODUCTION

The traditional wisdom of many cultures relates that what women eat while they are pregnant or lactating can have long-lasting effects on their children. Many of these food traditions evolved to protect and provide strength to both mother and child because of the high mortality that was typically associated with pregnancy and the early postpartum period [1]. Although there is little science-based evidence to support their efficacy, these practices, which revolve around a cuisine that combines local foods and spices, continue to be passed down from one generation to the next because they are often deeply rooted in religious and traditional beliefs [1].

We highlight here some examples of pregnancy-related food beliefs and taboos that are evident around the world. This is not intended to be a recommendation for such practices, but serves to acknowledge that they are shaped by a variety of social, cultural, economic,

From: *Nutrition and Health: Handbook of Nutrition and Pregnancy*
Edited by: C.J. Lammi-Keefe, S.C. Couch, E.H. Philipson © Humana Press, Totowa, NJ

and psychological factors and are often important for a people to maintain their cultural distinctiveness. Such practices can be observed in Shao, a rural village of Nigeria, where healers discourage pregnant women from eating meats because they believe the behavioral characteristics of the animals consumed will be imparted to the fetus [2]. Similarly, some women avoid eating foods such as strawberries or chicken because they believe that over-indulged food cravings can cause birthmarks or congenital deformities in the baby [3–5]. In South Africa, Zulu healers often prescribe *isihlambezo*, an herbal concoction of many different local plants to promote a healthy pregnancy and facilitate quick uncomplicated labor [6]. Depending on availability, additional ingredients to this herbal tonic may include fish heads, lizard or snakeskin, dried hyrax urine, mercury, clay, or sand. Such reliance on local resources is also observed in Mexico, where women eat more local fruit during pregnancy and lactation because they crave and prefer the tastes of these foods, or because their physicians, mothers, or other women with ascendance advise them to do so [7].

Similar food traditions are evident throughout lactation as there is a strong belief that the mother can optimize the quality and quantity of her milk to meet the needs of her child through her own diet and psychological well-being [7]. In Egypt, foods that are considered to increase the quantity of milk (galactagogues) include juices, fenugreek tea and milk, green leafy vegetables, halva (a sesame-based sweet), and yogurt [8]. In some parts of Mexico, women drink *pulque*, a low-alcoholic beverage made of the fermented juice of a local fruit *Agave atrovirens*, because they believe that it will enhance their milk supply [9], whereas others consume milk and gruels to thicken their breast milk [7]. It is important to note that no scientific evidence supports any of these claims, and some of these practices may be nutritionally unsound [10–13]. From the perspective of early flavor learning, however, emerging scientific evidence suggests that when women adhere to these cultural food practices they in a sense "educate" children to the flavor principles of their culture, because amniotic fluid and breast milk acquire the flavors of the foods consumed by mothers throughout pregnancy and lactation, respectively [14, 15].

In this chapter, we first discuss how dietary cravings may alter maternal diets over the course of pregnancy. We then review the ontogeny of olfaction and taste, which suggests that by the third trimester, the fetus is capable of detecting and learning about the flavors of the mother's diet. Finally, we discuss evidence that suggests that these early pre- and postnatal experiences program later food preferences, which may serve to facilitate the transition to solid foods commonly eaten within the infants' cultures.

4.2 DIETARY CHANGES THROUGHOUT PREGNANCY

In the United States, it is estimated that approximately 50–90% of pregnant women experience food cravings during the course of pregnancy [16–18]. Despite its prevalence, the etiology of pregnancy-related cravings is not well understood. Whereas some hypothesize that cravings are a function of cognitive characteristics of the individual [19], others claim that cravings may represent "wisdom of the body" [20]. For example, pregnant women may crave certain foods to overcome nutritional deficiencies, thereby ensuring that they consume a varied diet with enough calories to support the growth of a healthy fetus [21]. This is analogous to the embryo protection hypothesis [22–24], which proposes that nausea and vomiting during pregnancy evolved to prevent pregnant women from ingesting toxic foods that may harm the fetus.

Although the specific foods women crave may be a function of their culture or geographic location [25, 26], pregnant women generally tend to crave and eat more foods that are sweet and/or sour, with fruits and fruit juices being most commonly consumed [18, 22, 27]. Whether these dietary changes are related to taste [28] and olfactory changes [29] during pregnancy has been the focus of a few experimental studies. In the 1990s, Duffy and colleagues published one of the only prospective studies (i.e., the Yale Pregnancy Study) on taste changes during pregnancy [28]. Women were tested before they became pregnant and then during each trimester throughout their pregnancy. During each test session, women were asked to rate the intensity and hedonic value of sodium chloride (salt), citric acid (sour), quinine (bitter), and sucrose. The results indicated that bitter sensitivity increased during the first trimester, a finding that coincides with previous work [30]. These data suggest that avoidance of bitter-tasting foods such as green vegetables during early pregnancy [31] may be due, in part, to this initial hypersensitivity to bitter stimuli. As pregnancy progressed, women's sensitivity to bitter and salt tastes decreased, and their liking for bitter, salt, and sour tastes increased [28].

Regardless of the mechanisms underlying taste and presumably dietary changes during pregnancy, from the perspective of the fetus, such changes coincide with important developmental milestones. As will be discussed, this dietary information is passed through the mothers' amniotic fluid, providing the fetus with important orosensory stimulation that will modulate later food and flavor preferences.

4.3   ONTOGENY OF TASTE, SMELL, AND FLAVOR PERCEPTION

Flavor as an attribute of foods and beverages, is an integration of multiple sensory inputs including taste and retronasal olfaction in the oral and nasal cavities. Emerging scientific research reveals that the taste and olfactory systems are well developed before birth (see [32] for review). The apparatus needed to detect taste stimuli make their first appearance around the 7th or 8th week of gestation, and by 13 to 15 weeks, the taste bud begins to morphologically resemble the adult bud, except for the cornification overlying the papilla [33]. Taste buds are capable of conveying gustatory information to the central nervous system by the last trimester of pregnancy, and this information is available to systems organizing changes in sucking, facial expressions, and other affective behaviors.

Likewise, the olfactory bulbs and receptor cells needed to detect olfactory stimuli have attained adult-like morphology by the 11th week of gestation. Olfactory marker protein, a biochemical correlate of olfactory receptor functioning in fetal rats [34], has been identified in the olfactory epithelium of human fetuses at 28 weeks of gestation [35]. Because the epithelial plugs that obstruct the external nares resolve between gestational weeks 16–24, there is a continual turnover of amniotic fluid through the nasal passages, such that by the last trimester of pregnancy, the fetus swallows significant amounts of amniotic fluid, and inhales more than twice the volume it swallows. Even in air-breathing organisms, volatile molecules must penetrate the aqueous mucus layer covering the olfactory epithelium to reach receptor sites on the cilia. Thus, there is no fundamental distinction between olfactory detection of airborne and waterborne stimuli.

4.4   EARLY FLAVOR LEARNING

4.4.1   Amniotic Fluid

The environment in which the fetus lives, the amnion, can indeed be odorous. Its odor can indicate certain disease states, such as maple syrup disease or trimethylaminuria [36, 37]. In 1985, a case study report was published describing four infants who presented with peculiar body odors on delivery. Although each infant tested negative for syndromes that are associated with peculiar body odors, all were born to women who had ingested a spicy meal (e.g., cumin, fenugreek, curry) prior to delivery [14].

In the mid-1990s, an experimental study was conducted that revealed that the diet of the mother could alter the odor of amniotic fluid in humans [38]. Amniotic samples were collected from women undergoing routine amniocentesis. The women were randomized to one of two groups, in which they consumed either essential oil of garlic or placebo capsules approximately 45 min prior to the amniocentesis. The amniotic fluid from a portion of the sample was then evaluated by a trained sensory panel of adults who were screened for normal olfactory functioning. The results were unequivocal. Panelists judged the odor of the amniotic fluid of the women who consumed the garlic capsules as smelling stronger and more garlic-like than the amniotic fluid samples from the women who consumed the placebos. Since odor is an important component of flavor perception, these data provided the first experimental evidence that the amniotic fluid may provide infants with their first exposure to flavors within the mothers' diets.

That these flavor changes in amniotic fluid are perceived by fetuses and bias their preferences after birth was later demonstrated in a study conducted in Northern Ireland [39]. The response to the odor of garlic was assessed in two groups of infants: one group had mothers who consumed garlic-containing foods on a regular basis during the last month of pregnancy, whereas the other group did not. Between 15 and 24 h after birth, newborns were given a two-choice test between a cotton swab that contained garlic and an unadulterated cotton swab. The infants whose mothers consumed garlic before their birth oriented their head slightly more toward the cotton swab that smelled like garlic, whereas the infants whose mothers avoided garlic expressed their aversion for the garlic odor by orienting their heads more to the unadulterated swab than to the garlic swab.

A similar study was later conducted in France [40], but here the response to anise odors was assessed in infants whose mothers either regularly consumed anise-flavored foods and sweets, or those who did not consume anise-flavored foods. In this study, newborns of mothers who regularly consumed anise mouthed more and spent more time orienting toward a swab containing anise odor relative to the unadulterated swab and displayed fewer facial responses of distaste (e.g., brow lowering, cheek raising, nose wrinkling, gaping) toward the anise odor when compared with the infants whose mothers did not consume such flavored foods and sweets.

Taken together, these data suggest that neonates can respond positively to flavor volatiles that are experienced prenatally. However, experimental studies in which subjects are randomized to different treatment groups are considered the gold standard in research because they control for all extraneous variables, thereby permitting cause–effect inferences [41]. To this end, the first experimental study on how experience with flavors in amniotic fluid and mothers' milk affects infants' responses to these flavors is presented in the next section. But first, we review the evidence that reveals that like amniotic fluid, human milk provides the potential for a rich source of varying chemosensory experiences.

4.4.2 Breast Milk

Over the past 15 years, psychophysical research studies have revealed that like the milk of other mammals, human milk changes as a function of the dietary choices of the mother (for review, see [42]). Using a within-subjects design, milk samples were obtained from lactating women at fixed intervals before and after they ingested a particular food or beverage on one testing day and placebo during the other. These milk samples were placed individually in plastic squeeze bottles to minimize any visual differences in the milk samples, and all possible pairs of samples were presented to trained sensory panelists who were blind to the experimental condition. Using a forced-choice procedure, the panelists were asked to indicate which bottle of the pair "smelled stronger" or like the flavor under study. In general, panelists indicated that significant increases in the intensity of the milk odor occurred within a half hour to an hour after the mother consumed the flavor under study, with the intensity of the flavor decreasing thereafter. No such changes occurred on the days the mothers consumed the placebos. To date these psychophysical studies have revealed that a wide variety of volatiles either ingested (e.g., alcohol [43], garlic [44], vanilla [45], carrot [46]) or inhaled (i.e., tobacco [47]) by the lactating mother are transmitted to her milk.

Not only can infants detect these flavor changes in the milk, but other experimental studies revealed that they develop preferences for flavors experienced in amniotic fluid or mother's milk [46]. Pregnant women who planned on breastfeeding their infants were randomly assigned to one of three groups. The women consumed either 300 ml of carrot juice or water for 4 days per week for three consecutive weeks during the last trimester of pregnancy and then again during the first 2 months of lactation. The mothers in one group drank carrot juice during pregnancy and water during lactation, mothers in a second group drank water during pregnancy and carrot juice during lactation, whereas those in the control group drank water during both pregnancy and lactation. Approximately 4 weeks after the mothers began complementing their infants' diet with cereal, and before they had ever been fed foods or juices containing the flavor of carrots, the infants were videotaped as they were fed, in counterbalanced order, cereal prepared with water during one test session and cereal prepared with carrot juice during another.

Similar to the results of previous studies [39, 40], infants who had exposure to the flavor of carrots in amniotic fluid behaved differently in response to that flavor in a food base than did nonexposed control infants. Specifically, previously exposed infants displayed fewer negative facial expressions while eating the carrot-flavored cereal when compared with the plain cereal. They were also perceived by their mothers as enjoying the carrot-flavored cereal more when compared with the plain cereal. Postnatal exposure had similar consequences, thus highlighting the importance of a varied diet for both pregnant and lactating women. These findings provide the first experimental demonstration that prenatal or postnatal exposure to a flavor enhances the acceptance and enjoyment of that flavor in a food during weaning in humans [46].

The finding of enhanced acceptance of a flavor experienced in amniotic fluid and mothers' milk is not unique to humans, since similar findings have been observed in a wide variety of mammals such as dogs [48], rabbits [49], lambs [50], and rodents [51, 52]. The redundancy of dietary information transmitted during pregnancy and lactation may be important biologically because it provides complementary routes for the animal to learn about the types of foods available in the environment, should the mother's diet change between pregnancy and lactation. At weaning, young animals are faced with learning what to eat and how to forage. Exposure to dietary flavors in amniotic fluid and mother's milk may be one of several ways that mothers teach their young what foods are "safe." Consequently, young animals tend to choose a diet similar to that of their mothers when faced with their first solid meal. Such flavor learning is adaptive since these flavors tend to be associated with nutritious foods, or at least the foods the mother has access to, and hence they will likely be the foods to which young animals will have the earliest exposure during weaning.

4.5   FLAVOR VARIETY

Another advantage to breastfeeding is that it provides infants with rich and varied sensory experiences. That is, in contrast to formula-fed infants who become familiar with only a small set of invariant flavors, breastfed infants have extensive exposure to a wide range of flavors within their mothers' milks. As a result of these early and varied flavor experiences, breastfed infants are more willing to eat similarly flavored foods at weaning [53, 54], a finding that is consistent with research in a wide variety of mammals [49, 55].

In a recent study [54], breastfed infants showed greater liking of a fruit than did formula-fed infants, as did their mothers who reported eating more fruits in general when compared with mothers who formula fed. Similar findings were not observed among formula-fed infants despite their mothers eating more of a particular food. Although it remains unknown how much exposure the baby needs to enhance acceptance, our recent studies indicate that breastfeeding confers an advantage on initial acceptance of a food, but only if mothers eat the food regularly. Such flavor experiences during breastfeeding may serve to reduce food neophobia over the long term. In a study of 192, 7-year-old children, researchers found that girls who were breastfed for at least 6 months were less likely to be picky eaters [56]. One explanation for these findings is that breastfeeding provided varied sensory experiences, which in turn enhanced children's acceptance of a variety of flavors. In other words, exposure to flavor variety during breastfeeding may represent an important adaptive mechanism that facilitates food acceptance and diet diversity throughout life.

The importance of exposing infants to a variety of flavor experiences at weaning has been demonstrated experimentally [57, 58]. However, whether the experience with variety modified acceptance of a food depended on the flavors of foods experienced, and whether the novel food was a fruit or vegetable. Those infants who were fed a variety of vegetables that differed in taste, smell and texture subsequently ingested more of an orange vegetable (i.e., carrots) and a novel meat (i.e., chicken) after a 9-day exposure period when compared with those infants who were exposed to another vegetable (i.e., potatoes) [57]. Similarly, infants repeatedly fed a variety of fruits were more accepting of a novel fruit. However, the preference that developed appeared to be specific to the flavors experienced, since repeated exposure to a variety of fruits did not modify acceptance of a green vegetable [58]. Since flavor variety is often related to greater variety in the nutritive content of foods, preferences for varied flavors should ultimately enhance the range of nutrients consumed and thus increase the likelihood that a well-balanced diet is achieved. In this manner, the variety effect may reflect an important adaptive mechanism in the regulation of food intake.

4.6   LONG-TERM CONSEQUENCES OF EARLY FLAVOR LEARNING

Significant traces of the effects of early feeding experiences may remain as children age. In an 8-year longitudinal study of 70 white mother–child dyads living in Tennessee [59], interviews were conducted to determine whether food-related experiences at 2–24 months predicted dietary variety when children were between the ages of 6–8 years. Although vegetable variety in school-aged children was weakly correlated with mothers' vegetable preferences, 25% of the variance in school-aged children's fruit variety was predicted by breastfeeding duration and early fruit variety experience. Similar findings were reported in another longitudinal study from France [60] and a retrospective survey study conducted in England [61]. It is important to note that much of the research showing relationships between food habits in childhood and later in life are correlational in nature and consequently inconclusive regarding cause and effect relationships. The generality of such findings may be limited since all tested children were from predominately white, middle-class families. Moreover, it is possible that other important variables, such as genetic differences in taste (e.g., bitter) sensitivity, may contribute to individual differences in food preferences [62].

4.7   CONCLUSION

Over the course of pregnancy and lactation, a variety of factors interact to determine the food choices of mothers. While many of their food choices are driven by internal factors such as cravings and aversions, others may be influenced by environmental factors such as their cultural food practices and beliefs. Regardless of why women consume particular foods during pregnancy, emerging evidence reveals that such food choices can be detected by the fetus and young infant because of flavor changes in amniotic fluid and mother's milk. Although more research is needed to understand the mechanisms involved in early flavor learning, repeated exposure to flavors in amniotic fluid, mothers' milk, as well as to actual foods familiarizes infants to a wide range of flavors that influence their acceptance of foods and flavors at weaning. In other words, pre- and early-postnatal exposure, at the least, predisposes the young infant to accept the now-familiar flavor and facilitates the transition from fetal life through the breastfeeding period to the initiation of a varied solid food diet.

To be sure, many continue to learn and develop preferences for flavors and foods experienced later in life. The data reviewed in this chapter reveal that the development of preferences for culture-specific flavors has its beginnings during gestation and breastfeeding. It is the first, but not the only, way in which children learn about what foods are acceptable and preferred by their mothers. Strong adherence to cultural practices and beliefs during pregnancy, lactation and early childhood, helps to ensure that their children will learn to appreciate and prefer the flavors typical of their culture and will in turn pass on these cherished food practices to the next generation.

REFERENCES

1. Ahlqvist M, Wirfalt E (2000) Beliefs concerning dietary practices during pregnancy and lactation. A qualitative study among Iranian women residing in Sweden. Scand J Caring Sci 14:105–111
2. Ebomoyi E (1988) Nutritional beliefs among rural Nigerian mothers. Ecol Food Nutr 22:43–52
3. Frankel B (1977) Childbirth in the ghetto. R & E Research, San Francisco, Calif.

4. Kay MA (1977) Health and illness in a Mexican-American barrio. In: Spicer EH (ed) Ethnic Medicine in the Southwest. University of Arizona Press, Tucson, Ariz. pp 96–164
5. Snow LF, Johnson SM (1978) Folklore, food, female reproductive cycle. Ecol Food Nutr 7:41–49
6. Varga CA, Veale DJ (1997) Isihlambezo: utilization patterns and potential health effects of pregnancy-related traditional herbal medicine. Soc Sci Med 44:911–924
7. Mennella JA, Turnbull B, Ziegler PJ, Martinez H (2005) Infant feeding practices and early flavor experiences in Mexican infants: an intra-cultural study. J Am Diet Assoc 105:908–915
8. Harrison GG, Zaghloul SS, Galal OM, Gabr A (1993) Breastfeeding and weaning in a poor urban neighborhood in Cairo, Egypt: maternal beliefs and perceptions. Soc Sci Med 36:1063–1069
9. Backstrand JR, Goodman AH, Allen LH, Pelto GH (2004) Pulque intake during pregnancy and lactation in rural Mexico: alcohol and child growth from 1 to 57 months. Eur J Clin Nutr 58:1626–1634
10. Fikree FF, Azam SI, Berendes HW (2002) Time to focus child survival programmes on the newborn: assessment of levels and causes of infant mortality in rural Pakistan. Bull World Health Organ 80:271–276
11. Odebiyi AI (1989) Food taboos in maternal and child health: the views of traditional healers in Ile-Ife, Nigeria. Soc Sci Med 28:985–996
12. Veale DJH, Havlik I, Katsoulis LC, Kaido T, Arangies NS, Olive DW, Dekker T, Brookes KB, Doudoukina OV (1998) The pharmacological assessment of herbal oxytocics used in South African traditional medicine. Biomed Environ 2:216–222
13. Jelliffe DB (1968) Child nutrition in Developing Countries. US Department of Health, Education and Welfare, Washington, D.C.
14. Hauser GJ, Chitayat D, Berns L, Braver D, Muhlbauer B (1985) Peculiar odours in newborns and maternal prenatal ingestion of spicy food. Eur J Pediatr 144:403
15. Hepper PG (1988) Adaptive Fetal Learning: prenatal exposure to garlic affects postnatal preferences. Anim Behav 36:935–936
16. Taggart N (1961) Food habits in pregnancy. Proc Nutr Soc 20:35–40
17. Tierson FD, Olsen CL, Hook EB (1986) Nausea and vomiting of pregnancy and association with pregnancy outcome. Am J Obstet Gynecol 155:1017–1022
18. Bayley TM, Dye L, Jones S, DeBono M, Hill AJ (2002) Food cravings and aversions during pregnancy: relationships with nausea and vomiting. Appetite 38:45–51
19. Posner L, McCottry C, Posner A (1957) Pregnancy craving and pica. Obstet Gynecol 9:270–272
20. Wickham S (2005) Nutrition and the wisdom of craving. Pract Midwife 8:33
21. Weingarten HP, Elston D (1991) Food cravings in a college population. Appetite 17:167–175
22. Hook EB (1978) Dietary cravings and aversions during pregnancy. Am J Clin Nutr 31:1355–1362
23. Profet M (1995) Protecting your baby-to-be: preventing birth defects in the first trimester. Addison-Wesley, New York, N.Y.
24. Profet M (1988) The evolution of pregnancy sickness as protection to the embryo against Pleistocene teratogens. Evol Theory 8:177–190
25. Rozin P (1984) The acquisition of food habits and preferences. In: Mattarazzo HD, Weiss SM, Herd JA, Miller NE, Weiss SM (eds) Behavioral health: a handbook of health enhancement and disease prevention. Wiley, New York, N.Y, pp 590–607
26. Rozin P (1996) Sociocultural influences on human food selection. In: Elizabeth Capaldi (ed) Why we eat what we eat: the psychology of eating. American Psychological Association, Washington, D. C., pp 233–263
27. Pope JF, Skinner JD, Carruth BR (1992) Cravings and aversions of pregnant adolescents. J Am Diet Assoc 92:1479–1482
28. Duffy VB, Bartoshuk LM, Striegel-Moore R, Rodin J (1998) Taste changes across pregnancy. Ann N Y Acad Sci 855:805–809
29. Nordin S, Broman DA, Olofsson JK, Wulff M (2004) A longitudinal descriptive study of self-reported abnormal smell and taste perception in pregnant women. Chem Senses 29:391–402
30. Bhatia S, Puri R (1991) Taste sensitivity in pregnancy. Indian J Physiol Pharmacol 35:121–124
31. Flaxman SM, Sherman PW (2000) Morning sickness: a mechanism for protecting mother and embryo. Q Rev Biol 75:113–148

32. Ganchrow JR, Mennella JA (2003) The ontogeny of human flavor perception. In: Doty RL (ed) Handbook of olfaction and gustation, 2nd edn. Dekker, New York, N.Y., pp 823–846

33. Bradley RM (1972) Development of taste bud and gustatory papillae in human fetuses. In: Bosma JF (ed) The third symposium on oral sensation and perception: the mouth of the infant. Charles C. Thomas, Springfield, Ill., pp 137–162

34. Gesteland RC, Yancey RA, Farbman AI (1982) Development of olfactory receptor neuron selectivity in the rat fetus. Neuroscience 7:3127–3136

35. Chuah MI, Zheng DR (1987) Olfactory marker protein is present in olfactory receptor cells of human fetuses. Neuroscience 23:363–370

36. Lee CW, Yu JS, Turner BB, Murray KE (1976) Trimethylaminuria: fishy odors in children. N Engl J Med 295:937–938

37. Menkes JH, Hurst PL, Craig JM (1954) A new syndrome: progressive familial infantile cerebral dysfunction associated with an unusual urinary substance. Pediatrics 14:462–467

38. Mennella JA, Johnson A, Beauchamp GK (1995) Garlic ingestion by pregnant women alters the odor of amniotic fluid. Chem Senses 20:207–209

39. Hepper P (1995) Human fetal "olfactory" learning. Int J of Prenat Perinat Psych Med 7:147–151

40. Schaal B, Marlier L, Soussignan R (2000) Human foetuses learn odours from their pregnant mother's diet. Chem Senses 25:729–737

41. Trochim WMK (2002) Experimental design. Research Methods Knowledge Base. Available via http://www.socialresearchmethods.net/kb/desexper.htm)

42. Mennella JA (1995) Mother's milk: a medium for early flavor experiences. J Hum Lact 11:39–45

43. Mennella JA, Beauchamp GK (1991) The transfer of alcohol to human milk. Effects on flavor and the infant's behavior. N Engl J Med 325:981–985

44. Mennella JA, Beauchamp GK (1993) The effects of repeated exposure to garlic-flavored milk on the nursling's behavior. Pediatr Res 34:805–808

45. Mennella JA, Beauchamp GK (1996) The human infants' response to vanilla flavors in mother's milk and formula. Inf Behav Dev 19:13–19

46. Mennella JA, Jagnow CP, Beauchamp GK (2001) Prenatal and postnatal flavor learning by human infants. Pediatrics 107:E88

47. Mennella JA, Beauchamp GK (1998) Smoking and the flavor of breast milk. N Engl J Med 339:1559–1560

48. Hepper PG, Wells DL (2006) Perinatal olfactory learning in the domestic dog. Chem Senses 31:207–212

49. Bilko A, Altbacker V, Hudson R (1994) Transmission of food preference in the rabbit: the means of information transfer. Physiol Behav 56:907–912

50. Nolte DL, Provenza FD, Balph DF (1990) The establishment and persistence of food preferences in lambs exposed to selected foods. J Anim Sci 68:998–1002

51. Galef BG Jr, Sherry DF (1973) Mother's milk: a medium for transmission of cues reflecting the flavor of mother's diet. J Comp Physiol Psychol 83:374–378

52. Hepper PG (1988) Adaptable fetal learning: prenatal exposure to garlic affects postnatal preferences. Anim Behav 36:935–936

53. Mennella JA, Beauchamp GK (1997) Mothers' milk enhances the acceptance of cereal during weaning. Pediatr Res 41:188–192

54. Forestell CA, Mennella JA (2007) Early determinants of fruit and vegetable acceptance. Pediatrics 120:1247–1254

55. Nolte DL, Provenza FD, Callan R, Panter KE (1992) Garlic in the ovine fetal environment. Physiol Behav 52:1091–103

56. Galloway AT, Lee Y, Birch LL (2003) Predictors and consequences of food neophobia and pickiness in young girls. J Am Diet Assoc 103:692–698

57. Gerrish CJ, Mennella JA (2001) Flavor variety enhances food acceptance in formula-fed infants. Am J Clin Nutr 73:1080–1085

58. Mennella JA, Nicklaus S, Jagolino AL, Yourshaw LM (2007) Variety is the spice of life: strategies for promoting fruit and vegetable acceptance in infants. Physiol Behav (in press)

59. Skinner JD, Carruth BR, Bounds W, Ziegler P, Reidy K (2002) Do food-related experiences in the first 2 years of life predict dietary variety in school-aged children? J Nutr Educ Behav 34:310–315

60. Nicklaus S, Boggio V, Chabanet C, Issanchou S (2005) A prospective study of food variety seeking in childhood, adolescence and early adult life. Appetite 44:289–297
61. Cooke LJ, Wardle J, Gibson EL, Sapochnik M, Sheiham A, Lawson M (2004) Demographic, familial and trait predictors of fruit and vegetable consumption by pre-school children. Public Health Nutr 7:295–302
62. Mennella JA, Pepino MY, Reed DR (2005) Genetic and environmental determinants of bitter perception and sweet preferences. Pediatrics 115:e216–e222

II

NUTRIENT NEEDS AND FACTORS RELATED TO HIGH-RISK PREGNANCY

5 Obesity and Pregnancy

Sarah C. Couch and Richard J. Deckelbaum

Summary Obesity in pregnancy is associated with numerous maternal and neonatal complications including difficulty conceiving, increased risk of miscarriage, fetal anomalies and mortality, higher rates of gestational hypertension, gestational diabetes and preeclampsia, and an increased risk of cesarean section and delivery related complications. Nevertheless, more women are entering pregnancy with excessive weight and are gaining weight above the Institute of Medicine (IOM) recommendations during pregnancy. Weight loss is not recommended during pregnancy; however, overweight and obese women should be advised to aim for a moderate weight loss prior to conception and during the postpartum period. Strategies for achieving moderate pregestational and postpartum weight loss include a low-calorie, low-fat diet and at least 45 min of daily physical activity. Benefits to mother and child are achieved with even a moderate weight loss. Importantly, health care professionals should counsel women on gaining an appropriate amount of weight during pregnancy. More research is needed on effective intervention approaches to promote optimal weight status before and after pregnancy and to support optimal weight gain during pregnancy.

Keywords: Obesity, Overweight, Pregnancy, Gestational weight gain, Maternal and neonatal complications, Weight management strategies

5.1 INTRODUCTION

The prevalence of overweight and obesity in the United States has reached epidemic proportions. Nearly two thirds of adults >20 years of age have a body mass index (BMI) $\geq 25\,kg/m^2$ and are considered overweight; of these, a third have a BMI $\geq 30\,kg/m^2$ and are considered obese [1]. From 1999 to 2002, 29.1% of women of childbearing age (20–39 years) were obese, with the highest prevalence in non-Hispanic black women (49%), followed by Mexican-American (38.9%) and non-Hispanic white women (31.3%) [1]. Notably, more women are entering pregnancy with excess weight. Recent US data collected from nine states from the Pregnancy Risk Assessment Monitoring System showed a 69% increase in prepregnancy obesity from 1993 to 2003 [2]. These alarming statistics raise questions regarding the potential health implications of pregestational obesity for both mother and infant. This chapter defines overweight and obesity in pregnancy and examines the short- and long-term risks associated with excessive weight for

From: *Nutrition and Health: Handbook of Nutrition and Pregnancy*
Edited by: C.J. Lammi-Keefe, S.C. Couch, E.H. Philipson © Humana Press, Totowa, NJ

women during the childbearing years. Current recommendations for gestational weight gain for overweight women will be examined from the standpoint of adherence and complications resulting from poor compliance to guidelines. Finally, weight management strategies will be discussed and recommendations provided for use by the obstetrician in counseling women regarding achieving a healthy weight and rate of weight change before, during, and after pregnancy.

5.2   OBESITY: DEFINITION

BMI is considered the gold standard when determining weight status, and provides the basis for current weight gain recommendations from the IOM. It is calculated by dividing weight in kilograms by height in meters squared. BMI is not valid while pregnant, and so should be measured pre- and postgestation. In the case where pregestational weight is unknown, the first weight measured at prenatal clinic is generally used to calculate prepregnancy BMI [3]. Pregestation and postpartum obesity is most commonly defined according to the World Health Organization's (WHO) definition: underweight, BMI < $18.5 \, kg/m^2$; normal weight, BMI = $18.5–24.9 \, kg/m^2$; overweight, BMI = $25–29.9 \, kg/m^2$; and obese, BMI > $30 \, kg/m^2$ [4]. Less often, obesity is defined using the exact weight in kilograms or pounds. Since the definition of obesity is often not consistent in studies using absolute weight rather than BMI, this review focuses on those studies using weight classification based on BMI.

5.3   CONSEQUENCES OF PREGESTATIONAL OBESITY

5.3.1   *Infertility and Risk of Miscarriage*

Adverse effects of obesity on natural conception and assisted reproductive therapy in women are well documented in the literature [5–8] (see Table 5.1). For example, Rich-Edwards et al. [7] found that, among ~2,500 married, infertile nurses, those with pregestational obesity experienced more frequent anovulation and had longer mean time to pregnancy than did normal-weight women. Also, higher rates of early miscarriage have been found among obese women as compared with women of normal weight. In a case-control study of 1,644 obese women compared with 3,288 normal weight, age-matched controls, Lashen et al. [9] found an increase risk of first trimester and recurrent miscarriage associated with pregestational obesity. Similarly, a Swedish population–based cohort study of over 800,000 women showed that obesity was associated with a twofold greater risk of spontaneous abortion compared with normal weight mothers [10].

Table 5.1
Potential Reasons for Infertility in Obese Women

Menstrual irregularities
Hyperandrogenism
Oligo-/amenorrhea
Chronic anovulation
Decreased conception rates after assisted reproductive techniques
Increased risk of miscarriage

In overweight women conceiving after in vitro fertilization (IVF) or intracytoplasmic sperm injection, miscarriage rate is also reportedly higher in obese compared with lean or average weight women. A systematic review of the literature by Maheshwari et al. [11] found that when compared with women with a BMI < 25 kg/m^2, women with a BMI ≥ 25 kg/m^2 had a 29% lower likelihood of pregnancy and a 33% higher risk of miscarriage following IVF. In this same study, obese women were found to have a reduced number of oocytes retrieved despite requiring higher doses of gonadotropins. Mechanisms for the relationship between obesity and infertility are unknown. Suggested roles of hyper-androgenism, insulin resistance, high leptin levels, and polycystic ovarian syndrome are currently under investigation [12]. Regardless of mechanism, these data suggest that obesity may delay or prevent conception in women who want to become pregnant. Of some consolation, weight loss before infertility therapy may improve a women's likelihood of conceiving. Notably, over a dozen studies have documented improvement in reproductive parameters and fertility outcome following moderate weight loss [13–24].

5.3.2 Neural Tube Defects and Congenital Malformations

Several reports suggest an increased risk of congenital malformations, particularly neural tube defects (NTD), for infants born to obese mothers [25–28]. In a case-control study by Waller et al. [25] of 499 mothers of infants with NTDs, 337 mothers of infants with other major birth defects and 534 mothers of infants without birth defects (*n* = 534), women who were obese before pregnancy (BMI > 31 kg/m^2) were significantly more likely to have an infant with an NTD, e.g., spina bifida, compared with normal weight mothers. Results were adjusted for age, race, education, and family income. In a comparison of 604 fetuses and infants with NTDs and 1,658 fetuses and infants with other major malformations, Werler et al. [26] found a significant association between NTD and maternal pregestational weight, independent of folic acid intake. Ray et al. [27] examined antenatal maternal screening data from over 400,000 women in Canada to determine whether the risk of NTDs was lower after flour fortification with folic acid. Surprisingly, these researchers found that before fortification, greater maternal weight was associated with a modest increase risk in NTD (OR 1.4, 95% CI 1.0–1.8); after flour fortification, the risk actually increased (OR 2.8, 95% CI 1.2–6.6). While these results do not preclude folic acid supplementation as a means of reducing risk of NTDs in overweight and obese women, they do raise questions about the potential for adverse effects of excessive pregestational weight on folic acid bioavailability/metabolism in mother and fetus. Research is needed to examine these effects and to determine whether obese women may benefit from higher doses of folic acid, as recommended in mothers with diabetes [3].

Several other malformations have been associated with maternal obesity including defects of the heart, central nervous system, ventral wall, and other intestinal defects [25] (Table 5.2). Because higher risk of these same congenital anomalies often occurs with pregestational diabetes, the question is often raised as to whether undiagnosed type 2 diabetes mellitus could be the underlying cause. To rule out this factor, early and repeat screening for diabetes should be considered for women entering pregnancy with excessive weight.

Importantly, significant impairment of sonographic visualization of the fetal anatomy has also been reported in obese women. Hendler et al. [29] reported a decrease in visualization

Table 5.2
Possible Maternal and Fetal Complications Associated with Obesity in Pregnancy

Prior to pregnancy and early gestation
    Infertility
    Miscarriage
    Neural tube defects
    Heart and craniospinal defects
    Ventral wall and other intestinal defects

Late gestation
    Chronic hypertension
    Gestational diabetes
    Thromboembolic disease

Labor and delivery
    Increased incidence of cesarean section
    Increase incidence of preterm labor
    Postsurgical wound infection
    Poor lactation outcomes

Neonatal
    Neonatal death
    Macrosomia
    Shoulder dystocia/birth trauma

of fetal organs in obese versus nonobese women that was most marked for cardiac and craniospinal structures. Advancing gestation is the best predictor of visualization of fetal abnormalities in nonobese women, while among obese women Wolfe et al. [30] found no improvement with advancing gestation or duration of the examination. As this obese population is at higher risk of fetal abnormalities, this is cause for concern. At the very least, the use of advanced ultrasound equipment for these women, with anomaly scans done by an ultrasonographer with an appropriate level of expertise should be routinely recommended.

5.3.3 *Preeclampsia and Gestational Diabetes*

While the normal pregnancy is characterized by maternal hemodynamic changes and an insulin resistant state, obesity in pregnancy appears to complicate these expected physiological adaptations to pregnancy. Accordingly, the risk for hypertensive disorders and gestational diabetes (GDM) is reportedly higher in obese and morbidly obese women compared to women who are not obese. In a prospective, multicenter study of more than 16,000 women, Weiss et al. [31] observed a 2.5-fold greater risk of gestational hypertension, and a 2.6-fold greater risk of GDM among obese versus nonobese women. Risk for these conditions was even greater in a morbidly obese subset, e.g., 3.2- and 4-fold respectively. Similarly, these researchers found the risk for developing preeclampsia was 1.6 and 3.3 times more likely to develop in obese and morbidly obese women, respectively. Results from this study have been confirmed by others [32, 33] and found to be independent of other related factors including age, parity, ethnicity, and family history of chronic diseases.

Frequently, GDM and preeclampsia go hand in hand. Several studies suggest that obesity may be at the "metabolic core" of these conditions. For example, regardless of treatment type or degree of glucose control, Yogev et al. [34] reported that the risk for developing preeclampsia in women with GDM was significantly greater in obese (10.8%) versus normal weight women (8.2%). Notably, in this study the risk of preeclampsia escalated in obese women with poor glucose control (14.9%), suggesting that tighter glucose control in women with GDM may decrease risk. Barden et al. [35] found that late-onset preeclampsia in women with GDM was more likely to develop in women who were not only obese but had preexisting hypertension, more severe insulin resistance, subclinical inflammation, and a family history of diabetes and hypertension. Similar to the "metabolic syndrome" in the nonpregnant state, this clustering of risk factors suggests that obese women with GDM and preeclampsia may be at greater risk for cardiovascular disease and type 2 diabetes in later life.

In clinical practice, consideration should be given to screening obese women for GDM and hypertension as soon as possible, preferably upon presentation or during the first trimester. Screening for these conditions should be repeated later in pregnancy if the initial results are negative. Importantly, postpartum follow-up should include advice on achieving a healthful weight and modifying cardiovascular risk factors if they are present.

5.3.4   Thromboembolic Complications

In the United States, thromboembolic disease is the leading cause of death in pregnant women [36]. Obesity is a documented risk factor for thromboembolism in pregnancy. As evidence, in a retrospective study [37] comparing 683 obese women with 660 normal weight women (all had singleton live births), the risk of thromboembolic disease was twofold greater among obese versus normal weight women. The risk of developing thromboembolic disease is increased for about 6–8 weeks after delivery and is much greater after a cesarean section than after vaginal delivery [36]. Postpartum heparin therapy is often recommended for patients thought to be at high risk for venous thromboembolism [38]. Also, obese pregnant women may warrant prophylaxis measures against venous thromboembolism, such as compression stockings or heparin therapy, especially if exposed to other risk factors (e.g., bed rest).

5.3.5   Preterm Delivery, Cesarean Section, and Operative Complications

Obesity has been independently associated with an increased risk of a number of obstetric complications including preterm delivery, cesarean section, and post–cesarean section infectious morbidity. With respect to preterm delivery, BMI on both ends of the weight spectrum, e.g., BMI $\geq 40\,kg/m^2$ and BMI $\leq 18.5\,kg/m^2$, has been observed to increase risk of preterm delivery in comparison to normal BMI (18.5–$24.5\,kg/m^2$). In the multicenter study of Weiss et al [31], morbidly obese women had a 1.5 times greater risk of preterm delivery in comparison with a normal weight control group; underweight women had a 6.7-fold greater risk. Preterm delivery, particularly before 32 weeks gestation, is cause for concern, because it places the infant at increased risk of morbidity and mortality [39]. Reasons for the greater risk of preterm delivery in underweight and obese women may differ and have not been clearly defined. In obese women, underlying medical and obstetric issues may be the dominant cause.

Cesarean section rates are higher among obese women compared to nonobese women. In Washington State, Baeten et al. [40] examined delivery data from over 96,000 mother–infant pairs. These researchers found that obese women had nearly a threefold greater risk of having a cesarean section than women who were not obese. After adjusting for gestational hypertension, GDM, and preeclampsia, the risk decreased nominally, to 2.7-fold. In another study, Brost et al. [41] found that for each 1-kg/m^2 increase in prepregnancy BMI, the odds of having a cesarean section increased by 7%. Obesity is associated with a reduced likelihood of vaginal birth after cesarean section (VBAC) [42], and a lower success rate for VBAC compared with normal weight women (68 versus 79.9%, respectively [42, 43]). Operative and postoperative complications associated with cesarean surgery in obese women, especially the morbidly obese, are many including greater risk of excessive blood loss, prolonged operative time, higher rates of anesthesia, difficulty placing regional anesthesia, and greater risk of wound infection compared to nonobese women who have had a cesarean section. [41] Given this litany, it is not surprising that postoperative hospital stay is reportedly longer and delivery associated medical costs higher in obese versus nonobese women [12].

5.3.6 Macrosomia, Shoulder Dystocia, and Fetal Death

Several large US population–based cohort studies have shown a significant relationship between pregestational BMI and macrosomia, which is defined as a birth weight >4,000 g or above the 90th percentile [40, 44–46]. Risk for having a macrosomic infant appears to increase in mothers with degree of excess weight. For example, in the multicenter study by Weiss et al. [31], the incidence of macrosomia was 8.3% in nonobese women, 13.3% in obese women, and 14.6% in morbidly obese women. In the study by Baeten et al. [40], the odds of having a macrosomic infant were 1.2 in women who were normal weight, 1.5 in women who were overweight, and 2.1 in women who were obese. The analysis excluded women with chronic hypertension, pregestational and gestational diabetes, and preeclampsia. National and international trends in North America [47] and Europe [48, 49] report an increase in incidence of large for gestational age infants, and implicate rising trends in maternal obesity and diabetes, and declining trends in maternal smoking as causal factors.

Macrosomia is a well-established risk factor for shoulder dystocia and birth trauma. The risk is directly related to birth weight and increases substantially with birth weight >4,500 g. Injury to the brachial plexus is reportedly rare but increases substantially (tenfold) with birth weights >4,500 g [50]. Bassaw et al. [51] conducted a 9-year review of over 100 cases of shoulder dystocia from among ~47,000 vaginal deliveries. As compared with infants weighing between 3,500 and 3,999 g, these researchers found a 2.2% higher frequency of shoulder dystocia in infants weighing 4,000–4,499 g at birth. The frequency of shoulder dystocia increased by 7.1% with birth weights >4,500 g. Obesity was the most important identifiable predisposing factor for shoulder dystocia and occurred in 35.9%.

Risk for late-gestation fetal demise is also greater among obese women compared with their normal weight counterparts. Population-based studies in England [44], Sweden [52], Norway [53], and Canada [54] showed a significant relationship between obesity and risk of late fetal death (stillbirth occurring after 28 weeks of gestation) even after adjusting for gestational diabetes, gestational hypertension, preeclampsia, maternal age, and parity. Given the dire consequences of pregestational obesity, intervention strategies for helping women achieve a healthy prepregnant BMI are urgently needed.

5.4 WEIGHT GAIN RECOMMENDATIONS AND CONSEQUENCES OF NONCOMPLIANCE

In 1990, the IOM issued recommendations for weight gain during pregnancy based on prepregnancy weight status [3]. The goal of these recommendations was to optimize neonatal birth weight to between 3 and 4 kg and prevent the morbidity and mortality associated with low birth weight (LBW). According to these recommendations, an underweight woman (based on WHO BMI criteria above) should gain 28–40 lb (12.5–18 kg), a normal weight woman should gain 25–35 lb (11.5–16 kg), an overweight woman should gain 15–25 lb (7–11.5 kg), and an obese woman should gain less than or equal to 15 lb (7 kg). Recently, these recommendations have been criticized for being too liberal and not making allowances for women who gain excessive amounts of weight during pregnancy.

Evidence is mounting that significant numbers of women, particularly overweight and obese women, are not adhering to IOM guidelines. In an investigation of over 120,000 women enrolled in Women, Infants, and Children (WIC) clinics over a 6-year period, Schieve et al. [55] found that the percentage of women reporting a pregnancy weight gain greater than the IOM recommendations increased significantly from 41.5 to 43.7%. In 2005, Jain et al. [56] examined data from the New Jersey Pregnancy Risk Assessment Monitoring System ($n = 7,661$) and found that nearly 64% of overweight women and 78% of obese women were noncompliant with IOM recommendations (e.g., overgained).

This lack of adherence to weight gain recommendations during pregnancy should be cause for concern. Excessive gestational weight gain has been shown to be a risk factor for maternal and neonatal complications, independent of prepregnancy BMI. For example, among the 7661 pregnant women in New Jersey examined by Jain et al. [56], women who gained greater than 35 lb during pregnancy increased their risk of macrosomia and delivery by cesarean section by 60–180% and had lower rates of breastfeeding by 30%. In another study, Hilson et al. [57] showed that mothers who exceeded IOM weight gain recommendations failed to initiate and/or sustain breastfeeding in all categories of prepregnancy BMI.

Excessive gestational weight gain may lead to child adiposity. In a recent prospective study of over 1,000 mother–child pairs, mothers with greater gestational weight gain had children with greater BMI and skin fold thicknesses (triceps and subscapular) at 3 years of age [58]. This association was independent of parental BMI, maternal glucose intolerance, breastfeeding duration, gestational age at delivery, and birth weight. Children of mothers who gained more weight also had higher systolic blood pressure, a cardiovascular risk factor that has been shown to track into adulthood.

Poor adherence to weight gain recommendations may also have serious ramifications for a woman's health in midlife. Weight gain during pregnancy, weight loss at 6 months postpartum, and prepregnancy BMI all predicted BMI 15 years later [59]. Data from the National Maternal and Infant Health Survey showed that among the women who gained more than the recommended weight during pregnancy, greater than 30% retained an average of 2.5 kg at 10–18 months postpartum as opposed to retention of 1 kg among women who gained at the recommended level [60]. In a longer-term study, Rooney et al. [61] followed 540 women for ~8 years after childbirth and found that women who gained more than the IOM recommended weight during pregnancy retained 2 kg above their prepregnancy weight more at 8 years postpartum compared with those who complied with weight guidelines. These researchers also showed that at the mean age of 42 years, obese women weighed 34 lb more than when they became pregnant. Given that obesity is a risk factor for chronic diseases, these findings suggest that interventions are needed to improve adherence to IOM recommendations and to help women achieve a healthful weight postpartum.

5.5 MANAGEMENT

5.5.1 *Weight Loss Strategies for Prepregnancy and Postpartum*

Weight loss is not recommended during pregnancy; however, overweight and obese women should be advised to aim for a moderate weight loss prior to conception and postpartum. To help motivate women in their efforts, health care providers and obstetricians should clearly define the risks associated with overweight and obesity in pregnancy and beyond. Consultation with a registered dietitian (RD) should be considered, as these individuals can provide assistance in the assessment of current eating habits and in formulating approaches for healthful weight reduction. Varieties of approaches exist for the management of overweight and obesity among women. These have been reviewed in Chap. 13 ("Popular Diets"). For women who are overweight or obese, the American Dietetic Association recommends a low-calorie (1,000–1,500 kcal/day), low-fat (25–30% of energy) diet with generous amounts of protein (15–25% of energy), and regular exercise as a first-line approach [62]. The rate of weight loss recommended is no more than 1.5 to 2 lb per week; this equates to a calorie deficit of 750–1,000 kcal/day. During the postpartum period, a slower rate of weight loss is advised for breastfeeding women (no more than 1 lb per week) to ensure adequate energy intake to support lactation [63]. The physical activity goal to lose weight and maintain a healthy weight after weight loss is at least 45 min of moderate physical activity [64, 65]. As reported in most weight-loss studies, self-monitoring of food intake has been significantly associated with weight loss [62]. Women are likely to benefit from consistent support and advice from health care professionals. As data from numerous weight loss studies will attest, significant long-term weight loss is difficult to achieve and maintain [66]. However, even a moderate degree of weight loss of 10 lb can reduce the risk of GDM among obese women [67, 68]. Encouraging breastfeeding postpartum can help weight loss efforts after delivery. Twelve weeks of breastfeeding was associated with lower BMI later in life [59].

5.5.2 *Considerations for Bariatric Surgery*

Given the growing number of women with severe obesity, it is not surprising that the number of women who are seeking extreme measures to lose weight, e.g., bariatric surgery, is increasing. As discussed in detail in Chap. 6 ("Pregnancy and Weight Loss Surgery"), surgical interventions to lose weight, unless expertly planned, are not without potential consequences for mother and infant. In addition to promoting weight loss, malabsorptive type surgeries such as gastric bypass have resulted in suboptimal maternal absorption of calcium, iron, folic acid, and vitamin B_{12} [69]. While these surgeries have had a positive impact on reducing maternal risk for GDM and hypertensive disorders, case reports of intrauterine growth restriction, premature birth, and NTDs have been described [70]. Because of these limitations, the laparoscopic adjustable gastric banding procedure is being used more frequently as a means of restricting stomach volume, decreasing intake, and promoting weight loss [71]. The adjustability of banding also allows for adaptations to altered requirements of pregnancy. Early reports on follow-up of pregnant women who have had this type of procedure are encouraging and indicate reduced risk of malabsorption, GDM, gestational hypertension, and preterm deliveries [71]. To ensure optimal pregnancy outcome and minimize maternal and fetal risks, the American College of Obstetricians and Gynecologists recommends that women delay pregnancy for 12–18 months after surgery to avoid pregnancy during the rapid weight loss phase [67]. Vitamin supplementation is also advised if nutritional deficiencies occur.

5.5.3   Improving Compliance to IOM Recommendations

In 2000, Abrams et al. [72] conducted a systematic review of available observational data published between 1990 and 1997 on weight gain and maternal and fetal outcomes. Not surprising, this review showed that pregnancy weight gain within the IOM recommended range was associated with the best outcome for both mothers and infants. However, this review also found that most women were noncompliant with these guidelines; many women were gaining excessive amounts of weight. Researchers speculated many reasons for these findings, including environmental temptations, inactivity, and prepregnancy restrictive dieting. They also reported that many women were not given appropriate targets for weight gain. The Women and Infants Starting Healthy study also found that from pregnant women studied in the San Francisco Bay area (excluding women with preterm birth, multiple gestation, or maternal diabetes), 50% of obese women were given advice by their physician to overgain, 35% of underweight women were given advice to undergain, and 87% of women with normal weight were given advice to gain an appropriate amount of weight [73]. The proportion of women who received no advice at all was 33%. This suggests that some providers are not aware of BMI-specific weight gain guidelines and may be advising all women to gain within the same range. Greater public health efforts should be made to ensure that all clinicians and pregnant women are educated on appropriate weight gain targets in pregnancy. In clinical practice, prepregnancy BMI should be ideally recorded at the first visit in the first trimester, followed by regular monitoring of gestational weight gain throughout pregnancy. An appropriate care plan should be developed and implemented should rate of weight gain veer significantly from established guidelines.

5.5.4   Successful Interventions

So what works to control excessive weight gain in pregnancy? Unfortunately, few studies have been done to answer this question. A Medline search using the keywords pregnancy, intervention, weight gain, revealed three intervention trials for healthy pregnant women in this area. The most recent study [74] investigated whether individual counseling on diet and physical activity during pregnancy could increase diet quality and leisure time physical activity and prevent excessive weight gain among healthy pregnant primiparas. The study was conducted in six maternity clinics in Finland. Women in the treatment group received one 30-min counseling session on diet and physical activity and three 10-min booster sessions on the same until the 37th gestation week. Weight gain, diet, and physical activity guidelines appropriate for pregnancy were recommended as part of counseling. The control group received standard maternity care. Results showed that, while participants in the treatment group improved diet quality compared to standard care, the more extensive counseling was unable to prevent excessive gestational weight gain. A similar study conducted in Pennsylvania [74] compared a behavioral nutrition intervention with standard hospital-based care for women during pregnancy. Women (BMI > 19.8 kg/m^2) assigned to the behavioral intervention received written material on weight gain guidelines, exercise, and a healthy diet plan for pregnancy. Women were weighed monthly, and if weight gain was within established limits for stage of pregnancy, then positive feedback was given. If weight gain was outside of limits, additional nutrition information and behavioral counseling was provided. The control group received written material on healthy diet planning during pregnancy. Results showed that normal-weight women who participated in the intervention group were more successful than controls were in keeping weight within recommended limits. However, overweight women in both groups overgained (~75% per group). In a third trial, Olson et al. [75] compared weight gain in women who participated in a mail-based nutrition education intervention versus those who did not participate. The intervention group was mailed materials including weight gain guidelines, healthy diet planning, and appropriate exercises for pregnancy. Controls received no mailed materials. Results showed that the number of women who overgained did not differ between groups; however, low-income women, regardless of prepregnancy weight, were more

Table 5.3
Recommendations to Reduce Risks Associated with Pregestational and Postpartum Obesity and Excess Weight Gain in Pregnancy

Prepregnancy

- Counsel women regarding the maternal and fetal risks associated with overweight and obesity in pregnancy
- Provide guidance on healthy eating and exercise habits as part of routine care; if overweight or obese, referral to a registered dietitian is encouraged
- Counsel women to quit smoking, avoid alcohol, and consume adequate calcium, iron, and folic acid
- Screen for diabetes and hypertension

Prenatal

- Discuss recommended weight gain based on BMI
- Provide support and advice if rate of weight gain is outside suggested limits
- Screen for gestational diabetes and monitor blood pressure

Postpartum

- Provide guidance on healthy eating and exercise habits to help women return to healthy weight
- Encourage breastfeeding

likely to stay within weight gain limits as a result of the intervention. Although limited, results from these studies suggest that for normal weight and low-income women, low-intensity nutrition education programs may assist them in adhering to IOM weight gain recommendations. For overweight and obese women, more intensive behavioral nutrition interventions may be necessary to improved compliance.

5.6 CONCLUSION

As rates of obesity escalate, more women are entering pregnancy obese and gaining weight above established recommendations. Excess weight in pregnancy can adversely affect both mother and infant. Limited data are available on successful approaches to improve compliance to established weight gain recommendations. Importantly, women should be provided with appropriate resources and support from health care professionals to achieve a healthy weight before conception, maintain a healthy rate of weight gain during pregnancy, and attain a healthy BMI postpartum. Table 5.3 summarizes considerations for healthcare professionals regarding obesity and pregnancy.

REFERENCES

1. Hedley AA, Ogden CL, Johnson Dl, Carroll MD, Curin LR, Flegal KM (2004) Prevalence of overweight and obesity among US children, adolescents, and adults, 1999–2002. J Am Med Assoc 291:2847–2850
2. Kim SY, Dietz PM, England L, Morrow B, Callaghan WM (2007)Trends in prepregnancy obesity in nine states, 1993–2003. Obesity 25:986–993
3. Institute of Medicine (1990) Nutrition During Pregnancy. Part I. Weight Gain. Part II. Dietary Supplements. Committee on Nutritional Status during Pregnancy and Lactation. National Academy Press, Washington, D.C.
4. World Health Organization (1997) Obesity: preventing and managing the global epidemic. Report of a WHO consultation presented at the World Health Organization: 3–5 June, Geneva, Switzerland. Publication WHO/NUT/NCD/98.1
5. Lake JK, Power C, Cole TJ (1997) Women's reproductive health: the role of body mass index in early and adult life. Int Obes Rel Metab Disord 21:432–438
6. Foreyt JP, Poston WS (1998) Obesity: a never-ending cycle? Int J Fertil Womens Med 43:111–116
7. Rich-Edwards JW, Manson JE, Goldman MD (1992) Body mass index at age 18 years and the risk of subsequent ovulatory infertility. Am J Epidemiol 5:247–250
8. Howe G, Westhoof C, Vessey M, Yeates D (1985) Effects of age, cigarette smoking, and other factors on fertility: findings in a large prospective study. Br Med J 290:1697–1700
9. Lashen H, Fear K, Sturdee DW (2004) Obesity is associated with increased risk of first trimester and recurrent miscarriage: matched case control study. Hum Reprod 19:1644–1646
10. Cedergren MI (2004) Maternal morbid obesity and the risk of adverse pregnancy outcome. Obstet Gynecol 103:219–224
11. Maheshwari A, Stofberg l, Bhattacharya S (2007) Effect of overweight and obesity on assisted reproductive technology—a systematic review. Hum Reprod Update 13:433–434
12. Sarwer DB, Allison KC, Gibbons LM, Markowitz JT, Nelson DB (2006) Pregnancy and obesity: A review and agenda for future research. J Womens Health 15:720–733
13. Guzick DS, Wing R, Smith D, Berga SL, Winters SJ (1994) Endocrine consequences of weight loss in obese, hyperandrogenic, anovulatory women. Fertil Steril 61:598–604
14. Norman RJ, Noakes M, Wu R, Davies MJ, Moran L, Wang JX (2004) Improving reproductive performance in overweight/obese women with effective weight management. Hum Reprod Update 10:267–280
15. Grenman S, Ronnemaa T, Irjala K, Kaihola HL, Gronroos M (1986) Sex steroid, gonadotropin, cortisol, and prolactin levels in healthy, massively obese women: Correlation with abdominal fat cell size and effect of weight reduction. J Clin Endocrinol Metab 63:1257–1261

16. Moran LJ, Norman RJ (2002) The obese patient with infertility: a practical approach to diagnosis and treatment. Nutr Clin Care 5:290–97

17. Stamets K, Taylor DS, Kunselman A, Demers LM, Pelkman CL, Legro RS (2004) A randomized trial of the effects of two types of short term hypocaloric diets on weight loss in women with polycystic ovary syndrome. Fertil Steril 81:630–637

18. Clark AM, Thornley B, Tomlinson L, Galletley C, Norman RJ (1998) Weight loss in obese, infertile women results in improvement in reproductive outcome for all forms of fertility treatment. Hum Reprod 13:1502–1505

19. Bates GW, Whitworth NS (1982) Effect of body weight reduction on plasma androgens in obese, infertile women. Fertil Steril 38:406–409

20. Kiddy DS, Hamilton-Fairley D, Bush A (1992) Improvement in endocrine and ovarian function during dietary treatment of obese women with polycystic ovary syndrome. Clin Endocrinol 36:105–111

21. Jones GE, Nalley WB (1959) Amenorrhea: a review of etiology and treatment in 350 patients. Fertil Steril 10:461–479

22. Hollmann M, Runnebaum B, Gerhard I (1996) Effects of weight loss on the hormonal profile in obese, infertile women. Hum Reprod 11:1884–1891

23. Pasquali R, Anenucci D, Casimirri F (1989) Clinical and hormonal characteristics on obese amenorrheic hyperandrogenic women before and after weight loss. J Clin Endocrinol Metab 68:173–179

24. Crosignani PG, Colombo M, Vegetti W, Somigliana E, Gessati A, Ragni G (2003) Overweight and obese anovulatory patients with polycystic ovaries: parallel improvements in anthropometric indices, ovarian physiology and fertility rate induced by diet. Hum Reprod 18:1928–1932

25. Waller DK, Mills JL, Simpson JL, Cunningham, GC, Conley MR, Lassman MR (1994) Are obese women at higher risk for producing malformed offspring? Am J Obstet Gyncol 170:541–548

26. Werler MM, Louik C, Shapiro S, Mitchell AA (1996) Prepregnant weight in relation to risk of neural tube defects. J Am Med Assoc 275:1089–1092

27. Ray JG, Wyatt PR, Vermeulen MJ, Meier C, Cole DE (2005) Greater maternal weight and the ongoing risk of neural tube defects after folic acid flour fortification. Obstet Gynecol 105:261–265

28. Watkins ML, Rasmussen SA, Honein MA, Botto LD, Moore CA (2003) Maternal obesity and risk for birth defects. Pediatrics 111:1152–1158

29. Hendler I, Blackwell SC, Bujold E (2004) The impact of maternal obesity on midtrimester sonographic visualization of fetal cardiac and craniospinal structures. Int J Obes Relat Metab Disord 28:1607–1611

30. Wolfe HM, Gross TL, Sokol RJ (1990) Maternal obesity: a potential source of error in sonographic prenatal diagnosis. Obstet Gynecol 76:339–342

31. Weiss JL, Malone FD, Emig D, Ball RH, Nyberg DA, Comstock CH (2004) Obesity, obstetric complications and cesarean delivery rate—a population-based screening study. Am J Obstet Gynecol 190:1091–1097

32. Xiong X, Saunders LD, Wang FL, Demianczuk NN (2001) Gestational diabetes mellitus: prevalence, risk factors, maternal and infant outcomes. Int J Gynecol Obstet 75:221–228

33. O'Brien TE, Ray JG, Chan WS (2003) Maternal body mass index and the risk of preeclampsia: a systematic overview. Epidemiology 14:368–374

34. Yogev Y, Xenakis EM, Langer O (2004) The association between preeclampsia and the severity of gestational diabetes: the impact of glycemic control. Am J Obstet Gynecol 191:1655–1660

35. Barden A, Singh R, Walters BN, Ritchie J, Roberman B, Beilin LJ (2004) Factors predisposing to preeclampsia in women with gestational diabetes. J Hyperten 22:2371–2378

36. Colman-Brochu S (2004) Deep vein thrombosis in pregnancy. Am J Matern Child Nurs 29:186–192

37. Edwards LE, Hellerstedt WL, Alton IR, Story M, Himes JH (1996) Pregnancy complications and birth outcomes in obese and normal-weight women: effects of gestational weight change. Obstet Gynecol 87:389–394

38. Bates SM, Greer IA, Hirsh J, Ginsberg JS (2004) Use of antithrombotic agents during pregnancy: the Seventh ACCP Conference on Antithrombotic and Thrombolytic Therapy. Chest 126:627S–644S

39. March of Dimes (2002) Nutrition today matters tomorrow: a report of the March of Dimes Task Force on Nutrition and Optimal Human Development. March of Dimes Fulfillment Center, Wilkes-Barre, Pa.

40. Baeten JM, Bukusi EA, Lambe M (2001) Pregnancy complications and outcomes among overweight and obese nulliparous women. Am J Public Health 91:436–440

41. Brost BC, Goldenberg RL, Mercer BM, Iams JD, Meis PJ, Moawad AH (1997) The Preterm Prediction Study: association of cesarean delivery with increases in maternal weight and body mass index. Am J Obstet Gynecol 177:333–337
42. Juhasz G, Gyamfi C, Gyamfi P, Tocce K, Stone JL (2005) Effect of body mass index and excessive weight gain on success of vaginal birth after cesarean section. Obstet Gynecol 106:741–746
43. Chauhan PS, Magann EF, Carroll CS, Berrileaux PS, Scardo JA, Martin JN Jr (2001) Mode of delivery for the morbidly obese with prior cesarean delivery: vaginal versus repeat cesarean section. Am J Obstet Gynecol 185:349–354
44. Sebire NJ, Jolly M, Harris JP, Wadsworth J, Joffe M, Beard RW (2001) Maternal obesity and pregnancy outcome: a study of 287,213 pregnancies in London. Int J Obes 25:1175–1182
45. Jensen DM, Damm P, Sorenson B, Molsted-Pedersen L, Westergaard JG, Ovesen P (2003) Pregnancy outcome and prepregnancy body mass index in 2,459 glucose-tolerant Danish women. Am J Obstet Gynecol 189:239–244
46. Galtier-Dereure F, Montpeyroux F, Boulot P, Bringer J, Jaffiol C (1995) Weight excess before pregnancy: complications and cost. Int J Obes 19:443–449
47. Ananth CV, Wen SW (2002) Trends in fetal growth among singleton gestations in the United States and Canada, 1985 through 1998. Semin Perinatol 26:260–267
48. Surkan PJ, Hsieh CC, Johansson Al, Dickman PW, Cnattingius S (2004) Reasons for increasing trends in large for gestational age births. Obstet Gynecol 104:720–726
49. Orskou J, Kesmodel U, Henriksen TB, Secher NJ (2001) An increasing proportion of infants weigh more than 4,000 grams at birth. Acta Obstet Gynecol Scand 80:931–936
50. Mehta SH, Blackwell SC, Chadha R, Sokol RJ (2007) Shoulder dystocia and the next delivery: Outcomes and management. J Matern Fetal Neonatal Med 20:729–733
51. Bassaw B, Roopnarinesingh S, Mohammed N, Ali A, Persad H (1992) Shoulder dystocia: an obstetric nightmare. West Indian Med J 42:158–159
52. Stephannsson O, Dickman PW, Johansson A, Cnattingius S (2001) Maternal weight, pregnancy weight gain, and the risk of antepartum stillbirth. Am J Obstet Gynecol 184:463–469
53. Froen JF, Arnestad M, Frey K, Vege A, Saugstad OD, Stray-Pedersen B (2001) Risk factors for sudden intrauterine unexplained death: epidemiologic characteristics of singleton cases in Oslo, Norway 1986–1995. Am J Obstet Gynecol 184:694–702
54. Huang DY, Usher RH, Kramer MS, Yang H, Morin L, Fretts RC (2000) Determinants of unexplained antepartum fetal deaths. Obstet Gynecol 95:215–221
55. Schieve LA, Cogswell ME, Scanlon KS (1998) Trends in pregnancy weight gain within and outside ranges by the Institute of Medicine in a WIC population. Matern Child Health J 2:111–116
56. Jain NJ, Denk CE, Kruse LK, Dandolu V (2007) Maternal obesity: can pregnancy weight gain modify risk of selected adverse pregnancy outcomes? Am J Perinatol 24:291–298
57. Hilson JA, Rasmussen KM, Kjolhede CL (2006) Excessive weight gain during pregnancy is associated with earlier termination of breast-feeding among white women. J Nutr 136:140–146
58. Oken E, Taveras EM, Kleinman KP, Rich-Edwards, JW, Gillman MW (2007) Gestational weight gain and child adiposity at age 3 years. Am J Obstet Gynecol 196:322.e1–e8
59. Rooney B, Schauberger CW, Mathiason MA (2005) Impact of perinatal weight change on long-term obesity and obesity-related illnesses. Obstet Gynecol 106:1349–1356
60. Keppel KG, Taffel SM (1993) Pregnancy-related weight gain and retention: implications of the 1990 Institute of Medicine guidelines. Am J Public Health. 83:1100–1103
61. Rooney BL, Schauberger CW (2002) Excess pregnancy weight gain and long-term obesity: one decade later. Obstet Gynecol 100:245–252
62. Cummings S, Parham ES, Strain GW (2002) Position of the American Dietetic Association: Weight management. J Am Diet Assoc 102:1145–1155
63. Institute of Medicine (1991) Nutrition during lactation: summary, conclusions and recommendations. National Academy Press, Washington, D.C.
64. United States Department of Agriculture, United States Department of Health and Human Services (2000) Nutrition and your health: dietary guidelines for Americans, 5th ed,. Home and Garden bulletin no. 232
65. National Institutes of Health (1996) Physical activity and cardiovascular health. NIH Consensus Development Panel on Physical Activity and Cardiovascular Health. J Am Med Assoc 276:241–246 (review)

66. Douketis JD, Macie C, Thabone L, Williamson DF (2005) Systematic review of long-term weight loss studies in obese adults: clinical significance and applicability to clinical practice. Int J Obes 29:1153–1167
67. American College of Obstetricians and Gynecologists (2005) Committee opinion paper no. 315. Obesity in pregnancy. Obstet Gynecol 106:671–675
68. Glazer NL, Hendrickson AF, Schellenbaum GD, Mueller BA (2004) Weight change and the risk of gestational diabetes in obese women. Epidemiology 15:733–737
69. Gurewitsch ED, Smith-Levitin M, Mack J (1996) Pregnancy following gastric bypass surgery for morbid obesity. Obstet Gynecol 88:658–661
70. Catalano P (2007) Management of obesity in pregnancy. Obstet Gynecol 109:419–33
71. Dixon JB, Dixon ME, O'Brien PE (2005) Birth outcomes in obese women after laparoscopic adjustable gastric banding. Obstet Gynecol 106:965–972
72. Abrams B, Altman SC, Pickett KE (2000) Pregnancy weight gain: still controversial? AJCN 71:1233–1241
73. Stotland NE, Haas JS, Brawarsky P, Jackson RA, Fuentes-Afflick E, Escobar GJ (2005) Body mass index, provider advice, and target gestational weight gain. Obstet Gynecol 105:633–638
74. Polley BA, Wing RR, Sims CJ (2002) Randomized controlled trial to prevent excessive weight gain in pregnant women. Int J Obes Relat Metab Disord 26:1494–1502
75. Olson CM, Strawderman MS, Reed RG (2004) Efficacy of an intervention to prevent excessive gestational weight gain. Am J Obstet Gynecol 191:530–536

6 Pregnancy and Weight Loss Surgery

Daniel M. Herron and Amy Fleishman

Summary The dramatic increase in the incidence of obesity has resulted in an overwhelming increase in the number of bariatric, or weight loss, operations performed in the United States. These operations induce long-term weight loss through a combination of volume restriction and malabsorption. As a result, bariatric surgery patients may suffer from nutritional deficiencies over the long term and need to be followed extremely closely before, during, and after pregnancy. Bariatric patients are given regimens of nutritional supplementation that are specific for their operation. This chapter describes the different types of bariatric surgery and the nutritional disturbances associated with each one. Additionally, the standard recommendations for supplementation and follow up are reviewed. Alterations to these regimens during pregnancy are discussed. Pregnancy outcomes after bariatric surgery are reviewed.

Keywords: Bariatric surgery, Weight loss surgery, Gastric bypass, Gastric band, Lap Band, Biliopancreatic diversion, Duodenal switch, Pregnancy, Nutrition

6.1 INTRODUCTION

In 2006, greater than 50% of American adults were obese [1]. Despite the fact that more than 365,000 deaths per year are caused by obesity, current treatment options are limited [2]. The single intervention with demonstrated long-term efficacy for the severely obese is bariatric surgery. A consensus statement from the National Institutes of Health in 1991 recognized bariatric surgery as a safe and effective treatment option for the severely obese patient [3]. Following the publication of this report, the national interest in bariatric surgery increased dramatically. In 2003, it was estimated that over 102,000 bariatric operations were performed in the United States [4].

This chapter addresses the indications for bariatric surgery, the types of surgical procedures available, and the impact of bariatric surgery on pregnancy. Additionally, we review the common side effects and nutritional sequelae of bariatric surgery. Finally, we address the nutritional recommendations for pregnant women who have undergone bariatric procedures.

6.2 CONSIDERATIONS FOR BARIATRIC SURGERY

Bariatric surgery is reserved for individuals who are severely obese as defined by body mass index, or BMI. The BMI is calculated by dividing a patient's weight in kilograms by the square of their height in meters. Alternatively, BMI equals the patient's weight

From: *Nutrition and Health: Handbook of Nutrition and Pregnancy*
Edited by: C.J. Lammi-Keefe, S.C. Couch, E.H. Philipson © Humana Press, Totowa, NJ

in pounds divided by the square of their height in inches and multiplied by 703. BMI is measured in units of kg per m². A BMI between 18 and 25 kg/m² is considered normal. Individuals are considered candidates for bariatric surgery when their BMI is greater than 40 kg/m², or greater than 35 kg/m² with one or more comorbidities including severe hypertension, sleep apnea, or diabetes. For most patients, this BMI corresponds to being approximately 45 kg (100 lb) or more above ideal body weight [5].

6.3   TYPES OF BARIATRIC SURGERY

A number of different bariatric procedures are currently performed in the United States. Currently, Roux-en-Y gastric bypass (RYGB) is the most common operation (Fig. 6.1). In the RYGB, a surgical stapler is used to divide the stomach into a small upper pouch and a large gastric remnant. The upper pouch, only 15 to 30 ml in volume, causes the patient to feel full after eating a small meal. The small intestine is reconnected in a Y shape to the gastric pouch in such a manner that the ingested food bypasses the stomach, duodenum, and proximal jejunum. In addition to the volume restriction caused by the small pouch, the RYGB causes weight loss by decreasing absorption (malabsorption) and altering levels of hormones involved in weight maintenance, such as insulin and ghrelin [6].

The adjustable gastric band (AGB) is the second most common bariatric operation in the United States (Fig. 6.2). In this operation, commonly referred to as the Lap-Band® (Allergan, Irvine, Calif.), a small adjustable ring made of silicone rubber is wrapped around the upper portion of the stomach, creating a pouch of 15- to 20-ml volume [7].

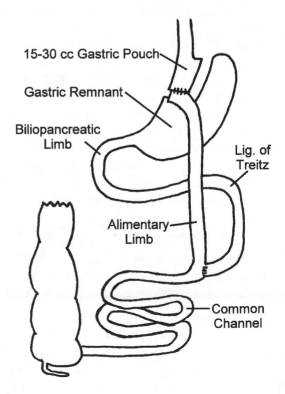

Fig. 6.1. Diagram of the Roux-en-Y gastric bypass operation. (Image ©2005 Daniel M. Herron, reprinted with permission)

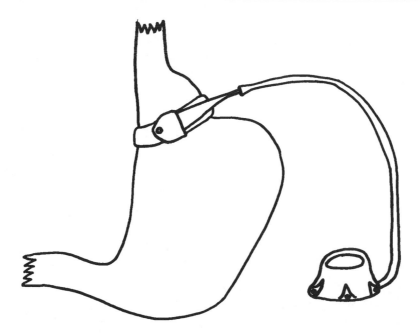

Fig. 6.2. Diagram of the adjustable gastric band. (Image ©2005 Daniel M. Herron, reprinted with permission)

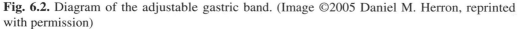

The band is connected, via a thin flexible tube, to an access port placed underneath the skin on the abdominal wall. By injecting or withdrawing saline from the access port, the band can be tightened or loosened and the amount of restriction adjusted. Unlike the RYGB, the AGB is a purely restrictive operation.

Vertical banded gastroplasty, also known as VBG or "stomach stapling," was at one time the most common bariatric operation, but has lost favor recently due to its poor long-term results [8]. Like the adjustable gastric band, the VBG is a purely restrictive operation that works by decreasing the volume of food that a patient can eat at one sitting. Unlike the gastric band, the VBG cannot be adjusted. VBG is now an uncommon operation.

One of the most recently developed bariatric operations is the sleeve gastrectomy (SG), a more modern variant of the VBG (Fig. 6.3) [9]. In this technically straightforward operation, the entire left side of the stomach is surgically removed, resulting in a small, banana-shaped stomach. For superobese patients in whom a complex operation like the RYGB may present excessive technical difficulty, the SG can be used as the first component of a two-staged approach. The SG will result in a weight loss of 50 kg or more, after which the patient can be safely taken to the operating room for conversion to a more definitive operation like the RYGB [10]. SG without a second stage may also be used as a purely restrictive procedure.

The least common and most complex bariatric operation performed in the United States is the biliopancreatic diversion with duodenal switch (BPD-DS, Fig. 6.4) [11]. The BPD-DS consists of a SG combined with the bypass of a substantial portion of the small intestine. The first portion of the duodenum is divided and reconnected to the

Fig. 6.3. Diagram of the sleeve gastrectomy. (Image ©2005 Daniel M. Herron, reprinted with permission)

Fig. 6.4. Diagram of the biliopancreatic diversion with duodenal switch (BDP-DA). (Image ©2005 Daniel M. Herron, reprinted with permission)

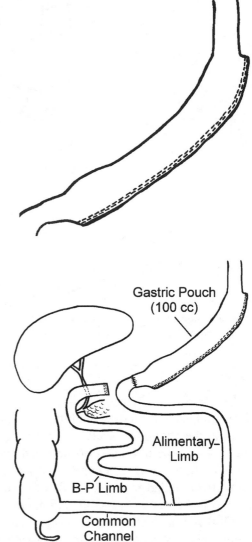

distal 250 cm of small intestine. Additionally, bile and pancreatic secretions are diverted to the distal ileum. The BPD-DS results in moderate volume restriction and significant malabsorption. While providing the best long-term weight loss of any bariatric operation, the BPD-DS causes the most nutritional disturbance.

6.4 WEIGHT LOSS AFTER SURGERY AND POSTOPERATIVE RECOMMENDATIONS FOR PREGNANCY

The rate of weight loss after surgery varies with the type of procedure. A large meta-analysis of surgical interventions for weight loss reported a mean weight loss regardless of operation of 61.2% [12]. Specifically, excess body weight loss was 47.5% for patients who underwent AGB, 61.6% for those who underwent RYGB, and 70.1% for those who had BPD-DS.

With the RYGB and BPD-DS, the most rapid weight loss occurs during the first 3 weeks after surgery, when patients typically lose 1 lb per day or more. The rate of weight loss gradually decreases until weight stabilizes, about 12 to 18 months after surgery [13]. Weight loss after AGB occurs at a slower rate, but may continue for 2 to 3 years after surgery. Most bariatric surgeons recommend that female patients avoid pregnancy for a period of 18 months or more after their operation or until their weight has stabilized.

6.5   NUTRITION DEFICIENCIES AFTER WEIGHT LOSS SURGERY

Deficiencies in vitamins and other nutrients are common after bariatric surgery, particularly with RYGB and BPD-DS, since these operations result in decreased intestinal surface area and bypass the duodenum (Fig. 6.5). Since BPD-DS results in more significant malabsorption than does RYGB, there are more nutrient deficiencies reported among BPD-DS patients. Although not as prevalent, nutritional deficiencies have also been reported after AGB and SG, primarily because of decreased food intake and the avoidance of certain nutrient-rich foods because of individual intolerances.

In order to better understand what the postoperative nutrition needs are for pregnant women who have had bariatric surgery, it is important to first understand the nutritional deficiencies that commonly accompany these procedures. The main deficiencies reported among postoperative patients are protein, iron, vitamin B_{12}, folate, calcium, vitamin D, and fat-soluble vitamins [15]. Below is a brief review of studies that have been carried out as well as the assessments that are recommended as a check for nutrient deficiencies following bariatric surgery.

- **Protein.** In a prospective randomized study of patients with a BMI greater than 50 kg/m^2, 13% of the patients who underwent distal RYGB experienced protein deficiency 2 years after surgery [16]. Protein deficiency occurred more frequently after BPD-DS than RYGB due to the more severe malabsorption caused by this operation. It is recommended that total serum protein and albumin be assessed on a regular basis after bariatric surgery to measure protein stores, typically 3, 6, and 12 months after surgery, then annually.
- **Iron.** In a study of RYGB patients before surgery and up to 5 years after the procedure, iron deficiency was identified in 26% of patients preoperatively, in 39% at 4 years postoperatively, and in 25% of those 5 years postoperatively [17]. The anatomic changes resulting from RYGB reduce the exposure of iron-containing food to the acidic environment in the stomach, which is required for the release of iron from protein and conversion into its absorbable ferrous form [18]. It is recommended that hemoglobin, hematocrit, iron, ferritin, and total iron binding capacity be evaluated for diagnosis of iron deficiency or anemia.
- **Vitamin B_{12} (cobalamin) and folate.** Deficiencies of vitamin B_{12} and folate are common in bariatric surgery patients. Halverson studied patients 1 year after RYGB and found 33% of patients had a vitamin B_{12} deficiency, and 63% had a folate deficiency [19]. As with iron digestion after RYGB, the absence of an acidic environment prevents the release of vitamin B_{12} from food [17]. In addition, intrinsic factor (IF), secreted from parietal cells of the stomach, is responsible for the absorption of vitamin B_{12}. Therefore, after bariatric surgery, inadequate IF secretion or function is a possible mechanism for vitamin B_{12} deficiency [20]. It is recommended that vitamin B_{12} and folate be assessed regularly. Blood levels of >300 pg/ml for B_{12} are considered normal.

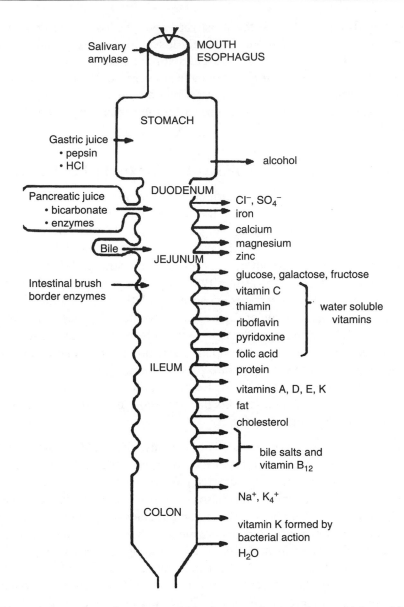

Fig. 6.5. Sites of absorption of nutrients within the gastrointestinal tract. (Adapted from: Mahan and Escott-Stump: Krause's Food, Nutrition and Diet Therapy, 9/e, p 13, ©1996, with permission from Elsevier)

- **Calcium and vitamin D.** Calcium and vitamin D are usually assessed together since vitamin D promotes the intestinal absorption of calcium. Brolin et al. found a 10% incidence of calcium deficiency and 51% incidence of vitamin D deficiency in patients who had distal RYGB [16]. Parathyroid hormone (PTH) levels may be a more sensitive indicator of calcium deficiency [18]. If PTH is elevated, then calcium deficiency is presumed. As for vitamin D, it is important to check 25(OH) vitamin D levels rather than 1,25(OH)$_2$ vitamin D. Although the normal range of vitamin D is variable depending on the lab, it is usually recommended that serum 25(OH) vitamin D to be >20 ng/ml.

- **Other fat-soluble vitamins: A, E, K.** Malabsorption of these vitamins is most commonly seen after BPD-DS. Slater et al. studied 170 patients following BPD and BPD-DS and reported that 69% were deficient in vitamin A and 68% were deficient in vitamin K 4 years after surgery [21]. Dolan et al. showed that 5% of patients had low levels of vitamin E an average of 28 months after BPD and BPD-DS [22]. Therefore, vitamin A, vitamin E (tocopherol), and INR (the International Normalized Ratio, used to measure clotting and indirectly assess vitamin K deficiency) should be assessed at least annually.

6.6   COMMON POSTOPERATIVE PROBLEMS

In addition to nutrient deficiencies after bariatric surgery, there are other common side effects that are worth mentioning because they may be confused with symptoms of pregnancy.

- **Vomiting.** Emesis may occur after surgery if the patient eats too much at one time, eats too quickly, or does not chew solid food thoroughly. Food that is not chewed well may get stuck in the narrow anastomosis between the stomach pouch and jejunum. Scarring and stricture may also narrow the outlet and lead to vomiting. Excessive vomiting in patients who have had the AGB requires removal of some saline from their band to reduce the degree of restriction [23]. It is important to correct the problem because persistent vomiting may lead to malnutrition and dehydration, which are harmful during pregnancy.
- **Constipation.** Constipation may occur because postoperative patients focus on high-protein foods as a dietary mainstay and reduce their overall quantity of food intake, so their fiber intake may be suboptimal. If the patient is not drinking an adequate amount of fluid, then this may exacerbate the constipation. In addition, the iron that is recommended for bariatric patients and for pregnancy is also known to cause constipation. A stool softener or fiber supplement such as Metamucil® can be suggested.
- **Dumping syndrome.** This occurs after RYGB when there is consumption of simple sugars. The patient may feel nauseated, and suffer tachycardia, syncope, and diarrhea. The syndrome may be averted by instructing patients to avoid concentrated sweets. Importantly, when screening for gestational diabetes mellitus (GDM) during pregnancy, a glucose tolerance test will likely cause dumping syndrome. An alternative approach to assess for GDM would involve measuring fasting serum glucose periodically, since an elevation usually correlates with elevated postprandial blood glucose [24]. The overall risk of GDM is significantly lower in bariatric patients than in morbidly obese women [25].

6.7   PRECONCEPTION CARE

The best pregnancy is a planned pregnancy, especially after bariatric surgery, because the patient is able to take preventive measures against postnatal nutrient deficiencies. In addition to meeting with their obstetrician, women who have undergone a bariatric operation should schedule a follow-up visit with their bariatric surgeons and dietitians. At this visit, the surgeon will check a complete laboratory assessment for nutrient status. If any levels are low, then there is adequate time to correct them. Deficiencies in iron, calcium, vitamin B_{12}, and folate can result in maternal complications, such as anemia, and in fetal complications, such as neural tube defects. Even if all lab values are normal, the patient should be encouraged to continue her bariatric prenatal supplement regimen.

Assessments should be carried out before becoming pregnant and frequently (every 2 to 3 months) during gestation, especially if deficiencies were present initially. In addition to reviewing blood work, the dietitian will ensure the patient is eating an adequate amount of protein, drinking enough fluid, and eating a healthy diet. The main goals of medical nutrition therapy during pregnancy are to facilitate adequate weight gain to promote fetal growth and development, to provide appropriate vitamin and mineral supplementation to prevent or correct deficiencies, and to assess nutrition education needs.

6.8 STANDARD SUPPLEMENT RECOMMENDATIONS

Dietary supplementation before and during pregnancy should be based on laboratory findings as well as the type of bariatric surgery. General practice includes adding a prenatal vitamin to the current supplement regimen for the bariatric surgery, not giving it in lieu of the standard postoperative regimen.

The standard prenatal vitamin contains 1 mg of folic acid, which is sufficient to reduce the risk of neural tube defect in the fetus (1 mg of folic acid per day should be taken prior to pregnancy for maximum benefit). For RYGB and BPD-DS, it is important that the calcium source be citrate because citrate does not require an acidic environment in order to be broken down. It is best that vitamin B_{12} be taken sublingually in crystalline form for better absorption [18]. When prescribing iron, ferrous fumarate is best tolerated [25]. For all patients who take iron and calcium, to maximize absorption the two should not be taken at the same time. If BPD-DS patients are compliant with taking fat-soluble vitamins but remain deficient in vitamin A, then additional vitamin A should be given as beta-carotene, since vitamin A may have teratogenic potential [27]. While general recommendations are given below, each patient should be treated as an individual and may require a slightly different regimen (Table 6.1).

Table 6.1
Standard Prenatal Supplementation for Women who have Undergone Bariatric Surgery

Procedure	*Supplement and dose*
Laparoscopic adjustable gastric banding or sleeve gastrectomy	Prenatal vitamin with iron Calcium citrate 1200 mg/day plus vitamin D 800 IU/day
Roux-en-Y gastric bypass	Prenatal vitamin with iron Calcium citrate 1200 mg/day plus vitamin D 800 IU/day Vitamin B_{12}: 500 mcg/day Elemental iron: 300 mg/day (ferrous and polysaccharides complex)
Biliopancreatic diversion with duodenal switch	Prenatal vitamin with iron Calcium citrate 2000 mg/day plus vitamin D 800 IU/day Vitamin B_{12}: 500 mcg/day Elemental iron: 300 mg/day (ferrous and polysaccharides complex) ADEK: three times per day

6.9   PROTEIN RECOMMENDATIONS

Protein is the most important macronutrient for the bariatric patient. For this reason, patients should be instructed to consume protein at the beginning of a meal to ensure adequate intake if the patient becomes sated prematurely. During pregnancy, protein needs are 1.1 g/kg/day. There are no published protein guidelines for bariatric patients during pregnancy. Therefore, protein recommendations may vary among institutions. Our program recommends 60–80 g/day for the RYBG, 80–120 g/day for the BPD-DS, and 0.8–1 g/kg/day of adjusted body weight for AGB in nonpregnant women. Each pregnant bariatric patient should be encouraged to meet the upper end of the protein range specified for her type of surgery. If needed, protein intake may be supplemented with sugar-free protein shakes.

Patients are instructed to avoid pregnancy during the first year after surgery, since this is the period of most rapid weight loss. If there is an unplanned pregnancy during this time, protein requirements should be increased to 1.5 g/kg/day of ideal body weight.

6.10   CALORIE RECOMMENDATIONS AND WEIGHT GAIN DURING PREGNANCY

Calorie recommendations for the pregnant bariatric patient include approximately 300 kcal/day above maintenance guidelines for bariatric surgery. As with protein, calorie recommendations may vary between institutions. Typically, 1 year after surgery, individuals consume approximately 1,200 kcal/day, so this would result in a caloric recommendation of 1,500 kcal/day for pregnant bariatric patients. These are general guidelines, and each patient should be monitored for appropriate weight gain during pregnancy to ensure she is getting adequate caloric intake.

Weight gain during pregnancy after bariatric surgery is variable, as with any pregnancy. There are no published guidelines for pregnancy weight gain in bariatric patients. Therefore, the guidelines set forth by the Institute of Medicine should be used (Table 6.2) [28]. The postoperative BMI should be used to determine the appropriate weight category.

Even when weight gain is normal and expected during pregnancy, some patients may have an emotional response to this gain since they worked so hard to lose weight. It should be reinforced to patients that pregnancy is not a time to lose weight. Inadequate weight gain can result in intrauterine growth retardation and fetal abnormalities [18]. Patients may also be nervous about losing their pregnancy weight after the birth of their infants. It may be helpful to refer the patient to a psychologist to help sort out these issues.

Table 6.2
Recommendations for Weight Gain during Pregnancy

Weight category	Recommended weight gain (lb)
Underweight: BMI < 19.8	28–40
Normal weight: BMI = 19.8–25	25–35
Overweight: BMI = 26–29	15–25
Obese: BMI > 29	15

From [29]

6.11   OTHER RECOMMENDATIONS DURING PREGNANCY

General pregnancy recommendations should be stressed to the bariatric patient, such as avoiding alcohol, drugs, and undercooked meat, as well as increasing intake of omega-3 fatty acids and avoiding fish that are high in mercury. Patients should be advised to do the following:

- Eat slowly and chew very well. It is best to try to take a half hour to eat each meal. If the patient does not stop eating when she is full, then there is an increased chance that vomiting will follow. This may lead to malnutrition and dehydration.
- Do not skip meals. Have three to five small, protein-rich meals every day. Skipping meals may lead to malnutrition.
- Avoid concentrated sugars. RYGB patients may experience dumping syndrome. For the patient with a purely restrictive procedure, it is important to avoid concentrated sugars because they are high in calories and will contribute to excessive weight gain.
- Limit fat intake. Fat is high in calories, and like sugar, will contribute to excess weight gain. Some patients may also experience nausea after eating a high fat meal.
- Maintain adequate hydration. The patient should drink six to eight 8-ounce, noncaloric, noncarbonated, noncaffeinated drinks per day. Patients should avoid drinking 15 min before, during, or 1 h after meals. Drinking immediately before a meal may cause the stomach pouch to fill up with fluid, and there will not be room to eat an adequate amount of protein. Drinking during the meal may cause the food to move quicker through the stomach, and therefore decrease a feeling of fullness. Drinking 1 h after may cause vomiting because there will not be any room for the fluid with all the food that was eaten.
- Increase intake of fiber to prevent or treat constipation.

6.12   OUTCOMES OF PREGNANCY AFTER WEIGHT LOSS SURGERY

A number of published studies have addressed the issue of outcomes after bariatric surgery. One of the largest studies to date evaluated the perinatal outcome of 159,210 deliveries occurring in Israel between 1988 and 2002. Of these deliveries, 298 were from women who had previously undergone bariatric surgery [30, 31]. Although there was a higher rate of caesarean delivery in the bariatric surgery group (25.2 vs. 12.2%), no difference was found in perinatal mortality, congenital malformations and Apgar scores at 1 and 5 min.

6.13   CONCLUSION

It is important for the obstetrician and bariatric surgeon to work together to care for the bariatric surgery patient. In addition, the patient will benefit from attending bariatric surgery support groups where she can share her experiences, ask for advice, and help others with the challenges they encounter. Encourage the bariatric patient to attend these support groups because they will find help in dealing with nutrition before and after pregnancy.

REFERENCES

1. Centers for Disease Control and Prevention (2004) Health, United States, 2004. Available via http://www.cdc.gov/nchs/hus.htm
2. Mokdad AH, Marks JS, Stroup DF, Gerberding JL (2005) Correction: actual causes of death in the United States, 2000. J Am Med Assoc 293:293–429

3. National Institutes of Health (2000) The practical guide: identification, evaluation, and treatment of overweight and obesity in adults, National Institutes of Health, National Heart, Lung, and Blood Institute, and North American Association for the Study of Obesity, NIH publication number 00-4084, Bethesda, Md.

4. Santry HP, Gillen DL, Lauderdale DS (2005) Trends in bariatric surgical procedures. J Am Med Assoc 294:1909–1917

5. Herron DM (2004) The surgical management of severe obesity. Mt Sinai J Med 71:63–71

6. Cummings DE, Weigle DS, Frayo RS, Breen PA, Ma MK, Dellinger EP, Purnell JQ (2002) Plasma ghrelin levels after diet-induced weight loss or gastric bypass surgery. N Engl J Med 346:1623–1630

7. Parikh MS, Fielding GA, Ren CJ (2005) US experience with 749 laparoscopic adjustable gastric bands: intermediate outcomes. Surg Endosc 19:1631–1635

8. Balsiger BM, Poggio JL, Mai J, Kelly KA, Sarr MG (2000) Ten and more years after vertical banded gastroplasty as primary operation for morbid obesity. J Gastrointest Surg 598–605

9. Regan JP, Inabnet WB, Gagner M, Pomp A (2003) Early experience with two-stage laparoscopic Roux-en-Y gastric bypass as an alternative in the super-super obese patient. Obes Surg 13:861–864

10. Aggarwal S, Kini SU, Herron DM (2007) Laparoscopic sleeve gastrectomy for morbid obesity: a review. Surg Obes Relat Dis 3:189–194

11. Herron DM (2004) Biliopancreatic diversion with duodenal switch vs. gastric bypass for severe obesity. J Gastrointest Surg 8:406–407

12. Buchwald H, Avidor Y, Braunwald E, Jenson MD, Pories W, Fahrbach K, Schoelles K (2004) Bariatric surgery—a systematic review and meta-analysis. J Am Med Assoc 292:1724–1737

13. Sugerman HJ, Starkey JV, Birkenhauer R (1987) A randomized prospective trial of gastric bypass versus vertical banded gastroplasty for morbid obesity and their effects on sweets versus non-sweets eaters. Ann Surg 205:613–624

14. Ponce J, Paynter S, Fromm R (2005) Laparoscopic adjustable gastric banding: 1,014 consecutive cases. J Am Coll Surg 201:529–535

15. Bloomberg RD, Fleishman A, Nalle JE, Herron DM, Kini S (2005) Nutritional deficiencies following bariatric surgery: what have we learned? Obes Surg 15:145–154

16. Brolin RE, LaMarca LB, Kenler HA, Cody RP (2002) Malabsorptive gastric bypass in patients with superobesity. J Gastrointest Surg 6:195–203; discussion 4–5

17. Skroubis G, Sakellaropoulos G, Pouggouras K, Mead N, Nikiforidis G, Kalfarentzos F (2002) Comparison of nutritional deficiencies after Roux-en-Y gastric bypass. Obes Surg 12: 551–558

18. Woodard CB (2004) Pregnancy following bariatric surgery. J Perinat Neonat Nurs 4:329–340

19. Halverson JD (1986) Micronutrient deficiencies after gastric bypass for morbid obesity. Am Surg 52:594–598

20. Marcuard SP, Sinar DR, Swanson MS, Silverman JF, Levine JS (1989) Absence of luminal intrinsic factor after gastric bypass surgery for morbid obesity. Dig Dis Sci 34:1238–1242

21. Slater GH, Ren CJ, Siegel N, Williams T, Barr D, Wolfe B, Dolan K, Fielding GA (2004) Serum fat-soluble vitamin deficiency and abnormal calcium metabolism after malabsorptive bariatric surgery. J Gastrointest Surg 8:48–55; discussion 4–5

22. Dolan K, Hatzifotis M, Newbury L, Lowe N, Fielding G (2004) A clinical and nutritional comparison of biliopancreatic diversion with and without duodenal switch. Ann Surg 240: 51–56

23. Weiss HG, Nehoda H, Labeck B, Hourmont K, Marth C, Aigner F (2001) Pregnancies after adjustable gastric banding. Obes Surg 11:303–306

24. Carroll MF, Izard A, Riboni K, Burge MR, Schade DS (2002) Fasting hyperglycemia predicts the magnitude of postprandial hyperglycemia. Diabetes Care 25:1247–1248

25. Wittgrove AC, Jester L, Wittgrove P, Clark GW (1998) Pregnancy following gastric bypass for morbid obesity. Obes Surg 8:461–464

26. Deitel M, Ternamian AM, Noor SS (1997) Intravenous nutrition in obstetrics and gynecology. J Soc Obstet Gynecol Can 19:1171–1178

27. Miller RK, Hendrickz AG, Mills JL, Hummler H, Wiegand UW (1998) Periconceptional vitamin A use: how much is teratogenic? Repr Toxicol 12:75–88

28. Food and Nutrition Board, Institute of Medicine (1990) Nutrition During Pregnancy. National Academy Press, Washington, D.C.

29. Parker JD, Abrams B (1992) Prenatal weight gain advice: an examination of the recent prenatal weight gain recommendations of the Institute of Medicine. Obstet Gynecol 79:664–669
30. Sheiner E, Levy A, Silverberg D, Menes TS, Levy I, Katz M, Mazor M (2004) Pregnancy after bariatric surgery is not associated with adverse perinatal outcome. Am J Obstet Gynecol 190:1335–1340
31. Dixon JB, Dixon ME, O'Brien PE (2005) Birth outcomes in obese women after laparoscopic adjustable gastric banding. Obstet Gynecol 106:965–972

7 Nutrition in Multifetal Pregnancy

Elliot H. Philipson

Summary All multifetal pregnancies can be considered high risk due to frequent obstetrical complications associated with this type of pregnancies. Nutritional assessment with current dietary recommendations and specialized antenatal care are important for a good outcome. Maternal weight and weight gain are also important factors, but nutritional supplements, assessment of calorie intake, and adjustments as needed are crucial as well. Confounding variables must be considered, particularly when examining outcomes. This chapter addresses some of the important components that can contribute to a healthy outcome in multifetal pregnancy. Future research and knowledge is needed in this challenging area. A multidisciplinary approach to antepartum and even preconceptional care will help optimize the outcome.

Keywords: Multifetal pregnancy, Nutrition, Maternal weight, Prenatal care, Supplements

7.1 INTRODUCTION

In 2004 there were more than 139,000 multiple or multifetal births in the United States [1, 2]. The birth rate for twin gestation, which accounts for more than 95% of these multifetal pregnancies, continues to rise, with the latest birth rate in 2004 at a record of 32.2 twins per 1,000 live births. The birth rate of triplets and higher-order multifetal births has been relatively stable since its peak in 1998. Almost all pregnancy complications are more frequent in multifetal pregnancies [3, 4]. The most recent American College of Obstetricians and Gynecologists (2004) published their practice guidelines, that describes many of these complications [4]. However, there is no mention in this document of any nutritional assessment or treatment for this high-risk group.

Nutrition information for clinicians who provide antenatal care for women with twins or triplets often has been based on scientific data obtained from singleton pregnancies and then extrapolated and applied to multifetal pregnancy. A search of Ovid Medline for all English-language articles related to multiple pregnancy in humans and nutrition or diet shows 331 articles published on this subject in the past 30 years. Almost 50% of these articles (151) have been published since 1996, indicating that interest in and knowledge of this specific area of high-risk pregnancy is increasing. The purposes of this chapter are to: (1) evaluate some of the most relevant and important new clinical information regarding nutrition and multiple pregnancy, (2) describe well-documented

From: *Nutrition and Health: Handbook of Nutrition and Pregnancy*
Edited by: C.J. Lammi-Keefe, S.C. Couch, E.H. Philipson © Humana Press, Totowa, NJ

differences in nutrition needs between singleton and multiple pregnancies, (3) provide practical antenatal nutritional guidelines for the clinician and nutritionist managing multiple pregnancies, and (4) stimulate further research and knowledge for this interesting and challenging aspect of obstetrics.

7.2 WEIGHT

The first aspect of nutrition to be examined relates to weight. Much attention has been directed toward this aspect of nutrition because weight is measured easily and can be followed over time. Measurement of weight has been examined by prepregnancy maternal weight, maternal weight at different times during the pregnancy, weight gain based on body mass index (BMI), and neonatal birth weight [5–8]. Maternal weight gain and patterns of this weight gain have been shown to be important predictors of a good perinatal outcome defined in various ways, but usually by birth weights greater than 2,500 g [5–8].

Specific guidelines have been established; clinicians have recommended that women with a normal weight (BMI) (20–25) gain approximately 25 lb for a singleton pregnancy and an additional 10 lb for multiple pregnancies [9, 10]. Women with triplet pregnancies should gain at least 50 lb [10]. These recommendations are based on the neonatal birth weight that is considered appropriate.

Luke et al. [11] examined the maternal weight gain stratified by BMI as it relates to the optimal fetal growth and weight in twins. In this historical cohort study of 2,324 twin pregnancies, optimal rates of fetal growth and birth weights were associated with varying rates of maternal weight gain, depending on the pregravid BMI and the period of gestation. These data were obtained over a 20-year period, from 1979–1999, in four different locations, with no specific information on meal plans, dietary interventions, or changes in dietary habits. The authors concluded that the rates of maternal weight gain in twin pregnancies are best viewed as guidelines that can be used antenatally.

The results of an individualized intervention program in twin pregnancy demonstrated that nutritional intervention, that went beyond measuring body weights, can significantly improve pregnancy outcome [12, 13]. In this study, the Higgins Nutrition Intervention program, developed at the Montreal Diet Dispensary, was used first to assess each pregnant woman's risk profile for adverse birth outcomes and to adjust the diet using an individualized nutrition program. The individualized program included education about food consumption patterns to meet individual dietary requirements and allow for preexisting food habits. Regular follow-up visits at 2- to 4-week intervals with the same dietitian were included. Other features of this intervention program included supplementation with milk and eggs, an additional 1,000 kcal/day, a 50-g protein allowance for each fetus, and smoking cessation. Significantly, the group of patients in this study was at high risk, not only because of the twin gestation, but also because they were economically disadvantaged. The results of this study demonstrated a lower rate of preterm delivery, lower neonatal mortality, and lower maternal mortality. There also was a small (80 g) increase in twin birth weight.

In 2003, a specialized, prospective intervention program in multiple pregnancy at one institution reported its effect not only on maternal, but also on neonatal and early childhood outcomes [14]. In addition to regular prenatal care, this study provided twice-monthly prenatal visits with a dietetian and nurse practitioner, additional maternal nutrition education, modification of maternal activity, multivitamin supplementation, and individualized dietary prescription with serial monitoring of nutritional status during the pregnancy. Specific dietary recommendations and weight gain goals based on BMI were provided and monitored. The initial dietary assessment was based on a 24-h recall. The diet was adjusted to 3,000–4,000 kcal per day, depending on the pregravid BMI. Dietary assessment was made at each subsequent visit, and changes were made. The plan was to provide three meals and three snacks with 20% of the calories from protein, 40% from carbohydrates, and 40% from fat. A multivitamin containing 100% of the Recommended Daily Allowances (RDAs) for the nonpregnant woman was provided for daily consumption and then increased to twice daily after 20 weeks gestation. In addition, a daily mineral supplement with calcium, magnesium, and zinc in divided doses was provided. Over a 6-year period, participation in this program improved pregnancy outcomes as assessed by the frequency of preeclampsia, preterm premature rupture of the membranes, delivery less than 36 weeks gestation, and an increased birth weight (220 g). There was also less neonatal morbidity, as assessed by retinopathy of prematurity, necrotizing enterocolitis, intraventricular hemorrhage, and ventilator support. Through 3 years of age, children whose mothers participated in the program were less likely to be hospitalized or to be developmentally delayed.

These results form the basis for the most current dietary recommendations and specialized antenatal care during multiple pregnancy.

As guidelines, maternal weight gain in twin pregnancy would be based on the gestational period. The underweight gravida (BMI < 20) should gain between 1.25 and 1.75 lb per week in the early (0–20 weeks) gestational period, 1.5–2 lb per week in the mid (20–28 weeks) period, and 1.25 lb per week in the late (>28 weeks to delivery) period, for a total weight gain of 47–61 lb. A normal-weight gravida (BMI = 20–25) should gain 1–1.5 lb per week in the early period, 1.25–2 lb per week in the mid period, and 1 pound per week in the late period, for a total weight gain of 38–54 lb. The overweight gravida (BMI = 26–30) should gain 1–1.25, 1–1.5, and 1 lb per week in the early, mid, and late periods, respectively. The total weight gain should be 36–45 lb. The obese gravida (BMI > 30) should gain 0.75–1.25, 1, and 0.75 lb per week in the early, mid, and late periods respectively. The total weight gain should be 29–39 lb.

As demonstrated by the study of Luke et al. [11], maternal weight, weight gain based on BMI, and the specific weight at each gestational period represent only some of the aspects of a nutritional management plan for multifetal pregnancy. Another aspect relates to the basic metabolic rate, as measured by resting energy expenditure, which is about 10% higher in the third trimester of twin pregnancy compared with women with singleton pregnancies [15]. Therefore, an increase in caloric intake, frequent monitoring, and adjustment of the dietary plan, maternal activity considerations, educational and instructional assessment, and nutritional supplements are also crucial to a successful outcome.

7.3   NUTRITION SUPPLEMENTS

7.3.1   *Iron*

Daily iron supplementation is recommended by the National Academy of Science for pregnant women because the iron content in most American diets, and the nonheme iron stores of American woman are insufficient to provide for the increased iron requirements of pregnancy [16–18] as further discussed in Chap. 16 ("Iron requirements and Adverse

Outcomes"). In multifetal pregnancy, as in singleton pregnancies, supplementations should include iron and folic acid. The majority of studies report an iron deficiency in multifetal pregnancy [10, 18, 19].

7.3.2 Minerals

There are no known trials using mineral supplements in twin gestation [20]. However, magnesium, calcium, and zinc have been recommended to reduce pregnancy complications and improve outcome [10, 21, 22]. In the study mentioned previously [14], 3 g of calcium, 1.2 g of magnesium, and 45 mg of zinc were added to the specialized diet. Again, there is no scientific proof that this supplementation is effective, but there are some data suggesting that it may be beneficial. However, various studies have been inconclusive, and mineral supplementation remains an area that needs good clinical trials, particularly in multifetal pregnancy.

7.3.3 Fish Oil

Fish oil supplementation has been shown to play an important role in pregnancy, parturition, lactation, and even childhood development [23–26]. A recent study in rats showed that dietary docosahexaenoic acid (DHA) can suppress the indices of premature labor and shortened gestation [26]. An older study in rats also reported that gestational age is related positively to high dietary intake of n-3 fatty acids [27]. In human pregnancy, daily supplements in the third trimester of pregnancy using capsules containing 2.7 g of n-3 fatty acids have been shown to prolong pregnancy [28]. In a randomized, double-blind, controlled clinical trial, the long-chain omega-3 fatty acid DHA obtained from egg ingestion beginning at 24–28 weeks until delivery has been shown to significantly increase the length of gestation [29]. Other studies have supported the association between an increased intake of DHA and longer gestations, increased birth weight, head circumference, and birth length [25]. Low-birth-weight babies also have lower levels of DHA than full-term infants [24]. Recently, maternal consumption of DHA during pregnancy has been shown to benefit infant performance on problem solving at nine months and infant visual acuity at 4 months of age [30, 31]. It is important to note that none of these studies has shown any detrimental effects on the growth of the fetus, course of labor, or neonatal outcome. Therefore, it appears that DHA supplementation could help prolong pregnancy. In fact, educational materials have been developed for Women, Infants, and Children (WIC) programs to increase DHA intake during pregnancy. This program is called the "Omega-3 for Baby and Me" [32].

In a multicenter, randomized clinical trial including nineteen hospitals in Europe, fish oil supplementation reduced the recurrence risk of preterm delivery from 33 to 21% [33]. However, this study reported that fish oil supplements had no effect on preterm delivery in twin pregnancies. In this study, 579 twin gestations received prophylactic fish oil, beginning at a mean gestational age of 21+ weeks. Some limitations of this study include different recruitment rates between the centers, little available clinical information regarding these pregnancies, and the fact that no other complications of twin gestation other than intrauterine growth rate (IUGR) and pregnancy-induced hypertension (PIH) were reported. Were there any inventions or services provided for this high-risk group of patients that could have influenced the outcome? Should the amount of fish oil be increased with multiple pregnancies? In spite of these shortcomings, this is the only study that this author was able to find that specifically examined the issue of fish oil supplementation and multiple births. More research is needed in this area.

7.4 CONFOUNDING VARIABLES

There are many confounding variables that can contribute to birth weight and early neonatal outcomes in multifetal pregnancies. All multifetal pregnancies can be considered high risk because of the increase of many obstetrical complications, compared with those of singleton pregnancies. Maternal complications include preeclampsia, hypertension, gestational diabetes, placenta previa, abruptio placenta, cesarean birth, and maternal mortality. For the fetuses, the risks include prematurity, low birth weight, birth asphyxia, cerebral palsy, and neonatal death.

These complications associated with multifetal pregnancies should be considered when examining pregnancy outcomes. Some other important variables can contribute to the outcome of a multifetal pregnancy.

First, pregnancy complications such as preeclampsia, hypertension, diabetes, abruptio placenta, or placenta previa can be very problematic for the clinician and can result in medical or surgical interventions or even early delivery [34]. Some of these pathologic conditions are known to influence intrauterine growth. Would bed rest with reduced maternal activity and stress reduction alter interuterine fetal growth? Can overdistension of the uterus or an increased amniotic fluid precipitate early labor and delivery? All of these maternal factors and pregnancy complications need to be considered if birth weight is considered as a crude marker for nutrition.

Second, discordancy could account for a difference in fetal growth and birth weights with multifetal pregnancy. Growth discordancy, usually more than 20 or 25%, offers a unique challenge for clinicians. Discordancy occurs at a higher frequency and severity in triplet pregnancies [35]. Fetal growth, discordant or not, depends on many factors. One factor relates to the function of the placenta or uteroplacental unit, as nutrients cross from the mother to the fetus [36]. Umbilical cord insertion and fusion of the placentas has been shown to influence birth weight in twin gestation [37]. Fetal growth also seems to be influenced by plurality, that is, the mean birth weight decreases as the number of fetuses increases [38]. Fetal intrauterine growth depends on the time in gestation, as "growth curves" for singletons, twins, and triplets are similar until 28 weeks' gestation. After 28 weeks, the curves of the multifetal pregnancies deviate from the singleton pregnancy. Uterine adaptation and volume also may influence fetal growth [39, 40]. Therefore, when considering fetal growth or neonatal birth weight as a marker for nutrition, not only is the length of gestation critical, but pregnancy complications and other factors that influence fetal growth also must be considered. In fact, the monochorionic type of placentation in multifetal pregnancies has been reported as a risk factor for increased discordance [41, 42]. Nutritional status of multifetal pregnancies and outcome has not been well evaluated by the type of placentation or the other placental complications that occur more frequently in multifetal pregnancies.

Third, second-born twins have a greater risk for perinatal morbidity than first-born twins [43, 44]. Is this finding due to difference in mode of delivery, intrauterine or neonatal growth, or perinatal nutrition? The author was unable to find any literature that evaluates or links nutrition to the birth order.

Finally, carbohydrate metabolism in pregnancy can be characterized by the phenomenon of accelerated starvation; that is, the fasting glucose decreases as pregnancy advances with an accelerated insulin response to meals [45]. Pregnant women with twin gestation have an accelerated response compared with singleton pregnancy [46]. This indicates that lower glucose in the fasting state can increase the depletion of glycogen stores resulting in metabolism of fat. Ketones may result, and ketonuria has been associated with preterm delivery. Compared with the recommended diet for women whose pregnancies are complicated by carbohydrate intolerance (diabetes), changes in diet composition for women with multiple pregnancies would be a lowering (40%) of the carbohydrates to avoid more hyperglycemic peaks and an increase in the percent of fats (40%) to provide more substrate. This adjustment in the distribution of the macronutrients may be important not only for nutritional value, but also for prevention of preterm labor. Carbohydrate intolerance, particularly in the absence of longstanding vascular disease, is a well-recognized risk factor for fetal or neonatal macrosomia.

7.5   CONCLUSION

In summary, multifetal pregnancy offers many challenges for both physician and patient. Maternal, fetal, and neonatal complications are more common and make mothers with multifetal pregnancies a high-risk pregnancy group. These patients usually receive their clinical care from well-trained obstetricians or maternal–fetal medicine specialists, and yet, the clinicians do not commonly request nutrition consultation. Certainly, consults are not as common as when pregnancies are complicated by diabetes or hypertension. A reasonable approach would be to obtain a nutritional assessment every trimester. There are good data to support a recommendation for an increase in caloric intake and dietary management, based on BMI. Maternal weight gain, however followed clinically, is only one part of the algorithm for good nutritional care, as other variables are also crucial. Additional supplements with micronutrients and fish oil seem to be supported by the literature. Neonatal birth weight, as a marker for nutrition, must also be considered in its entirety, as many confounding variables can influence birth weight. Nutritional guidelines as part of an overall program to improve perinatal outcome in mutifetal pregnancy are important. Further study and research into nutrition during multifetal pregnancy is needed, as many of the complications in the group are increased. Hence, the program for this group should be multidisciplinary and provide close monitoring throughout the pregnancy, even preconceptionally.

REFERENCES

1. National Center for Health Statistics (2006) National vital statistics reports, births: final data for 2004, vol.55, no. 1, 29 September 2006
2. Russell RB, Petrini, JR, Mattison DR, and Schwarz RH (2003) The changing epidemiology of multiple births in the United States. Obstet Gynecol 101:129–135
3. Multifetal gestation (2005) In: Williams' obstetrics, 22nd edn. McGraw-Hill, New York, N.Y., pp 911–948
4. American College of Obstetricians and Gynecologists (ACOG) Practice Bulletin. Multiple gestation: complicated twin, triplet, and high-order multifetal pregnancy. October, no 56. ACOG, Washington, D.C.

5. Lantz ME, Chez RA, Rodriquez A, Porter KB et al (1996) Maternal weight gain patterns and birth weight outcome in twin gestations. Obstet Gynecol 87:551–556
6. Carmichael S, Abrams B, Selvin S (1997) The pattern of maternal weight gain in women with good pregnancy outcomes. Am J Public Health 87:1984–1988
7. Luke B, Min S-J, Gillespie B, Avni M, Witter FR, Newman RB, Mauldin JG, Salman FA, O'Sullivan MJ (1998) The importance of early weight gain in the intrauterine growth and birth weight of twins. Am J Obstet Gynecol 179:1155–1161
8. Bracero LA, Byrne DW (1998) Optimal maternal weight gain during singleton pregnancy. Gynecol Obstet Invest 1998:46:9–16
9. Abrams B, Altman SL, Picket KE (2000) Pregnancy weight gain: still controversial. Am J Clin Nutr 71:1233–1241
10. Brown JE, Carlson M (2000) Nutrition and multifetal pregnancy. J Am Diet Assoc 100:343–348
11. Luke B, Hediger ML, Nugent C, Newman RB, Mauldin JG, Witter FR, O'Sullivan MJ (2003) Body mass index–specific weight gains associated with optimal birth weights in twin pregnancies. J Reprod Med 48:217–224
12. Dubois S, Dougherty C, Duquette MP, Hanley JA, Moutquin JM (1991) Twin pregnancy: the impact of the Higgins Nutrition Intervention Program on maternal and neonatal outcomes. Am J Clin Nutr 53:1397–1403
13. Higgins, AC, Moxley JE, Pencharx PB, Mikolainis D, Dubois S (1989) Impact of the Higgins Nutrition Intervention Program on birth weight: results of a within-mother analysis. J Am Diet Assoc 89:1097–2003
14. Luke B, Brown MB, Misiunas R, Anderson E, Nugent C, van de Ven C, Burpee B, Gogliotti S (2003) Specialized Prenatal care and maternal and infant outcomes in twin pregnancy. Am J Obstet Gynecol 189:934–938
15. Shinagawa S, Suzuki S, Chihara H, Otsubo Y, Takeshita T, Araki T (2005) Maternal basal metabolic rate in twin pregnancy. Gynecol Obstet Invest 60:145–148
16. Institute of Medicine (1992) Nutrition services in prenatal care, 2nd edn. National Academy Press, Washington, D.C.
17. Institute of Medicine (1992) Nutrition during pregnancy and lactation: an implementation guide. subcommittee for a clinical application guide, Committee on Nutritional Status during Pregnancy and Lactation, Food and Nutrition Board. National Academy Press, Washington, D.C.
18. Allen LH (2000) Anemia and iron deficiency; effects on pregnancy outcome. Am J Clin Nutr 71(Suppl):1280–1284
19. Blickstein I, Goldschmit R, Lurie S (1995) Hemoglobin levels during twin vs. singleton pregnancies. J Reprod Med 40:47–50
20. Luke B (2004) Improving multiple pregnancy outcomes with nutritional interventions. Clin Obstet Gynecol 47:146–162
21. Villar J, Gulmezoglu M, de Oris M (1998) Nutritional and anti-microbial interventions to prevent preterm birth: an overview of randomized controlled trials. Obstet Gynecol Surv 53:575–585
22. Roem K (2003) Nutritional management of multiple pregnancies. Twin Res 6:514–519
23. McGregor JA, Allen KGD, Harris MA, Reece M, Wheeler M, French JI, Morrison J (2001) The omega-3 story: nutritional prevention of preterm birth and other adverse pregnancy outcomes. Obstet Gynecol Surv 56:1–13
24. Al MDM, van Houwelingen AC, Hornstra G (2000) Long-chain polyunsaturated fatty acids, pregnancy, and pregnancy outcome. Am J Clin Nutr 71:285s–291s
25. Foreman-van Drognelen MM, van Houwelingen AC, Kester AD, Hasaart TH, Blanco CE, Hornstra G (1995) Long-chain Polyunsaturated fatty acids in preterm infants: status at birth and its influence on postnatal levels. J Pediatr 126:611–618
26. Perez, MA, Hansen, RA, Harris MA, Allen KGD (2005) Dietary docosahexaenoic acid alters pregnant rat reproductive tissue prostaglandin and matrix metalloproteinase production. J Nutr Biochem 17:446–453
27. Olsen SF, Hansen HS, Jensen B (1990) Fish oil versus arachis oil food supplementation in relation to pregnancy duration in rats. Prostaglandins Leukot Essent Fatty Acids 40:255–260

28. Olsen SF, Sorensen JD, Secher NJ, Hedegaard M, Henriksen TB, Hansen HS, Grant A (1992) Randomized controlled trial of effect of fish-oil supplementation on pregnancy duration. Lancet 339:1003–1007

29. Smuts CM, Huang M, Mundy D, Plasse T, Major S, Carlson S (2003) A randomized trial of docosahexaenoic acid supplementation during the third trimester of pregnancy. Obstet Gynecol 101:469–479

30. Judge MP, Harel O, Lammi-Keefe CJ (2007) Maternal consumption of a docosahexaenoic acid-containing functional food during pregnancy: benefit for infant performance on problem solving but not recognition memory tasks at age 9 mo. Am J Clin Nutr 85:1572–1577

31. Judge MP, Lammi-Keefe C (2007) A docosahexaenoic acid functional food during pregnancy benefits infant visual acuity at four but not six months of age. Lipids 42:117–122

32. Troxell H, Anderson J, Auld G, Marx N, Harris M, Reece M, Allen K (2005) Omega-3 for baby and me: maternal development for a WIC intervention to increase DHA intake during pregnancy. Maternal Child Health J 9:189–197

33. Olsen SF, Secher NJ, Tabor A, Weber JW, Gludd C (2000) Randomized clinical trials of fish oil consumption in high-risk pregnancies. Br J Obstet Gynaecol 107:382–395

34. Sibai, BM, Hauth J, Caritis S, Lindheimer MD, MacPherson C, Klebanoff M, VanDorsten JP, Landon M, Miodovnik M, Paul R, Meis P, Thurnau G, Dombrowski M, Roberts J, McNellis D (2000) Hypertension disorders in twin versus singleton gestations. national institute of child health and human development network of maternal- fetal medicines units. Am J Obstet Gynecol 182:938–942

35. Jones JS, Newman RB, Miller MC (1991) Cross-sectional analysis of triplet birth weight. Am J Obstet Gynecol 164:135–140

36. Blickenstein I (2005) Growth aberration in multiple pregnancy. Obstet Gynecol Clin N Am 32:39–54

37. Loos RJ, Derom C, Derom R et al (2001) Birthweight in liveborn twins: the influence of the umbilical cord insertion and the fusion of the placentas. Br J Obstet Gynaecol 108:943–948

38. Alexander GR, Kogan M, Martin J et al (1998) What are the fetal growth patterns of singleton, twins, and triplets in the United States? Clin Obstet Gynecol 41:114–125

39. Blickstein I, Goldman RD (2003) Intertwin birth weight discordance as a potential adaptive measure to promote gestational age. J Reprod Med 48:449–454

40. Blickstein I, Jacques DL, Keith LG (2003) Effect of maternal height on gestational age and birth weight in nulliparous mothers of triplets with a normal pregravid body mass index. J Reprod Med 48:335–338

41. Victoria A, Mora G, Arias F (2001) Perinatal outcome, placental pathology, and severity of discordance in monochorionic and dichorionic twins. Obstet Gynecol 97:310–315

42. Gonzalez-Quintero VH, Luke B, O'Sullivan MJ et al (2003) Antenatal factors associated with significant birth weight discordancy in twin gestations. Am J Obstet Gynecol 189:813–817

43. Armson BA, O'Connell C, Persad V et al (2006) Determinants of perinatal morbidity and serious neonatal morbidity in the second twin. Obstet Gynecol 108:556–564

44. Malone FD, Kaufman GE, Chelman et al (1998) Maternal morbidity associated with triplet pregnancy. Am J Perinatal 15:73–77

45. Freinkel N, Dooley SL, Metzger BE (1985) Care of the pregnant woman with insulin dependent diabetes mellitus. N Engl J Med 96:313

46. Casele HL, Dooley SL, Metzger BE (1996) Metabolic response to meal eating and extended overnight fast in twin gestation. Am J Obstet Gynecol 175:917–921

8 Adolescent Pregnancy: Where Do We Start?

Linda Bloom and Arlene Escuro

Summary Adolescent pregnancy is associated with many high-risk conditions. Some of these are amenable to nutritional support and interventions to reduce risk. Nutrients that have specific implications for adolescents are energy, protein, iron, calcium, folate, and fluids. This chapter addresses assessment and interventions specific to the adolescent including the above elements; weight gain in pregnancy; the use of groups to promote behavior change; Women, Infants, and Children nutrition program; and Internet and print resources that have been identified as helpful.

Keywords: Adolescent, Teen, Anemia, Birth weight, Group prenatal care, Weight gain in pregnancy, Motivational negotiation

8.1 INTRODUCTION

Teen pregnancy is recognized as a high-risk condition because it is associated with obstetric complications such as preeclampsia, preterm delivery, low-birth-weight infants, and neonatal death, especially in very young teens [2–4]. This was recognized as early as 1856 by Alcott [1] and is true across cultures and continents [2]. Adolescent pregnancy is a significant public health issue in the United States and in other nations [1–3]. Adolescents have been establishing pregnancies throughout recorded history and presumably before. Throughout history, depending on food availability and the health of the population, menarche occurred earlier or later. With sexual maturity came sexual activity and pregnancy. In colonial America, teen pregnancy was possibly the norm, and young marriage was certainly acceptable. The values associated with marital versus premarital sex have fluctuated based on economics of the time and the cultural and religious beliefs of an individual community. In the twentieth century, the average age of marriage and childbearing in the United States and Europe increased as formal education and the need for extended preparation for careers increased. The highest rate of teen pregnancy in the United States was in 1957, but it was not identified as a "problem" by the government until the 1970s, when the rate was already headed downward [1]. The teen pregnancy rate, which in the United States in 1995 was 57 per 1,000, is however much higher than it is in other developed countries such as the United

From: *Nutrition and Health: Handbook of Nutrition and Pregnancy*
Edited by: C.J. Lammi-Keefe, S.C. Couch, E.H. Philipson © Humana Press, Totowa, NJ

Kingdom (28/1000), Europe (France: 10/1,000, Italy 7/1,000) and Japan (4/1,000) and compares to developing nations [2, 3].

There is evidence that nutrition and appropriate weight gain in the adolescent pregnancy have a relationship to obstetric risks [5, 6] and thus may be amenable to intervention. There is also modest indication in the available literature that nutritional interventions for pregnant teens can positively affect pregnancy outcome [7]. The young pregnant woman, however, has many obstacles impinging on her ability and/or desire to eat a healthy, adequate diet. A multitude of social problems such as substance abuse, incest, truancy, sexually transmitted infections, poverty, and dysfunctional families may influence her [3, 9, 10]. This chapter reviews these issues and possible complications that can result from adolescent pregnancy. Also, important nutritional considerations for maintaining optimal health of the teenager during pregnancy are discussed as well as assessment and interventions specific to the adolescent mother.

8.2 SCOPE OF TEEN PREGNANCY

Data from the Centers for Disease Control (CDC) underline a decline in 2004 in the teenage birth rate in the US, with 41.2 births per 1,000 females aged 15–19 years [11]. Rates increased slightly in 2004 for girls aged 10–14 years [11]. These data are worrisome, as these young women are nearer to menarche, and still growing themselves, with increased nutritional needs. These very young women also have increased risk for maternal death [12]. The younger-aged group received the lowest rate of timely prenatal care, highest rates of late or no prenatal care, and was at highest risk of pregnancy-associated hypertension. Among the youngest cohort, pregnancy outcome was poor, e.g., infants were more likely to be preterm, to be born with low birth weight, and to die in their first year at a rate that was three times the overall rate of 15.4 per 1,000 [13].

8.3 DEVELOPMENTAL CONSIDERATIONS FOR THE ADOLESCENT

Despite the high risk for pregnant teens, there is a paucity of research on effective interventions for their care and nutrition [7]. Providers can extrapolate from the research on adolescents in general to identify strategies to motivate and guide these young women to nourish themselves and their babies. To develop an approach for pregnant adolescents it is first appropriate to examine theories of development for this age group.

Erickson proposed the developmental tasks of teenagers as accepting body image, determining and internalizing sexual identity, developing a personal value system, preparing for productive function, achieving independence from parents and, finally, developing an adult identity [8]. Pregnancy, planned or unplanned, has a tremendous impact on all of these tasks, both positive and negative. How does a young woman accept a body image that is changing every week? How does she deal with a rapid weight gain in a culture that is fraught with "skinny" images and advertisements for weight loss products? It has been reported in the literature that lesbian and bisexual young women are at higher risk for pregnancy compared with their peers, although it is unclear why this is the case; how do their attempts to internalize their sexual identity impact on this risk [10]? Some young women may plan their pregnancy to demonstrate their complete rejection of their parents' value system (thus stating their own) and achieving independence through their actions. And, some may become pregnant as acceptance of a value system where parenthood is a defining necessity of adulthood.

Against this background, providers struggle to provide the interventions that will assist young women to successfully navigate into adult roles and maintain health for themselves and their infants. Appropriate nutrition is one aspect of these interventions for their patients. The first step is to establish the individual needs of the pregnant adolescent in light of the current recommendations.

8.4   NUTRIENT NEEDS OF THE PREGNANT ADOLESCENT

Nutrition screening and assessment is a cornerstone for comprehensive prenatal care for all pregnant adolescents. Adolescence is a period of rapid physical growth, with heightened nutritional requirements to support growth and development. The additional energy and nutrient demands of pregnancy place adolescents at nutritional risk [13]. The physiological and psychosocial immaturity of the teen compounds the potential for obstetric risks and complications [13]. Nutrition screening and counseling should be aimed at alleviating the risks and promoting optimal maternal and fetal outcomes.

Important assessment data that need to be collected and evaluated to comprehensively develop educational approaches for pregnant adolescents can be categorized as follows: (1) determining the quality, quantity, and rate of weight gain in pregnancy; (2) evaluating current dietary intake to determine the adequacy of nutrient and energy intake during pregnancy; and (3) assessing dietary issues that may affect intake, e.g., food allergies or vegetarianism [13]. Data derived from these assessments can provide a focus for discussions with all adolescents throughout pregnancy. Adolescents, especially those younger than 15 years of age, are at high risk for inappropriate maternal weight gain, anemia, and more serious complications such as lung and renal disease. Maternal weight gain is reportedly more influential than age of mother on fetal birth weight [11, 14]. Given that fetal birth weight < 3,000 g is related to increased infant morbidity and mortality, optimizing maternal weight gain should be central to any intervention efforts for the pregnant teen and adult.

Pregnancy places the adolescent at high nutritional risk because of the increased energy and nutrient demands of pregnancy. Data regarding nutrient requirements of pregnant adolescents are extremely limited. In general, however, the closer a teen is to menarche (younger teens with incomplete growth) at conception, the greater her need is for energy and nutrients above the normal requirements for pregnancy [15, 16]. The Institute of Medicine (IOM) Dietary Reference Intakes (DRIs) provide recommendations for nutrient and energy needs during pregnancy by trimester of pregnancy. Adequate energy intake should be a primary consideration for adolescent pregnancy; if energy needs are not met, then available protein, vitamins, and minerals cannot be used effectively in various metabolic functions [17]. Energy requirements may be greater for adolescents who begin pregnancy underweight, are still growing, or who are physically active [13]. The additional energy needs during the second and third trimesters of pregnancy are approximately 300 kcal per day in adults and older adolescents and 500 kcal per day in younger adolescents (aged 14 years and younger) [18, 19].

Protein needs are increased during pregnancy. The additional 25 g of protein required each day during pregnancy are generally not a problem to obtain for most adolescents in industrialized countries, given that many teens consume twice their recommended daily protein intake [20]. Routine ingestion of high-protein powders and specially formulated high-protein supplements and beverages are not routinely needed and may be potentially harmful (increasing risk of preterm birth) during pregnancy [17, 21]. These supplements should be avoided during pregnancy. Rather, increased use of food sources of protein is recommended, such as milk and flesh (meat) foods, as part of a well-balanced diet, especially because these foods are also rich sources of vitamins and minerals [21]. A careful assessment of dietary protein intake of a pregnant adolescent is important. About two thirds of total protein should be of high biologic quality, such as the protein that comes from eggs, milk, meat, or other animal sources [17].

Iron deficiency is one of the most common nutritional problems among both pregnant and nonpregnant adolescent females and occurs in all socioeconomic groups [17]. The need for iron increases during adolescence due to increased deposition of lean body mass, heightened synthesis of red blood cells, and onset of menses [17]. Iron requirements are further increased during pregnancy due to the rapidly expanding blood volume, which far exceeds the expansion of red blood cells, and results in decreased hemoglobin concentration [13]. The DRI for pregnant women is 27 mg per day, a level almost twice that for nonpregnant adolescents. The CDC recommends a routine low-dose iron supplement (30 mg per day) beginning at the first prenatal visit [22]. Although conclusive evidence for the benefits of universal supplementation is lacking, the CDC advocates this position because many women have difficulty maintaining iron stores during pregnancy [23]. Given the typically low intake of iron in most adolescent diets [17], supplemental sources of iron may be even more important for pregnant teens. Liquid and chewable forms of iron are available if teenagers have trouble swallowing tablets or capsules. To ensure adequate absorption, iron supplements should be taken at bedtime or between meals with water or juice; milk, tea, or coffee should be avoided, as these block iron absorption [24]. Pregnant adolescents should be encouraged to consume iron-rich foods such as lean red meat, fish, poultry, dried fruits, and iron-fortified cereals. Iron from animal sources is well absorbed. Iron from plant sources is poorly absorbed, but absorption is enhanced by simultaneous intake of vitamin C, meat, fish, and poultry [17]. If iron deficiency anemia develops, then iron supplementation is typically increased to 60 to 120 mg per day until the anemia is resolved. A multivitamin–mineral supplement supplying 15 mg of zinc and 2 mg of copper is also recommended because the therapeutic dosage of iron may impair the absorption or utilization of these nutrients [20].

Calcium is another nutrient of concern during pregnancy, especially among adolescents. The DRI for calcium for adolescents is 1,300 mg per day, yet 12- to 19-year-old females in the United States have average calcium intakes of about 800 mg per day [25]. Consumption of low-calcium beverages like soft drinks and fruit drinks displace milk and likely accounts for the suboptimal intakes reported [26]. The DRI for calcium does not increase during pregnancy, but adolescents may potentially have increased needs secondary to continued skeletal growth and consolidation of bone mass during the adolescent years [13]. Health professionals should counsel pregnant teens on good dietary sources of calcium and ways to meet the recommended intake. Dairy products and fortified foods including orange juice are high-quality calcium sources to recommend during counseling [17]. For pregnant teens that do not consume milk products (due to milk allergy or other reason) or calcium-fortified foods, a calcium and vitamin D supplement may be needed [23]. The IOM recommends taking a separate calcium supplement supplying 600 mg of elemental calcium per day. Most calcium supplements have comparable absorption rates, and two common forms include calcium citrate and calcium carbonate. Pregnant teens should be advised not to take supplements from bone meal calcium, dolomite, and calcium carbonate made from oyster shells, as these may contain lead [13]. For optimal absorption, supplements should be taken with meals and in doses no greater than 600 mg at a time [13].

Folate, essential for nucleic acid synthesis, is required in greater amounts during pregnancy, because of maternal and fetal tissue growth and red blood cell formation [15]. Folate deficiency during pregnancy may result in intrauterine growth restriction, congenital anomalies, neural tube defects, or spontaneous abortion [21, 23, 29]. The DRI for folate during pregnancy is 600 mcg per day [28]. The major natural sources of dietary folate are legumes, green leafy vegetables, liver, citrus fruits and juices, and whole wheat bread. Compared to naturally occurring folate in foods, the folic acid contained in fortified foods and supplements is almost twice as well absorbed, so that 1 mcg from these sources is equivalent to 1.7 mcg of dietary folate equivalents [23]. Because an adolescent's diet tends to be low in foods naturally high in folate, such as fruits and vegetables, nutrition counseling should focus on ways to incorporate both fortified foods and fruits and vegetables on a daily basis [13]. Most women are not able to meet the recommended intake of folate without supplements, and the general recommendation is to supplement with 400 mcg of folic acid per day both prior to and during pregnancy [13].

Fluid needs increase during pregnancy because of increased blood volume [17]. Adequate water helps the body maintain proper temperature, transports nutrients and waste products, moistens the digestive tract and tissues, and cushions and protects the developing fetus [17]. At least eight 8-oz cups of noncaffeinated fluids should be consumed each day. Water is the best choice, as the body absorbs it rapidly. Adolescents may obtain appreciable amounts of caffeine through consumption of soft drinks and coffee or tea beverages. Caffeinated beverages can increase urinary output, contributing to fluid depletion, and should be consumed in limited amounts [17]. High-caffeine intakes have been linked to low birth weight and increased risk of spontaneous abortion [23]. Prudent advice would be to discourage caffeine intake above 300 mg/day [23]. To translate that level into servings, this equates to the amount of caffeine in about two 8-oz cups of brewed coffee (135 mg/cup), three 8-oz cups of instant coffee (95 mg/cup), and six 8-oz cups of leaf/bag tea (50 mg/cup). The caffeine in a 12-oz soft drink ranges from 23 to 71 mg [17]. Newer "energy drinks" are often higher in caffeine and reading the labels is important.

Weight gain during pregnancy is a good assessment of the adequacy of an adolescent's dietary intake. How much should she gain? The IOM recommendations are 28–40 lb (12.5–18 kg) for women with body mass index (BMI) < 19.8, 25–35 lb (11.5–16 kg) for BMI = 19.8–26, 15–25 lb (7–11.5 kg) for BMI = 26–29, and at least 15 lbs (7 kg) for BMI > 29. Adolescents should gain at the upper end of these ranges. These recommendations were established in 1990, and several authors have suggested that they be revisited to evaluate long-term effects on infant and child health. [11] There is agreement, however, that weight gain should be based on prepregnancy BMI (kg/m^2) to promote healthy outcomes, avoid postpartum weight retention, and reduce risk of chronic diseases in childhood and beyond [11, 13, 23]. The current recommendations by IOM are supported by the American College of Obstetricians and Gynecologists (ACOG). Maternal weight gain strongly influences fetal growth, infant birth weight, and length of gestation [21].

The recommendation for obese women to gain at least 15 lb is to reduce risk for delivering small for gestational age infants [13]. The American Dietetic Association (ADA) suggests that efforts should also be focused on assisting the postpartum adolescent to return to a healthy weight to reduce risk to future pregnancies [23]. There is no benefit in weight gain above the range suggested, and some indication of harm to mother and baby has been reported (see Chap. 5, "Obesity in Pregnancy"). Very young teens may be the exception to this for reasons discussed earlier regarding difficulties meeting nutrient recommendations. However, more studies are needed to determine whether changes to the IOM recommendations are warranted for the adolescent.

The overall issue of weight gain may be problematic for teens responding to the "skinny" image presented in pervasive media. Croll in *Guidelines for Adolescent Nutrition Services* [30] presents an entire chapter dedicated to body image issues and tools to assist teens to establish a healthy appreciation for their unique appearance. She suggests that routine patient counseling should include assessment for body image concerns, and if present, teens should be provided with appropriate resources to address these issues. In her book, Croll provides specific questions to use in assessing body image, and suggests several strategies and tools to use with teens and their parents on body distortion, dieting, and media literacy. The same source [30] also has a chapter by Alton on eating disorders and offers diagnostic criteria and treatment information for these psychiatric syndromes with disturbed body images.

8.5 ADOLESCENT BARRIERS TO HEALTHY EATING

Practically speaking, knowledge of nutrients and their value is often not sufficient to encourage teenagers to consume the appropriate foods. Motivational approaches are often necessary to enable diet-related behavior change and to assist the teen in overcoming environmental obstacles and other barriers to healthy eating. Teens are frequently not involved in either food purchase or preparation in the household. Their financial situation may preclude the purchase of fruits and vegetables. For those who obtain supplemental food from the Women, Infants, and Children (WIC) program, often purchases are expended before the month is over. Also, school lunch choices for those still attending school are widely varied from system to system, and the "optional" food line may be overwhelmed with high-fat content items, providing relatively few healthy choices. For example, in northeast Ohio, one local school system provides a basic healthy lunch (based on the US Food and Drug Administration [FDA] guidelines), and then an á la carte menu of French fries smothered in cheese sauce, pepperoni or sausage pizza, fried-chicken sandwiches, cheeseburgers, pretzels, baked potatoes (with sour cream, butter or cheese sauce as requested) and bags of chips (personal observation). Teens frequent fast food restaurants and eat "junk" food at home at a high rate, obtaining high numbers of calories and little in the way of necessary nutrients. These issues need to be acknowledged and explored with each teen.

8.6 PROGRAMS AND RESOURCES FOR PREGNANT ADOLESCENTS

8.6.1 *Community-Based Nutrition Programs*

Each community will have a unique response to the need for adolescent services and for others in need of nutrition support. There may be churches or food pantries that assist those with food insecurity. Health care professionals should become familiar with these local resources (or with the social worker, nurse, or person in the office answering the phones who knows these resources), understand how they can be utilized to assist the adolescent, and know their limits. Networking with community health care providers and health service employees is a skill that can be modeled for these young learners.

In addition, each community should have some way to access the WIC program. Many adolescent patients will be eligible for WIC services. To be eligible, the pregnant adolescent must meet specific residency, income, and nutritional risk criteria as specified by WIC. To qualify based on nutritional risk, the adolescent must have a medically-based risk such as anemia, be underweight (less than 100 pounds) or overweight (over 200 pounds), have a history of pregnancy complications or poor pregnancy outcomes, or have other dietary risks such as failure to meet the dietary guidelines for any food groups or have inappropriate nutrition practices, e.g., pica.

The WIC fact sheet states in part [31]: "In most WIC state agencies, WIC participants receive checks or vouchers to purchase specific food each month that are designed to supplement their diets. A few WIC state agencies distribute the WIC foods through warehouses or deliver the foods to participants' homes. The foods provided are high in one or more of the following nutrients: protein, calcium, iron, and vitamins A and C. These are the nutrients frequently lacking in the diets of the program's target population. Different food packages are provided for different categories of participants.

"WIC foods include iron-fortified infant formula and infant cereal, iron-fortified adult cereal, vitamin C-rich fruit or vegetable juice, eggs, milk, cheese, peanut butter, dried beans/peas, tuna fish, and carrots. Special therapeutic infant formulas and medical foods may be provided when prescribed by a physician for a specified medical condition."

The WIC program was started in 1974. Since its inception, numerous studies have been performed to evaluate its effectiveness, and it has earned a reputation of being one of the most effective nutrition intervention programs in the United States [32]. Various studies performed by Food and Nutrition Services (FNS) of the US Department of Agriculture (USDA) and other entities have demonstrated benefits for WIC participants including longer pregnancy duration, less low-birth-weight infants, reduced infant morbidity and mortality, improved infant feeding practices, and higher diet quality of pregnant and postpartum mothers [32].

The WIC program is a proven effective resource that should be pursued for any pregnant adolescent who qualifies. The general guidelines for the program are nationwide, but each individual state has flexibility in implementing the program. Contacting the local agency implementing the program and obtaining details of community specifics will allow tailoring of interventions to build around the existing programs. The webpage www.fns.usda.gov/wic/Contacts/ContactsMenu.HTM has agency contacts listed by state. Providing a list for your patients of the locations of WIC offices along with other local resources such as food banks, churches with meals, etc should be considered.

8.6.2   Group Programs

Educational and support groups have been used successfully to promote healthy behaviors in various populations with chronic diseases such as diabetes [33]. Several different group programs have been described specifically for the pregnant adolescent [34–36]. Each program addresses nutrition differently, however. The Pregnancy Aid Center (PAC) located in Maryland has been providing care to pregnant adolescents since 1995, and was described by Bowman and Palley [35]. As part of this program, participants meet with social workers in groups; social workers then assist these individuals with direct service provision including transportation to and assistance with registration in government programs, including Medicaid and WIC. Bowan and Palley tracked 40 adolescent participants for a year and reported improved mean maternal weight gain and birth weights over the background statistics for the area they served. These authors postulated that the individual contact with social workers, who provided flexible services to participants, helped to reduce program-related anxiety. Other reported positive outcomes of the program included an increase in the number of participants who received regular prenatal care, and a reported 100% participation rate in WIC services for their study population [35].

Another program designed to deliver prenatal care with an emphasis on nutrition is the CenteringPregnancy program developed by Sharon Rising. This program uses a structured curriculum with clearly defined nutrition objectives. The content of the curriculum is based on assessment, education, and skills building, and support [36]. CenteringPregnancy uses the benefits of group interaction, emphasizes individual responsibility for health, and combines it with well-designed information on important topics such as prenatal nutrition and infant feeding [37]. Although preliminary, outcome data suggest that this program has promise for reducing the incidence of preterm birth and low birth weight in a cost effective manner [33, 38, 39].

Nutrition experts at the University of Minnesota as part of their Leadership, Education and Training Program in Maternal And Child Nutrition have proposed a method of counseling pregnant adolescents in individual and group sessions based on motivational negotiation [40]. This technique has been described as "a dance," in which the provider is the expert in knowledge, and the client is the expert in her abilities, struggles, and motivation. Motivational negotiation is a fairly new counseling technique that has shown promise in changing adult and adolescent dietary and health-related behaviors. The technique and its applications for the pregnant adolescent are described in more detail and can be accessed from the webpage www.epi.umn.edu/let/nutri/pregadol/index.shtm.

8.6.3 *Internet-Based Programs and Resources*

Our lives are increasingly influenced by the World Wide Web, which is now accessible in our schools, workplaces, libraries, and homes. This technology has even greater impact on the young as they have grown up with computers. Teenagers routinely access health information on the Web. Therefore, computers can be a useful way of facilitating teenage acquisition of health-related information. Importantly, the adolescent must be educated on where to find accurate Web-based information.

The following is a list of websites that provide adolescents and their parents with important, reliable information on prenatal care and related considerations. These websites will change as the Web grows, but these are sites to explore and to share with your adolescent patients.

1. http://www.marchofdimes.com/pnhec/159_153.asp. The March of Dimes provides good factual information on weight gain, a printable chart, a game to play about what to put in your grocery cart and other information presented at an appropriate reading level.
2. http://www.nlm.nih.gov/medlineplus/pregnancy.html#cat11. Medline has several topics for nutrition in pregnancy and gives links to other good sources including Environmental Protection Agency recommendations for fish and shellfish in pregnancy.
3. http://www.MyPyramid.gov/. This site presents the new food pyramid, with a chance to register for an individualized record of food and activity. At present, it does not have accommodation for pregnancy, and has a specific disclaimer for pregnancy and

lactation, but a dietitian could provide extra information to personalize the pyramid. See Chap. 1 ("Nutrient Recommendations and Dietary Guidelines for Pregnant Women") for a discussion of how the nutrients in the six food intake patterns, within the recommended energy intake ranges for pregnancy, of the MyPyramid Food Guidance System compare with the RDAs for other nutrients in pregnancy.

4. http://win.niddk.hig.gov/publications/two.htm. Fit for Two has tips for pregnancy. This is not specifically for teens, but has appropriate reading level, good pictures, and graphics, very appealing.

5. http://www.nal.usda.gov/fnic/pubs/bibs/gen/vegetarian.htm presents a list of resources for vegetarians that are useful for providers and consumers.

6. http://www.vrg.org/nutrition/protein.htm has information on protein for vegetarians, including vegans. Sample menus are provided.

7. http://www.clevelandclinic.org/health/health-info/docs/1600/1674.asp?index=4724 provides information on appropriate diet for vegetarians.

8. http://teamnutrition.usda.gov/library.html is a general resource on nutrition for young teens, not specific to pregnancy, but it is a useful source for colorful materials. There is also information about school lunch programs.

9. http://www.fns.usda.gov/wic/ provides information about the WIC program, including how to register.

10. http://www.nal.usda.gov/wicworks/ is a link to a 12-unit training module for WIC staff. It has continuing education credits for nurses and dietitians. The material covered is useful for any provider of nutrition education and includes information and guidance for leading group discussions on nutrition topics.

Other sources to explore include the state agriculture department, dairy industry materials, and local university extension service fact sheets. The Web makes information from other states just as accessible as your own state or local university. For example, at www.nal.usda.gov/fnic/pubs/bibs/topics/pregnancy/pregcon.html there is a bibliography of available resources for nutrition, some specific to adolescents and includes ordering information for many materials.

These websites will give you a place to begin developing resources. Ask your teen patients to let you know when they find a site they like; use them as sources for Web news. Assist them in evaluating the information presented on different websites; encourage them to screen the information carefully before implementing recommendations. Some questions to ask include:

- Who wrote the pages and is the author an expert?
- What does the author say is the purpose of the site?
- When was the site created and last updated?
- What is the source of information? Can it be confirmed?
- Why is the information useful for my purpose? [41, 42]

8.6.4 Other Resources

Story and Stang's *Nutrition and the Pregnant Adolescent: A Practical Reference Guide 2000* [13] is a very useful text for any program providing nutrition education. There are handouts, suggestions for topics, and techniques and many hints for this area. This text is available in downloadable form at www.epi.umjn.edu/let/pubs/nmpa.shtm.

The *Journal of Midwifery and Women's Health* provides several nutritional topics in their Share with Women series. You can go to their website at www.jmwh.org and print out information on weight gain during pregnancy, folic acid, eating safely during pregnancy, and others. These are provided courtesy of the American College of Nurse Midwives.

The March of Dimes produces many printed resources, in addition to their online materials, for nutrition education, including an easy to read booklet, video on healthy pregnancy for teens, and other fact sheets and materials. These may be accessed by going to the professional section on the website www.marchofdimes.org.

The Philadelphia Department of Public Health has produced a beautiful brochure supported by Title V, called "Healthy Foods, Healthy Baby." It has great illustrations of two teenagers, and takes them through pregnancy and the decision to breastfeed, in 28 pages. There are sections on eating out, grocery shopping, weight gain, and nausea and vomiting. It includes weight chart and graph. The new nutrition pyramid information is included in the 2007 version. It is available for order at: http://www.phila.gov/health/units/mcfh.

For those interested in implementing CenteringPregnancy (described above), the association provides handbooks, self-assessment surveys, and printed information along with recommendations for teaching aids as part of its initial training (www.centering-pregnancy.com).

A larger concern throughout the provision of services to adolescents may be the attitudes of providers. Are they viewed as "problems" since their pregnancy presents risks? Kenneth R. Ginsberg proposes that professionals look to the patient's strengths, build on them, and reframe the issue in terms of what opportunities are present in the situation [43]. This framework is consistent with Motivational Negotiation [39] and the techniques presented in CenteringPregnancy [36–38]. For example, pregnancy gives a young woman and her partner the chance to expand their knowledge of nutrition so they can produce the healthiest child in the neighborhood. They can use that protein to make brain cells, and consume those vitamins to make a healthy immune system. Ginsberg [43] suggests that we consider resilience theory and move beyond the "problem" to recognize the competency and strengths of the young woman.

8.7   CONCLUSION

Adolescents are a special challenge in nutritional interventions due to their developmental stage, their increased nutritional requirements, and their perceptions of weight gain. There are no "proven" techniques to address all these issues but the methods summarized below show promise.

8.7.1   Recommendations to Health Care Providers Regarding Nutrition and the Pregnant Adolescent

1. Evaluate each adolescent's nutritional status at her initial visit and consider doing it each trimester (if this is onerous for you, refer to a dietitian each trimester and follow up on the client's attendance at the visits).
2. Prescribe a good prenatal vitamin and evaluate the compliance; if she is unable to swallow or hates the taste, find another choice (Flintstones vitamins with iron works well!).

3. Prescribe iron supplements (especially to vegetarian teens).
4. Recommend calcium supplements (Tums is easy!).
5. Provide easy-to-read nutrition resources.
6. Send your patients to WIC at their first visit.
7. Establish group opportunities for young mothers for education and support (or find ones already existing in your community, possibly the WIC office).
8. Graph weight gain for each young woman (visual learning) or better yet, provide her a chart to graph her own weight (see above for sources).
9. Provide material to explain the purpose of all weight gain (i.e. baby, placenta, fluids, etc.); ask body image questions.
10. Use your dietitian as a resource, s/he has lots of materials and ideas that may be useful.
11. Enjoy the interaction with the young!

A striking issue that becomes apparent in the review of materials for adolescent care is the lack of research in this area. There are obvious problems in doing randomized, controlled trials with pregnant adolescents since not just one, but two patients (mother and child) are affected, and both pose issues for informed consent. As providers, we need to seek methods to overcome these barriers to develop effective tools and processes to ensure the health and well-being of these individuals now and in the future. As we discover these answers, we can promote adolescents' strengths, their new knowledge, and the confidence they build by successfully navigating this life passage.

REFERENCES

1. Vinovskis MA (2003) Historical perspectives on adolescent pregnancy and education in the United States. Hist Fam 8:399–421
2. Conde-AgudeloA, Belizan JM, Lammers C (2005) Maternal-perinatal morbidity and mortality associated with adolescent pregnancy in Latin America: cross-sectional study. Am J Obstet Gynecol 192:342–349
3. Rondo PH, Souza MR, Moraes F, Nogueira F (2004) Relationship between nutritional and psychological status of pregnant adolescents and non-adolescents in Brazil. J Health Popul Nutr 22:34–45
4. Fraser AM, Breckert JE, Ward RH (1995) Association of young maternal age with adverse reproductive outcomes. N Engl J Med 332:1113–1117
5. Herrmann TS, Siega-Riz A, Aurora C, Dunkel-Schetter C (2001) Prolonged periods without food intake during pregnancy increase risk for elevated maternal cortiocotrophin-releasing hormone concentrations. Am J Obstet Gynec 185:403–412
6. Siega-Riz A, Herrmann T, Savitz D, Thorp J (2001) Frequency of eating during pregnancy and its effect on preterm delivery. Am J Epidemiol 153:647–652
7. Nielsen JN, Gittelsohn J, Anliker J, O'Brien K (2006) Interventions to improve diet and weight gain among pregnant adolescents and recommendations for future research. J Am Diet Assoc 106:1825–1840
8. Erikson E (1968) Identity: youth and crisis. Norton, New York, N.Y.
9. Rome ES, Rybicki LA, Durant RH (1998) Pregnancy and other risk behaviors among adolescent girls in Ohio. J Adolesc Health 22:50–55
10. Saewyc EM, Bearinger L, Blum RW, Resnick MD (1999) Sexual intercourse, abuse and pregnancy among adolescent women: does sexual orientation make a difference? Fam Plan Perspect 31:127–131
11. Lenders CM, McElrath TF, Scholl TO (2000) Nutrition in adolescent pregnancy. Curr Opin Pediatr 12:291–296
12. Otterblad Olausson P, Haglund B, Ringback Weitoft G, Chattingius S (2004) Premature death among teenage mothers. Br J Obstet Gynaecol 111:793–799
13. Stang, J, Story M, Feldman S (2005) Nutrition in adolescent pregnancy. Int J Childbirth Educ 20:4–11
14. Rees LM, Lederman SA, Kiely JL (1996) Birth weight associated with low neonatal mortality: infants of adolescent and adult mothers. Pediatrics 98:1161–1166

15. Story M, Alton I (1995) Nutrition issues and adolescent pregnancy. Nutr Today 30:142–151
16. Buschman NA, Foster G, Vickers P (2001) Adolescent girls and their babies: achieving optimal birthweight, gestational weight gain and pregnancy outcomes in terms of gestation at delivery and infant birth weight: a comparison between adolescents under 16 and adult women. Child Care Health Devel 2:163–171
17. Story M, Stang J (eds) (2000) In: Nutrition and the pregnant adolescent: a practical reference guide. Center for Leadership, Education, and Training in Maternal and Child Nutrition. University of Minnesota, Minneapolis, Minn., pp 37–46
18. National Academy of Sciences (1989) National Research Council. Recommended Dietary Allowances, 10th edn. National Academy Press, Washington, D.C.
19. Gutierrez Y, King JC (1993) Nutrition during teenage pregnancy. Pediatr Ann 22:99–108
20. Devaney BL, Gordon AR, Burghardt JA (1995) Dietary intakes of students. Am J of Clin Nutr 61(Suppl):205S–212S
21. Institute of Medicine (1990) Nutrition during pregnancy: Part I, Weight gain. Part II, Nutrient supplements. Committee on Nutrition Status During Pregnancy and Lactation. National Academy Press, Washington, D.C.
22. Recommendations to prevent and control iron deficiency in the United States (1998) Centers for Disease Control and Prevention. Morb Mortal Wkly Rep 47:1–36
23. American Dietetic Association (2002) Position of the American Dietetic Association: nutrition and lifestyle for a healthy pregnancy outcome. J Am Diet Assoc 102:1479–1490
24. Institute of Medicine (1992) National Academy of Sciences, Food and Nutrition Board. Nutrition during pregnancy and lactation: an implementation guide. National Academy Press, Washington, D.C.
25. US Dept of Health and Human Services, Centers for Disease Control and Prevention, National Center for Health Statistics (USDHHS CDCP NCHS) (1997) National Health and Nutrition Survey III, 1988–1994. USDHHS CDCP NCHS, Hyattsville, Md.
26. Harnack L, Stang J, Story M (1999) Soft drink consumption among US children and adolescents: nutritional consequences. J Am Diet Assoc 99:436–441
27. Ritchie LD, King JC (2000) Dietary calcium and pregnancy-induced hypertension: is there a relation? Am J Clin Nutr 71:1371S–1374S
28. Institute of Medicine (1988) Dietary Reference Intakes for thiamin, riboflavin, niacin, vitamin B_6, folate, vitamin B_{12}, pantothenic acid, biotin, and choline. National Academy Press, Washington, D.C.
29. Goldenburg R, Tamura T, Cliver S, Cutter G, Hoffman H, Cooper R (1992) Serum folate and fetal growth retardation: a matter of compliance? Obstet Gynecol 79:719–723
30. Stang J, Story M (eds) (2005) Guidelines for adolescent nutrition services. Available via http://www.epi.umn.edu/let/pubs/adol_book.shtm
31. US Department of Agriculture (USDA), Food and Nutrition Services (FNS) (2007) WIC—the special supplemental nutrition program for women, infants and children. Nutrition Program Facts, Food and Nutrition Services, USDA. Available via www.fns.usda.gov/wic/WIC-Fact-Sheet.pdf
32. US Department of Agriculture (USDA), Food and Nutrition Services (FNS) (2007) About WIC—how WIC helps. Available via http://www.fns.usda.gov/wic/aboutwic/howwichelps.htm.
33. Klima C (2003) Centering Pregnancy: a model for pregnant adolescents. J Midwifery Womens Health 48 :220–225
34. Rothenberg A, Weissman A (2002) The development of programs for pregnant and parenting teens. Soc Work in Health Care 35:65–83
35. Bowman EK, Palley HA (2003) Improving adolescent pregnancy outcomes and maternal health: a case study of comprehensive case managed services. J Health Soc Policy 18:15–42
36. Carlson NS, Lowe NK (2006) Centering pregnancy: a new approach in prenatal care. Am J Matern Child Nurs 31:218–222
37. Ickovics JR, Kershaw TS, Westdahl C, Schindler Rising S, Klima C, Reynolds H, Magriples, U (2003) Group prenatal care and preterm birth weight: results from a matched cohort study at public clinics. Obstet Gynecol 102:1051–1057
38. Grady MA, Bloom K (2004) Pregnancy outcomes of adolescents enrolled in a CenteringPregnancy Program. J Midwifery Womens Health 49:412–420

39. Nutrition curricula, University of Minnesota (2007) Leadership, Education and Training Nutrition Module. Connecting with the Adolescent. Available via www.epi.umn.edu/let/nutri/pregadol/fla_mod3.shtm
40. Schrock K (2007) Critical evaluation of a website. Available via http://school.dicovery.com/schrock-guide/pdf/o7–01-cic.pdf
41. University of Southern Maine (2007) Checklist for evaluating Web resources. Available via http://library.usm.maine.edu/researchguides/webevaluating.html
42. Ginsburg K (2003) Developing our future: seeing and expecting the best in youth. J Midwifery Womens Health 48:167–169

9 Anorexia Nervosa and Bulimia Nervosa During Pregnancy

Sharon M. Nickols-Richardson

Summary Anorexia nervosa (AN) and bulimia nervosa (BN) present high-risk situations during pregnancy. These conditions have been associated with poor energy and nutrient intakes, notably total energy; folate; vitamins B_6, B_{12}, and A; calcium; iron; and zinc. Electrolyte imbalances are also of concern. Inadequate or excessive weight gain, spontaneous abortion, intrauterine growth restriction, preterm delivery, and low birth weight, among other adverse outcomes, have been reported in pregnant women with AN or BN and their offspring. Screening and assessment of women for these eating disorders during prenatal clinic visits is recommended. An interdisciplinary approach to care during pregnancy, the postpartum period, and beyond is critical to the successful management of AN or BN and optimal pregnancy outcomes.

Keywords: Anorexia nervosa, Binge eating, Bulimia nervosa, Compensatory behavior, Purging

9.1 INTRODUCTION

During periods of severe caloric deprivation, reproduction becomes a nonessential life function. For example, approximately half of all women of childbearing age experienced amenorrhea during the Dutch Winter Famine of 1944–1945 [1]. Reproduction requires a nutritionally replete woman at conception and the availability of energy, macronutrients, and micronutrients throughout pregnancy (see Chap. 1, "Nutrient Recommendations and Dietary Guidelines for Pregnant Women"). A supply of energy that is balanced to support eumenorrhea, implantation, and growth and development of the placenta and other maternal and fetal tissues is critical for optimal pregnancy outcomes. Anorexia nervosa (AN), bulimia nervosa (BN), eating disorder not otherwise specified (EDNOS), and binge eating disorder (BED) represent conditions of energy, macronutrient, and micronutrient imbalances for individuals with such eating disorders.

When occurring before, during, or after pregnancy, such a disorder may impact maternal and fetal outcomes. Treatment prior to conception is ideal; however, screening for eating disorders during pregnancy is important to facilitate early intervention that may optimize maternal and fetal health.

From: *Nutrition and Health: Handbook of Nutrition and Pregnancy*
Edited by: C.J. Lammi-Keefe, S.C. Couch, E.H. Philipson © Humana Press, Totowa, NJ

9.2 EATING DISORDERS DEFINED

The American Psychiatric Association defines AN, BN, EDNOS, and BED based on diagnostic criteria for each eating disorder [2]. Selected criteria are displayed in Table 9.1. Very little is known about the incidence and outcomes of EDNOS and BED during pregnancy. Thus, this discussion focuses on AN and BN.

9.2.1 Anorexia Nervosa

Characterized by extreme voluntary weight loss due to self-starvation or binge eating followed by purging, AN occurs in 0.5–3% of the female population [3, 4]. Clinical signs and symptoms of AN include an emaciated appearance, prepubertal features, lethargy, lanugo, alopecia, acrocyanosis, hypothermia, swollen joints, pitting edema, and bradycardia and hypotension. Biochemical evaluation often shows fluid and electrolyte disturbances and hypercarotenemia as well as endocrine and hematologic abnormalities such as hypothyroidism and anemia, respectively. Several cardiovascular irregularities develop along with a host of gastrointestinal complications, particularly in those with the binge eating–purging type of AN. Osteoporosis and skeletal fractures are common in persons with AN. Some may experience peripheral neuropathy and seizures. Mortality is as high as 22% in women with long-term AN [5].

9.2.2 Bulimia Nervosa

Individuals with BN engage in binge eating episodes, followed by compensatory behaviors to prevent any increases in body weight. Purging behaviors include self-induced vomiting or self-prescribed use of enemas, laxatives, or diuretics. Nonpurging behaviors include fasting and excessive exercise. While clinically diagnosed BN occurs in approximately 5% of the female population, up to 20% of women have reported bulimic behaviors in their lifetimes [6, 7]. Clinical features of BN include Russell's sign, dental enamel erosion, dental caries, and enlargement of the parotid glands in those who use self-induced vomiting as a purging behavior. Use of enemas, laxatives, and diuretics as well as vomiting can lead to electrolyte imbalances, cardiac dysfunction, and other neurologic disorders. Gastrointestinal symptoms may range from constipation to esophageal or gastric rupture. In those who engage in nonpurging behaviors, electrolyte imbalances, renal and cardiac dysfunction, and gastrointestinal disorders are common. Mortality occurs in less than 10% of individuals with BN [8].

9.3 ETIOLOGY

Several models for the etiology of AN and BN have been developed. Genetic foundations may explain up to 76 and 83% of the variance in AN and BN, respectively [9]. Candidate genes include the serotonergic neurotransmitter system (5-HT$_{2A}$), agouti related melanocortin-4 receptor, uncoupling protein-2/-3, and the estrogen-beta-receptor. Alterations in these genes would impact appetite regulation, mood, energy utilization, and body weight [10, 11]. Other variables included in explanatory models involve biologic, psychologic, sociologic, and behavioral factors. While multifactorial in nature, AN and BN have relatively clear clinical manifestations and anticipated outcomes. Dire consequences of AN and BN are expected in untreated and long-term conditions. Because of these complications, pregnancy places the woman with AN or BN and her fetus at high risk for adverse outcomes.

Table 9.1
Diagnostic Features of Eating Disorders [2]

Feature	Anorexia nervosa (AN)	Bulimia nervosa (BN)	Eating disorder not otherwise specified (EDNOS)	Binge eating disorder (BED)
Body weight	• < 85% of expected • Refusal to achieve or maintain > 85% of expected weight • Intense fear of weight gain	• 90–110% of expected (normal range)	• Generally of normal range • Intense fear of weight gain	• Generally > 100% of expected
Eating pattern	• Restricting • Binge eating–purging	• Recurrent binge eating with compensatory behavior (≥ twice per week for ≥ 3 months) • During a binge episode, intake of larger than typical quantity of food within 2 hours • Sensation of no control over intake during binge eating episode	• Recurrent binge eating with compensatory behavior (< twice per week for < 3 months) • During a binge episode, intake of larger than typical quantity of food within 2 hours • Sensation of no control over intake during binge eating • Chewing and spitting out of large quantities of food	• Binge eating without compensatory behavior (≥ twice per week for ≥ 6 months) • During a binge episode, intake of larger than typical quantity of food within 2 hours • Sensation of no control over intake during binge eating
Body image	• Disturbance • Dissatisfaction • Distortion	• Disturbance • Dissatisfaction • Distortion	• Disturbance • Dissatisfaction • Distortion	• Disturbance • Dissatisfaction
Menstrual function	• Amenorrhea for ≥ 3 consecutive cycles	• Eumenorrhea in 50%; oligomenorrhea or amenorrhea in 50%	• Generally normal	• Generally normal
Type	• Restricting • Binge eating–purging	• Purging as compensatory behavior[a] • Nonpurging but with other compensatory behaviors[a]	• Based on primary characteristics of AN or BN	• Not applicable

[a] Independent of AN

117

9.4 ANOREXIA AND BULIMIA NERVOSA DURING PREGNANCY

AN and BN are typically manifested in the early postpubertal to young adult years [12] and continue throughout the reproductive years [13]. Amenorrhea is a diagnostic criteria for AN, suggesting that pregnancy is of little concern in a woman with this eating disorder. However, approximately 10% of women who sought treatment in an infertility clinic presented with AN or BN [14]. Moreover, 60% of women with oligomenorrhea had eating disorders [14], indicating the desire for fertility despite any dysmenorrhea associated with AN or BN.

Studies confirm that women with AN may become pregnant [15, 16], particularly those women who are undergoing active treatment or are in remission [17]. While pregnancies in women with active BN are more common than in women with active AN, women with histories of AN and BN have similar pregnancy rates compared to the general population of women of childbearing age [17, 18].

Many factors are important to a successful course and outcome of pregnancy. Yet, prepregnancy weight and weight gain by the mother during pregnancy (see Chap. 2, "Optimal Weight Gain") are the two most salient indicators of infant outcome, including birth weight [19, 20]. Common characteristics of AN and BN are body image dissatisfaction or disturbances and desire to prevent weight gain. Thus, pregnancy presents a pivotal life cycle stage for a woman with AN or BN, because body weight and shape transform gradually over the course of pregnancy and abruptly upon delivery.

9.5 NUTRITIONAL CONCERNS

Adequate dietary intake is essential to meet the energy demands of pregnancy as well as to provide micronutrients that are critical to the growth and development of the woman and her fetus. Specific nutrient requirements for pregnancy are presented in Chap. 1, ("Nutrient Recommendations and Dietary Guidelines for Pregnant Women"). Nutrients of special concern in the woman with AN or BN are discussed here (Table 9.2).

9.5.1 Energy and Macronutrients

Energy needs increase in the last two trimesters to support the maternal and fetal products of pregnancy as well as spare protein to build these new tissues. Weight gain serves as a proxy indicator that these tissues have developed normally (see Chap. 2, "Optimal Weight Gain"). What is unique in AN is the controlled intake of food energy in those with restricting type. Intakes of 200–700 kcal per day, typical of an individual with restricting-type AN, are simply inadequate to supply the energy required for most successful pregnancies. In binge eating–purging-type AN and purging-type BN, adequate and even overly abundant kilocalories may be consumed—but are purged before the body has the opportunity to either fully digest or absorb nutrients. With nonpurging-type BN, adequate energy may be consumed; however, laxative, diuretic, and/or enema use as well as excessive exercise may result in malabsorption, excessive excretion, or altered utilization of nutrients such that the stream of nutrients is inadequate during pregnancy.

To provide an adequate amount of glucose and nonprotein kilocalories, over 175 g of carbohydrate per day are needed in pregnancy. If composed solely of carbohydrate, 700 kcal would meet this minimum need; however, most intakes of women with active restricting type AN do not contain adequate carbohydrate levels. Conversely, foods

Table 9.2

Nutrients of Special Concern in Women with Anorexia Nervosa (AN) or Bulimia Nervosa (BN) During Pregnancy

Nutrient	Concern in AN or BN	Dietary Reference Intake[a] during Pregnancy	Role during Pregnancy
Energy and macronutrients			
• Energy	• Energy severely restricted in AN (200–700 kcal per day) or excessive (1,500–9,000 kcal per binge episode, followed by compensatory behavior) with limited energy availability	• +340 kcal per day in second trimester and +452 kcal per day in third trimester	• Energy to supply production and growth of maternal and fetal tissues of pregnancy
• Carbohydrate	• Severely restricted in AN or binged but purged in BN	• Minimum of 175 g per day	• Glucose availability and non-protein energy needs for mother and fetus
• Protein	• Adequate proportion relative to energy intake, but total intake limited in AN	• 71 g per day	• Amino acid supply for maternal and fetal tissue production, maternal blood volume expansion and fluid balance
• Fat (lipids)	• Intake generally avoided or purposefully restricted	• 13 g per day of linoleic acid and 1.4 g per day of alpha-linolenic acid	• Growth, development, and function of fetal nerve and brain tissue, cell membranes, and organs
Vitamins			
• Folate	• Poor intake and subclinical deficiency	• 600 mcg per day	• Fetal neural tube formation
• Pyridoxine (B$_6$)	• Poor intake and subclinical deficiency	• 1.9 mg per day	• Coenzyme for maternal energy metabolism
• Cobalamin (B$_{12}$)	• Poor intake (especially in vegans) and subclinical deficiency	• 2.6 mcg per day	• Required for maternal folate metabolism and DNA and RNA synthesis for fetal tissues
• Vitamin A	• Hypercarotenemia in AN due to catabolism	• 770 mcg retinal activity equivalents per day	• Cellular differentiation for fetal tissue development
Minerals			
• Calcium	• In AN, skeletal calcium stores may be compromised	• 1,000 mg per day	• Fetal skeletal mineralization
• Iron	• Poor intake (especially in vegans)	• 27 mg per day	• Hemoglobin synthesis; support of maternal blood volume expansion

(continued)

Table 9.2
Nutrients of Special Concern in Women with Anorexia Nervosa (AN) or Bulimia Nervosa (BN) During Pregnancy

Nutrient	Concern in AN or BN	Dietary Reference Intake[a] during Pregnancy	Role during Pregnancy
• Zinc	• Poor intake (especially in vegans)	• 11 mg per day	• DNA and RNA synthesis and cofactor for enzymes
• Potassium	• Hypokalemia due to purging and other compensatory behaviors	• 4.7 g per day	• Transmission of nerve impulses; major intracellular cation
• Sodium	• Hyponatremia due to purging and other compensatory behaviors	• 1.5 g per day	• Transmission of nerve impulses; major extracellular cation
• Chloride	• Hypochloremia due to purging and other compensatory behaviors	• 2.3 g per day	• Part of hydrochloric acid in stomach; transmission of nerve impulses; major extracellular anion

Compiled from Food and Nutrition Board, Institute of Medicine (1998) Dietary Reference Intakes for thiamin, riboflavin, niacin, vitamin B_6, folate, vitamin B_{12}, pantothenic acid, biotin, and choline. National Academies Press, Washington, D.C.; Food and Nutrition Board, Institute of Medicine (2001) Dietary Reference Intakes for vitamin A, vitamin K, arsenic, boron, chromium, copper, iodine, iron, manganese, molybdenum, nickel, silicon, vanadium, and zinc. National Academies Press, Washington, D.C.; Food and Nutrition Board, Institute of Medicine (1997) Dietary Reference Intakes for calcium, phosphorus, magnesium, vitamin D, and fluoride. National Academies Press, Washington, D.C.; and Food and Nutrition Board, Institute of Medicine (2004) Dietary Reference Intakes for water, potassium, sodium, chloride, and sulfate. National Academies Press, Washington, D.C.

[a]From [28]

ingested during binge eating episodes have been shown to contain high amounts of carbohydrate [21, 22]. Yet, if purged, malabsorbed, or used to support increased energy expenditure of exercise, this carbohydrate is not readily available to support growth and development of the mother and fetus. Low-carbohydrate diets do not meet the minimum need for carbohydrate during pregnancy (see Chap. 13, "Popular Diets"). Moreover, these diets result in mild ketosis [23], which may pose harm to the fetus [24].

Protein intake is crucial to supply the amino acids needed for production of new tissues and to support blood volume expansion and fluid balance. Restricted intake and purging and nonpurging behaviors do not allow for the provision of adequate protein during pregnancy. Vegetarianism is common in women with eating disorders [25]. Dietary protein and specific amino acid deficiencies may be of concern in women with AN or BN who are also vegetarians (see Chap. 15, "Vegetarian Diets in Pregnancy).

A small portion of total kilocalories as dietary fat is needed to supply the essential fatty acids—linoleic acid and alpha-linolenic acid. These fatty acids are crucial, however, to the growth, development, and function of nerve and brain tissues, cell membranes, and organs. Docosahexaenoic acid (DHA) plays a role in cognitive development and visual acuity. Avoidance of dietary fat in women with AN or BN has been documented [26, 27]. This has implication for overall energy intake as well as absorption, metabolism, and utilization of fat-soluble vitamins.

Balanced energy intake is critical to adequate weight gain and micronutrient availability during pregnancy. In general, total energy intake should be made up of 45–65% carbohydrate, 10–35% protein, and 20–35% lipids or dietary fat [28].

9.5.2 Micronutrients

9.5.2.1 VITAMINS

The B-complex vitamins—folate, pyridoxine (B_6), and cobalamin (B_{12})—are of special concern in pregnant women with AN or BN. The metabolic needs for these vitamins do not appear to be greater in pregnant women with AN or BN compared with pregnant women without these eating disorders. However, due to past and present eating behaviors, women with AN or BN may have subclinical deficiencies prior to pregnancy and poor dietary intakes of these nutrients during pregnancy.

Folic acid supplementation prior to and in early pregnancy has been shown to reduce the risk and incidence of neural tube defect (NTD) in the infant (see Chap. 17, "Folate: a Key to Optimal Pregnancy Outcome"). The risk for NTD incidence was 1.7 times greater in women with eating disorders compared with controls [29]. Moreover, NTD incidence was 2.7 times higher in women who used diuretics compared with control women [29]. Folate intake is generally suboptimal in women with active AN and BN [30].

Pyridoxine and B_{12} intakes are of concern in women with AN and BN as eating pattern data show limited intakes of meat and whole-grain foods [25, 30], the primary sources of these two nutrients. Vitamin B_6 is required for serotonin synthesis. Serotonin is a neurotransmitter that controls satiety and mood, and serotonin deficiency has been implicated in AN, BN, and depression. A lack of dietary B_6 limits the conversion of the amino acid tryptophan to serotonin, leading to poor functioning of the serotonergic transmitter system. Moreover, in niacin deficiency, dietary tryptophan is competitively converted to niacin. Thus, in certain phenotypes, inadequate dietary B_6 intake may precipitate or exacerbate AN or BN and associated mood disorders.

Because a vegetarian pattern of eating is relatively common in AN and BN, consumption of meat, chicken, fish, milk products, and eggs varies widely. Cobalamin deficiency may result in pernicious anemia for the mother and may impair deoxyribonucleic acid (DNA) synthesis and nerve function in the offspring. Inadequate B_{12} intake has been reported in women with AN [30].

Hypercarotenemia is common in AN [31] and results in the yellowish skin tone sometimes observed clinically. Elevated blood carotene is due to catabolism of lipid stores and not excessive intake. During pregnancy, vitamin A needs, inclusive of carotenoids and retinoids, increase slightly to support cellular differentiation and tissue and organ formation. Excessive consumption should be avoided (see Chap. 14, "Dietary Supplements during Pregnancy: Need, Efficacy, and Safety").

9.5.2.2 MINERALS

Notably in the last trimester of pregnancy, maternal skeletal calcium is mobilized and calcium absorption is upregulated to meet the calcium demands of the fetus [32]. An adequate supply of calcium is necessary to support mineralization of the fetal skeleton (see Chap. 14). Osteoporosis risk is significantly greater in women who have experienced amenorrhea due to low estrogen concentration, as in AN. Therefore, women with AN may begin their pregnancies with poor skeletal calcium reserves [33]. At least 1,000 mg of dietary calcium per day are needed during pregnancy for adult women; however, this requirement may be higher in the woman with AN who also has osteopenia or osteoporosis. Dietary calcium intake in those with BN is generally adequate, with the exception of vegan or fruitarian diets (see Chap. 15, "Vegetarian Diets in Pregnancy").

The physiologic need for iron decreases in a woman with AN in relation to the duration of amenorrhea and degree of catabolism. However, during pregnancy, the iron requirement increases by 50%. A woman with AN may have difficulty meeting this requirement during pregnancy. Moreover, in those women with AN who also present with iron-deficiency anemia prior to or during pregnancy, supplemental iron may be necessary to meet the increased need.

Zinc intake is poor in women with AN and BN [30, 34], particularly in vegetarians. This mineral is required for DNA and ribonucleic acid (RNA) synthesis, protein production, and as a cofactor for enzymatic activity.

Electrolyte balance may be moderately to severely disturbed in those who engage in purging, laxative and diuretic use, and excessive exercise. Self-induced vomiting can lead to hypokalemia and hypochloremic alkalosis. Laxative and diuretic abuse is associated with hypokalemia. Excessive exercise may result in hyponatremia and hypokalemia. These compensatory behaviors lead to dehydration, elevation in blood urea nitrogen (BUN), and retardation of glomerular filtration rate. Alterations in maternal kidney function due to AN or BN may impair the efficiency with which fetal waste products are excreted during pregnancy.

9.6 MATERNAL AND FETAL RISKS AND OUTCOMES

Pregnancy represents a life cycle stage during which energy balance is critical to optimal outcomes for both mother and fetus. Body weight and shape changes that occur during pregnancy and serve as indicators of appropriate energy, macronutrient, and micronutrient intakes to support fetal growth and development must be accepted by the

Table 9.3
Adverse Findings during Pregnancy in Women with Anorexia Nervosa (AN) or Bulimia Nervosa (BN)

Maternal outcomes	*Fetal outcomes*
• Inadequate or excessive body weight gain	• Intrauterine growth restriction
• Spontaneous abortion	• Preterm delivery
• Hyperemesis gravidarum	• Low birth weight
• Relapse in AN or BN symptoms	• Small for gestational age
• Worsening of AN or BN symptoms	• Microcephaly
• Cesarean section	• Short body length
• Death	• Neural tube defect
• Postpartum depression	• Other birth defects
	• Poor Apgar scores

woman with AN or BN. A variety of risks and adverse outcomes have been documented in women with AN or BN during pregnancy (Table 9.3).

9.6.1 Findings Related to Anorexia Nervosa

In studies that investigated only AN in pregnant women, inadequate weight gain [35, 36], spontaneous abortion [17], and delivery by cesarean section [17, 36] were common. Other adverse events included hyperemesis gravidarum, intrauterine growth restriction (IUGR), preterm delivery, and low birth weight (LBW) [35–41]. Unexpected maternal death occurred in one pregnant woman with AN [42], and pregnancy and poor postpartum coping appeared to have precipitated AN in another woman [43].

9.6.2 Findings Related to Bulimia Nervosa

Individual case studies and case series reports show a range of maternal and fetal outcomes, largely due to the inconsistencies in use of diagnostic criteria, length of BN history, treatment interventions during pregnancy, and lack of comparison to appropriate controls [44–49]. One consistent finding in these studies, however, was inadequate [45, 46, 48] or excessive [46, 49] weight gain.

Adverse outcomes in larger studies include inappropriate weight gain [50], spontaneous abortion [24, 50, 51], hyperemesis gravidarum, LBW, small for gestational age (SGA), and poor Apgar scores [24, 50, 52].

9.6.3 Findings Related to Anorexia and Bulimia Nervosa

In those studies in which women with AN or BN were investigated together, risk and incidence of inappropriate weight gain [53–55], hyperemesis gravidarum [56], cesarean section [57], preterm delivery [58], LBW [56, 58, 59], SGA [56, 58], small head circumference or microcephaly [56], short body length [59], NTD [29], and other birth defects [57] were high. In general, women who entered pregnancy in remission from their AN or BN had optimal maternal and fetal outcomes [50, 60], while women with active eating disorders prior to conception and during pregnancy fared less well [24, 58].

9.7 CHANGES IN BEHAVIORS

A relapse in eating disorder symptoms in women who were previously in remission may occur during pregnancy [56]. In active AN or BN, body dissatisfaction and low body esteem may worsen during pregnancy [61] in addition to an increased frequency of restricting, binge eating–purging, and nonpurging behaviors [36, 44, 46, 51, 58]. Conversely, AN or BN symptoms and behaviors improved during pregnancy in women receiving treatment [45, 46, 53, 54] and not currently receiving treatment [47, 49, 55, 61–63]. Yet, postpartum resumption of AN and BN behaviors occurred with some regularity [45, 46, 48, 49, 54, 55, 62, 63].

Postpartum depression (PPD) requires assessment in women with AN or BN as this mood disorder is tightly linked to eating disorders [64, 65]. While most studies report an increased incidence of PPD in women with AN or BN [24, 57, 64, 65], one study reported fewer symptoms of depression in women with treated BN who delivered infants compared to women with treated BN who had not given birth [66].

9.8 NUTRITION CARE OF WOMEN WITH ANOREXIA OR BULIMIA NERVOSA DURING PREGNANCY

The first step in the nutrition management of the pregnant woman with AN or BN is identification of the eating disorder. Assuming that prenatal care is sought, many women with AN or BN do not disclose their conditions at any of their prenatal visits [42, 46, 49]. In addition, most obstetricians do not inquire about eating disorders in their patients. For example, only 18% of obstetricians in prenatal clinics questioned their pregnant patients about AN and BN [67]. The secrecy of these disorders and lack of inquiry lead to suboptimal care of these pregnant women.

9.8.1 Assessment

Clinicians may pose several questions to their patients to identify preexisting or newly developed AN or BN (Table 9.4). Once such screening suggests the coexistence of pregnancy and an eating disorder, medical nutrition therapy (MNT) can be applied. As part of MNT, a full nutritional assessment involves systematic collection and evaluation of anthropometric, biochemical, clinical, and dietary intake data. In addition, functional and behavioral status may be evaluated based on responses to screening questions (Table 9.4).

9.8.1.1 ANTHROPOMETRIC DATA

An easily obtained parameter of adequate dietary intake and fetal growth is maternal body weight. Body weight should be measured at each prenatal visit, recognizing that this assessment may make a woman with AN or BN uncomfortable. Some women may even refuse to have body weight measured. In those women who may increase eating disorder behaviors with body weight gain [46, 48, 50], nondisclosure of weight changes may be appropriate. Alternatively, in women who relax their eating disorder behaviors during pregnancy [45–47, 49, 55], discussion of weight changes may provide positive reinforcement of healthy behaviors. Inadequate [35, 36, 45, 46, 48] and excessive [46, 49, 50] weight gain must be tracked and compared to the recommended weight gain based on prepregnancy body mass index (BMI) (see Chap. 2, "Optimal Weight Gain") and any needed nutritional repletion in AN.

Table 9.4
Screening for an Eating Disorder During Pregnancy

Body weight

1. What is your current weight? Current height?
2. What is your usual weight?
3. What was your highest weight as an adult?
4. What was your lowest weight as an adult?
5. Have you had any changes in weight in the last month?
6. Does your weight fluctuate very often? If yes, by how much and how often?
7. How much weight do you think that you will gain during this pregnancy?
8. How much weight do you want to gain during this pregnancy?
9. Do you think that your body shape will change during this pregnancy?
10. What do you think about any body shape changes?

Weight control tactics

Have you previously (or are you currently):

1. Gone on (On) a diet to lose weight?
2. Fasted (Fasting) for more than or equal to eight hours (other than during sleep)?
3. Made yourself vomit (Vomiting frequently)?
4. Used (Using) laxatives? If yes, how often?
5. Used (Using) diuretics? If yes, how often?
6. Eaten (Eating) very large amounts of food in a short period of time? If yes, how often?
7. Restricted (Restricting) the amount of food or beverages that you consume(d)? If yes, how often?
8. Eaten (Eating) in private or in secret?
9. Avoided (Avoiding) certain types of food?
10. Engaged (Engaging) in exercise? If yes, what type, frequency, duration, and intensity?

Dietary intake

1. How frequently do you eat foods and drink beverages?
2. Do you ever skip meals?
3. Do you have any food allergies?
4. Do you have any food cravings?
5. Do you have any food aversions?
6. Have you had morning sickness?
7. Do you drink fluids in place of solid foods or meals?
8. Do you take any vitamin and/or mineral supplements?
9. Do you take any other supplements such as herbal products?
10. Do you use sugar substitutes or fat substitutes?

General health

1. Were your menstrual cycles regular prior to this pregnancy?
2. Have you experienced constipation or diarrhea?
3. Have you experienced heartburn?
4. Do you feel "stressed" or anxious?
5. Do you think that you are "retaining fluid"?
6. Have you felt weak or light-headed?
7. Are you experiencing frequent urination?
8. Are you taking any over-the-counter or prescription medications?
9. Do you plan to breastfeed your infant?
10. How does this pregnancy compare to your previous pregnancy (pregnancies)?

Body weight and weight gain may be affected by the patient's hydration status, glycogen stores (in AN), and changes in lean and fat mass. Body composition testing during pregnancy may be performed by bioelectrical impedance analysis but is significantly affected by hydration status. Skin fold and circumferential measurements will be affected by differential changes in maternal body fat deposition. Dual-energy X-ray absorptiometry should not be performed during pregnancy. Pitting edema may be a sign of BN.

9.8.1.2 BIOCHEMICAL OR LABORATORY DATA

Biochemical or laboratory values are generally normal in women with AN or BN. During semistarvation in AN, catabolic and compensatory mechanisms mobilize tissue stores, releasing nutrients to the serum pool. As a result, hypercarotenemia is often found in women with moderate to severe AN. Yet when serum concentrations of nutrients are low, severe AN is likely. At this point, several B vitamins, including B_6 and B_{12}, and minerals, such as zinc, will show signs of depletion. Dehydration in either AN or BN may falsely normalize or elevate several biochemical markers of nutritional status, such as serum albumin and iron. Thus, establishing normal hydration is important for accurate nutrition assessment. Vitamin and mineral supplement use is common in AN and BN and may mask nutrient deficiencies. Elevated blood lipids may be noted in the majority of women, due to liver and hypothalamic dysfunction in AN and inappropriate intake of dietary fats or lipids in AN or BN during binge eating.

9.8.1.3 CLINICAL DATA

Upon clinical assessment, signs and symptoms of AN or BN will be present (see Sect. 9.2 and Table 9.1). Bleeding gums or sensitive teeth may present as new symptoms or worsened conditions. Assessment instruments such as the Eating Disorders Examination [68] may also be useful when evaluating the full clinical picture and relate to the functional and behavioral aspects of AN or BN.

9.8.1.4 DIETARY INTAKE DATA

Methods designed to gather dietary intake information include dietary history, food frequency questionnaire, 24-h recall, and food diary or record. A variety of techniques for data collection, including written, computerized, and Web based have been used with validated instruments. Dietary intake data may be evaluated for energy, macronutrient, and micronutrient intakes; food patterns; food groups; and/or food variety. Comparison to established standards is important, as is comparison to the woman's previous intake. Such evaluation will identify foods, nutrients, and/or eating behaviors of concern as well as areas where improvements have been made.

Inquiry about eating behaviors may also uncover related issues such as food cravings or aversions, timing and triggers of intake, and fasting and ritualistic behaviors. These may be linked to dental problems, morning sickness, hyperemesis gravidarum, gastrointestinal symptoms, and mood changes during pregnancy.

9.8.2 Nutrition Diagnosis

Based on the nutrition assessment, the registered dietitian can establish a nutrition diagnosis (or diagnoses). Such statements identify nutrition problems (diagnostic labels), related etiology, and distinguishing characteristics (signs and symptoms) and are important to document in the medical record. These nutrition diagnoses serve as the foundation for nutrition interventions, monitoring and evaluation, and anticipated

Table 9.5

Examples of Nutrition Diagnostic Statements for Women with Anorexia Nervosa or Bulimia Nervosa during Pregnancy

- Inadequate energy intake (problem) related to restriction of food intake (etiology) as evidenced by mean daily dietary intake of less than 1,000 kcal per day and 7-lb weight loss during the past 6 weeks (signs)
- Frequent stool output (problem) related to thrice daily use of laxatives and low–dietary fiber intake (etiology) as evidenced by increased diarrhea, dehydration, and estimated average daily dietary intake of fiber of less than 2 g per day (signs and symptoms)
- Suboptimal folate intake (problem) related to avoidance of leafy green vegetables, orange fruits, and fortified grain products (etiology) as evidenced by red blood cell folate of 2.7 nmol/l (sign)
- Hypokalemia (problem) related to increased frequency of self-induced vomiting (etiology) as evidenced by change in self-reported behaviors and serum potassium of 2.9 mmol/l (sign)
- Rapid weight gain (problem) related to relaxation of compensatory behaviors (etiology) as evidenced by self-reported change in behaviors, weight gain of 12 lb from the 14th to 18th week of gestation, and average dietary intake of approximately 1,000 kcal per day beyond estimated energy needs (signs)

outcomes. Table 9.5 presents examples of nutrition diagnostic statements for AN or BN during pregnancy.

9.9 NUTRITION INTERVENTION

Working from nutrition diagnoses, areas for nutrition intervention that will positively alter behaviors, reduce risks, and improve and/or promote the health of the mother and fetus can be identified. Pregnancy can provide a unique opportunity to improve AN or BN behaviors if interventions focus on fetal nutritional requirements [45, 47, 55], fetal growth and development [45, 46], and relationships among maternal body weight gain, shape changes, and fetal growth [47]. Planning individualized, patient-focused care, activities, and expected outcomes is essential. The overall goal of nutrition intervention is to "promote the consumption of foods that will best meet the nutritional requirements of pregnancy, essential for fetal growth and development, within the context of the woman's often uncontrolled [or overly restricted] eating" [60, p 452].

The primary objective for AN is to gradually increase energy intake to support a positive energy balance to allow repletion of the mother while meeting fetal energy demands. An intake of 130% of estimated energy needs is initially recommended. Reaching this goal should be attained through incremental increases of 100–200 kcal per day approximately twice per week. In the first trimester, additional kilocalories are not needed to support fetal growth and development; however, maternal weight gain of one to two pounds per week may be expected due to repletion of maternal energy stores. During the second and third trimesters, energy intake should increase beyond maternal repletion needs to supply requirements of the fetus (see Table 9.2). Frequent recalculation of estimated energy needs is necessary to adjust for changes in body composition, basal metabolic rate, and energy expenditure, including physical activity.

The primary objective for BN is to disrupt binge eating–purging episodes and eating restraint so that intake becomes more consistent, and to stop other compensatory behaviors to achieve a stable energy and nutrient supply. In the first trimester, when additional energy is not required, body weight stabilization is critical. Approximately 100–130% of estimated energy needs are recommended, depending on prepregnancy BMI, weight fluctuations, and energy expenditure of physical activity. Through the second and third trimesters, additional energy intake should match recommended increases. In those women with BN who are also overweight or obese, dietary recommendations specific to these conditions should also be considered when setting energy intake levels (see Chap. 5, "Obesity and Pregnancy").

As stated above, macronutrient distribution of total energy in both AN or BN should be made up of 45–65% carbohydrate, 10–35% protein, and 20–35% dietary fat or lipids. Adjustments may be needed based on food aversions, gastrointestinal complaints, continued binge eating–purging episodes, or other issues.

Vitamin and mineral supplementation is warranted in pregnant women with AN or BN. A prenatal supplement that meets but does not exceed 100% of the Dietary Reference Intake for micronutrients for adult women is suggested to allow for consumption of food-based nutrients and to avoid excessive intakes that may potentially occur from binge eating. A thorough discussion of prenatal supplements is found in Chap. 14.

The registered dietitian should involve the patient in menu planning and food selection. Emphasis on specific micronutrient intake is important to stress the relationship of these nutrients to optimal fetal growth and development.

Nutrition education is a vital intervention component. Most women with eating disorders are well versed in nutrition facts and knowledge. However, they may be less aware of nutrition needs for healthy pregnancies. Discussion of micronutrient requirements and roles of these nutrients in fetal growth and development may redirect the mother's preoccupation with body weight and shape to fetal needs for intrauterine health. Other important nutrition education topics are listed in Table 9.6.

Table 9.6
Nutrition Education Topics during Pregnancy in Women with Anorexia Nervosa or Bulimia Nervosa

- Body weight gain: where does this weight go?
- Pregnancy outcomes with maternal malnutrition: what are the risks?
- Behavioral strategies to improve eating and intake
- Menu planning, food choices, and portion sizes
- Differences between nutrients from foods and from prenatal supplements
- Folate intake and neural tube defect
- Meeting fetal nutrient needs with a vegetarian diet
- Alcohol intake: what are the effects?
- Caffeine intake: what are the effects?
- Distinguishing morning sickness from self-induced vomiting
- Severe vomiting, prolonged morning sickness, and hyperemesis gravidarum
- Managing constipation, diarrhea, hemorrhoids, and heartburn
- Pica practices
- Exercise recommendations
- Postpartum body weight loss: what can be expected?
- Planning for lactation
- Energy and nutrient needs of lactation
- Postpartum eating: maintaining healthy habits

9.10   MONITORING AND EVALUATION

At each prenatal visit, eating disorders screening may be conducted (see Table 9.4) along with measurement and documentation of parameters or outcomes related to nutrition interventions and diagnoses. Body weight and rate of weight gain should be tracked and evaluated. Adjustments in energy intake should be based on appropriateness of weight changes. Eating behaviors and dietary intake should be examined at each prenatal visit to assess the adequacy of dietary composition and patterns of intake. Changes in purging and nonpurging behaviors should be noted and addressed. Fingersticks to check hematocrit and glucose may be useful in the monitoring of iron status and hypoglycemia or hyperglycemia. In women with established eating disorders, urinalysis may detect starvation or dehydration as noted by urinary ketones, elevated specific gravity, and alkaline urine. Vital signs will show any change in general health status. Glucose tolerance testing should be conducted in the 24th to 28th week of pregnancy to screen for gestational diabetes mellitus (see Chap. 10, "Diabetes and Pregnancy"). Resolution of any nutrition diagnoses should be documented and any new issues addressed.

More aggressive and intensive inpatient care may be warranted if monitoring and evaluation shows a worsening of the eating disorder, IUGR, or other fetal growth and development problems. In AN or BN, a reduction in body weight to less than 75% of expected; hypokalemia, hyponatremia, or hypochloremic alkalosis; dehydration; hyperemesis gravidarum; cardiovascular changes; prolonged fasting; uncontrolled binge eating–purging cycles; severe depression; suicidal ideation; and any obstetrical complication are justification for hospitalization.

9.11   PLANNING FOR POSTPARTUM CARE

Relapses in eating disorders often occur in the postpartum period [46–50, 55]. Moreover, the rate of PPD in women with eating disorders is high (see Chap. 19, "Postpartum Depression and the Role of Nutritional Factors"). Changes in estrogen status and estrogen-beta-receptor function or other gene–nutrient interactions may be responsible for observed relapses. The registered dietitian should work closely with the patient toward the end of pregnancy to set realistic goals for dietary intake, weight loss, eating behaviors, and expectations during lactation.

9.11.1   Interdisciplinary Care

Nutrition care is but one part of treatment for AN or BN. These complex disorders require multidisciplinary and integrated care, due to the multifactorial etiology and wide scope of signs and symptoms. The obstetrician, nurse practitioner, psychologist or psychiatrist, dietitian, dentist, social worker, family therapist, occupational therapist, pharmacist, certified exercise physiologist, and other allied health care professionals must openly and cohesively interact with one another and most importantly with the patient to provide effective treatment. Cognitive-behavioral therapy is used to modify anorexic and bulimic behaviors. Medications may be used in treatment, but a risk–benefit assessment for use during pregnancy should be completed (Table 9.7). An increased frequency of prenatal visits is warranted in these high-risk conditions. Monitoring of fetal heart rate and more frequent ultrasounds may shift the center of attention from the mother's AN or BN

Table 9.7
Selected Medications Used in the Treatment of Anorexia Nervosa or Bulimia Nervosa

Medication	Classification	Drug–nutrient interactions	Food and Drug Administration pregnancy category*
• Desipramine	• Antidepressant, tricyclic	• Limit caffeine • Increase riboflavin intake • Avoid alcohol • Incompatible with lactation	B
• Fluoxetine	• Antidepressant, antibulimic, selective serotonin reuptake inhibitor	• Avoid tryptophan supplements • Avoid alcohol • Incompatible with lactation	C
• Nortriptyline	• Antidepressant, tricyclic	• Limit caffeine • Increase riboflavin intake • Avoid alcohol • Incompatible with lactation	C
• Paroxetine	• Antidepressant, selective serotonin reuptake inhibitor	• Avoid tryptophan supplements • Avoid alcohol • Incompatible with lactation	C
• Phenelzine	• Antidepressant, monoamine oxidase inhibitor	• Avoid high-tyramine-containing foods, such as aged cheeses, avocados, grapes, prunes raisins, beef liver, soy sauce, nuts, chocolate, and Chianti wine, among other foods • Limit caffeine • Avoid tryptophan supplements • Increase pyridoxine (B_6) intake • Avoid alcohol • Incompatible with lactation	C
• Tranylcypromine	• Antidepressant, monoamine oxidase inhibitor	• Avoid high-tyramine-containing foods • Limit caffeine • Avoid tryptophan supplements • Avoid alcohol • Incompatible with lactation	C

*Category A includes drugs which were shown to have no increased risk of fetal abnormalities in well-controlled studies including pregnant women; Category B includes drugs which failed to demonstrate any risk to the fetus in well-controlled studies including pregnant women, although animal studies demonstrated an adverse effect or animal studies resulted in no harm to the fetus, but well-controlled studies including pregnant women were not available; Category C includes drugs which resulted in harm to the fetus in animal studies and well-controlled studies including pregnant women were not available or no animal and well-controlled studies including pregnant women have been conducted; Category D includes drugs which resulted in risk to the fetus in well-controlled or observational studies including pregnant women; however, the benefits of the drug may outweigh risk of harm to the fetus; Category X includes drugs which produced fetal abnormalities in well-controlled studies including pregnant women or animals, and the use of the drug is not recommended during pregnancy or by women who may become pregnant.

behaviors to the growing fetus. An informal or formal support network that includes friends, family members, and possibly other patients can provide more constant reassurance, advice, assistance, and positive reinforcement, often valued by women with AN or BN.

9.12 CONCLUSION

Women with active AN or BN during pregnancy are at high risk for adverse outcomes. Ideally, treatment of the AN or BN should occur prior to conception. If not feasible, screening for and assessment of eating disorders during prenatal visits is critical. If an eating disorder is detected, then interdisciplinary care is vital to address all medical issues of the mother and developing fetus. Nutrition requirements of both the mother and fetus must be addressed, and eating patterns and behaviors that optimize a consistent and appropriate stream of nutrients to mother and fetus are key components of care. Treatment of the woman with AN or BN during pregnancy should not end at delivery, but rather, must continue into the postpartum period and beyond.

9.13 CASE STUDY: BULIMIA NERVOSA DURING PREGNANCY

T.J. is a 32-year-old Caucasian, married woman, gravida 2, para 1, seeking prenatal care in the 11th week of gestation. Medical history reveals current BN, the onset of which occurred in the third month postpartum of her previous pregnancy. Since the onset of BN at age 27, T.J. has engaged in binge eating–purging cycles at least twice per day, consuming approximately 2,200 kcal of high-fat, high-carbohydrate snack-type foods during each binge with subsequent vomiting. She reports "problems with my teeth" and "frequent heartburn." T.J. denies laxative, diuretic, or enema use, but admits to moderate exercise of "fast-paced walking" of up to 2 h per day. She was dissatisfied with her body shape and inability to quickly lose weight after her first pregnancy and is fearful that she will lose control of her body weight during this pregnancy. She gained 47 lb during her first pregnancy. T.J. currently weighs 145 lb and is 5′ 7″. Laboratory values are within normal limits. She reports having the "baby blues" after her first delivery and "frustration" with her husband who "travels too much to be of any help with our child." T.J. has not confided in her husband regarding her BN and engages in binge eating–purging episodes "in secret."

1. Calculate T.J.'s body mass index and determine an appropriate weight gain for T.J. for her current pregnancy.
2. Estimate T.J.'s energy needs for weight maintenance and weight gain during pregnancy.
3. With T.J., plan a 7-day menu that includes appropriate food choices to meet nutrient needs of pregnancy and strategies to avoid binge eating.
4. Identify potential adverse outcomes for T.J. and her fetus if BN continues during this pregnancy.
5. Discuss the impact of T.J.'s exercise habits on her energy needs and course of pregnancy.
6. Establish criteria to monitor and evaluate T.J.'s BN during pregnancy on an outpatient basis. Identify key indicators that will be used to determine if inpatient care is needed.
7. List all of the health professionals and others who should be involved in T.J.'s prenatal care and provide reasons for their involvement.

REFERENCES

1. Stein AD, Ravelli AC, Lumey LH (1995) Famine, third-trimester pregnancy weight gain, and intrauterine growth: the Dutch Famine Birth Cohort Study. Hum Biol 67:135–150
2. American Psychiatric Association (1994) Diagnostic and statistical manual of mental disorders, 4th ed. American Psychiatric Association, Washington, D.C.
3. Hoek HW, van Koeken D (2003) Review of the prevalence and incidence of eating disorders. Int J Eat Disord 34:383–396
4. Marcus MD, Levine MD (1998) Eating disorder treatment: an update. Curr Opin Psychiatry 11:159–163
5. Steinhausen HC (2002) The outcome of anorexia nervosa in the 20th century. Am J Psychiatry 159:1284–1293
6. Muscari ME (1998) Screening for anorexia and bulimia. Am J Nurs 98:22–24
7. James DC (2001) Eating disorders, fertility, and pregnancy: relationships and complications. J Perinat Neonat Nurs 15:36–48
8. Quadflieg N, Fichter MM (2003) The course and outcome of bulimia nervosa. Eur Child Adolesc Psychiatry 12[Suppl 1]:99–109
9. Bulik CM, Sullivan PF, Wade TD, Kendler KS (2000) Twin studies of eating disorders: a review. Int J Eat Disord 27:1–20
10. Gorwood P, Kipman A, Foulon C (2003) The human genetics of anorexia nervosa. Eur J Pharmacol 480:163–170
11. Park RJ, Senior R, Stein A (2003) The offspring of mothers with eating disorders. Eur Child Adolesc Psychiatry 12[Suppl 1]:110–119
12. Wolfe BE (2005) Reproductive health in women with eating disorders. J Obstet Gynecol Neonatal Nurs 34:255–263
13. Patel P, Wheatcroft R, Park RJ, Stein A (2002) The children of mothers with eating disorders. Clin Child Fam Psychol Rev 5:1–19
14. Stewart DE, Robinson E, Goldbloom DS, Wright C (1990) Infertility and eating disorders. Am J Obstet Gynecol 163:1196–1199
15. Finfgeld DL (2002) Anorexia nervosa: analysis of long-term outcomes and clinical implications. Arch Psychiatr Nurs 16:176–186
16. Rome ES (2003) Eating disorders. Obstet Gynecol Clin North Am 30:353–377
17. Bulik CM, Sullivan PF, Fear JL, Pickering A, Dawn A, McCullin M (1999) Fertility and reproduction in women with anorexia nervosa: a controlled study. J Clin Psychiatry 60:130–135
18. Crow SJ, Thuras P, Keel PK, Mitchell JE (2002) Long-term menstrual and reproductive function in patients with bulimia nervosa. Am J Psychiatry 159:1048–1050
19. Johnson AA, Knight EM, Edwards CH, Oyemade UJ, Cole OJ, Westney OE, Westney LS, Laryea H, Jones S (1994) Dietary intakes, anthropometric measurements and pregnancy outcomes. J Nutr 124[Suppl 6]:936S–942S
20. Abrams BF, Laros RK (1986) Prepregnancy weight, weight gain, and birth weight. Am J Obstet Gynecol 154:503–509
21. Wallin G van der Ster, Norring C, Holmgren S (1994) Binge eating versus nonpurged eating in bulimics: is there a carbohydrate craving after all? Acta Psychiatr Scand 89:376–381
22. Hadigan CM, Kissileff HR, Walsh BT (1989) Patterns of food selection during meals in women with bulimia. Am J Clin Nutr 50:759–766
23. Coleman MD, Nickols-Richardson SM (2005) Urinary ketones reflect serum ketone concentration but do not relate to weight loss in overweight premenopausal women following a low-carbohydrate/high-protein diet. J Am Diet Assoc 105:608–611
24. Abraham S (1998) Sexuality and reproduction in bulimia nervosa patients over 10 years. J Psychosom Res 44:491–502
25. Bakan R, Birmingham CL, Aeberhardt L, Goldner EM (1993) Dietary zinc intake of vegetarian and nonvegetarian patients with anorexia nervosa. Int J Eat Disord 13:229–233
26. Fernstrom MH, Weltzin TE, Neuberger S, Srinivasagam N, Kaye WH (1994) Twenty-four-hour food intake in patients with anorexia nervosa and in healthy control subjects. Biol Psychiatry 36:696–702

27. Rolls BJ, Andersen AE, Moran TH, McNelis AL, Baier HC, Fedoroff IC (1992) Food intake, hunger, and satiety after preloads in women with eating disorders. Am J Clin Nutr 55:1093–1103

28. Food and Nutrition Board, Institute of Medicine (2005) Dietary Reference Intakes for energy, carbohydrate, fiber, fat, fatty acids, cholesterol, protein, and amino acids. National Academies Press, Washington, D.C.

29. Carmichael SL, Shaw GM, Schaffer DM, Laurent C, Selvin S (2003) Dieting behaviors and risk of neural tube defects. Am J Epidemiol 158:1127–1131

30. Hadigan CM, Anderson EJ, Miller KK, Hubbard JL, Herzog DB, Klibanski A, Grinspoon SK (2000) Assessment of macronutrient and micronutrient intake in women with anorexia nervosa. Int J Eat Disord 28:284–292

31. Boland B, Beguin C, Zech F, Desager JP, Lambert M (2001) Serum beta-carotene in anorexia nervosa patients: a case-control study. Int J Eat Disord 30:299–305

32. Cross NA, Hillman LS, Allen SH, Krause GF, Vieira NE (1995) Calcium homeostasis and bone metabolism during pregnancy, lactation, and postweaning: a longitudinal study. Am J Clin Nutr 61:514–523

33. Ward A, Brown N, Treasure J (1997) Persistent osteopenia after recovery from anorexia nervosa. Int J Eat Disord 22:71–75

34. Gendall KA, Sullivan PE, Joyce PR, Carter FA, Bulik CM (1997) The nutrient intake of women with bulimia nervosa. Int J Eat Disord 21:115–127

35. Milner G, O'Leary MM (1988) Anorexia nervosa occurring in pregnancy. Acta Psychiatr Scand 77:491–492

36. Sengupta-Giridharan R, Settatree RS, Jones A (2003) Complex long-term eating disorder, Bartter's syndrome and pregnancy: a rare combination. Aust N Z J Obstet Gynaecol 43:384–385

37. Hart T, Kase N, Kimball CP (1970) Induction of ovulation and pregnancy in patients with anorexia nervosa. Am J Obstet Gynecol 108:580–584

38. Ho E (1985) Anorexia nervosa in pregnancy. Nurs Mirror 160:40–42

39. Strimling BS (1984) Infant of a pregnancy complicated by anorexia nervosa. Am J Dis Child 138:68–69

40. Treasure JL, Russell GF (1988) Intrauterine growth and neonatal weight gain in babies of women with anorexia nervosa. Br Med J (Clin Res Ed) 296:1038

41. Weinfeld RH, Dubay M, Burchell RC, Millerick JD, Kennedy AT (1977) Pregnancy associated with anorexia and starvation. Am J Obstet Gynecol 129:698–699

42. Ahmed S, Balakrishnan V, Minogue M, Ryan CA, McKiernan J (1999) Sudden maternal death in pregnancy complicated by anorexia nervosa. J Obstet Gynaecol 19:529–531

43. Benton-Hardy LR, Lock J (1998) Pregnancy and early parenthood: factors in the development of anorexia nervosa? Int J Eat Disord 24:223–226

44. Conrad R, Schablewski J, Schilling G, Liedtke R (2003) Worsening of symptoms of bulimia nervosa during pregnancy. Psychosomatics 44:76–78

45. Feingold M, Kaminer Y, Lyons K, Chaudhury AK, Costigan K, Cetrula CL (1988) Bulimia nervosa in pregnancy: a case report. Obstet Gynecol 71:1025–1027

46. Hollifield J, Hobdy J (1990) The course of pregnancy complicated by bulimia. Psychotherapy 27:249–255

47. Lacey JH, Smith G (1987) Bulimia nervosa. The impact of pregnancy on mother and baby. Br J Psychiatry 150:777–781

48. Price WA, Giannini AJ, Loiselle RH (1986) Bulimia precipitated by pregnancy. J Clin Psychiatry 47:275–276

49. Ramchandani D, Whedon B (1988) The effect of pregnancy on bulimia. Int J Eat Disord 7:845–848

50. Stewart DE, Raskin J, Garfinkel PE, MacDonald OL, Robinson GE (1987) Anorexia nervosa, bulimia, and pregnancy. Am J Obstet Gynecol 157:1194–1198

51. Mitchell JE, Seim HC, Glotter D, Soll EA, Pyle RL (1991) A retrospective study of pregnancy in bulimia nervosa. Int J Eat Disord 10:209–214

52. Conti J, Abraham S, Taylor A (1998) Eating behavior and pregnancy outcome. J Psychosom Res 44:465–477

53. Lemberg R, Phillips J (1989) The impact of pregnancy on anorexia nervosa and bulimia. Int J Eat Disord 8:285–295

54. Namir S, Melman KN, Yager J (1986) Pregnancy in restricter-type anorexia nervosa: a study of six women. Int J Eat Disord 5:837–845

55. Willis DC, Rand CSW (1988) Pregnancy in bulimic women. Obstet Gynecol 71:708–710
56. Kouba S, Hallstrom T, Lindholm C, Hirschberg AL (2005) Pregnancy and neonatal outcomes in women with eating disorders. Obstet Gynecol 105:255–260
57. Franko DL, Blais MA, Becker AE, Delinsky SS, Greenwood DN, Flores AT, Ekeblad ER, Eddy KT, Herzog DB (2001) Pregnancy complications and neonatal outcomes in women with eating disorders. Am J Psychiatry 158:1461–1466
58. Sollid CP, Wisborg K, Hjort J, Secher NJ (2004) Eating disorder that was diagnosed before pregnancy and pregnancy outcome. Am J Obstet Gynecol 190:206–210
59. Waugh E, Bulik CM (1999) Offspring of women with eating disorders. Int J Eat Disord 25:123–133
60. Morrill ES, Nickols-Richardson SM (2001) Bulimia nervosa during pregnancy: a review. J Am Diet Assoc 101:448–454
61. Crow SJ, Keel PK, Thuras P, Mitchell JE (2004) Bulimia symptoms and other risk behaviors during pregnancy in women with bulimia nervosa. Int J Eat Disord 36:220–223
62. Blais MA, Becker AE, Burwell RA, Flores AT, Nussbaum KM, Greenwood DN, Ekeblad ER, Herzog DB (2000) Pregnancy: outcome and impact on symptomatology in a cohort of eating-disordered women. Int J Eat Disord 27:140–149
63. Morgan JF, Lacey JH, Sedgwick PM (1999) Impact of pregnancy on bulimia nervosa. Br J Psychiatry 174:135–140
64. Abraham S, Taylor A, Conti J (2001) Postnatal depression, eating, exercise, and vomiting before and during pregnancy. Int J Eat Disord 29:482–487
65. Mazzeo SE, Slof-Op't Landt MCT, Jones I, Mitchell K, Kendler KS, Neale MC, Aggen SH, Bulik CM (2006) Associations among postpartum depression, eating disorders, and perfectionism in a population-based sample of adult women. Int J Eat Disord 39:202–211
66. Carter FA, McIntosh VVW, Joyce PR, Frampton CM, Bulik CM (2003) Bulimia nervosa, childbirth, and psychopathology. J Psychosom Res 55:357–361
67. Abraham S (2001) Obstetricians and maternal body weight and eating disorders during pregnancy. J Psychosom Obstet Gynecol 22:159–163
68. Wilfley DE, Schwartz MB, Spurrell EB, Fairburn CG (2000) Using the eating disorder examination to identify the specific psychopathology of binge eating disorder. Int J Eat Disord 27:259–269

ADDITIONAL RESOURCES

Katz MG, Vollenhoven B (2000) The reproductive endocrine consequences of anorexia nervosa. Br J Obstet Gynaecol 107:707–713
Mitchell-Gieleghem A, Mittelstaedt ME, Bulik CM (2002) Eating disorders and childbearing: concealment and consequences. Birth 29:182–191
Rocco PL, Orbitello B, Perini L, Pera V, Ciano RP, Balestrieri M (2005) Effects of pregnancy on eating attitudes and disorders: a prospective study. J Psychosom Res 59:175–179
Spear BA, Myers ES (2001) Position of the American Dietetic Association: nutrition intervention in the treatment of anorexia nervosa, bulimia nervosa, and eating disorders not otherwise specified (EDNOS). J Am Diet Assoc 101:810–819

The American Dietetic Association	www.eatright.org
The American Psychiatric Association	www.psych.org
Anorexia Nervosa and Related Eating Disorders, Inc.	www.anred.com
National Association of Anorexia Nervosa and Associated Disorders	www.anad.org
National Eating Disorders Association	www.edap.org

10 Diabetes and Pregnancy

Alyce M. Thomas

Summary Diabetes mellitus is the most common complication in pregnancy, affecting nearly 8% of all pregnancies. Nearly 90% of women with diabetes develop the condition during pregnancy; diabetes in the other 10% antedated the pregnancy. Since the discovery of insulin, perinatal mortality rates for women with diabetes have decreased, however, infant morbidity remains higher than in the nondiabetes pregnant population.

Diabetes in pregnancy can be classified as type 1 diabetes, type 2 diabetes, and gestational diabetes mellitus (GDM). Type 1 diabetes is characterized by insulin deficiency caused by autoimmune destruction of the pancreatic beta-cells. Type 2 diabetes is associated with insulin resistance and obesity rather than insulin deficiency. GDM is defined as glucose intolerance with onset or first recognition during pregnancy.

The risk for maternal and fetal complications decreases if the woman is in optimal blood glucose control during pregnancy. Women with preexisting diabetes should receive preconceptional counseling during their childbearing years to achieve and maintain glycemic control and to address medical conditions that could affect the pregnancy.

Self-management is the key to reducing the risks associated with diabetes and pregnancy. For women with preexisting diabetes, these include medical nutrition therapy (MNT), insulin therapy, self-monitoring of blood glucose and ketones, and physical activity. Current nutrition recommendations for the treatment of diabetes may be used for pregnant women with type 1 diabetes and type 2 diabetes. MNT is the cornerstone of treatment for women with GDM. Most women with GDM can control their blood glucose by following a carbohydrate modified meal plan that also provides sufficient energy and nutrients to promote maternal and fetal health. Occasionally, medication may be added to maintain optimal blood glucose control.

Breastfeeding is not contraindicated in GDM and should be encouraged. Women with GDM in a previous pregnancy are at risk of developing this condition in subsequent pregnancies and type 2 diabetes later in life. Women should be encouraged to develop healthy lifestyles to decrease their risk of developing diabetes-related conditions.

Keywords: Preexisting diabetes, Gestational diabetes mellitus, Congenital anomaly, Insulin resistance, Insulin sensitivity, Normoglycemia

From: *Nutrition and Health: Handbook of Nutrition and Pregnancy*
Edited by: C.J. Lammi-Keefe, S.C. Couch, E.H. Philipson © Humana Press, Totowa, NJ

10.1 INTRODUCTION

Diabetes mellitus affects 20.8 million or 7% of the United States' population, with 14.6 million diagnosed cases and 6.2 million unaware they have the disease [1]. It is the most common complication of pregnancy, estimated at 8% of all pregnancies or more than 200,000 cases annually [2]. Nearly 90% will develop diabetes during pregnancy [3]; the other 10% had diabetes that predated the pregnancy [2]. Although perinatal morbidity and mortality have decreased in the last 80 years, the prevalence of fetal complications in women with diabetes is higher than in women without diabetes. With intensive management and optimal glycemic control, prior to and throughout pregnancy, women with diabetes can reduce their risk of perinatal complications.

10.2 HISTORICAL BACKGROUND

Before 1921, women with diabetes were advised to avoid pregnancy or to abort if they conceived because of adverse perinatal outcomes. If the pregnancies advanced to the stage of fetal viability, the infants were often stillborn or were born with major malformations. Medical nutrition therapy was the primary method of management for pregnant women with diabetes prior to 1921; however, the diets were often severely restricted or nutritionally unbalanced. These dietary approaches varied from high carbohydrate-low protein, or high protein–high fat, to brief periods of starvation [4, 5]. Alcohol was often included because of its calming effect on the mother [6].

Although insulin injections revolutionized diabetes management, nutrition therapy remained virtually unchanged in the early years after its discovery. In 1937, Priscilla White, a physician at the Joslin Diabetes Center in Boston, Mass., developed a new meal plan, which consisted of 30 kcal/kg body weight, 1 g protein/kg actual body weight, and 180–250 g carbohydrate with the remainder as fat [7]. Other researchers used similar meal plans to achieve maternal blood glucose control [8, 9].

During the 1950–1960s, health care providers were concerned with the risk of macrosomia and hypertension in pregnancy. Weight gain and sodium were restricted to less than 15 lb and 2 g, respectively, in all pregnant women. After the publication of *Maternal Nutrition and the Course of Pregnancy* in 1970 [10], weight gain recommendations were increased to 22–30 lb, and the sodium restriction was discontinued. This comprehensive literature review found no evidence to support the restriction of weight or sodium in pregnancy. However, weight gain and sodium restrictions for pregnant women with diabetes continued until 1970, when the American Diabetes Association recommended the same regimen for pregnant women with diabetes as for the general pregnant population [11]. Today, pregnant women with and without diabetes follow the same weight gain recommendations.

10.3 CLASSIFICATION OF DIABETES

The American Diabetes Association defines diabetes mellitus as a group of metabolic diseases characterized by hyperglycemia resulting from defects in insulin secretion, insulin action or both [12]. The main classification of diabetes mellitus is type 1, type 2, and GDM.

Type 1 diabetes, formerly known as insulin-dependent or juvenile-onset diabetes, is characterized by autoimmune destruction of the pancreatic beta-cells and accounts for 5–10% of all diabetes cases. Type 1 diabetes requires exogenous insulin for survival and is diagnosed primarily in persons less than 30 years of age. Type 2 diabetes, which accounts for almost 90% of diabetes cases, was previously known as adult-onset or non-insulin dependent diabetes. Insulin resistance rather than insulin deficiency and obesity are associated with type 2 diabetes. GDM is defined as any degree of glucose intolerance with onset or first recognition during pregnancy. The definition applies if medication or MNT is used in treatment or the condition persists after pregnancy. It does not exclude the possibility that the diabetes may have existed prior to pregnancy.

Diabetes in pregnancy is classified whether the condition predated (type 1 diabetes or type 2 diabetes) or was diagnosed during pregnancy (GDM). Other classifications have been used to identify risk factors associated with diabetes in pregnancy, including age of onset, presence of preexisting complications and degree of metabolic control [13, 14].

10.4 PREEXISTING DIABETES

10.4.1 Pathophysiology of Normal Pregnancy

During pregnancy, the fetus receives nutrients from across the placenta, including glucose, amino acids, and fatty acids via either active transport or facilitated diffusion. In the first trimester, maternal glycogen storage and endogenous glucose production increase. Pregnancy hormones (human placental lactogen and cortisol), estrogen, progesterone, and the constant fetal demand of glucose lower fasting maternal blood glucose levels [15, 16]. The maternal appetite is stimulated resulting in consumption of additional calories. Fasting and postprandial glucose levels rise in response to the extra glucose required for fetal growth. Elevated hormonal levels increase insulin resistance and the beta-cells produce and secrete additional insulin as glucose is transported across the placenta. Insulin resistance peaks by the latter part of the third trimester, which is characterized by a three-fold increase in insulin production and secretion. After delivery, insulin production returns to prepregnancy levels.

Other hormones thought to affect insulin resistance include leptin, insulin-like growth factors, relaxin, and adiponectin [17–19]. Maternal insulin does not cross the placental barrier unless bound to insulin immunoglobulins. Fat is deposited and stored primarily in early pregnancy, then mobilized in the third trimester as fetal energy demands increase. Free fatty acids have been shown to contribute to insulin resistance in late pregnancy [20].

10.4.1.1 TYPE 1 DIABETES

Insulin is necessary for carbohydrate, fat, and protein metabolism. In type 1 diabetes, blood glucose levels remain elevated as insulin deficiency and the rise in free fatty acids lead to the formation of ketones and beta-hydroxybutyrate. The risk of diabetic ketoacidosis increases in the absence or lack of insulin. Women in optimal glycemic control may experience increased insulin sensitivity and decreased insulin requirements in the first trimester. During the second and third trimesters, elevated hormonal levels increase insulin resistance and additional insulin is necessary to maintain normal maternal glycemic levels and decrease fetal complications.

10.4.1.2   TYPE 2 DIABETES

Type 2 diabetes is associated with impaired insulin secretion, insulin insensitivity, and pancreatic beta-cell dysfunction. Women with type 2 diabetes tend to be older, heavier, and have higher insulin resistance than women with type 1 diabetes. The fetal pancreas is stimulated to secrete additional insulin in the presence of excessive glucose. Higher fetal insulin levels may result in macrosomic growth. Exogenous insulin may be necessary to maintain normoglycemia as insulin deficiency and insulin resistance increase.

10.4.2 Complications Associated with Preexisting Diabetes

Complications associated with diabetes can adversely affect both the woman and fetus. The incidence of fetal complications is correlated with maternal glycemic control and the trimester of pregnancy.

10.4.2.1   FETAL

Congenital malformations and spontaneous abortions are associated with maternal hyperglycemia in the first 12 weeks of gestation. The central nervous system, heart, lungs, gastrointestinal tract, kidneys, urinary tract, skeleton, and placenta are all vulnerable to adverse effects (Table 10.1) [21–23]. The frequency and severity of complications decrease if maternal normoglycemia is maintained throughout pregnancy.

Second- and third-trimester fetal complications include macrosomia, neonatal hypoglycemia, neonatal hypocalcemia, hyperbilirubinemia, polycythemia, respiratory distress syndrome, preterm delivery, and stillbirth. With the exception of stillbirth, other complications are more closely associated with infant morbidity than mortality.

Macrosomia is the most common complication associated with diabetes and pregnancy, estimated at 20–45%, depending on the population [24, 25]. The definition of macrosomia varies and ranges from 4,000 to 4,500 g [26]. Macrosomia is thought to occur if maternal glycemic levels are elevated in the third trimester. Pedersen hypothesized that maternal hyperglycemia leads to fetal hyperglycemia, which stimulates the fetal pancreas to produce excessive insulin and results in excess growth [27]. Macrosomic infants have disproportional large fetal trunks in relation to their head size, thereby increasing the risk of difficult delivery, shoulder dystocia, brachial plexus palsy, or facial nerve injury.

Neonatal hypoglycemia is a fetal serum glycemic level <35 g/dl in the first 12 h of life. Maternal glucose transport abruptly ceases when the umbilical cord is clamped. If fetal hyperinsulinemia continues, the infant will experience a rapid decrease in glycemic levels. The preferred method of treatment is oral feeding, preferably with breast milk, and frequent blood glucose monitoring within the first 4–6 h of life. Respiratory distress syndrome is caused by a deficiency of surfactant, necessary for fetal lung maturity. Neonatal hypocalcemia is serum calcium <7 mg/dl. Hyperbilirubinemia occurs when the serum bilirubin level of the neonate >13 mg/dl. Polycythemia, which is a hematocrit >65% at delivery, could lead to perinatal asphyxia. The risk of these conditions decreases if the mother maintains optimal glycemic control throughout pregnancy.

Advances in diabetes research and management have led to decreased risks of stillbirth in infants born to women with preexisting diabetes, though it remains higher than in the general pregnant population. Maternal vascular complications, poor blood glucose control, and inadequate or no prenatal care are associated with higher rates of stillbirths in women with diabetes prior to pregnancy.

Table 10.1
Congenital Anomalies Associated with Preexisting Diabetes and Pregnancy

Central Nervous System

- Neural tube defects (e.g., anencephaly, spina bifida, hydrocephalus)
- Microcephaly
- Dandy-Walker complex

Cardiovascular

- Coarctation
- Transportation of great vessels
- Truncus arteriosus
- Aortic stenosis

Gastrointestinal

- Duodenal atresia
- Anorectal atresia
- Gastroschisis

Genitourinary

- Renal agenesis
- Hydronephosis
- Cystic kidneys
- Anal/rectal atresia

Skeletal

- Caudal regression syndrome

From [21–23]

10.4.2.2 MATERNAL

Preconceptional maternal complications include nephropathy, neuropathy, retinopathy, hypertension, and diabetic ketoacidosis. Diabetic nephropathy is associated with other complications including preeclampsia, anemia, intrauterine growth restriction, fetal demise, and preterm delivery [28, 29]. If maternal glycemic levels are in optimal control before conception, the severity of complications and further renal deterioration during and after pregnancy are reduced. Pregnancy itself is not a risk factor for the development or progression of diabetic neuropathy. Gastroparesis, a condition in which the stomach's ability to empty its contents is impaired because of a possible disruption of nerve stimulation to the intestine, occurs more often in type 1 diabetes. Women with gastroparesis may experience nausea, vomiting abdominal discomfort and difficulty in controlling blood glucose. Few studies have been published on gastroparesis and pregnancy. One case report noted severe and intractable vomiting in two women with gastroparesis resulting in fetal demise in one of the pregnancies [30]. The effect of pregnancy on diabetic retinopathy depends on the severity of the condition, whether proteinuria or hypertension are present. In most cases, background retinopathy regresses after delivery. Proliferative retinopathy may progress if the condition was untreated prior to pregnancy [31, 32]. Laser photocoagulation is contraindicated in pregnancy, and the woman is advised to delay conception to avoid further eye damage. Obesity is a risk factor for hypertension and is primarily associated with type 2 diabetes [3, 34]. Diabetic ketoacidosis occurs more rapidly in pregnancy than in nonpregnancy because of increased insulin resistance and accelerated starvation ketosis. Factors that precipitate diabetic ketoacidosis include hyperemesis, gastroparesis, insulin pump failure, and certain medications, such as steroids [35].

Complications that develop during pregnancy include hypertensive disorders, polyhydramnios, preterm delivery, and cesarean section. Poor blood glucose control in early pregnancy is associated with the development of preeclampsia and pregnancy-induced hypertension [35]. Although the etiology of polyhydramnios (excessive amniotic fluid) is not well understood, it is associated with suboptimal blood glucose control. Macrosomia may warrant preterm or cesarean delivery.

10.5 MEDICAL NUTRITION THERAPY

There are no specific dietary guidelines for pregnant women with preexisting diabetes. Current guidelines for nutrition recommendations in pregnant women without diabetes may be used for pregnant women with type 1 diabetes and type 2 diabetes. Individualizing the meal plan is the key to providing adequate calories and nutrients to the woman and fetus. The meal plan works concurrently with the insulin regimen to achieve target blood glucose levels. The goals of MNT for pregnancy and diabetes are (1) to provide adequate nutrients for maternal-fetal nutrition, (2) to provide sufficient calories for appropriate weight gain, and (3) to achieve and maintain optimal glycemic control.

10.5.1 Weight Gain

Weight gain recommendations are based on the 1990 Institute of Medicine's publication, *Nutrition during Pregnancy*, according to the women's prepregnancy BMI (Table 10.2) [37]. The prepregnancy BMI and the amount of weight gained during pregnancy are two factors affecting perinatal outcome. Weight gain below the Institute of Medicine's recommendations is associated with low birth weight and small-for-gestational-age infants. Excessive weight gain may lead to macrosomia, cesarean section, and unnecessary postpartum weight retention. Overweight women with diabetes need to gain minimum weight to decrease the risk of macrosomia.

10.5.2 Energy Requirements

The estimated energy requirements (EER) during pregnancy are based on the 2002 Dietary Reference Intakes [38]. The EER for pregnancy are:

Table 10.2
Recommended Ranges of Total Weight Gain for Pregnant Women

Prepregnancy BMI	Recommended weight gain	Rate of gain/week (2nd and 3rd trimesters)
Underweight (<19.8)	28–40 lb (12.7–18.2 kg)	1.5 lb (0.7 kg)
Normal weight (19.8–26)	25–35 lb (11.2–15.9 kg)	1 lb (0.5 kg)
Overweight (26.0–29)	15–25 lb (6.8–11.3 kg)	0.6 lb (0.3 kg)
Obese (>29)	15 lb (6.8 kg)	Individualize
Twin gestation	35–45 lb (15.9–20.5 kg)	1 ½ lb (0.7 kg)
Triplet gestation	45–55 lb (20.5–25 kg)	2–2 ½ lb (0.9–1.1 kg)

From [37]

- First trimester: adult EER for women (no calorie increase)
- Second trimester: adult EER for women +160 kcal (8 kcal/week × 20 weeks) + 180 kcal
- Third trimester: adult EER for women +272 kcal (8 kcal/week × 34 weeks) + 180 kcal

The EER for adult women is based on age, height, weight, and physical activity level, which is higher than the previous recommendation of 300 extra kilocalories daily, beginning in the second trimester. Adequate calories are required to avoid starvation ketosis and ketoacidosis. A comprehensive nutrition history/questionnaire, food record/diary and blood glucose records, and regular monitoring of weight are used to develop individualized meal plans. Fluctuating blood glucose levels may necessitate frequent adjustments in the meal plan.

10.5.3 Macronutrients

The requirement for protein is 71 g/day or 1.1 g/kg/day for women over 18 years of age [38]. High-fat diets are not recommended and saturated fats are limited to less than 10% of total calories from fat.

The Recommended Daily Allowance for carbohydrate intake in pregnancy is 175 g/day to ensure sufficient glucose for fetal brain growth and development, estimated to be 33 g/day [38]. While there is no carbohydrate restriction for women with preexisting diabetes, adjustments may be necessary to maintain normoglycemia.

10.5.4 Micronutrients

Calcium, vitamin D, magnesium, iron, and folic acid are frequently consumed in inadequate amounts in pregnancy [39]. The fetus requires calcium throughout pregnancy, which is mostly deposited in the skeletal tissues in the third trimester. Vitamin D is required for calcium absorption and deposition in the fetal skeleton. Women with inadequate vitamin D intakes or limited sunlight exposure are at risk for vitamin D deficiency, increasing the risk of neonatal rickets. Magnesium deficiency may be associated with preeclampsia [40]. A low dose supplementation of 30 mg iron/day is recommended beginning in the second trimester. Folate is necessary for DNA synthesis and maternal and fetal cell proliferation. Folate deficiency is associated with maternal megaloblastic anemia, neural tube defects, spontaneous abortions, and low birth weight [41]. Folic acid supplementation should begin prior to conception and continue throughout pregnancy.

10.5.5 Nonnutritive Sweeteners

Five nonnutritive sweeteners are approved for use in pregnancy when used within the Acceptable Daily Intakes: saccharin, aspartame, acesulfame potassium, sucralose, and neotame [42].

10.6 MEAL-PLANNING APPROACHES

Various meal-planning approaches are used in diabetes management. The meal plan followed prior to conception may need only minor adjustments to account for fetal growth. Women with no previous MNT will need more intensive self-management education. The appropriate meal-planning tool selected depends on the woman's ability and motivation to follow the plan. Meal-planning approaches include menus, plate method, Food Guide Pyramid, or Exchange Lists for Meal Planning [43]. Carbohydrate counting is used more often today as clients learn the importance of employing amounts and food sources of carbohydrates, label reading, and food records. Pattern management, calculating insulin-to-carbohydrate ratios and correction factors are advanced forms of meal planning [43].

10.7   MEDICATIONS

Exogenous insulin therapy is used for women with preexisting diabetes. Although certain oral antidiabetic agents are used with GDM, there are limited studies on their use in pregnancy with type 2 diabetes.

Human insulin is recommended in pregnancy, as it is less allergenic and has a quicker absorption rate than animal-based insulin. Rapid-acting insulin analogs (Lispro, Aspart) are frequently used in pregnancy, yielding results similar to short-acting insulin [44–47]. Glargine and Detemir are long-acting, peakless insulin analogs. No clinical studies have been conducted on their use during pregnancy, though case reports have not shown teratogenic effects. Injectable therapies that have not demonstrated safety in pregnancy include incretin mimetic hormones (Pramlintide and Exenatide).

Multiple daily injections of rapid-acting insulin or short-acting insulin with an intermediate acting are the most frequently used insulin administrations in pregnancy (Table 10.3). Women who were on a fixed dose of insulin before conception are often switched to multiple daily injections because of the need for frequent insulin adjustments.

A common insulin regimen is rapid-acting or short-acting insulin before breakfast and dinner, or before each meal and intermediate-acting before breakfast and at bedtime. Intermediate-acting insulin is not usually injected before dinner because of possible nocturnal hypoglycemia.

Insulin requirements change during pregnancy as fetal growth continues and insulin resistance increases. First-trimester insulin regimen varies but is usually 0.7–0.8 units/kg actual body weight/day; second trimester: 0.8–1 unit/kg actual body weight/day; and 0.9–1.2 units/kg actual body weight/day in the third trimester. [35]. The requirements for obese women may be higher (1.5–2 units/kg actual body weight/day).

Insulin injection devices include syringes, pens, and continuous subcutaneous insulin pumps (insulin pump therapy). Pump therapy requires rapid-acting insulin with 50–60% of the dose as basal for continuous insulin and 40–50% as boluses before meals and snacks. Advantages to the insulin pump are flexibility with lifestyle and meal times, and improved glucose control. The disadvantages include cost, risk of interruption in insulin delivery or infection at the infusion site.

Table 10.3
Human Insulin

Insulin type	Onset (h)	Peak action (h)	Duration (h)
Rapid-acting (Lispro, Aspart, Glulisine)	5–15 min	1–2	4–6
Short-acting (Regular)	0.5–1	2–4	6–10
Intermediate-acting (NPH)	1–2	4–8	10–18 (long-acting)
Long-acting (Glargine, Detemir)	1–2	Peakless	Up to 24

From Messing C (ed) (2006) The art and science of diabetes self-management education: a desk reference for healthcare professionals. American Association of Diabetes Educators, Chicago, Ill., p 38

10.8   SELF-MANAGEMENT TOOLS

Medical nutrition therapy and insulin therapy are only two of the components for successful self-management. Food records will assist the registered dietitian to adjust the meal plan, when necessary. Other tools include sick-day rules, self-monitoring of blood glucose and ketones, and physical activity.

10.8.1   Sick-Day Rules

Hypoglycemia is a concern if the woman is ill and consuming inadequate calories. All pregnant women with preexisting diabetes should be aware of hypoglycemia symptoms, which range from sweating, blurred vision, nervousness, anxiety, headache, weakness, or in severe cases, seizures or unconsciousness. The treatment for hypoglycemia depends on the severity of the symptoms. Mild-to-moderate symptoms are treated with 15 g of carbohydrate if the blood glucose level is <60 mg/dl. This is repeated at 15 minutes later if the blood glucose level remains <60 mg/dl. If severe hypoglycemia occurs, either glucagons or intravenous glucose is used [48].

10.8.2   Self-Monitoring

Monitoring provides a necessary tool for adjusting food, medication, and physical activity in diabetes management. Women with preexisting diabetes need to monitor their blood glucose levels, using a glucose meter before and after meals. The blood glucose goals for diabetes and pregnancy are in Table 10.4. Urine ketone monitoring of the first morning specimen may be necessary if energy intake or weight gain is inadequate. Glycosylated hemoglobin (HbA1C), while not a self-monitoring tool, is used clinically to assess blood glucose levels in the preceding 6–8 weeks to determine metabolic control and treatment.

10.8.3   Physical Activity

No evidence has shown any beneficial effect of physical activity on glycemic control in women with type 1 diabetes. Unless contraindicated, women who were physically active before pregnancy are encouraged to continue, although the type and duration may change. Contraindications to exercising with diabetes in pregnancy include glycemic levels <100 mg/dl or >250 mg/dl [49].

Table 10.4
Glycemic Goals in Diabetes and Pregnancy

	mg/dl	mmol/l
Fasting	60–95	3.3–5.2
Preprandial	60–115	3.3–6.4
1 h postprandial	<140	<8.1
2 h postprandial	<120	<6.6
Nocturnal	60–135	3.3–5.6

From [35]

10.9 POSTPARTUM

Insulin requirements usually decrease after delivery and it is not uncommon for the woman to forego insulin for the first 1–2 days after delivery. Insulin adjustments are necessary to prevent hypoglycemia.

There are no contraindications to lactation for the woman with diabetes, and women should be encouraged to breastfeed. The meal plan is adjusted to include additional snacks to avoid hypoglycemia, which may be more frequent during lactation. Women with type 2 diabetes and choosing to breastfeed are advised to continue insulin therapy for the duration of lactation [35, 50]. Oral antidiabetic agents may resume once breastfeeding is terminated or if the woman chooses to formula feed her infant.

Family planning is an important topic to discuss with the woman with preexisting diabetes. The use of contraceptive agents will depend on whether cardiovascular disease is present [51]. Low-dose combinations of progestin and estrogen or progestin-only oral contraceptive agents are recommended for women with hyperlipidemia. Intrauterine devices and barrier methods do not affect blood glucose levels.

10.10 PRECONCEPTIONAL COUNSELING

Preconception counseling is essential for all women with preexisting diabetes in their childbearing years. Women with type 1 diabetes or type 2 diabetes should delay pregnancy until their glycosylated hemoglobin levels are <1% above the normal range prior to conception to decrease the risk of adverse perinatal outcomes [52]. Preconceptional care includes a complete physical examination to identify and treat any preexisting diabetes-related or other medical condition, an assessment of her nutritional status, and self-management education, including psychosocial assessment. A discussion of finances is also important because of the additional expense of more frequent testing or diabetes supplies.

Although all women with preexisting diabetes should receive preconceptional counseling, this often does not occur. Recent studies have shown that women with type 2 diabetes are not referred as frequently for preconceptional counseling as their type 1 diabetes counterparts [35, 53–56]. McElduff et al. found that only 12% of women with type 2 diabetes received preconceptional care compared with 27.8% of women with type 2 diabetes [54]. In another study, women with type 2 diabetes had a higher incidence of poor perinatal outcome, including fetal demise, congenital anomalies, and difficult deliveries than women with type 1 diabetes [53]. One reason for this higher prevalence of complications is the misconception that type 2 diabetes is not as severe a condition as type 1 diabetes. This myth must be dispelled and strategies developed, including intensive diabetes self-management, to improve outcomes.

10.11 GESTATIONAL DIABETES MELLITUS

It is estimated that 90% of cases of diabetes in pregnancy is GDM [35]. This includes women with possible undiagnosed type 1 or type 2 diabetes prior to conception. For most women who develop diabetes during pregnancy, normoglycemia returns following delivery.

10.11.1 Pathophysiology

GDM is similar to type 2 diabetes, as it is associated with insulin resistance and insensitivity. The exact mechanism responsible for the development of GDM is not fully understood; however, pancreatic beta-cell dysfunction may be responsible. As hormonal levels continue to rise in the second and third trimesters, beta-cells are unable to produce or secrete sufficient insulin for glucose regulation. Fasting blood glucose levels are elevated as insulin deficiency and resistance increase. Delayed insulin response, insulin resistance, and placental hormonal antagonism are responsible for postprandial glucose excursions. Human placental lactogen and cortisol block insulin receptors, which creates a deficiency in circulating insulin production, and results in increased glucose intolerance. In normal pregnancy, the beta-cells compensate by increasing insulin secretion; in GDM the decreased insulin response will result in elevated glycemic levels [57].

10.11.2 Complications

Maternal risks for gestational diabetes include hypertension, higher rates of caesarean sections and preterm deliveries [3, 58]. Congenital anomalies are rare in gestational diabetes. The exception would be the woman diagnosed with gestational diabetes early in the first trimester and in poor blood glucose control. Macrosomia is the most common complication in gestational diabetes. Other complications include neonatal hypoglycemia, neonatal hypocalcemia, neonatal hyperbilirubinemia, and polycythemia. The risk for respiratory distress syndrome decreases if delivery occurs at term.

10.11.3 Risk Factors for Gestational Diabetes Mellitus

There is considerable controversy over the screening and diagnosis of GDM. The American Diabetes Association recommends assessing all pregnant women for risk of GDM at their first prenatal visit. Risk factors for diabetes are categorized as low, average, and high [3]. Women in the low risk category must meet all of the following criteria and require. No further screening:

- Less than 25 years of age
- Normal BMI
- No first-degree family history of diabetes
- No history of glucose intolerance
- No history of poor perinatal outcome
- Not a member of a group with a high prevalence of diabetes, which includes those of African, Hispanic, Asian, Pacific Islander, or Native American descent

   Women at high risk must meet one or more of the following criteria:

- Obese
- Previous history of GDM
- Glycosuria
- Strong family history of diabetes
- Member of an ethnic group with a high prevalence of diabetes (see above)

   High-risk women are screened at their first prenatal visit. The test is repeated between 24 and 28 weeks of gestation if the initial screen was normal.

Women at average risk are those not at low or high risk. They are screened for GDM at 24–28 weeks of gestation.

10.11.4   Screening and Diagnosis of Gestational Diabetes Mellitus

Two approaches are used to screen and diagnose for GDM, the two-step and the one-step method [3]. The two-step method is used primarily in the United States. The first step is the oral glucose challenge test (OGCT). A solution containing 50 g glucose is consumed, and the plasma glucose level is checked 1 h later. If the test is ≥140 mg/dl, the second step, the oral glucose tolerance test (OGTT) is administered after 3 days of unrestricted carbohydrates (at least 150 g/day) and unlimited physical activity. The woman fasts for at least 8 h the night before the test. Blood is drawn for a fasting glucose level, followed by 100 g of glucose solution given orally and redrawn at 1, 2, and 3 h. The oral glucose tolerance test is discontinued if the fasting glucose is ≥126 mg/dl or a random glucose is ≥ 200 mg/dl. GDM is diagnosed if at least two of the values exceed the Carpenter and Coustan criteria (see Table 10.5).

The second method eliminates the 50 g OGCT. The one-step approach uses a 75-g glucose solution as the OGTT and the blood is drawn at fasting, 1 h, and 2 h. The criteria for the diagnosis of GDM are the same as the 3-h oral glucose tolerance test (Table 10.5). This method is used by the World Health Organization, but may also be more cost-effective in populations at high risk for GDM. One abnormal value on the oral glucose tolerance test is not a diagnosis for GDM; however, it may indicate adverse perinatal outcome compared to women with normal results [59].

10.11.5   Management of Gestational Diabetes

There are no universal guidelines in the management of GDM. A recent Australian randomized, controlled trial of 1,000 women with gestational diabetes showed that treating women with GDM reduced the risk of perinatal complications [60]. In this study by Crowther et al., the intervention group received MNT, self-monitored their blood glucose levels, and if indicated received insulin therapy. Perinatal complications were 1% in the intervention group and 4% in the group receiving routine care.

10.11.5.1   MEDICAL NUTRITION THERAPY

MNT is the cornerstone of treatment in the management of GDM. The American Diabetes Association and the American College of Obstetricians and Gynecologists recommend nutritional counseling by a registered dietitian and an individualized meal plan [3, 58]. The American Dietetic Association's evidence-based Nutrition Practice Guidelines have identified the following MNT goals for GDM: (1) to achieve and main-

Table 10.5
Diagnostic Criteria for Gestational Diabetes Mellitus

	100 g OGTT	75 g OGTT
Fasting	95 mg/dl (5.2 mmol/l)	95 mg/dl (5.2 mmol/l)
1 h	180 mg/dl (10 mmol/l)	180 mg/dl (10 mmol/l)
2 h	155 mg/dl (8.6 mmol/l)	155 mg/dl (8.6 mmol/l)
3 h	140 mg/dl (8.1 mmol/l)	

From [12], OGTT: Oral Glucose Tolerance Test

tain normoglycemia, (2) to provide sufficient calories to promote appropriate weight gain and avoid maternal ketosis, and (3) to provide adequate nutrients for maternal and fetal health [61]. The American Dietetic Association provides an algorithm for MNT for GDM (Fig. 10.1). The Institute of Medicine's recommendations are used to determine the appropriate weight gain for women with gestational diabetes (38). The EER are the same for pregnant women without diabetes. Monitoring weight gain, and reviewing blood glucose, food and if necessary, ketone records are other useful tools to determine diet adequacy.

The Dietary Reference Intakes do not provide a recommendation EER for obese women. Several studies used various calorie restrictions to determine minimum energy requirements, while avoiding ketonuria and ketonemia. A minimum of 1,700–1,800 kcal/day appears to improve glucose control without increasing ketone levels [3, 58, 61].

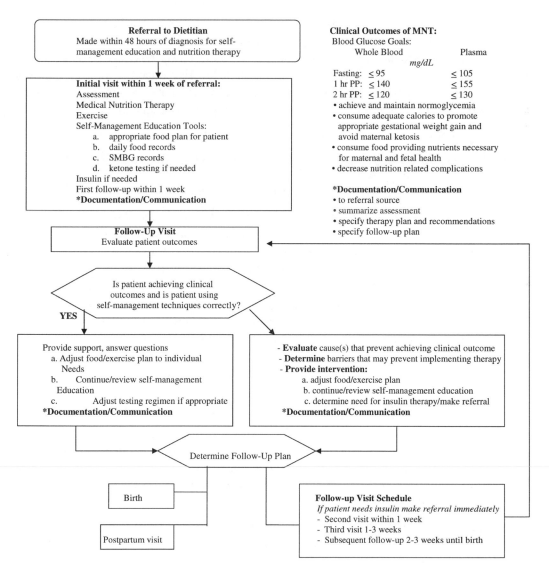

Fig. 10.1. Algorithm for MNT. (From [61], used with permission)

Carbohydrates are the main contributors of postprandial glucose excursions in GDM. The amount, source, and distribution of carbohydrates are determined in conjunction with blood glucose monitoring. The nutrition practice guidelines recommend restricting the carbohydrate content to 40–45% of total calories, but not less than the Dietary Reference Intakes recommendation of 175 g/day to achieve blood glucose goals. Carbohydrate sources include whole grains, dried beans and peas, and lentils, which are more nutrient dense and have a lower glycemic response than processed foods, such as instant products (e.g., cereals, rice, and potatoes) or highly refined grain products.

The distribution of carbohydrates into three meals and two to four snacks will help control postmeal blood glucose levels [61]. Carbohydrate intake is more restricted at breakfast than at other meals, as hormonal levels are higher in the morning. The total amount of carbohydrates at breakfast can range from 15 to 45 g. Breakfast cereals, milk, and fruit may need to be consumed at other meals or snacks. Carbohydrate distribution at lunch and dinner is usually 30–45 g or higher, depending on postprandial glycemic levels. The distribution of snacks is 15–45 g, with a smaller snack in the morning. An evening snack will help avoid overnight starvation ketosis.

Protein is not associated with postprandial glycemic elevations. The protein intake increases to 25–25% of total calories as the carbohydrate level is reduced, and usually exceeds the Dietary Reference Intakes of 71 g/day or 1.1 g/kg/day [61]. Fat makes up 35–40% of the total calories, with the majority as monosaturated and polyunsaturated fats [62, 63].

10.11.5.2 SELF-MANAGEMENT TOOLS

Self-management is important for improving perinatal outcome in GDM. Other self-management tools in addition to MNT include self-monitoring of blood glucose and ketone levels, physical activity, and the initiation of medication, when necessary.

10.11.5.2.1   Self-Monitoring of Blood Glucose. Although the American Diabetes Association and the American College of Obstetricians and Gynecologists recommend daily monitoring of blood glucose levels in GDM, there is no consensus on frequency of testing [3, 58]. Research has indicated fewer complications (macrosomia, cesarean section, birth injury, neonatal hypoglycemia) with daily use of blood glucose meters than with weekly laboratory testing of fasting and postprandial levels [3]. While preprandial testing is recommended with preexisting diabetes, postprandial monitoring of blood glucose levels have yielded better outcomes, e.g., decrease in fetal macrosomia, large-for-gestational-age infants, in GDM [64]. Optimal testing times for blood glucose levels have not been established. Studies comparing 1-h and 2-h postprandial monitoring have shown conflicting results [65, 66]. (See Table 10.4 for target blood glucose levels.)

10.11.5.2.2   Ketone Monitoring. In an effort to avoid insulin injections, women may consume fewer calories than recommended; however, this practice may increase their risk of developing ketones. Rizzo et al. found decreased intelligence scores correlated with ketonemia [67]. Urine ketone monitoring is not widely used in practice because it does not reflect the level of ketonemia, but it may be useful in detecting inadequate calorie or carbohydrate intake [3, 61].

10.11.5.2.3   Physical Activity. Physical activity may have a positive effect on glycemic control by increasing insulin sensitivity and obviating the need for insulin therapy [58, 68, 69]. Low-impact aerobics, such as walking, stair climbing, or swimming are acceptable. The activity should be performed after meals to improve glycemic levels. Pregnant women with diabetes should seek medical approval prior to beginning an exercise program.

10.11.5.2.4 Medication. Insulin therapy is used concurrently with MNT if normoglycemia is not consistently maintained with diet only. There is no consensus of opinion as to when insulin therapy should be instituted. The nutrition practice guidelines recommend beginning insulin therapy 2 weeks after MNT is implemented [61]. The American Diabetes Association and the American College of Obstetricians and Gynecologists use different glycemic cut-offs for initiating insulin (Table 10.6). Ultrasound measurement of the fetal abdominal circumference to determine macrosomic growth is also used to determine initiatiation of insulin therapy [70].

Human-based insulin is preferred over animal-based because it is less allergenic. The type, dosage, and regimen vary but are usually a combination of short-acting and intermediate-acting insulin. Calculation of the starting dose uses an approach that is similar to that employed for preexisting diabetes, and self-monitoring of blood glucose is used to guide the dose and timing of the regimen and subsequent adjustments. Insulin analogs are not yet approved by the Food and Drug Administration for use in GDM. Studies using insulin Lispro and insulin Aspart in GDM were not associated with adverse effects [71, 72].

Prior to 2000, oral antidiabetic agents were contraindicated during pregnancy. First-generation sulfonylureas crossed the placenta and were thought to cause fetal hyperinsulinemia or teratogenicity. A randomized trial by Langer et al. in which Glyburide, a second-generation sulfonylurea, was compared to insulin and reported no difference in the incidence of maternal or fetal complications, including preeclampsia, cesarean sections, macrosomia, or fetal anomalies [73]. Glyburide was also not detected in the cord serum. Four percent of the women on Glyburide did require insulin therapy. The American Diabetes Association and the American College of Obstetricians and Gynecologists have not recommended Glyburide in pregnancy, although both organizations have acknowledged its use in controlling blood glucose levels in GDM. The advantages of Glyburide in pregnancy, according to healthcare providers who advocate its use, are that it is cost-effective, non-invasive, and may result in better compliance than insulin injections [74]. Recent research in Glyburide therapy use in pregnancy demonstrated a failure rate of 12–20% [75, 76]. Women with high fasting blood glucose levels (≥ 110 mg/dl) were more likely to be switched to insulin therapy [76]. Further research is needed to determine the safety of other oral antidiabetic agents in pregnancy.

10.11.6 Postpartum

Women with GDM are at increased risk for developing type 2 diabetes after pregnancy and should be screened 6–12 weeks postpartum [3, 58]. The American Diabetes Association recommends a 75-g, 2-h oral glucose tolerance test to identify women with possible undiagnosed diabetes before conception, impaired glucose tolerance, or risk for

Table 10.6
Criteria for Initiating Insulin in GDM

	American Diabetes Association	American College of Obstetricians and Gynecologists
Fasting	≤ 105 mg/dl (5.8 mmol/l)	<95 mg/dl (5.2 mmol/l)
1 h postprandial	≤ 155 mg/dl (8.6 mmol/l)	<130 mg/dl (7.2 mmol/l)
2 h postprandial	≤ 130 mg/dl (7.2 mmol/l)	<120 (6.6 mmol/l)

From [3, 58]

Table 10.7
Criteria for Diagnosis of Diabetes Mellitus using a 75 g OGTT

	Normal values mg/dl (mmol/l)	Impaired fasting glucose mg/dl (mmol/l)	Impaired glucose tolerance mg/dl (mmol/l)	Diabetes mellitus mg/dl (mmol/l)
Fasting plasma glucose	<100 mg/dl (5.6 mmol/l)	≥100 to <126 mg/dl (≥ 5.6–7.8 mmol/l)	<100 mg/dl (< 5.6 mmol/l)	≥ 126 mg/dl (≥ 7.0 mmol/l)
75 g OGTT	<140 mg/dl (7.8 mmol/l)	<140 mg/dl (7.8 mmol/l)	≥140 to <200 mg/dl (≥ 7.8–11.1 mmol/l)	≥200 mg/dl 11.8 mmol/l)

From American Diabetes Association (2004) Screening for type 2 diabetes. Diabetes Care 27(Suppl 1): S11–S14, OGTT: Oral Glucose Tolerance Test

future diabetes (Table 10.7) [3]. If the oral glucose tolerance test is normal at 6–12 weeks postpartum, the woman should be reassessed every 3 years. Women with impaired fasting glucose or impaired glucose tolerance need to be tested annually for diabetes.

Breastfeeding, unless contraindicated, is recommended for women with GDM [3]. Lactation may improve glucose control, mobilize fat stores, promote weight loss, and protect against future risk of developing diabetes [36, 77]. Gradual weight loss (1–2 kg/month) is encouraged.

Oral contraceptive use in women with previous GDM is associated with thromboembolism, myocardium infarction, stroke, and increased insulin resistance [78, 79]. If an oral contraceptive agent is desired, a low potency dose of progestin and estrogen is prescribed to minimize the adverse effects of glucose intolerance and increased serum lipids [51].

Women with previous histories of GDM are also at risk of developing GDM in recurring pregnancies. Factors that increase the risk of GDM in a subsequent pregnancy are hip-to-waist ratio >0.84, weight gain >11 lb (5.0 kg) between pregnancies, and a fat intake >40% of the total calorie intake [80]. Women should be encouraged to adopt healthy lifestyles to lessen their risk of developing type 2 diabetes or GDM in subsequent pregnancies. Recommended lifestyle modifications include achieving and maintaining normal body weight, healthy eating habits, and consistent physical activity [61].

10.12 CONCLUSION

Advances in diabetes management have greatly improved pregnancy outcomes. For the woman with preexisting diabetes, optimal maternal blood glucose control must begin before conception and continue throughout the pregnancy. All women with type 1 diabetes and type 2 diabetes of childbearing age should be referred for preconceptional care to incorporate self-management strategies that can decrease perinatal morbidity and mortality. Self-management care includes MNT, self-monitoring of blood glucose, and if necessary, ketone testing, insulin therapy, and physical activity.

MNT is a key component in the management of GDM. An individualized meal plan should be designed to provide adequate energy and nutrients for maternal and fetal health and promote appropriate weight gain based on prepregnancy BMI. The registered dietitian will use food, blood glucose and, if necessary, ketone records to adjust the meal plan. After delivery, lifestyle modifications will be necessary to reduce the long-term risk of developing type 2 diabetes. These modifications should focus on diet, physical activity and achieving and maintaining a healthy weight.

REFERENCES

1. American Diabetes Association (2007) Total prevalence of diabetes and prediabetes. Available at http://www.diabetes.org/diabetes-statistics/prevalence.jsp
2. Martin JA, Hamilton BE, Sutton PD, Venture SJ, Menacker F, Munson ML (2004) Births: final data for 2002. Natl Vital Stat Rep 52:1–113
3. American Diabetes Association (2004) Position statement. Gestational diabetes mellitus. Diabetes Care 27(Suppl 1):S88–S90
4. Hadden DR (1998) A historical perspective on gestational diabetes. Diabetes Care 21(Suppl 2):B3–B4
5. Duncan JM (1982) On puerperal diabetes. Trans London Obstet Soc 24:256–285
6. Dey D, Hollingsworth DR Nutritional management of pregnancy complicated by diabetes: historical perspective (1981) Diabetes Care 4:647–655
7. White P (1937) Diabetes complicating pregnancy. Am J Obstet Gynecol 33:380–385
8. Duncan GG (1951) Diabetes mellitus: principles and treatment. Saunders, Philadelphia, Pa.
9. Reis RS, DeCosta EJ, Allweiss MD (1952) Diabetes and pregnancy. Charles C. Thomas, Springfield, Ill.
10. Committee on Maternal Nutrition/Food and Nutrition Board (1970) Maternal Nutrition and the Course of Pregnancy. National Academy of Sciences, National Research Council, Washington, D.C.
11. American Diabetes Association and American Dietetic Association (1979) Principles of nutrition and dietary recommendations for patients with diabetes mellitus. Diabetes Care 2:520–523
12. American Diabetes Association (2007) Position statement. Diagnosis and classification of diabetes mellitus. Diabetes Care 30(Suppl):S42–S47
13. White P (1949) Pregnancy complicating diabetes. Am J Med 7:609–616
14. Buchanan TA, Coustan DR (1994) Diabetes mellitus. In: Burrows GN, Ferris TN (eds) Medical complications during pregnancy, 4th edn. Saunders, Philadelphia, Pa. pp 29–61
15. Metzger BE, Phelps RL, Dooley SL (1997) The mother in pregnancies complicated by diabetes mellitus. In: Porte D, Sherwin RS (eds) Ellenberg and Rifkin's diabetes mellitus. 5th edn. Appleton and Lange, Stamford, Conn.
16. Catalano PM, Buchanan TA (2004) Metabolic changes during normal and diabetic pregnancies. In: Reece EA, Coustan DR, Gabbe SG (eds) Diabetes in women: adolescence, pregnancy and menopause. Lippincott Williams & Wilkins, Philadelphia, Pa., pp 129–145
17. Eidelman AI, Samueloff A (2002) The pathophysiology of the fetus of the diabetic mother. Sem Perinatol 26:232–236
18. Mazaki-Tovi S, Kanety H, Sivan E (2005) Adiponectin and human pregnancy. Curr Diab Rep 5:278–281
19. Krechowec SO, Vickers M, Gertler A, Breier BH (2006) Prenatal influences on leptin sensitivity and susceptibility to diet-induced obesity. J Endocrinol 5:355–363
20. Sivan E, Boden G (2003) Free fatty acids, insulin resistance and pregnancy. Curr Diabet Rep 3:319–322
21. Gabrielli S, Pilu G, Reese EA. Prenatal diagnosis and management of congenital malformations in pregnancies complicated by diabetes. In: Reece EA, Coustan DR, Gabbe SG (eds) Diabetes in women: adolescence, pregnancy and menopause. Lippincott Williams & Wilkins, Philadelphia, Pa., pp 299–319
22. Wren C, Birrell G, Hawthorne G (2003) Cardiovascular malformations in infants of diabetic mothers. Heart 89:1217–1220
23. Farrell T, Neale L, Cundy T (2002) Congenital anomalies in the offspring of women with type 1, type 2 and gestational diabetes mellitus. Diabet Med 19:322–326
24. Ehenberg HM, Durnwald CP, Catalano PM, Mercer BM (2004) The influence of obesity and diabetes on the risk of cesarean delivery. Am J Obstet Gynecol 191:969–974
25. Ballard JL, Rosenn B, Khy JC, Miodovnik M (1993) Diabetic fetal macrosomia: significance of disproportionate growth. J Pediatr 122:115–119

26. American College of Obstetricians and Gynecologists (2000) ACOG bulletin no 22. Fetal Macrosomia. ACOG, Washington, D.C.
27. Pedersen J (1977) The pregnant diabetic and her newborn, 2nd edn. Williams & Wilkins, Baltimore, Md.
28. Carr DB, Koontz GL, Gardella C, Holing EV, Brateng DA, Brown ZA, Easterling TR (2006) Diabetic nephopathy in pregnancy: suboptimal hypertensive control associated with preterm delivery. Am J Hypertens 19:513–519
29. Khy JC, Miodovnik M, LeMasters G, Sibai B (2002) Pregnancy outcome and progression of diabetic nephopathy. What's next? J Matern Fetal Neonat Med 1194:238–244
30. MacLeod AF, Smith SA, Sonksen PH, Lowry C (1990) The problem of autonomic neuropathy in diabetic pregnancy. Diabet Med 7:80–82
31. Lauszus FF, Klebe JG, Bek T, Flyvbjerg A (2003) Increased serum IGT-I during pregnancy is associated with progression of diabetic retinopathy. Diabetes 52:852–856
32. Lauszus FF, Gron PL, Klebe JG (1998) Pregnancies complicated by diabetic proliferative retinopathy, Acta Obstet Gynecol Scand 77:814–818
33. Kristensen J, Vestergaard M, Wisborg K, Kesmodel U, Secher NJ (2005) Pre-pregnancy weight and the risk of stillbirth and neonatal death. Br J Obstet Gynaecol 112:403–408
34. Cundy T, Gamble G, Townend K, Henley PG, MacPherson P, Roberts AB (2000) Perinatal mortality in type 2 diabetes mellitus. Diabet Med 17:33–39
35. American College of Obstetricians and Gynecologists (2005) Pregestational diabetes mellitus. Practice bulletin no. 60. Obstet Gynecol 105:675–685
36. American Diabetes Association (2007) Position statement. Evidence-based nutrition principles and recommendations for the treatment and prevention of diabetes and related complications. Diabetes Care 30(Suppl 1):S48–S65
37. National Academy of Sciences (1990) Nutrition during pregnancy. National Academy Press, Washington, D.C.
38. Institute of Medicine of the National Academies (2002) Dietary Reference Intakes: energy, carbohydrate, fiber, fat, fatty acids, cholesterol, protein, and amino acids. National Academies Press, Washington, D.C.
39. U.S. Department of Health and Human Services, Department of Agriculture (2005) Dietary guidelines for Americans. National Academies Press, Washington, D.C.
40. Frederick IO, Williams MA, Dashow E et al (2005) Dietary fiber, potassium, magnesium and calcium in relation to the risk of preeclampsia. J Reprod Med 50:332–344
41. Picciano MF (2003) Pregnancy and lactation: physiological adjustments, nutritional requirements and the role of dietary supplements. J Nutr 133:1997S–2002S
42. American Dietetic Association (2004) Position statement. Use of nutritive and nonnutritive sweeteners. J Am Diet Assoc 104:255–275
43. Green Pastors J, Waslaski J, Gunderson H (2005) Diabetes meal planning. In: Ross, TA, Boucher JL, O'Connell BS (eds) American Dietetic Association Guide to Medical Nutrition Therapy and Education. American Dietetic Association, Chicago, Ill., pp 201–217
44. Hirsch I (2005) Insulin analogues. New Eng J Med 352:174–183
45. Lapolla A, Dalfra MG, Fedele D (2005) Insulin therapy in pregnancy complicated by diabetes: are insulin analogs a new tool? Diabetes Metal Res Rev 21:241–252
46. Carr KJE, Idama TO, Masson EA, Ellis K, Lindow SW (2005) A randomized controlled trial of insulin lispro given before or after meals in pregnant women with type 1 diabetes —the effect on glycemic excursion. J Obstet Gynecol 24:382–386
47. Garg SK, Frias JP, Anil S, Gottlieb PA, MacKenzie T, Jackson WE (2003) Insulin Lispro therapy in pregnancies complicated by type 1 diabetes: glycemic control and maternal and fetal outcomes. Endocr Prac 9:187–193
48. Arnold MS (2005) Hypoglycemia and hyperglycemia. In: Ross, TA, Boucher JL, O'Connell BS (eds) American Dietetic Association guide to medical nutrition therapy and education. American Dietetic Association, Chicago, Ill., pp 128–145
49. American Diabetes Association (2004) Position statement. Physical activity/exercise and diabetes. Diabetes Care 27(Suppl 1):S58–S62
50. Reader D. Diabetes in pregnancy and lactation (2005) In: Ross, TA, Boucher JL, O'Connell BS (eds) American Dietetic Association guide to medical nutrition therapy and education. American Dietetic Association, Chicago, Ill., pp 189–197

51. Kjos SL, Buchanan TA (2004) Postpartum management, lactation and contraception. In: Reece EA, Coustan DR, Gabbe SG (eds) Diabetes in women: adolescence, pregnancy and menopause. Lippincott Williams & Wilkins, Philadelphia, Pa., pp 441–449

52. American Diabetes Association, Position statement (2004) Preconception care of women with diabetes. Diabetes Care 27(Suppl 1):S76–S78

53. Clausen TD, Mathiesen E, Ekborn P, Hellmuth E, Mandrup-Poulsen T, Damm P (2005) Poor pregnancy outcome in women with type 2 diabetes. Diabetes Care 128:323–328

54. McElduff A, Ross GP, Lagström, Champion B, Flack JR (2005) Lau SM, Moses RG, Seneratne S, McLean M, Cheung NW. Pregestational diabetes and pregnancy: an Australian experience. Diabetes Care 28:1260–1261

55. Vangen S, Stoltenberg C, Holan S, Moe N, Magnus P, Harris J, Stray-Pedersen B (2003) Outcome of pregnancy among immigrant women with diabetes. Diabetes Care 26:327–332

56. Dunne F, Brydin P, Smith K, Gee H (2003) Pregnancy in women with type 2 diabetes: 12 years outcome data 1990–2002. Diabet Med 20:734–738

57. Thomas AM, Gutierrez YM (2005) American Dietetic Association guide to gestational diabetes mellitus. American Dietetic Association Chicago, Ill.

58. American College of Obstetricians and Gynecologists (2001) Gestational diabetes mellitus. Practice bulletin no. 30. Obstet Gynecol 8:525–538

59. McLaughlin GB, Cheng YW, Caughey AB (2006) Women with one elevated 3-h glucose tolerance test value: are they at risk for adverse perinatal outcome? Am J Obstet Gynecol 194:e16–e19

60. Crowther CA, Hiller JE, Moss JR, McPhee AJ, Jeffries S, Robinson JS (2005) Effect of treatment of gestational diabetes mellitus on pregnancy outcomes. N Eng J Med 352:2477–2486

61. American Dietetic Association (2001) Medical nutrition therapy evidence-based guides for practice: nutrition practice guidelines for gestational diabetes mellitus [CD-ROM]. American Dietetic Association, Chicago, Ill.

62. Lauszus FF, Rasmussen OW, Henriksen JE, Klebe JG, Jensen L, Lauszus KS, Hermansen K (2001) Effect of a high monosaturated fatty acid diet on blood pressure and glucose metabolism in women with gestational diabetes mellitus. Eur J Clin Nutr 55:436–443

63. Wang Y, Storlien LH, Jenkins AB, Tapsell LC, Jin Y, Pan JF, Shao YF, Clavert GD, Moses RG, Shi HL, Zhu XX (2000) Dietary variables and glucose tolerance in pregnancy. Diabetes Care 23:460–464

64. deVeciana M, Major CA, Morgan MA, Asrat T, Toohey JS, Lien JM, Evans AT (1996) Postprandial versus preprandial blood glucose monitoring in women with gestational diabetes mellitus. N Engl J Med 333:1237–1241

65. Moses RG, Lucas EM, Knights S (1999) Gestational diabetes mellitus. At what time should the postprandial glucose level be monitored? Aust N Z J Obstet Gynaecol 39:457–460

66. Sivan E, Weisz B, Homko CJ, Reece EA, Schiff E (2001) One or two h postprandial glucose measurements: are they the same? Am J Obstet Gynecol 185:604–607

67. Rizzo T, Metzger BE, Burns WJ, Burns K (1991) Correlations between antepartum maternal metabolism and intelligence of offspring. N Engl J Med 325:911–916

68. Brankston GN, Mitchell BF, Ryan EA, Okun NB (2004) Resistance exercise decreases the need for insulin in overweight women with gestational diabetes mellitus. Am J Obstet Gynecol 190:188–193

69. Avery MD, Walker AJ (2001) Acute effect of exercise on blood glucose and insulin levels in women with gestational diabetes. J Matern Fetal Med 10:52–58

70. Buchanan TA, Kjos SL, Schaefer U, Peters RK, Xiang A, Byrne J, Berkowitz K, Montoro M (1998) Utility of fetal measurements in the management of gestational diabetes mellitus. Diabetes Care 21(Suppl):B99–B105

71. Pettit DJ, Kolaczynski JW, Ospina P et al (2003) Comparison of an insulin analog, insulin aspart and regular human insulin with no insulin in gestational diabetes mellitus. Diabetes Care 26:183–186

72. Jovanovic L, Ilic S, Pettit DJ, Hugo K, Gutierrez M, Bowsher RR, Bastyr EJ III (1999) Metabolic and immunologic effects of insulin Lispro in gestational diabetes. Diabetes Care 22:1422–1427

73. Langer O, Conway DL, Berkus MD, Xenakis EM, Gonzales O (2000) A comparison of glyburide and insulin in women with gestational diabetes mellitus. N Engl J Med 343:1134–1138

74. Goetzl L, Wilkins I (2001) Glyburide compared to insulin for the treatment of gestational diabetes mellitus: a cost analysis. J Perinatol 22:403–406

75. Jacobson GF, Ramos GA, Ching JY, Kirby RS, R Ferrara A, Field DR (2005) Comparison of glyburide and insulin for the management of gestational diabetes in a large managed care organization. Am J Obstet Gynecol 193:118–124

76. Conway DL, Gonzales O, Skiver D (2004) Use of glyburide for the treatment of gestational diabetes: the San Antonio experience. J Matern Fetal Med 15:51–55

77. Kjos SL, Henry O, Lee RM, Buchanan TA, Mishell DR (1993) The effect of lactation on glucose and lipid metabolism in women with recent gestational diabetes. Obstet Gynecol 82:451–455

78. Kjos SL, Peters RK, Xiang A, Thomas D, Schaefer U, Buchanan TA (1998) Contraception and the risk of type 2 diabetes mellitus in Latina women with prior gestational diabetes mellitus. J Am Med Assoc 280:533–538

79. Kjos SL (1996) Contraception in diabetic women. Obstet Gynecol Clin North Am 23:243–258

80. Jacob Reichelt AA, Ferraz TM, Rocha Oppermann ML, Costa E, Forti A, Duncan BB, Fleck Pessoa E, Schmidt MI (2002) Detecting glucose intolerance after gestational diabetes: inadequacy of fasting glucose alone and risk associated with gestational diabetes and second trimester waist-hip ratio. Diabetologia 45:455–457

11 Preeclampsia

*Lana K. Wagner, Larry Leeman,
and Sarah Gopman*

Summary Preeclampsia is a multi-organ disease that is specific to pregnancy and is characterized by the development of proteinuria and hypertension. It complicates 5–7% of pregnancies and specific criteria must be met for diagnosis. The exact etiology or pathophysiology of preeclampsia is poorly understood and as such, there are no well-established methods of primary prevention or of reliable and cost-effective screening. Calcium and aspirin may have a role in preventing preeclampsia in certain subpopulations, and research continues regarding these and other possible nutritional interventions. Preeclampsia is associated with increased maternal mortality and morbidity, and childbirth is the only known cure. Women with preeclampsia need to have regular surveillance. The associated hypertension may warrant treatment under certain conditions and magnesium sulfate is the drug of choice for the prevention and treatment of eclamptic seizures.

Keywords: Preeclampsia, Eclampsia, Hypertensive disorders of pregnancy, Gestational hypertension

11.1 INTRODUCTION

Preeclampsia is a multi-organ disease that occurs after 20 weeks gestation and is characterized by the development of proteinuria and hypertension. It is specific to pregnancy and complicates 5–7% of pregnancies [1]. It falls into the larger category of hypertensive disorders of pregnancy, which is addressed briefly within this chapter. The exact etiology or pathophysiology of this disorder is poorly understood. Additionally, there are no well-established methods of primary prevention or of reliable and cost-effective screening. Yet, preeclampsia is associated with increased maternal mortality and morbidity including placental abruption, acute renal failure, cerebrovascular and cardiovascular complications, and disseminated intravascular coagulation [2].

11.2 HYPERTENSIVE DISORDERS OF PREGNANCY

Complications from hypertension are a leading cause of pregnancy-related deaths, ranking third behind hemorrhage and embolism [2]. Hypertensive disorders that may be found during pregnancy include chronic hypertension, preeclampsia–

From: *Nutrition and Health: Handbook of Nutrition and Pregnancy*
Edited by: C.J. Lammi-Keefe, S.C. Couch, E.H. Philipson © Humana Press, Totowa, NJ

eclampsia, preeclampsia superimposed on chronic hypertension, and gestational hypertension [3].

Chronic hypertension is hypertension that exists outside of the pregnancy. As such, it will predate the pregnancy, be documented prior to 20 weeks, or will still be present 12 weeks after delivery [3]. Treatment of mild-to-moderate chronic hypertension during pregnancy has not been shown to prevent preeclampsia and has shown no proven fetal benefit [4–6].

Preeclampsia–eclampsia is the onset of hypertension with proteinuria that occurs after 20 weeks of pregnancy. Eclampsia, which occurs in less than one percent of women with preeclampsia [1], is the new onset of seizures during preeclampsia.

If a patient has chronic hypertension but develops new or worsened proteinuria, then this is preeclampsia superimposed on chronic hypertension [7]. Also, if there is an acute increase in the hypertension (assuming preexisting proteinuria) or if HELLP (hemolysis, elevated liver enzymes, low platelet count) syndrome develops, then this is also considered to be preeclampsia superimposed on chronic hypertension [7].

Gestational hypertension, which used to be known as "pregnancy induced hypertension," [3] is hypertension that develops in the absence of proteinuria. Gestational hypertension develops after 20 weeks of pregnancy and returns to normal within 12 weeks of delivery [7].

11.3 DIAGNOSIS OF PREECLAMPSIA

As mentioned previously, both proteinuria and hypertension after 20 weeks of gestation must be present for a diagnosis of preeclampsia to be made. The diagnostic criteria for preeclampsia are presented in Table 11.1. Blood pressures should be measured with an appropriately sized cuff, with the patient in an upright position [8]. Edema and blood pressure elevations above the patient's baseline are no longer included in diagnostic criteria [3, 7].

In severe preeclampsia, blood pressures may be higher and proteinuria more pronounced. The diagnostic criteria for severe preeclampsia are also presented in Table 11.1. Any signs or symptoms indicating end organ damage make the diagnosis of severe preeclampsia.

Although 24-h urine collections are the gold standard for measuring proteinuria, a random urinary protein-to-creatinine ratio can rule out significant proteinuria if the ratio is less than 0.19 [9]. The urine protein-to-creatinine ratio using the 0.19 cutoff has a sensitivity of 90%, specificity of 70%, and a negative predictive value of 87% [9].

11.4 PATHOPHYSIOLOGY AND RISK FACTORS

The etiology of preeclampsia remains poorly understood. Multiple theories have been proposed regarding the pathophysiology, and no single causal factor has been found [10].

Theories of pathophysiology include genetic predisposition [11–14], abnormal placental implantation [15, 16], angiogenic factors [17], exaggerated inflammatory responses [18], inappropriate endothelial activation [18], vasoconstriction [19], and coagulation cascade defects [19]. Although hypertension and proteinuria are the criteria by which preeclampsia is diagnosed, the pathophysiologic changes associated with preeclampsia affect virtually every organ system. Microthrombi from activation of the coagulation

Table 11.1
Diagnostic Criteria for Preeclampsia and Severe Preeclampsia [3, 7]

	Preeclampsia[a]	*Severe preeclampsia*[b]
Blood pressure[c]	140 mmHg or higher systolic	160 mmHg or higher systolic Or 90 mmHg or higher diastolic 110 mmHg or higher diastolic
Proteinuria	0.3 grams or more in a 24-h urine (this usually corresponds with a 1+ or greater on dipstick)	5 g or more in a 24-h urine Or 3+ or more on two random urine samples collected at least 4 h apart
Other features		Oliguria (less than 500 ml of urine in 24 h) Cerebral or visual disturbances Pulmonary edema or cyanosis Epigastric or right upper-quadrant pain Impaired liver function Thrombocytopenia Fetal growth restriction

[a] Both hypertension and proteinuria components must be present
[b] One or more must be present in addition to criteria for preeclampsia
[c] Taken on two occasions at least 6 h apart

Table 11.2
Risk Factors for Preeclampsia [20]

Increased maternal age (>40 years of age)
Nulliparity
Multiple gestation
Preeclampsia in a prior pregnancy
Elevated body mass index
Certain medical conditions:
Chronic hypertension
Chronic renal disease
Antiphospholipid syndrome
Diabetes mellitus

cascade, as well as systemic vasospasm, decrease blood flow to organs [19]. Perfusion is further compromised by vascular hemoconcentration and third spacing of intravascular fluids [18].

Risk factors for preeclampsia are presented above in Table 11.2 [20]. Note that young maternal age is no longer considered a risk factor, as this was not supported by a systematic review [20]. Women with preeclampsia should be counseled about the increased risk of recurrent preeclampsia in future pregnancies. The recurrence rate may be as high as 40% in nulliparous women with preeclampsia before 30 weeks gestation and even higher in multiparous women [3].

11.5 PREVENTION AND NUTRITION

As was stated earlier, there are no well-established methods of primary prevention for preeclampsia, although numerous supplements have been studied regarding their ability to impact its occurrence. Thus far, randomized controlled trials do not support routine prenatal supplementation with magnesium, omega-3 fatty acids, antioxidants (vitamins E and C), or calcium to prevent preeclampsia [21–24].

However, calcium and aspirin may have a role in preventing preeclampsia in certain subpopulations, though the optimal treatment regimens will require further research. Calcium supplementation in high-risk women and in women with low dietary calcium intakes reduced the risk of hypertension and preeclampsia [25]. Also, calcium supplementation has been shown to decrease the incidence of neonatal mortality and severe maternal morbidity due to hypertensive disorders when given to normotensive nulliparous women [26].

Low-dose aspirin was shown to have small to moderate benefits for prevention of preeclampsia within certain groups of women. A Cochrane analysis demonstrated that in women at increased risk for preeclampsia, 69 women would need to be treated with low-dose aspirin to prevent one case of preeclampsia [27]. However, in the subgroup of women at highest risk for preeclampsia (because of histories of previous severe preeclampsia, diabetes, chronic hypertension, renal disease, or autoimmune disease), only 18 would need to be treated with low-dose aspirin to prevent one case of preeclampsia [27].

Research continues regarding possible nutritional interventions for preeclampsia. While larger studies that are more reliable are needed to confirm results, diets high in fiber and potassium may reduce the risk of preeclampsia [28]. Additionally, diets high in calories, sucrose, and polyunsaturated fatty acids may increase the risk for preeclampsia [29].

11.6 MANAGEMENT OF PREECLAMPSIA

Childbirth is the cure for preeclampsia as the disease process usually resolves within days of delivery. Delivery is always preferable from the perspective of maternal health. However, decisions on induction of labor or cesarean delivery must include a consideration of prematurity-related neonatal risks and the severity of the preeclampsia. Women with mild preeclampsia should be carefully followed until they are close to term and delivered at 37–39 weeks [30]. Women with severe preeclampsia may be expectantly managed until 32–34 weeks, or delivered sooner based on maternal and fetal status [31]. Women with preeclampsia need to have regular surveillance of the fetus with nonstress testing and amniotic fluid volume assessment. Blood work should be checked periodically to detect renal or hepatic involvement, hemolysis, or thrombocytopenia.

The hypertension of preeclampsia only warrants treatment if the systolic blood pressure is above 160 mmHg or the diastolic blood pressure is above 110 mmHg [3]. If these pressures occur near term, then the blood pressure may be managed with intravenous hydralazine or labetalol until delivery [32]. Women with severe preeclampsia undergoing expectant management may have their blood pressure controlled with oral labetalol, methyldopa, or nifedipine [3]. Magnesium sulfate is the drug of choice for the prevention and treatment of eclamptic seizures [33]. All women with severe preeclampsia need intravenous magnesium in labor and for 24 h postpartum [34]. The use of magnesium sulfate in women with mild preeclampsia remains controversial, as 400 women may need to be treated to prevent one eclamptic seizure [35].

Neonatal morbidity and mortality is due to the risk of prematurity, uteroplacental insufficiency, or placental abruption. An ultrasound for estimated fetal weight should be done at the time of diagnosis to evaluate for possible intrauterine growth restriction secondary to uteroplacental insufficiency [3]. If delivery is required prior to term, then the birth should occur at an institution with a neonatal intensive care unit capable of caring for infants at the anticipated gestational age. Placental abruption is an unpredictable event, which can lead to fetal death or morbidity.

REFERENCES

1. Witlin AG, Sibai BM (1998) Magnesium sulfate therapy in preeclampsia and eclampsia. Obstet Gynecol 92:883–889
2. Mackay AP, Berg CJ, Atrash HK (2001) Pregnancy-related mortality from preeclampsia and eclampsia. Obstet Gynecol 97:533–538
3. National Heart Lung and Blood Institute (2000) National High Blood Pressure Education Program Working Group Report on High Blood Pressure in Pregnancy. Am J Obstet Gynecol 183:S1–S22
4. Abalos E, Duley L, Steyn DW, Henderson-Smart DJ (2001) Antihypertensive drug therapy for mild to moderate hypertension during pregnancy. Cochrane Database Syst Rev 2
5. Magee LA, Duley L (2003) Oral beta-blockers for mild to moderate hypertension during pregnancy. Cochrane Database Syst Rev 3
6. American College of Obstetricians and Gynecologists (2001) ACOG practice bulletin. Chronic hypertension in pregnancy, no. 29. Obstet Gynecol 98(Suppl):177–185
7. American College of Obstetricians and Gynecologists (2002) ACOG practice bulletin no. 33: Diagnosis and management of preeclampsia and eclampsia. Obstet Gynecol 99:159–167
8. US Preventative Services Task Force (1996) Guide to clinical preventative services, 2nd ed,. Williams and Wilkins, Baltimore, Md. Available via http://www.ahrq.gov/clinic/uspstfix.htm
9. Rodriquez-Thompson D, Lieberman ES (2001) Use of a random urinary protein-to-creatinine ratio for the diagnosis of significant proteinuria during pregnancy. Obstet Gynecol 185:808–811
10. Davison JM, Homuth V, Jeyabalan A, Conard KP, Karumanchi SA, Quaggin S, Dechend R, Luft FC (2004) New aspects in the pathophysiology of preeclampsia. J Am Soc Nephrol 15:2440–2448
11. Esplin MS, Fausett MB, Fraser A, Kerber R, Mineau G, Carrillo J, Varner MW (2001) Paternal and maternal components of the predisposition to preeclampsia. N Engl J Med 344:867–872
12. Morgan T, Ward K (1999) New insights into the genetics of pre-eclampsia. Semin Perinatol 23:14–23
13. Lin J, August P (2005) Genetic thrombophilias and preeclampsia: a meta-analysis. Obstet Gynecol 105:182–192
14. Migini LE, Latthe PM, Villar J, Kilby MD, Carroli G, Khan KS (2005) Mapping the theories of preeclampsia: the role of homocysteine. Obstet Gynecol 105:411–425
15. McMaster MT, Zhou Y, Fisher SJ (2004) Abnormal placentation and the syndrome of preeclampsia. Semin Nephrol 24:540–547
16. Merviel P, Carbillon L, Challier JC, Rabreau M, Beaufils M, Uzan S (2004) Pathophysiology of preeclampsia: links with implantation disorders. Eur J Obstet Gynecol Reprod Biol 115:134–147
17. Levine RJ, Thadhani R, Quian C, Lam C, Lim KH, Yu KF, Blink AL, Sachs BP, Epstein FH, Sibai BM, Sukhatme VP, Karumanchi SA (2005) Urinary placental growth factor and the risk of preeclampsia. J Am Med Assoc 293:77–85
18. Dekker GA, Sibai BM (1998) Etiology and pathogenesis of preeclampsia: current concepts. Am J Obstet Gynecol 179:1357–1375
19. Roberts JM, Cooper DW (2001) Pathogenesis and genetics of preeclampsia. Lancet 357:53–56
20. Milne F, Redman C, Walker J, Baker P, Bradley J, Cooper C, de Swiet M, Fletcher G, Jokinen M, Murphy D, Nelson-Piercy C, Osgood V, Robson S, Shennan A, Tuffnell A, Twaddle S, Waugh J (2005) The pre-eclampsia community guideline (PRECOG): how to screen for and detect onset of preeclampsia in the community. Brit Med J 330:576–580

21. Levine RJ, Hauth JC, Curet LB, Sibai BM, Catalano PM, Morris CB, DerSimonian R, Esterlitz JR, Raymond EG, Bild DE, Clemens JD, Cutler JA (1997) Trial of calcium to prevent pre-eclampsia. N Engl J Med 337:69–76

22. Sibai BM, Villar MA, Bray E (1989) Magnesium supplementation during pregnancy: a double blind randomized controlled clinical trial. Am J Obstet Gynecol 161:115–119

23. Salvig JD, Olsen SF, Secher NJ (1996) Effects of fish oil supplementation in late pregnancy on blood pressure: a randomized controlled trial. Br J Obstet Gynaecol 103:529–533

24. Poston L, Briley A, Seed P, Kelly F, Shennan A (2006) Vitamin C and vitamin E in pregnant women at risk for pre-eclampsia (VIP trial): randomised placebo-controlled trial. Lancet 367:1145–1154

25. Atallah AN, Hofmeyr GJ, Duley L (2002) Calcium supplementation during pregnancy for preventing hypertensive disorders and related problems. Cochrane Database Syst Rev 1

26. Villar J, Abdel-Aleem H, Merialdi M, Mathai M, Ali MM, Zavaleta N, Purwar M, Hofeyr J, Nguyen TN, Campodonico L, Landoulsi S, Carroli G, Lindheimer M, (2006) World Health Organization Calcium Supplementation for the Prevention of Preeclampsia Trial Group. Am J Obstet Gynecol. 194:639–649

27. Duley L, Henderson-Smart DJ, Knight M, King JF (2004) Antiplatelet agents for preventing pre-eclampsia and its complications. Cochrane Database Syst Rev 1

28. Frederick IO, Williams MA, Dashow E, Kestin M, Zhang C, Leisenring WM (2005) Dietary fiber, potassium, magnesium and calcium in relation to the risk of preeclampsia. J Reprod Med 50:332–344

29. Clausen T, Slott M, Solvoll K, Drevon CA, Vollset SE, Henriksen T (2001) High intake of energy, sucrose, and polyunsaturated fatty acids is associated with increased risk of preeclampsia. Am J Obstet Gynecol 185:451–458

30. Sibai BM (2003) Diagnosis and management of gestational hypertension and preeclampsia. Obstet Gynecol 102:181–192

31. Churchill D, Duley L (2002) Interventionist versus expectant care for severe pre-eclampsia before term. Cochrane Database Syst Rev 3

32. Duley L, Henderson-Smart DJ (2002) Drugs for treatment of very high blood pressure during pregnancy. Cochrane Database Syst Rev 4

33. Duley L, Gulmezoglu AM, Henderson-Smart DJ (2003) Magnesium sulfate and other anticonvulsants for women with pre-eclampsia. Cochrane Database Syst Rev 2

34. The Magpie Trial Collaborate Group (2002) Do women with pre-eclampsia, and their babies, benefit from magnesium sulphate? The Magpie Trial: a randomized placebo-controlled trial. Lancet 359:1877–1890

35. Sibai, BM (2004) Magnesium sulfate prophylaxis in preeclampsia: Lessons learned from recent trials. Am J Obstet Gynecol 190:1520–1526

12 AIDS/HIV in Pregnancy

Katherine Kunstel

Summary Women are among the fastest growing populations of those infected with HIV and AIDS, and most infected women are of childbearing age. Women who are both HIV-positive and pregnant are faced with a double burden both in terms of immunity and nutrition. The HIV-infected pregnant woman is at increased nutritional risk compared to the HIV-uninfected pregnant woman. HIV-infected pregnant women tend to gain less weight during pregnancy. Macronutrient needs are increased to cover the increased demands of both HIV infection and pregnancy and inadequate intake is common. Micronutrient deficiencies are also common in HIV-infected pregnant women and can have adverse outcomes for both the mother and the developing child. Other nutrition-related considerations for this population include symptom management, the consequences of antiretroviral therapy, risk of transmission of the virus through breastfeeding, food safety, and food security.

Nutrition assessment and counseling is a critical component of the overall care plan for HIV-infected pregnant women. Counseling regarding weight gain, adequate nutrient intake, management of HIV-related symptoms, drug-nutrient interactions, and the risks associated with breastfeeding should be made available to all HIV-infected pregnant women. The nutrition care plan must aim to promote the best outcomes for both the mother and the developing child.

Keywords: Nutrition, Pregnancy, HIV, AIDS, Maternal health, Breastfeeding, Pregnancy outcomes

12.1 INTRODUCTION

In the United States and Canada, one of the fastest growing populations in the profile of acquired immune deficiency syndrome (AIDS) cases is women [1], and most women who are currently infected with human immunodeficiency virus (HIV) and AIDS are of childbearing age [2]. These facts support the increasing need to consider the implications of HIV and AIDS on pregnancy and lactation. There is no evidence to indicate that pregnancy and lactation have a significant effect on hastening the progression of HIV disease [3, 4]. However, women who are infected with HIV and are pregnant face a double burden in terms of their immune function and face additional potential complications. For example, HIV-infected pregnant women have higher risks of fetal loss [5, 6],

From: *Nutrition and Health: Handbook of Nutrition and Pregnancy*
Edited by: C.J. Lammi-Keefe, S.C. Couch, E.H. Philipson © Humana Press, Totowa, NJ

low birth weight, preterm delivery, and intrauterine growth retardation (IUGR) [7]. Additionally, the HIV-infected pregnant woman is at a higher nutritional risk, owing to increased energy needs, tendency to achieve suboptimal weight gain during pregnancy, micronutrient deficiencies, and management of disease symptoms.

12.2 NUTRITIONAL STATUS OF THE MOTHER

The nutritional status of an HIV-infected woman prior to and during pregnancy influences both her own health and the health of her unborn child [8]. The nutritional challenges for the HIV-infected pregnant woman are threefold. First, during pregnancy, just as in the uninfected woman, maternal metabolism is altered by hormones in preference of the developing infant, and nutrients are directed to the placenta, the mammary gland, and the infant [9, 10]. Additionally, HIV infection can prompt micronutrient deficiencies and lean body mass depletion because of decreased nutrient intake, malabsorption, and increased utilization and excretion of nutrients resulting in undernutrition [11]. Finally, HIV infection affects nutritional status through an increase in resting energy expenditure (REE) [12–14]. For women who are malnourished, an energy–protein supplement during pregnancy may improve pregnancy outcomes by improving maternal weight gain and reducing the risk of perinatal mortality [15, 16]. The consequences of maternal malnutrition extend beyond the mother and increase risks for the developing infant as well. For example, poor maternal nutritional status may impair the integrity of the placental lining, creating a more opportune condition for transplacental transmission of HIV [17]. Optimal nutrition during pregnancy increases weight gain and improves pregnancy and birth outcomes [4].

12.3 WEIGHT GAIN

HIV-infected pregnant women tend to gain less weight during pregnancy compared with women who are not infected [4], putting them at higher risk for complications. Additionally, HIV infection is oftentimes associated with wasting and a progressive loss of body mass [18]. This can lead to adverse pregnancy outcomes. Weight loss or suboptimal weight gain during pregnancy is related to increased risk of intrauterine growth retardation (IUGR) [10], fetal death, preterm delivery, and low-birth-weight (LBW) infants [19, 20]. While overall weight gain is indicative of pregnancy outcomes, Villamor et al. showed that weight loss during the third trimester was more strongly associated with preterm delivery than weight loss during the second trimester, but weight loss during the second trimester was related to an increased risk of fetal death [19].

Weight gain goals during pregnancy for HIV-infected women are the same as those for uninfected women. This weight gain is representative of two entities, the products of conception (fetus, placenta, amniotic fluid) and maternal tissues (expansion of blood and extracellular fluid, uterus, mammary glands, and adipose tissue) [10]. Weight gain goals should be based on the woman's prepregnancy weight and height [21]. It is recommended that a woman with a normal prepregnancy weight for height, or a body mass index (BMI) of 19.8–26 kg/m^2 gain approximately 3.5 lb (1.6 kg) in the first trimester and then approximately 1 lb (~0.5 kg) per week during the second and third trimesters, for a total pregnancy weight gain goal of 25–35 lb (11–16 kg). Women who are underweight for height (BMI < 19.8 kg/m^2) prepregnancy should aim to gain approximately 5 lb in the first trimester and a little more than a pound per week in the second and third trimesters, for a total weight gain of 28–40 lb. Women who are overweight for height (BMI > 26–29 kg/m^2) prepregnancy should gain approximately 2 lb in the first trimester and slightly less than 1 lb per week in the second and third trimesters, for a total pregnancy weight gain of 15–25 lb. Obese women (BMI > 29 kg/m^2) are recommended to gain at least 15 lb during pregnancy [22] (Table 12.1).

12.4 NUTRIENT NEEDS

12.4.1 Macronutrients

HIV infection increases energy needs due to an increase in REE, as previously stated. This increased REE coupled with HIV-related infections and complications, such as anorexia, place HIV-infected pregnant women at greater nutritional risk than the uninfected woman [23, 24]. Current energy recommendations for HIV-infected pregnant and lactating women are an increase of 10% over baseline energy needs during the asymptomatic phase and an increase of 20–30% over baseline energy needs during the symptomatic phase [25]. Early symptomatic HIV infection is defined as the stage of viral infection caused by HIV when symptoms have begun, but before the development of AIDS. Symptoms may include but are not limited to mouth disorders (oral hairy leukoplakia, oral thrush, gingivitis), prolonged diarrhea, swollen lymph glands, prolonged fever, malaise, weight loss, bacterial pneumonia, joint pain, and recurrent herpes zoster. In addition, the World Health Organization (WHO) recommends an intake of an extra 300 kcal per day during all trimesters of pregnancy [26], while the National Research Council recommends an additional intake of 300 kcal per day during the second and third trimesters of pregnancy [27] (Table 12.2). Energy intakes that fall below established recommendations are likely to result in coinciding low intakes of micronutrients such as calcium, magnesium, zinc, vitamin B-6, and folate [19], which have potential consequences for both the mother and developing child.

Table 12.1
Recommended Weight Gain for Pregnancy Based on Prepregnancy Weight

	BMI (kg/m^2)	Recommended total weight gain (lb)
Underweight	<19.8	28–40 (13–18 kg)
Normal weight	19.8–26	25–35 (11–16 kg)
Overweight	26–29	15–25 (7–11 kg)
Obese	>29	At least 15 (At least 7 kg)

Table 12.2
Adjusted Energy Needs for HIV-Infected Pregnant Women

	Increase in energy needs	Plus	Additional energy for pregnancy
HIV-asymptomatic pregnant women	10% above normal needs	+	300 kcal/day
HIV-symptomatic pregnant women	20–30% above normal needs	+	300 kcal/day

Protein requirements are higher during pregnancy to support maternal protein synthesis for expansion of the blood volume, uterus, and breasts and to supply amino acids for synthesis of fetal and placental proteins [27]. The recommended dietary allowance (RDA) for protein for a normal pregnancy and lactation is 71 g per day [28] or approximately 1.1 g/kg/day based on current body weight. This compares to a RDA for protein of 0.8 g/kg/day for nonpregnant, healthy women [28]. The Institute of Medicine recommends 71 g of protein per day during pregnancy [29]. In the United States and other developed countries, adequate protein intake is not usually a problem. As with nonpregnant individuals, additional protein may be needed under conditions of stress, such as symptomatic HIV infection. However, there is not sufficient data to suggest that HIV infection in and of itself demands a higher protein intake by the infected individual [4].

The Acceptable Macronutrient Distribution Range (AMDR) for fat for normal pregnancy and lactation is 20–35% of kilocalories for all women from 18 to 50 years of age [28]. Structural and functional changes in the gastrointestinal tract in HIV often affect the absorption of fat, leading to fat malabsorption [16]. In this setting, the use of medium chain triglyceride (MCT) oils may be beneficial. MCT are more easily digested than long-chain triglycerides (LCT) and can be absorbed across the small intestinal mucosa in the absence of pancreatic enzymes [30]. MCT oil can be used as a supplement and added to foods. Also, many enteral feeding formulas designed for patients with fat malabsorption contain MCT oil.

12.4.2   Micronutrients

Micronutrient deficiencies are seen more frequently in HIV-infected pregnant women than in HIV-uninfected pregnant women [4]. Deficiencies may be due to physiologic losses, malabsorption, inadequate intake, and lack of knowledge regarding appropriate prenatal nutrition [9]. Poor maternal micronutrient status has consequences for both the mother and the developing infant. Micronutrient deficiencies may result in increased risk for opportunistic infections, more rapid disease progression [31–33], and an increased risk of vertical transmission of the virus as a result of compromised immune status of the mother [17]. Common micronutrient deficiencies that are seen in people with HIV include vitamin A, B-complex, vitamin C, vitamin E, selenium, and zinc, all of which play an important role in immune function and defense against infection [34–37 and Chap. 23, "Micronutrient Status and Pregnancy Outcomes in HIV-Infected Women"]. Iron and folate supplementation is especially important to promote positive pregnancy outcomes, while vitamin A supplementation has been shown to increase the risk of vertical HIV transmission. Zinc deficiency results in reduction of T-cell development and function, affecting immune response [38]. It has also been linked to low-birth-weight outcomes in infected pregnant women.

12.4.3   Iron and Folate

During pregnancy, there is an increase in blood volume as a result of expansion of plasma volume by approximately 45–50% and of red cell mass by approximately 15–20% in the third trimester [10]. Anemia is a common condition associated with pregnancy because the expansion of blood cell mass is less relative to plasma, and hemoglobin and hematocrit levels drop as a result [39]. Anemia during pregnancy is associated with complications such as low birth weight, preterm delivery, and increased perinatal mortality

[40]. An additional concern for the HIV-infected pregnant woman is bone marrow suppression and hemolysis because of infections; the end result is further depressed hemoglobin [41]. Iron deficiency per se is the prevailing cause of low hemoglobin concentrations [42], however vitamin A, folate, riboflavin, vitamin B_{12} [43, 44], and zinc are also essential to erythropoiesis [36].

Folate deficiency causes macrocytic anemia [45] as well as neural tube defects [46] and possibly IUGR and preterm delivery [47]. In a study by Friis et al. [36], HIV was a negative predictor of serum folate, most likely owing to reduced intake and absorption and increased catabolism.

The Dietary Reference Intakes (DRI) suggests an intake of 27 mg of iron daily and 600 mcg of folate daily to prevent anemia [48, 49]. WHO recommendations regarding supplementation include: 400 mcg of folate and 60 mg of iron daily during the last 6 months of pregnancy to prevent anemia and twice daily doses to prevent severe anemia [50]. The WHO recommendation is most applicable to women who reside in resource-limited settings because it takes into consideration the bioavailability of dietary iron, which tends to be limited in developing countries [9]. There are concerns about potential adverse effects of iron supplementation in the setting of HIV infection [51]. Iron is important for immune function [52] but it also serves as a substrate for enzymes involved in HIV replication [53, 54]. More research is needed to clarify the role of iron in HIV infection before current recommendations regarding intake are reconsidered.

12.4.4   Vitamin A

Maternal vitamin A deficiency may cause IUGR and other adverse pregnancy outcomes [55] as well as maternal morbidity and mortality [56]. Additionally, infections may contribute to low vitamin A status in women of reproductive age [57]. In the United States, however, overt vitamin A deficiency is not commonplace [58], and as a fat-soluble vitamin, there is the potential for excessive intake in the setting of supplementation. Additionally, in trials conducted in Malawi, South Africa, and Tanzania, Fawzi et al. [59] found that vitamin A supplementation resulted in a significant increased risk of vertical HIV transmission and a significant increase in lower genital viral shedding [60]. Vitamin A supplementation is not recommended for HIV-infected pregnant women.

Multivitamin supplementation during pregnancy for the HIV-infected woman has many benefits. Supplementation with a multivitamin (including B-complex, vitamin C, and vitamin E) but not vitamin A alone effectively reduces adverse pregnancy outcomes, such as fetal loss, low birth weight, and prematurity, while also improving maternal weight gain during pregnancy [19, 50]. In addition, micronutrient supplementation has been shown to improve body weight and body cell mass [61], specifically during the last trimester of pregnancy [19, 62], while reducing incidence of opportunistic infections [63] and hospitalizations [64]. Multivitamin supplementation has been shown to significantly decrease the risk of maternal weight loss and also improve hemoglobin concentrations [62]. Fawzi et al. reported that multivitamins had a significant beneficial effect on T-cell subset counts. There were increases seen in both CD4 and CD8 cells, which are the main cellular indicators of immunity in HIV infection [65]. Finally, use of multivitamins or B vitamins and higher dietary intakes of riboflavin, thiamin, and niacin were linked to slower progression of HIV disease [32, 66, 67].

12.4.5 Zinc

Zinc deficiency has been associated with impaired growth outcomes [68]. In one study of women in the United States with low serum zinc concentrations, there was a significant increase in fetal growth among the group receiving zinc supplementation [69]. Both the control and supplemented groups received a multivitamin preparation. Results from trials of zinc supplementation in uninfected pregnant women are mixed, and although some benefit has been shown, the data are not conclusive.

The RDA for zinc supplementation during pregnancy is 12 mg/day for 14- to 19-year-olds and 11 mg/day for women ages 19–50 [49]. However, in developing countries, where there is an increased likelihood of chronic zinc deficiency, prophylactic doses of zinc have been used. The WHO set the Upper Limit (UL) for zinc supplementation at 35 mg/day [70, 29].

12.5 MEDICATIONS

Nutrients and nutritional status of the HIV-infected individual can affect the absorption, use, elimination, and tolerance of antiretroviral medications [71–73]. Treatment of HIV-infected pregnant women is based on the premise that therapies of known benefit to women should not be withheld during pregnancy unless they have known adverse effects on the mother, the fetus, or the infant, and unless the adverse effects outweigh the benefit to the woman [74]. While some antriretroviral mediations are well tolerated, others have potential nutritional side effects. Zidovudine, a nucleoside reverse transcriptase inhibitor (NRTI) that is approved by the US Food and Drug Administration (FDA) for use during pregnancy to prevent vertical transmission of HIV to the newborn is associated with nausea and vomiting, as well as bone marrow suppression that may increase the severity of anemia [16, 23, 75]. Ritonavir, a protease inhibitor (PI) is also associated with gastrointestinal upset, while Indinavir, another PI is associated with hyperbilirubinemia, which is a concern for the newborn [76]. Timing of antiretroviral (ARV) medications with regard to meals is important in terms of medication efficacy and managing drug-related side effects. Indinavir should be taken 1 h before or 2 h after meals [77], but can be taken with light, low-fat meals if necessary. Ritonavir should be taken with food and liquids such as, chocolate milk, to lessen the bitter aftertaste [78] (See Table 12.3).

A phenomenon of fat redistribution, or lipodystrophy, is often seen in HIV-infected individuals on ARV therapy, specifically PI, NRTI, or both [79, 80]. This usually occurs as a result of long-term ARV therapy. The syndrome is characterized by loss of subcutaneous fat from the face, arms, and legs and oftentimes deposition of excess fat in the neck, upper back, and the trunk [79]. Additionally, metabolic complications such as insulin resistance, hypertriglyceridemia, low levels of serum low density lipoprotein cholesterol, and hyperglycemia are commonly seen.

12.6 SYMPTOM MANAGEMENT

Most women who are infected with HIV and become pregnant are asymptomatic [81]. However, for those who are experiencing symptoms, nutritional consequences may occur. Some of the symptoms that may affect nutritional status include anorexia, nausea, vomiting, diarrhea, and chewing and swallowing difficulties [1]. In a study by Kim et al. [82], the clinical symptom that was consistently associated with an inability to meet the RDA for nutrients was anorexia. Reduced intake is an important risk factor for weight

Table 12.3
Common Antiretroviral Medications Used During Pregnancy and their Nutritional Effects [74, 75]

	Possible side effects	*Administration*
NRTI		
Zidovudine	Nausea, vomiting, anemia, neutropenia	Can be taken with or without food
Lamivudine	Nausea, vomiting, peripheral neuropathy	Can be taken with or without food
NNRTI		
Nevirapine	Skin rash (most common)	Can be taken with or without food
PI		
Lopinavir/ Ritonavir	Diarrhea, headache, nausea, vomiting, weakness, rash	Should be taken with food
Nelfinavir	Diarrhea, nausea, abdominal pain, dizziness, flatus	Should be taken with food
Indinavir	Kidney stones, abdominal pain, nausea, headache	Should be taken on an empty stomach, 1 h before meals or 2 h after meals. Can be taken with light, low-fat meal if stomach upset occurs.
Ritonavir	Nausea, vomiting, diarrhea	Should be taken with food or liquids Chocolate milk may help to lessen bitter aftertaste

NRTI: nucleoside reverse transcriptase inhibitor; NNRTI: non-nucleoside reverse transcriptase inhibitor; PI: protease inhibitor

loss and malnutrition in this population, especially in the setting of opportunistic infections [16]. Strategies for symptom management are outlined in Table 12.4.

12.7   FOOD SAFETY

HIV-infected women are more susceptible to bacteria and viruses contaminating food and water [39]. Pregnant, HIV-positive women with negative toxoplasmosis titers should be counseled regarding avoidance of undercooked meats and foods contaminated by soil and animal feces [22]. Food poisoning can lead to weight loss and further compromise immunity to future infections. Proper hand washing, safe food handling and storage, and cooking foods to appropriate temperatures are especially paramount for the safety and health of the HIV-infected pregnant woman. See Table 12.5 for guidelines.

12.8   BREASTFEEDING

De Martino et al. [83] report that the risk of mother to child transmission (MTCT) of HIV through breastfeeding ranges from 4 to 14%, depending upon geographic location and whether feeding was sustained for greater than 1 year. Some risk factors for MTCT through breastfeeding include viral load in the blood [84], viral load in breast milk [85], the mother's immune status [86], the breast health of the mother [87], and the mother's nutritional status, including hemoglobin and serum retinol levels. [58, 88–89]. Mothers who are HIV-positive should be educated regarding the risks and benefits of different feeding options, including the risk of transmitting HIV through breastfeeding [1] as well as

Table 12.4
Nutrition and Dietary Management of Common HIV Symptoms

Anorexia

- Eat small, frequent meals.
- Capitalize on moments when you are hungry. Keep snacks readily available.
- Choose nutrient dense foods (peanut butter, cheese, yogurt).
- Avoid foods of low nutrient value (diet foods and beverages).

Nausea and vomiting

- Keep something in your stomach to curb nausea (i.e. crackers).
- Choose bland foods that are easy to digest such as toast, pasta, oatmeal, turkey, and pudding.
- Avoid greasy, high fat foods.
- Avoid spicy or highly seasoned foods.

Diarrhea

- Select binding foods such as bananas, toast, rice, and applesauce.
- Avoid high fat foods and lactose-containing foods.
- Drink adequate fluids to replace losses (water, sports drink, juices).

Sore mouth and throat

- Select smooth textured foods, such as pudding, yogurt, scrambled eggs, bananas, and applesauce.
- Avoid foods that are acidic, spicy, and that have rough edges and textures.
- Add moisture to foods with broths and gravies.
- Use a straw to drink liquids.

Table 12.5
Guidelines for Food Safety

- Ensure proper hand washing.
- Handle and store food in a safe manner (i.e. separate raw foods from prepared foods).
- Cook foods to appropriate temperatures.
- Store and keep foods at appropriate temperatures.
- Avoid raw or undercooked potentially hazardous foods (i.e. meat, poultry, fish, eggs).
- Use water from a clean and safe water supply.

an increased mortality rate among HIV-infected pregnant women who breastfeed [90]. The metabolic demands of lactation as well as the metabolic demands of HIV infection are thought to lead to nutritional impairment and subsequently, an increased risk of infant mortality [90].

It is recommended that mothers who can provide an alternate feeding method that is acceptable, affordable, sustainable, and safe are advised to do so [91]. The Centers for Disease Control and Prevention (CDC) recommend that HIV-infected pregnant women in the United States avoid breastfeeding [92].

For those mothers who choose to breastfeed, the quality of the breast milk is linked to the nutritional status of the mother, specifically the concentrations of both macronutrients and micronutrients as well as immunologic properties [17]. Mixed feeding, which is defined as breastfeeding coupled with other foods and liquids by mouth, appears to add to transmission risk in the first 6 months of life [93]. Longer durations of breastfeeding by HIV-infected mothers are associated with an increased risk of HIV transmission to their infants and early weaning from human milk (i.e., at 6 months of age) is recommended to limit the child's exposure to HIV-infected human milk [90].

12.9   GOALS OF NUTRITION CARE

The following goals for maternal nutritional care have been outlined by Lwanga [94]:

1. Improve nutritional status, maintain weight, prevent weight loss, and preserve lean body mass.
2. Ensure adequate weight gain during pregnancy.
3. Ensure adequate nutrient intake by improving dietary habits and encouraging appropriate macronutrient and micronutrient intake.
4. Prevent foodborne illness by promoting food and water safety.
5. Manage symptoms that affect nutrient intake to minimize the impact of secondary infections on nutritional status.

The general goal of nutritional assessment and interventions is to improve nutritional status, enhance quality of life, and prolong the survival of the mother [95].

12.10   APPROACHES TO NUTRITIONAL CARE

The following is a recommended approach to nutritional care for the HIV-infected pregnant woman and includes a baseline nutritional assessment, nutritional counseling, and follow-up nutritional care.

12.10.1   Nutritional Assessment

A baseline assessment should be made at the first prenatal visit and follow-up care should be provided at subsequent visits. The initial nutritional assessment should include baseline anthropometric measurement such as weight, height, BMI (height [cm]/kg [m²]), and mid–upper arm circumference (MUAC) [23]. The BMI will indicate whether the woman is at an appropriate weight or is underweight or overweight at onset of pregnancy. This information will enable the provider to make specific weight gain recommendations and allow for tracking of weight gain during pregnancy. Additionally, women who have a MUAC of <23 cm are at even greater nutritional risk [96, 97]; thus, this anthropometric marker may indicate the need for more aggressive nutritional intervention. Biochemical assessment measures including serum albumin, transferrin, hematocrit, creatinine, urea nitrogen, lipids, and micronutrients indicate disease prognosis and potential complications [98, 99]. These markers should be included in the initial assessment and followed throughout pregnancy.

The initial assessment takes into account the diet history of the woman as well as any symptoms or problems that might hinder adequate intake. Typical dietary, appetite, gastrointestinal symptoms (i.e., nausea, vomiting, diarrhea, and constipation), difficulty with chewing and swallowing, food allergies, ethnic and cultural food practices, and household food security should be considered and included in the assessment. Furthermore, all medications and supplements as well as complementary therapies should be investigated in order to determine possible drug-nutrient interactions.

12.10.2   Nutritional Counseling

- Weight gain goals should be set based on the woman's BMI at baseline. If underweight (BMI < 19.8 kg/m^2), then she needs to gain 28–40 lb during pregnancy. If at appropriate weight (BMI 19.8 – 26 kg/m^2), then she needs to gain 25–35 lb during pregnancy. If overweight (BMI > 26–29 kg/m^2), then she needs to gain 15–25 lb during pregnancy. If obese (BMI > 29 kg/m^2), then she needs to gain at least 15 lb during pregnancy (refer to Table 12.1).

- Energy needs are dependent on whether the woman is asymptomatic or symptomatic at the time of assessment. If asymptomatic, then her calorie needs are increased by 10% plus an additional 300 kcal per day on average during pregnancy (500 kcal per day during lactation). (see Chaps. 1 and 18, "Nutrient Recommendations and Dietary Guidelines for Pregnanat Women" and "Nutrition Issues During Lactation"). If symptomatic, then her calorie needs are increased by 20–30% plus an additional 300 kcal per day during pregnancy (500 kcal per day during lactation).

- Protein needs are estimated to be approximately 1 g/kg/day.

- Encourage a diet that is nutritionally adequate and varied. Additional multivitamin and mineral supplementation is also encouraged. Specific nutrients of importance are iron and folate. The DRI for iron is 27 mg/day and for folate is 600 mcg per day. While prenatal vitamins vary, most provide approximately 800–1,000 mcg of folic acid and 30 mg of iron per dose.

- Provide counseling regarding breastfeeding and risk of MTCT of HIV.

- Address food safety with regard to proper hand washing, safe food handling and storage, and safe cooking temperatures.

- Provide counseling for women on ARV treatment regarding meal planning, symptom management, and the metabolic changes associated with ARV therapy.

12.11   CONCLUSION

Nutrition assessment and counseling are critical components of the care plan for HIV-infected pregnant women. The compounding effects of pregnancy and HIV-infection place HIV-infected pregnant women at greater nutritional risk. Adequate weight gain and nutrient intake and symptom management are especially challenging for this population. Intervention in terms of care for HIV-infected pregnant women tends to be directed towards pregnancy outcomes and fetal health, more so than to maternal health and maternal morbidity and mortality. Given the growing population of HIV-infected women of childbearing age, there is a need for more research to determine the best nutrition and medical care plans for promoting both maternal and fetal health.

REFERENCES

1. American Dietetic Association and Dietitians of Canada (2004) Position of the American Dietetic Association and Dietitians of Canada: nutrition intervention in the care of persons with human immunodeficiency virus infection. J Am Diet Assoc 104:1425–1440
2. Seguardo AC, Miranda SD, Latorre M (2003) Evaluation of the care of women living with HIV/AIDS in Sao Paulo, Brazil. AIDS Patient Care and STDs 17:85–93
3. Tinkle MB, Amaya MA, Tamayo OW (1992) HIV disease and pregnancy, Part 1. Epidemiology, pathogenesis, and natural history. J Obstet Gynecol Nurs 21:86–93
4. World Health Organization (2005) Nutrition and HIV/AIDS: Report by the secretariat, Geneva, Switzerland, 12 May 2005
5. Gray RH, Wawer MJ, Serwadda et al (1998) Population-based study of women with HIV-1 infection in Uganda. Lancet 351:98–103

6. Urassa EJN, Kilewo C, Mtvangu SR, Mhalu FS, Mbena E, Biberfeld G (1992) The role of HIV in pregnancy wastage in Dar es Salaam, Tanzania. J Obstet Gynaecol Central Afr 10:70–72
7. Abrams EJ, Matheson PB, Thomas PA (1995) Neonatal predictors of infection status and early death among 322 infants at risk of HIV-1 infection monitored prospectively from birth. Pediatrics 96:451–458
8. Piwoz EG, Bentley ME (2005) Women's voices, women's choices: The challenge of nutrition and HIV/AIDS. J Nutr 135:933–937
9. Ladipo OA (2000) Nutrition in pregnancy: mineral and vitamin supplements. Am J Clin Nutr 72(Suppl):280S–290S
10. Picciano, MF (2003) Pregnancy and lactation: physiological adjustments, nutritional requirements and the role of dietary supplements. J Nutr 133:1997S–2002S
11. Keusch GT, Farthing MJ (1990) Nutritional aspects of AIDS. Annu Rev Nutr 10:475–501
12. Melchior JC, Salmon D, Rigaud D et al (1991) Resting energy expenditure is increased in stable, malnourished, HIV-infected patients. Am J Clin Nutr 53:437–441
13. Mulligan K, Tai V, Schambelan M (1997) Energy expenditure in human immunodeficiency virus infection. N Eng J Med 336:70–71
14. Grunfeld C, Pang M, Snimizu L, Shigenaga JK, Jensen P, Feingold KR (1992) Resting energy expenditure, caloric intake, and short-term weight change in human immunodeficiency virus and the acquired immunodeficiency syndrome. Am J Clin Nutr 55:455–460
15. Ceesay SM, Prentice AM, Cole TJ et al (1997) Effects on birth weight and perinatal mortality of maternal dietary supplements in rural Gambia: 5 year randomized control trial. Br Med J 315:836–848
16. Keithley JK, Swanson B, Nerad J (2001) HIV/AIDS. In: Gottschlich MM (ed) American Society for Parenteral and Enteral Nutrition's The Science and Practice of Nutrition Support. Kendall/Hunt, Des Moines, Iowa, pp 619–641
17. Fawzi W (2000) Nutritional factors and vertical transmission of HIV-1. Ann N Y Acad Sci 918:99–114
18. Wanke CA, Silva M, Ganda A et al (2003) Role of acquired immune deficiency syndrome-defining conditions in human immunodeficiency virus-associated wasting. Clin Infect Dis 37(Suppl 2):S81–S84
19. Villamor E, Dreyfuss ML, Baylin A, Msamanga G, Fawzi WW (2004) Weight loss during pregnancy is associated with adverse pregnancy outcomes among HIV-1 infected women. J Nutr 134:1424–1431
20. Brocklehurst P, French R (1998) The association between maternal HIV infection and perinatal outcome: a systematic review of the literature and meta-analysis. Br J Obstet Gynaecol 108:1125–1133
21. Institute of Medicine (1990) Nutrition during pregnancy. National Academy Press, Washington, D.C., pp 9–12
22. Hillhouse JH, Neiger R (2001) Pregnancy and Lactation in HIV/AIDS. In: Gottschlich MM (ed) American Society for Parenteral and Enteral Nutrition's The Science and Practice of Nutrition Support. Kendall/Hunt, Des Moines, Iowa, pp 301–321
23. World Health Organization (2004) Nutrition counseling, care, and support for HIV-infected women: Guidelines on HIV-related care, treatment and support for HIV-infected women and their children in resource-constrained settings. WHO Department of HIV/AIDS and Department of Nutrition for Health and Development. Geneva, Switzerland
24. Kotler DP, Grunfeld C (1996) Pathophysiology and treatment of the AIDS wasting syndrome. In: Voldberding P, Jacobson MA (eds) AIDS clinical review 1995/1996. Dekker, New York, N.Y., pp 229–275
25. Seumo-Fosso E, Cogill B (2003) Meeting nutritional requirements of HIV-infected persons. FANTA Project, Academy for Educational Development, Washington, D.C.
26. World Health Organization (1985) Energy and protein requirements. Report of a joint FAO/WHO/UNU expert consultation. Technical report series 724. Geneva, Switzerland
27. National Research Council (1989) Recommended Dietary Allowances, 10th edn. National Academy Press, Washington, D.C.
28. The National Academies (2002) Dietary Reference Intakes for energy, carbohydrate, fiber, fatty acids, cholesterol, protein, and amino acids. National Academy Press, Washington, D.C.
29. Institute of Medicine (1990) Nutrition during pregnancy. National Academy Press, Washington, D.C.
30. DeLegge MH, Ridley C (2001) HIV/AIDS. In: Gottschlich MM (ed) American Society for Parenteral and Enteral Nutrition's The Science and Practice of Nutrition Support. Kendall/Hunt, Des Moines, Iowa, pp 1–16

31. Semba RD, Graham NMH, Caiaffa WT, Clement L, Vlahov D (1993) Increased mortality associated with vitamin A deficiency during human immunodeficiency virus type-1 infection. Arch Intern Med 153:2149–2154
32. Tang AM, Graham NMH, Saah AJ (1996) Effects of micronutrient intake on survival in human immunodeficiency virus type 1 infection. Am J Epidemiol 143:1244–1256
33. Tang AM, Graham NMH, Ranjit KC, Saah AJ (1997) Low serum vitamin B_{12} concentrations are associated with faster human immunodeficiency virus type 1 (HIV-1) disease progression. J Nutr 127:345–351
34. Kupka R, Fawzi WW (2002) Zinc nutrition and HIV infection. Nutr Rev 60:69–79
35. Semba RD, Tang AM (1999) Micronutrients and the pathogenesis of human immunodeficiency virus infection. Br J Nutr 81:181–189
36. Friis H, Gomo E, Koestel et al (2001) HIV and other predictors of serum folate, serum ferritin, and hemoglobin in pregnancy: a cross-sectional study in Zimbabwe. Am J Clin Nutr 73:1066–1073
37. Friis H, Gomo E, Koestel P et al (2001) HIV and other predictors of serum beta-carotene and retinol in pregnancy: a cross-sectional study in Zimbabwe. Am J Clin Nutr 73:1058–1065
38. Yoshida SH, Keen CL, Ansari AA, Gershwin ME (1999) Nutrition and the immune system. In: Shils ME, Olsom JA, Shike M, Ross AC (eds) Modern nutrition in health and disease. 9th edn. Williams and Williams, Baltimore, Md., pp 725–750
39. Bothwell TH (2000) Iron requirements in pregnancy and strategies to meet them. Am J Clin Nutr 72(suppl):257S–264S
40. Academy for Educational Development (2004) HIV/AIDS: a guide for nutritional care and support, 2nd ed. Food and Nutrition Technical Assistance Project, Academy for Educational Development, Washington D.C.
41. Tomkins A, Watson F (1989) malnutrition and infection. A review. ACC/SCN, Geneva, Switzerland
42. Cook JD, Skikne BS, Baynes RD (1994) Iron deficiency: the global perspective. Adv Exp Med Biol 356:219–228
43. van den Broek NR, Letsky EA (2000) Etiology of anemia in pregnancy in south Malawi. Am J Clin Nutr 72(Suppl):247S–256S
44. Hercberg S, Galan P (1992) Nutritional anaemias. Baillieres Clin Haematol 5:143–168
45. Scott J. Folate (folic acid) and vitamin B12 (2000) In: Garrow JS, James WPT, Ralph A (eds) Human nutrition and dietetics. Churchill Livingstone, Edinburgh, UK, pp 271–280
46. Daly LE, Kirke PN, Molloy A, Weir DG, Scott JM (1995) Folate levels and neural tube defects. Implications for prevention. J Am Med Assoc 274:1698–1702
47. Scholl TO, Johnson WG (2000) Folic acid: influence on the outcome of pregnancy. Am J Clin Nutr 71(Suppl):1295S–1303S
48. The National Academies (2004) Dietary Reference Intakes for individuals: vitamins. National Academy of Sciences, Washington, D.C.
49. The National Academies (2004) Dietary Reference Intakes for individuals: minerals. National Academy of Sciences, Washington, D.C.
50. Fawzi W, Msamanga G, Spiegelman D, Hunter D (2005) Studies of vitamins and minerals and HIV transmission and disease progression. J Nutr 135:938–944
51. Clark TD, Semba RD (2001) Iron supplementation during human immunodeficiency virus infection: a double-edged sword? Med Hypotheses 57:476–479
52. Beisel WR (1982) Single nutrients and immunity. Am J Clin Nutr 35:417–468
53. Weinberg ED (1996) Iron withholding: a defense against viral infections. Biometals 9:393–399
54. Georgiou NA, van der BT, Oudshoorn M, Nottet HS, Marx JJ, van Asbeck BS (2000) Inhibition of human immunodeficiency virus type 1 replication in human mononuclear blood cells by the iron chelators deferoxamine, deferiprone, and bleomycin. J Infect Dis 181:484–490
55. Stoltzfus RJ (1994) Vitamin A deficiency in the mother-infant dyad. SCN News 11:25–27
56. West KP Jr, Katz J, Khatry SK et al (1999) Double blind, cluster randomized trial of low dose supplementation with vitamin A or beta carotene on mortality related to pregnancy in Nepal. The NNIPS-2 Study Group. Brit Med J 318:570–575.
57. Friis H, Gomo E, Koestel P et al (2001) HIV and other predictors of serum beta carotene and retinol in pregnancy: a cross-sectional study in Zimbabwe. Am J Clin Nutr 73:1058–1065

58. Rothman KJ, Moore LL, Singer MR, Nguyen US, Mannino S, Milunsky A (1995) Teratogenicity of high vitamin A intake. N Eng J Med 23:1369–1373

59. Fawzi W, Msamanga GI, Hunter D et al (2002) Randomized trial of vitamin supplements in relation to transmission of HIV-1 through breastfeeding and early childhood mortality. AIDS 16:1935–1944

60. Fawzi W, Msamanga GI, Antelman G et al (2004) Effect of prenatal vitamin supplementation on lower genital levels of HIV type 1 and interleukin type 1β at 36 weeks gestation. Clin Infect Dis 38:716–722

61. Shabert JK, Winslow C, Lacey JM, Wilmore DW (1999) Glutamine-antioxidant supplementation increases body cell mass in AIDS patients with weight loss: a randomized, double-blind controlled trial. Nutrition 15:860–864

62. Villamor, Msamanga, Spiegelman et al (2002) Effect of multivitamin A supplements on weight gain during pregnancy among HIV-1-infected women. Am J Clin Nutr 76:1082–1090

63. Mocchegiana E, Muzzioli M (2000) Therapeutic application of zinc in human immunodeficiency virus against opportunistic infections. J Nutr 130:1424S–1431S

64. Burbano X, Miguez-Burbano MJ, McCollister K et al (2002) Impact of selenium chemoprevention clinical trial on hospital admissions of HIV-infected participants. HIV Clin Trials 3:483–491

65. Fawzi WW, Msamanga GI, Spiegelman D et al (1998) Randomised trial of effects of vitamin supplements on pregnancy outcomes and T cell counts in HIV-1-infected women in Tanzania. Lancet 351:1477–1482

66. Kanter AS, Spencer DC, Steinberg MH, Soltysik R, Yarnold PR, Graham NM (1999) Supplemental vitamin B and progression to AIDS and death in black South African patients infected with HIV. J Acquir Immune Defic Syndr Hum Retrovirol 21:252–253

67. Tang AM, Graham NMH, Kirby AJ, McCall AD, Willett WC, Saah AJ (1993) Dietary micronutrient intake and risk progression to acquired immunodeficiency syndrome (AIDS) in human immunodeficiency virus type 1 (HIV-1)-infected homosexual men. Am J Epidemiol 138:1–15

68. Fall, C, Yajnik C, Davies A, Brown N, Farrant H (2003) Micronutrients and fetal growth. J Nutr 133:1474S–1756S

69. Goldenberg RL, Tamura T, Neggers YH, Copper RL, Johnston KE, DuBard MB, Hauth MD (1995) Effect of zinc supplementation on pregnancy outcome. J Am Med Assoc 274:463–468

70. Gibson RS, Ferguson EL (1998) Nutrition intervention strategies to combat zinc deficiency in developing countries. Nutr Res Rev 274:463–468

71. Spada C, Treitinger A, Reis M et al (2002) An evaluation of antiretroviral therapy associated with alpha-tocopherol supplementation in HIV-infected patients. Clin Chem Lab Med 40:456–459

72. Damle BD, Yah JH, Behr D, O'Mara E, Nichola P, Kaul S, Knupp C (2002) Effect of food on the oral bioavailability of didanosine from encapsulated enteric-coated beads. J Clin Pharmacol 42:419–427

73. Kenyon CJ, Brown F, McClelland GR, Wilding IR (1998) The use of pharmacoscintigraphy to elucidate food effects observed with a novel protease inhibitor (saquinavir). Pharm Res. 15:417–422

74. Minkoff H, Augenbraun M (1997) Antiretroviral therapy for pregnant women. Am J Obstet Gynecol 176:478–489

75. Frascino RJ (1995) Changing face of HIV/AIDS care: mother–fetal and maternal–child HIV transmission. West J Med 163:368–369

76. Augenbaum M, Minkoff HL (1997) Antiretroviral therapy in the pregnant woman. Obstet Gynecol Clin North Am 24(4):833–854

77. Augenbaum M, Minkoff HL (1997) Antiretroviral therapy in the pregnant woman. Obstet Gynecol Clin North Am 24:833–854

78. Antimicrobial Therapy, Inc. (2001)The Sanford Guide to HIV/AIDS Therapy, 10th edn. Antimicrobial Therapy, Inc., Hyde Park, Vt.

79. Mosby (1999) Mosby's GenRx. 9th edn. Mosby, St. Louis, Mo.

80. Temple ME, Koryani KI, Nahata MC (2003) Lipodystrophy in HIV-infected pediatric patients receiving protease inhibitors. Ann Pharmaco 37:1214–1218

81. El Beitune P, Duarte G, Quintana SM, Figueiro-Filho EA (2004) HIV-1: maternal prognosis. Rev Hosp Clin Fac Med S Paulo 59:25–31

82. Kim JH, Spiegelman D, Rimm E, Gorbach SL (2001) The correlates of dietary intake among HIV-positive adults. Am J Clin Nutr 74:852–861

83. de Martino M, Tovo PA, Tozzi AE et al (1992) HIV-1 transmission through breast-milk: appraisal of risk according to duration of feeding. AIDS 6:991–997

84. Richardson BA, John-Stewart GC, Hughes JP et al (2003) Breast-milk infectivity in human immunodeficiency virus type-1 infected mothers. J Infect Dis 187:736–740

85. Rousseau CM, Nduati RW, Richardson BA et al (2003) Longitudinal analysis of human immunodeficiency virus type 1 RNA in breast milk and its relationship to infant infection and maternal disease. J Infect Dis 187:741–747

86. Leroy V, Karon JM, Alioum A et al (2002) Twenty-four month efficacy of a maternal short-course zidovudine regimen to prevent mother-to-child transmission of HIV-1 in West Africa. AIDS 16:631–641

87. Semba RD, Kumwenda N, Hoover DR (1999) Human immunodeficiency virus load in breast milk, mastitis, and mother-to-child transmission of human immunodeficiency virus type 1. J Infect Dis 180:93–98

88. Joao EC, Cruz ML, Menezes JA et al (2003) Vertical transmission of HIV in Rio de Janeiro, Brazil. AIDS 17:1853–1855

89. Ioannidis JP, Abrams EJ, Ammann A et al (2001) Perinatal transmission of human immunodeficiency virus type 1 by pregnant women with RNA virus loads <1,000 copies/ml. J Infect Dis 183:539–545

90. Read JS, and Committee on Pediatric AIDS (2003) Human milk, breastfeeding, and transmission of human immunodeficiency virus type 1 in the United States. Pediatrics 112:1196–1205

91. World Health Organization (2001) New data on the prevention of mother-to-child transmission of HIV and their policy implications: Conclusions and recommendations. WHO Technical Consultation on behalf of the UNFPA/UNICEF/WHO/UNAIDS Inter-Agency Task Team on Mother-to-Child Transmission of HIV. Geneva, Switzerland: World Health Organization, report no. WHO/RHR/01.28

92. Centers for Disease Control and Prevention (1985) Recommendations for assisting in the prevention of perinatal transmission of human T-lymphotropic virus type III/lymphadenopathy-associated virus and acquired immunodeficiency syndrome. Morb Mortal Wkly Rep 34:721–726, 731–732

93. Coutsoudis A (2000) Promotion of exclusive breastfeeding in the face of the HIV pandemic. Lancet 356:1620–1621

94. Lwanga D (2001) Clinical care of HIV-infected women in resource poor settings: Nutritional care and support. Johns Hopkins Program on International Education for Obstetrics and Gynecology (JHPIEGO), Baltimore, Md., CD-ROM tutorial

95. American Dietetic Association and Dietitians of Canada (2000) Position of the American Dietetic Association and Dietitians of Canada: nutrition intervention in the care of persons with human immunodeficiency virus infection. J Am Diet Assoc 100:708–717.

96. James WP, Mascie-Taylor GC, Norgan NC et al (1994) The value of arm circumference measurements in assessing chronic energy deficiency in Third World adults. Euro J Clin Nutr 48:883–894

97. Gartner A, Marie B, Kameli Y et al (2001) Body composition unaltered for African women classified as "normal but vulnerable" by body mass index and mid-upper arm circumference criteria. Euro J Clin Nutr 55:393–399

98. Feldman JG, Goldwasser P, Holman S, DeHovitz, Minkoff H (2003) C-reactive protein is an independent predictor of mortality in women with HIV-1 infection. J Acquir Immune Defic Syndr 32:210–214

99. Salomon J, de Truchis P, Melchoir JC (2002) Body composition and nutritional parameters in HIV and AIDS Patients. Clin Chem Lab Med 40:1329–1333

III

SPECIAL DIETS, SUPPLEMENTS, AND SPECIFIC NUTRIENTS DURING PREGNANCY

13 Popular Diets

Nancy Rodriguez and Michelle Price Judge

Summary Popular diets are a constant in the lives of women. And while some restricted-calorie plans recommending extreme differences in the contribution of calories from fat, carbohydrate, and protein may promote weight loss in some women, they should not be undertaken during pregnancy. Unfortunately, the scientific literature documenting the prevalence of use of popular diets in pregnant women, or more importantly the effects of such dietary behavior on pregnancy outcomes, is, for all practical purposes, nonexistent. Energy balance is central to the regulation of body weight. During pregnancy, the intent is for energy balance to be positive such that fetal growth and development is well supported by adequate calories and essential nutrients. The coexistence of a state of negative energy balance for the purpose of weight loss during pregnancy will not support optimal gestation. An appreciation of the basic principles of the most established popular diets in the context of nutrition and pregnancy will provide a foundation for the clinician and other healthcare professionals who treat and counsel pregnant women and women of childbearing age. This chapter provides a framework for practitioners from which thoughtful and appropriate discussions can result when approached with questions regarding the potential concerns associated with the practice of such diet behaviors during pregnancy. Without question, the best approach to healthful weight management during pregnancy combines a well-balanced diet with reasonable, routine physical activity and exercise. Habitual consumption of a well planned, energy sufficient diet containing the recommended amount of the macro- and micronutrients for the duration of pregnancy will minimize weight-related issues and other metabolic problems from conception to delivery. Clinicians and practitioners should be familiar with the premise of popular diets so they are prepared to offer appropriate counsel when they encounter dieting behaviors during their patients pregnancy.

Keywords: Popular diets, Weight management, exercise, pregnancy

13.1 INTRODUCTION

Women are most susceptible to the seduction of weight loss promised by the overwhelming number of popular diets promoted by the media. Although essentially no scientific literature documenting the prevalence of use of popular diets in pregnant women exists, clinicians and practitioners should be familiar with the premise of these

From: *Nutrition and Health: Handbook of Nutrition and Pregnancy*
Edited by: C.J. Lammi-Keefe, S.C. Couch, E.H. Philipson © Humana Press, Totowa, NJ

diets in the event they encounter dieting behaviors during pregnancy in their patients. This chapter highlights the basic principles of the most established popular diets in the context of nutrition and pregnancy to provide a foundation for the clinician and other healthcare professionals who treat and counsel pregnant women and women of child-bearing age. The intent is not to promote the use of popular diets by these women, but to provide a foundation for practitioners from which thoughtful and appropriate discussions can result when approached with questions regarding whether such diet behaviors are appropriate during pregnancy. Although less enticing, the best approach to healthful weight management during pregnancy combines a well-balanced diet with reasonable, as well as routine, physical activity and exercise. A well-planned diet that provides adequate energy and meets macronutrient and micronutrient requirements for the duration of pregnancy is the basis for minimizing weight-related issues and other metabolic problems from conception to delivery. Because energy balance (i.e., energy in and energy out), as well as diet composition (i.e., percent of calories provided by carbohydrate, protein and fat), is central to the discussion of popular diets, points specific to contemporary diet plans are integrated throughout the chapter.

13.2 RECOMMENDATIONS FOR WEIGHT GAIN DURING PREGNANCY

Guidelines for weight gain during pregnancy aim to promote adequate, but not excessive, weight gain for optimal fetal development. Weight gain is highly correlated with infant birth weight making optimal weight gain during pregnancy important to fetal outcomes [1]. For a thorough discussion of optimal weight gain for pregnancy, the reader is referred to Chap. 2, "Optimal Weight Gain," in Part 1 of this book. In brief, the Institute of Medicine (IOM) developed guidelines for maternal weight gain based on aggregate data examining fetal outcomes and associated maternal conditions [1]. These guidelines, adapted by both the American College of Obstetrics and Gynecology (ACOG) and the American Dietetic Association (ADA), use maternal body mass index (BMI, kg/m²) prior to conception (Tables 13.1, 13.2) as a starting point for recommended weight gain during pregnancy [1–4]. Although these guidelines are available to women during pregnancy, educational programs regarding how to follow these guidelines appear to be lacking as evidenced by several studies documenting that only 30–40% of American women meet the IOM guidelines for weight gain [1, 3].

Table 13.1
ACOG Guidelines for Weight Gain during Pregnancy [1]

Condition prior to pregnancy	Weight gain guideline (lb)
Underweight	28–40 (12.7–18.2 kg)
Normal weight	25–35 (11.4–16 kg)
Overweight	15–25 (6.8–11.4)
Obese	15 (6.8 kg)
Twins/triplets	35–45 (16–20.5 kg)

Table 13.2
ADA Position Paper: Guidelines for Weight Gain during Pregnancy

Body mass index (BMI)	Recommended weight gain	Weight gain/week after 12 weeks
BMI <19.8	12.5–18 kg (28–40 lb)	0.5 kg (1 lb)
BMI 19.8–26	11.5–16 kg (25–35 lb)	0.4 kg (0.9 lb)
BMI > 26–29	7–11.5 kg (15–25 lb)	0.3 kg (0.7 lb)
BMI > 29	At least 7 kg (15 lb)	
Twin pregnancy	15.9–20.4 kg (34–45 lb)	0.7 kg (1.5 lb)
Triplet pregnancy	Overall gain of 22.7 kg (50 lb)	

While managing a healthy weight throughout pregnancy may be a challenge for many women, it is important for women to embrace the fact that they are pregnant and need to gain weight in order to ensure having a healthy baby. Weight reduction during pregnancy is discouraged and has been associated with neuropsychological abnormalities and low birth weight in the infant [2]. Conversely, excess weight gain can place women at higher risk for complications. Women should not use pregnancy as an excuse to eat excessively.

Although diet is a key component to weight maintenance, exercise, rest, and lifestyle are also highly important. During pregnancy, women should strive to eat a varied diet that encompasses all of the nutrients essential for fetal development. This varied diet should be well balanced and not too high or low in any one of the macronutrients.

13.3   RECOMMENDATIONS FOR HEALTHY EATING

A complete approach to meeting nutrient and health needs during normal pregnancy is presented in Part 1 of this book. Specific aspects of this information in the context of weight management are presented here.

13.3.1   Energy

Activity level, age, height, and weight prior to pregnancy are all factors that are considered when determining an individual's energy requirements. Although energy requirements vary from woman to woman, most women's energy needs range from approximately 2,500 to 2,700 kcal daily [4]. Caloric requirements during the second and third trimesters of pregnancy are estimated to be 300 kcal/day (500 kcal/day for adolescents <14 years of age) above caloric requirements prior to pregnancy. The prepregnancy energy requirements used as a basis for this caloric estimation during pregnancy should account for age, activity level, and prepregnancy weight [4]. The Dietary Reference Intakes (DRIs) for pregnancy take into account increasing needs in support of fetal growth and appropriate maternal weight gain [5–7]. Butte and King [8] estimated energy costs of pregnancy using respiratory calorimetry throughout pregnancy and found that the daily cost of pregnancy increased by trimester. Weight gain during pregnancy should be individualized relative to prepregnancy BMI such that pregnancy outcome is improved, postpartum weight retention is minimized, and risk for adult chronic disease is reduced in the child [1]. In special situations such as adolescent pregnancy in which additional calories are necessary for the adolescent's own growth energy, requirements may be higher to maintain weight gain goals [2]. Pregnant adolescents less than 14 years of age require an additional 500 kcal/day [4, 9]. Likewise, women carrying twins need to consume 500 calories daily above energy needs for pregnancy as outlined above [7, 10]. Conversely, women who are overweight or obese prior to pregnancy will require fewer calories due to the availability of stored energy. In any case, it is important that positive energy balance (energy in vs. energy out) exists such that a pregnant woman is in an anabolic state (energy in > energy out). The degree of positive energy balance should be determined relative to an individual woman's prepregnancy body weight (Tables 13.1, 13.2). Theoretically, an overweight woman may eat fewer calories when she becomes pregnant but still achieve an anabolic state because the degree of positive energy balance (calorie intake in excess of energy expended), while less than that prior to the pregnancy will generally be sufficient to support fetal growth and development.

Once energy needs are established, it is important that the pregnant woman understands how to translate her calorie needs into appropriate food choices to support a healthy pregnancy. Requirements for some nutrients, such as protein, iron, and calcium, are increased during pregnancy. Therefore, pregnant women should focus on nutrient dense foods [foods that provide a lot of nutrients relative to the number of calories]. For example, one egg will contribute high-quality protein, essential fat, as well as a variety of vitamins and minerals, for approximately 75 calories. Too often patients are given calorie levels that may be specific to their needs without adequate instruction on how to incorporate these guidelines into their daily routines. Twenty-four-hour diet recalls conducted and evaluated by a registered dietitian, in combination with appropriate nutrition education materials, can be very useful in assisting the individual in translating their usual diet into meal plans that are consistent with recommendations for appropriate and healthy weight gain throughout pregnancy.

13.3.2 Diet Composition

The amount of energy contributed by the macronutrients—carbohydrate, protein, and fat—does not vary substantially during normal pregnancy. The role of each of these nutrients in normal physiology and metabolism remains intact with a heightened importance for some functions in the context of fetal growth and development. For that reason, it is critical that pregnant women do not self-impose diet restrictions or extremes during pregnancy.

13.3.2.1 CARBOHYDRATE

Carbohydrate is the brain's main fuel and the nutrient that fuels muscles for daily tasks and exercise. The growing fetus relies on the mother's carbohydrate supply. A diet too low in carbohydrate can affect the energy level of the mother. During normal pregnancy, care should be taken to ensure that 50–60% of daily calories are provided as carbohydrate. Diets low in carbohydrate should not be attempted during pregnancy, as the effects of such a diet on fetal development are not known. Carbohydrate source should be well planned to ensure that the majority of carbohydrates are complex, with a limited consumption of refined sugar or simple carbohydrate. Examples of complex carbohydrates include bread, rice, beans, pasta, and potatoes. When grains are refined, they are stripped of many important nutrients, including fiber, which are important in pregnancy. Pregnant women should be advised to consume whole-grain bread, cereal, and pasta products. Fruits, vegetables, and whole-grain products are good sources of dietary fiber that are beneficial in preventing constipation during pregnancy. Foods with simple sugar like candy, soft drinks, and desserts should be limited during pregnancy as they are high in calories and low in nutritional value. These fat and sugar-laden foods can displace other more nutritious foods and contribute to accelerated weight gain. A woman requiring 2,500 kcal daily would need 275–330 g of carbohydrate daily. One slice of bread, 1/2 cup of cooked pasta, 1/2 cup dry cereal or a serving of fruit all provide ~15 g of carbohydrate per serving.

13.3.2.2 PROTEIN

Adequate protein intake during pregnancy is important to maintain maternal health during pregnancy as well as provide important building blocks for fetal growth and development. Protein provides the structural framework for the body, is integral to the immune system, transports substances throughout the body, is the basis for many hormones and enzymes, and maintains fluid balance. Pregnant women need a minimum of 60 g of protein daily [2]. Good sources of dietary protein include meat, poultry, fish, dairy products, legumes, beans, and nuts. One ounce of meat, poultry, fish, or cheese provides 7 g of protein. One 8-oz. glass of skim milk provides 9 g of protein. Two 3-oz. servings of meat, poultry, or fish and three 8-oz glasses of milk will provide the protein necessary to meet the protein needs of pregnancy.

13.3.2.3 FAT

Fat is more energy dense than carbohydrate or protein (9 kcal/g vs. 4 kcal/g for fat and carbohydrate and protein, respectively) when consumed within recommended guidelines, fat is beneficial to maternal health and fetal development. Fats are important in maintaining skin health, as a structural component of cells, for absorption of vitamins A, D, E, and K as regulatory messengers (hormones), hormone, and for immune function. Recent evidence suggests that omega-3 fatty acids consumed during pregnancy are beneficial to cognitive development in infancy and childhood [11–14]. Pregnant women should consume 25–30% of their daily energy as fatty acids. A woman requiring 2,500 kcal/day would need to consume 69–83 total grams of fats daily. These fat requirements should be met using vegetable-based oils made up of unsaturated fat rather than animal and plant-based saturated fats that can be more problematic to health [15].

Sources of unsaturated fats include olive oil, canola oil, peanut oil, sunflower oil, flax seed oil, and fish oil. Canola oil and flax seed oil are sources of the essential fatty acid α-linolenic acid. Dietary oils either derived or obtained directly from fish are particularly beneficial during pregnancy as fish contains a preformed source of docosahexaenoic acid (DHA) which is the metabolic end product of α-linolenic acid in the body [16] and preferentially transferred to fetal tissue during pregnancy [17–19]. Women should consume 300 mg of DHA daily during pregnancy [20]. Many pregnant women are concerned about eating fish during pregnancy due to potential contamination; however, women need to be educated about safe fish intake during pregnancy to ensure consumption of these important omega-3 fatty acids (http://www.cfsan.fda.gov/~comm/haccp4.html). Omega-6 fatty (linoleic) acid should make up 2% of total energy during pregnancy. This recommendation does not differ from the recommendation for the general adult population [20].

Table 13.3
Micronutrient Deficiency Risks Associated with Various Popular Diets

Popular diet type:	Micronutrient
Fat restrictive	Fat soluble vitamins (A, D, E, K), essential fatty acids
Protein restrictive	Iron, vitamin B_{12}, zinc, magnesium, essential amino acids
Non-dairy	Calcium, vitamin D
Carbohydrate restrictive	B vitamins, vitamins C, A, K and D, potassium, fiber

Sources of saturated fat include whole milk, beef, cheese, lard, shortening, and palm and coconut oil. No more than 10% of daily calories should come from a saturated fat source. A woman who requires 2,200 kcal daily should consume no more than 24 g of saturated fat daily.

13.3.3 Micronutrients

Pregnancy is a time to ensure adequate micronutrient intake in addition to sufficient energy and macronutrients for optimal fetal development. The key to ensuring dietary adequacy of the micronutrients is a varied diet that includes multiple foods from all food groups. Factors that place pregnant women at risk for micronutrient deficiencies include diets that restrict energy, diets that omit one or more major food groups, food insecurity, food intolerances, allergy or food aversions. Dietary plans should be developed according to risk factors that include alternative dietary sources. Attention should also be given to potential micronutrient deficiencies associated with the respective types of diets. Several micronutrients are of critical importance to the fetus's growth and development (Table 13.3).

13.4   ACCEPTABLE PHYSICAL ACTIVITY AND EXERCISE PLANS

A plan for regular exercise is another key component of weight management [21–25]. Women need to dispel the myth that women should "take it easy" during pregnancy. Chapter 3, "Physical Activity and Exercise in Pregnancy," provides a thorough discussion regarding physical activity and exercise during pregnancy. In brief, pregnant women should consult their doctor before initiating an exercise program or modifying their existing regimen to rule out complications [24]. Light-to-moderate physical activity does not negatively influence fetal development for a normal, uncomplicated pregnancy [21, 25]. A properly designed exercise program during pregnancy is beneficial and can contribute to healthy weight management at this time [22, 24].

In addition to adequate calorie and nutrient intake, and appropriate exercise and physical activity, various lifestyle factors should be considered when planning for appropriate weight gain during pregnancy. Occupation, leisure activities, stress level, and habitual dietary behaviors (i.e., eating out, eating cues, binge eating) are important considerations for weight management programs. Behavior modification strategies may need to be implemented for women who have problems with habitual unhealthy dietary behaviors (See Chap. 9, "Anorexia Nervosa and Bulimia Nervosa during Pregnancy"). All of these factors should be taken into consideration in consultation with a registered dietitian and in collaboration with the supervising physician.

13.5 POPULAR DIETS: IMPLICATIONS FOR PREGNANCY

In general, it is important for women to understand that the goal during pregnancy is a healthy baby and not weight loss. Pregnant women should not fall prey to fad diets or diet plans that limit the types of foods that can be eaten. With the ongoing exposure to the written and visual media directed at body image, it is not surprising that some pregnant women might become fearful of weight gain during pregnancy and consider very meticulous and restricted eating plans. These plans are likely to parallel the popular diets being marketed at the time. The challenge to practitioners is to stay abreast of current diet trends so they are prepared to educate women regarding potential pitfalls of these weight loss plans specific to pregnancy. Fortunately for doctors, nurses, dietitians, and other allied health professionals working with pregnant women, the principles and recommendations for healthy weight gain during pregnancy are straightforward allowing for easy identification of nutrient- or metabolic-specific flaws of popular diets in the context of pregnancy. In addition, although not particularly seductive, the most effective approach to weight management is based on the simplicity of weight management: energy balance. To gain weight, energy intake must exceed that expended so the body is able to invest energy in the anabolic reactions critical to the deposition of musculoskeletal tissues and the development of organs and the nervous system. Weight loss or maintenance will occur when energy intake is less than or equal to that expended, respectively. Popular diets typically lead the consumer away from this concept with false promises most often based on changes in the composition of the calories eaten, rather than the total amount of calories consumed relative to what is needed.

Weight management plans focusing on the macronutrient composition of the diet have become the cornerstone of popular diets most likely to be considered by a pregnant woman. Plans that reduce carbohydrate intake in exchange for increasing protein and fat may be particularly troublesome during pregnancy [26, 27]. The following discussion addresses various versions of popular diets that practitioners working with pregnant women may encounter. Because the scientific literature is essentially void of research in this area, points are made within the context of what is known regarding optimal nutrition for a healthy pregnancy and how these particular diet practices might undermine these recommendations.

13.5.1 Low-Carbohydrate Diets

Low-carbohydrate diets are perhaps the most prolific of all weight loss plans. In 1992, Dr. Robert Atkins published *Dr. Atkins' New Diet Revolution* [28]. This was eventually followed by the publication of two subsequently revised versions of the book. The premise of this plan is to restrict carbohydrate intake so severely that ketoacidosis ensues. The plan, which provides only 20 g of carbohydrate during the "Induction Phase," allows for extremely high-fat (particularly saturated fat) and protein intakes [28] relative to what is generally recommended as healthful for the general population [29]. It should be readily apparent to the practitioner that this amount of carbohydrate is inadequate for supporting pregnancy [1, 6, 29, 30]. More importantly, a state of ketosis during pregnancy is not consistent with a normal metabolic profile for gestation and can be problematic for both the mother and fetus [26, 31]. In addition, the amount of calories that is ultimately provided by this plan will not meet the energy needs of pregnancy.

The immediate physiological response to the Atkins diet is a significant weight loss within 7–10 days of following the diet plan. A large component of weight loss is due to fluid losses that are associated with ketosis and needed for excretion of nitrogenous metabolites consequent to excessive dietary protein. Neither of these scenarios is desirable for the pregnant woman.

β-Hydroxybutyrate has been shown to cross the placenta in high amounts in sheep and produce significant decreases in fetal oxygenation, lactate, and fetal heart rate [32]. Although it is currently unknown how ketoacidosis impacts fetal outcome, in human, pregnancy complicated with diabetes places the developing fetus at risk for spontaneous abortion, stillbirth, congenital malformation, and macrosomia [33]. Given these negative outcomes associated with uncontrolled diabetes during pregnancy and subsequent ketoacidosis, women should avoid ketogenic diets during pregnancy.

Normal hydration is important for body functions and physical performance and concerns regarding the impact of excessive nitrogen secretion on renal function remain an issue for healthy men and nonpregnant women [34, 35]. Although various studies have been published to document potential benefits of the Atkins Diet to reducing risk factors for cardiovascular disease [36], there remains absolutely no premise for induction of ketosis during pregnancy. Even Dr. Atkins himself warned that his diet is not an appropriate eating plan for pregnancy [28].

The public's fascination with the resurgence of the Atkins diet in the twentieth century resulted in the publication of two other books that promoted higher protein intakes for weight loss, *The Zone Diet* and *The South Beach Diet* [37, 38]. These books exploited the concept of high-protein intakes for the purpose of weight loss. However, these higher protein intakes are not at levels that carbohydrate intake is restricted to the extent that ketosis occurs. Rather, these plans intended to slightly exceed recommended protein intakes for maintaining muscle mass during weight loss, keep fat within recommended ranges of approximately 30% of total calories, and promote a level of carbohydrate intake that prevents extreme fluctuations in blood glucose thereby keeping insulin secretion at bay [37, 38].

The drawback to these plans was the lack of controlled scientific studies in support of this approach to weight management. However, both authors do an exceptional job reviewing and distilling the existing scientific literature in developing the hypotheses for their respective weight loss programs. As a result, the arguments presented, along with the anecdotal reports of successful weight loss efforts sustained the popularity of these approaches to weight management. Eventually the scientific community was challenged to design and execute investigations that would either support or refute claims put forth by these weight loss programs. And indeed, it would appear that reduced calorie diets for which protein intakes are slightly above the recommended amount (1.5 g/kg vs. 0.8 g/kg) are quite effective in eliciting weight loss when combined with regular exercise [23, 39–42].

Of significance to this chapter, however, is that this work was carried out in healthy, overweight men and women, not pregnant women. An evidence-based analysis by the Cochrane group documented that balanced energy and protein intakes as recommended for pregnancy are an appropriate guide for normal, healthy fetal growth and development [43]. The report further states, "high-protein or balanced-protein supplementation alone is not beneficial and may be harmful to the fetus" [43]. That is, protein intakes in excess of the current RDA for pregnancy are of no benefit during pregnancy and may be detrimental to the fetus. This comprehensive report clearly illustrates the importance of balanced nutrition for pregnancy, particularly the critical relation between energy and protein for the purpose of weight gain during pregnancy and subsequent growth and development of the fetus. This report is consistent with current guidelines for nutrition during pregnancy and provides a sound, scientific-based rationale for counseling women to avoid diet plans that promote protein intakes in excess of the current recommendation whether for weight loss or management during pregnancy.

Adequate energy is needed to spare protein for the purpose of body protein synthesis, maintenance, and repair. This relationship is critical during pregnancy. Therefore, dieting behaviors that promote reduced energy intake, with or without additional protein, for the purpose of weight loss during pregnancy are not appropriate. In addition, consumption of protein intakes in excess of the RDA may be problematic for the mother and the fetus.

13.5.2   Low-Fat Plans

Low-fat diet plans should be cautiously evaluated. As mentioned previously, severe dietary fat restriction raises the concern for fat-associated micronutrient deficiency. Indeed, essential fatty acid deficiency has been reported during pregnancy when the woman self-imposed a dietary fat restriction [44]. Although a reduction in dietary fat intake can be paralleled by a reduction in energy intake, many "low-fat" and "no-fat" products are actually quite calorie dense due to substitution of fat with carbohydrate, specifically sugar, to overcome taste deficits when fat is removed. Women must be educated regarding the critical role that fats, specifically essential fatty acids, play in growth and development of the fetus [12, 17].

13.5.3   Very Low–Calorie Plans

By definition, very low–calorie diet plans provide approximately 400–900 kcal daily. Clearly, the aforementioned discussion, coupled with the information provided in Part 1 of this book, demonstrates the inappropriate nature of such an approach to weight management during pregnancy. Even the most proficient registered dietitian could not design a nutritional care plan that provided all of the essential nutrients required for a normal, healthy pregnancy with so few calories. Pregnant women must be cautioned against undertaking any type of reduced calorie plan that compromises energy and nutrient intake to this extent. If such behaviors were to persist, the practitioner must consider further evaluation or referral for eating disordered behaviors (see Chap. 9, "Anorexia Nervosa and Bulimia Nervosa during Pregnancy").

13.6   WEIGHT LOSS SUPPLEMENTS

13.6.1   Liquid Formulas and Meal-Replacement Bars

The marketplace is replete with a variety of nutritional supplements provided as liquid formulas and bars. For the purpose of weight loss, these products are not usually integrated into a balanced menu plan but used as substitutes for the entire meal, thereby reducing total calorie intake. While the literature presents a variety of supplementation and pregnancy studies, most deal with individual or mixed nutrient (i.e., vitamins or minerals) supplementation rather than energy or protein or both. Two supplementation studies appeared pertinent to the discussion of popular diets and pregnancy [45–47]. Both studies provided liquid formulas to pregnant women to improve energy and nutrient intake with the intent of improving pregnancy outcomes. Although the supplements did contribute additional calories to the total intake for some mothers, many of the women used the supplements as food and meal replacements [47]. The former option, drinking a nutritional supplement or eating a nutrition bar to ensure adequate calorie intake during pregnancy is a reasonable option in practice if a woman cannot consume sufficient calories and nutrients from whole foods because appetite is lacking or time is limiting. However, using a liquid nutritional supplement or a nutrient dense meal replacement bar as a substitute for a meal for the purpose of weight loss or weight management is not acceptable and should be discouraged in pregnancy.

13.6.2 *Pharmacological Diet Aids*

Herbal dietary supplements should be avoided during pregnancy due to the lack of clinical trials investigating safety and efficacy [4]. Herbal and botanical products are not regulated and therefore can contain a number of compounds that may be unsafe for consumption during pregnancy. Of particular concern would be herbal dietary supplements marketed for weight loss. The same risks associated with consumption of these products in the general population would likely be exaggerated during pregnancy. The governmental website http://vm.cfsan.fda.gov provides information regarding safety of, and adverse reactions associated with, herbal dietary supplements.

Caffeine acts to accelerate fat burning and is often used by dieters to assist weight loss attempts. During pregnancy, caffeine has been found to affect fetal heart rate and breathing [48] and increase risk for spontaneous abortion or low birth weight [49]. However, results regarding caffeine intake and risk for preterm labor are equivocal as more recent reports have not demonstrated a reduced risk of preterm labor associated with modest decreases in caffeine consumption [50, 51]. Given the potential for adverse impacts on the developing infant, it is recommended that pregnant women not consume more than 300 mg of caffeine daily [4].

13.7 BREASTFEEDING

Although pregnancy is a time when women should be discouraged from caloric restriction, concerns regarding weight gain provide an excellent opportunity to promote breastfeeding. Women concerned about weight gain may be particularly interested in learning how breastfeeding can help shed pounds gained during pregnancy [52]. Dewey and colleagues [53] found that weight loss is enhanced in the postpartum period if lactation is continued beyond 6 months. Rather than pursue weight loss by restricting calorie, and consequently nutrient, intake, women should be educated regarding energy balance in the context of breastfeeding. Specifically, by understanding the high-energy demands of lactation, women can apply the concept of negative energy balance subsequent to breastfeeding to potential weight loss during the postnatal period. The needs for successful lacatation are discussed further in Chap. 18, ("Nutrition Issues during Lactation").

13.8 SUMMARY AND RECOMMENDATIONS

Appropriate weight gain, not weight loss, should be the objective in place for achieving the goal of a healthy baby during pregnancy. Pregnant women should receive sound nutrition education regarding the interaction of healthy eating and routine physical activity for favorable pregnancy outcomes. While it may be tempting to restrict food consumption to minimize weight gain, pregnant women must be guided such that they do not fall prey to fad diets. Personal weight reduction goals should be deferred until after the baby is born. Health professionals working with pregnant women should be knowledgeable regarding the basic principles of energy balance so that information can be extended to practical application and education regarding healthy weight gain during pregnancy. It is recommended that practitioners be aware of the present onslaught of popular diets. Hence, physicians, dietitians, and nurses need to acknowledge the likelihood that pregnant women may want to minimize or avoid weight gain during pregnancy and do so by undertaking undesirable dietary behaviors. A basic understanding of the premise of popular diets and their impact on the ability for women to adequately meet energy and nutrient requirements for pregnancy will serve as a foundation for sound discussions to direct women to healthy diet and lifestyle behaviors at this important time.

REFERENCES

1. Food and Nutrition Board (1990) Nutrition during pregnancy, weight gain and nutrient supplements. Report of the Subcommittee on Nutritional Status and Weight Gain during Pregnancy, Committee on Nutritional Status During Pregnancy and Lactation, Food and Nutrition Board. National Academy Press, Washington, D.C.
2. Chicago Dietetic Association SSDA, Dietitians of Canada (2000) Pregnancy. In: Berzy D, Mehta A (eds) Manual of clinical dietetics, 6th edn. American Dietetic Association, Chicago, Ill.
3. Abrams B, Altman, SL, Pickett KE (2000) Pregnancy weight gain: still controversial. Am J Clin Nutr 71(Suppl):1233S–1241S
4. Kaiser LL, Allen, L (2002) Position of the American Dietetic Association: Nutrition and lifestyle for a healthy pregnancy outcome. J Am Diet Assoc 102:1479–1490
5. Brown JE, Sugarman Isaacs J, Murtaugh MA, Sharbaugh C, Stang J, Wooldridge NH (2005) Nutrition through the life cycle, 2nd edn. Wadsworth, Belmont, Calif.
6. American Dietetic Association (2002) Position of the American Dietetic Association: Nutrition and lifestyle for a healthy pregnancy outcome. J Am Diet Assoc 102:1479–90
7. American Dietetic Association (2006) ADA nutrition care manual. ADA, Chicago, Ill.
8. Butte NF, King JC (2005) Energy requirements during pregnancy and lactation. Public Health Nutr 8:1010–1027
9. Gutierrez Y, King JC (1993) Nutrition during teenage pregnancy. Pediatr Ann 22:99–108
10. Klein L (2005) Nutritional recommendations for multiple pregnancy. J Am Diet Assoc 105:1050–1052
11. Colombo J, Kannass KN, Shaddy DJ, Kundurthi S, Maikranz JM, Anderson CJ, Blaga OM, Carlson SE (2004) Maternal DHA and the development of attention in infancy and toddlerhood. Child Development 75:1254–1267
12. Helland IB, Smith L, Saarem K, Saugstad OD, Drevon CA (2003) Maternal supplementation with very-long-chain n-3 fatty acids during pregnancy and lactation augments children's IQ at 4 years of age. Pediatrics 111:39–44
13. Judge MP, Harel O, Lammi-Keefe CJ (2007) Maternal consumption of a DHA-functional food during pregnancy: comparison of infant performance on problem-solving and recognition memory tasks at 9 months of age. Am J Clin Nutr 85:1572–1577
14. Judge MP, Harel O, Lammi-Keefe CJ (2007) A docosahexaenoic acid-functional food during pregnancy benefits infant visual acuity at four but not six months of age. Lipids 42:117–122
15. Khoury J, Henriksen T, Seijeflot I, Mekrid L, Froslie K, Tonstad S (2007) Effects of an antiatherogenic diet during pregnancy on markers of maternal and fetal endothelial activation and inflammation: the CARRDIP study. Br J Obstet Gynecol 114:279–288
16. Sprecher H (1999) An update on the pathways of polyunsaturated fatty acid metabolism. Curr Opin Clin Nutr Metab Care 2:135–138

17. Haggarty P (2004) Effect of placental function on fatty acid requirements during pregnancy. Eur J Clin Nutr 58:1559–1570
18. Haggarty P (2002) Placental regulation of fatty acid delivery and its effect on fetal growth—a review. Placenta 23:S28–S38
19. Dutta-Roy AK (2000) Transport mechanisms for long-chain polyunsaturated fatty acids in the human placenta. Am J Clin Nutr 71:315–322
20. Simopoulos AP, Leaf A, Salem N (2000) Workshop statement on the essentiality and recommended dietary intakes for omega-6 and omega-3 fatty acids. Prostaglandins Leukot Essent Fatty Acids 83:119–121
21. American College of Obstetrics and Gynecology (2002) Opinion no. 267: Exercise during pregnancy and the postpartum period. Obstet Gynecol. 99:171–173
22. Clapp JF III, Little KD (1995) Effect of recreational exercise on pregnancy weight gain and subcutaneous fat deposition. Med Sci Sports Exerc 27:170–175
23. Layman DK, Evans E, Baum JI, Seyler J, Erickson DJ, Boileau RA (2005) Dietary protein and exercise have additive effects on body composition during weight loss in adult women. J Nutr 135:1903–1910
24. McArdle WD, Katch FI, Katch VL (2001) Exercise physiology: energy nutrition and human performance, 5th edn. Lippincott Williams & Wilkins, New York, N.Y.
25. Weissgerber T, Wolfe L, Davies G, Mottola M (2006) Exercise in the prevention and treatment of maternal-fetal disease: A review of the literature. Appl Physiol Nutr Metab 31:661–674
26. Romon M, Nuttens M, Vambergue A, Verier-MIne O, Biausque S, Lemaire C, Fontaine P, Salomez J, Beuscart R (2001) Higher carbohydrate intake is associated with decreased incidence of newborn macrosomia in women with gestational diabetes. J Am Diet Assoc 101:897–902
27. Jannette K, Henriksen T, Christophersen B, Tonstad S (2005) Effect of a cholesterol-lowering diet on maternal, cord, and neonatal lipids, and pregnancy outcome: a randomized clinical trial. Am J Obstet Gynecol 193:1292–1301
28. Atkins RC (2002) Dr. Atkins' new diet revolution. Evans, New York, N.Y.
29. Institute of Medicine (2002) Dietary Reference Intakes for energy, carbohydrates, fiber, fat, fatty acids, cholesterol, protein, and amino acids. National Academy Press, Washington, D.C.
30. American College of Obstetricians and Gynecologists (2000) Planning your pregnancy and birth. Editorial Task Force of the American College of Obstetricians and Gynecologists, Washington, D.C.
31. Potter J, Reckless J, Cullen D (1982) Diurnal variations in blood intermediary metabolites in mild gestational diabetic patients and the effect of a carbohydrate restricted diet. Diabetologia 22:68–72
32. Miodovnik M, Skillman C, Hertzberg V, Harrington D, Clark K (1986) Effect of maternal hyperketonemia in hyperglycemic pregnant ewes and their fetuses. Am J Obstet Gynecol 154:394–401
33. Siddiqui F, James D (2003) Fetal monitoring in type 1 diabetic pregnancies. Early Human Devel 72:1–13
34. Martin WF, Armstrong LE, Rodriguez NR (2005) Dietary protein intake and renal function. Nutr Metab 2:25
35. Martin WF, Cerundolo LH, Pikosky MA, Gaine PC, Maresh CM, Armstrong LE, Bolster DR, Rodriguez NR (2006) Effects of dietary protein intake on indexes of hydration. J Am Diet Assoc 106:587–589
36. Gardner CD, Kiazand A, Alhassan S, Kim S, Stafford RS, Balise RR, Kraemer HC, King AC (2007) Comparison of the Atkins, Zone, Ornish, and LEARN diets for change in weight and related risk factors among overweight premenopausal women: the A to Z Weight Loss Study: a randomized trial. J Am Med Assoc 297:969–977
37. Sears B (1995) The Zone. Harper Collins, New York, N.Y.
38. Agatston A (2003) The South Beach Diet. Rodale, Emmaus, Pa.
39. Layman DK (2004) Protein quantity and quality at levels above the RDA improves adult weight loss. J Am Coll Nutr 23:631S–616S
40. Layman DK, Baum JI (2004) Dietary protein impact on glycemic control during weight loss. J Nutr 134:968S–973S
41. Layman DK, Boileau RA, Erickson DJ, Painter JE, Shiue H, Sather C, Christou DD (2003) A reduced ratio of dietary carbohydrate to protein improves body composition and blood lipid profiles during weight loss in adult women. J Nutr 133:411–417
42. Layman DK, Shiue H, Sather C, Erickson DJ, Baum J. Increased dietary protein modifies glucose and insulin homeostasis in adult women during weight loss. J Nutr 133:405–410
43. Kramer M, Kakuma R (2007) Energy and protein intake in pregnancy. Wiley, New York, N.Y.

44. Tsai EC, Brown JA, Veldee MY, Anderson GJ, Chait A, Brunzell JD (2004) Potential of essential fatty acid deficiency with extremely low fat diet in lipoprotein lipase deficiency during pregnancy: a case report. BMC Pregnancy Childbirth [electronic resource] 4:27

45. Kusin JA, Kardjati S, Houtkooper JM, Renqvist UH (1992) Energy supplementation during pregnancy and postnatal growth. Lancet 340:623–626

46. Adams SO, Barr GD, Huenemann RL (1978) Effect of nutritional supplementation in pregnancy. I. Outcome of pregnancy. J Am Diet Assoc 72:144–147

47. Adams SO, Huenemann RL, Bruvold WH, Barr GD (1978) Effect of nutritional supplementation in pregnancy. II. Effect on diet. J Am Diet Assoc 73:630–634

48. Briggs GG, Freeman RK, Yaffe SJ (1994) Drugs in pregnancy and lactation. Williams & Williams, Baltimore, Md.

49. Fernandes O, Sabharwai M, Smiley T, Pastuszak A, Koren G, Einarson T (1998) Moderate to heavy caffeine consumption during pregnancy and relationship to spontaneous abortion and abnormal fetal growth: a meta-analysis. Reprod Toxicol 12:435–444

50. Grosso LM, Triche EW, Belanger K, Benowitz NL, Holford TR, Bracken MB (2006) Caffeine metabolites in umbilical cord blood, cytochrome P-450 1A2 activity, and intrauterine growth restriction. Am J Epidemiol 163:1035–1041

51. Bech BH, Obel C, Henriksen TB, Olsen J (2007) Effect of reducing caffeine intake on birth weight and length of gestation: randomized controlled trial. Brit Med J 334:7590

52. Kramer FM, Stunkard AJ, Marshall KA, McKinney S, Liebschultz J (1993) Breastfeeding reduces maternal lower-body fat. J Am Diet Assoc 93:429–433

53. Dewey KG, Heinig, MJ, Nommsen LA (1993) Maternal weight-loss patterns during prolonged lactation. Am J Clin Nutr 58:162–166

14 Dietary Supplements During Pregnancy: Need, Efficacy, and Safety

Mary Frances Picciano
and Michelle Kay McGuire

Summary National surveys indicate that as many as 97% of women living in the United States are advised by their health care providers to take multivitamin, multimineral (MVMM) supplements during pregnancy, and 7–36% of pregnant women use botanical supplements during this time. Although there is evidence of benefit from some of these preparations, efficacy has not been established for most of them. This chapter reviews some of the most commonly used prenatal supplements in terms of the evidence for their need, efficacy, and safety. Specifically, MVMM, folate, vitamin B6, vitamin A, vitamin D, iron, zinc, magnesium, and iodine are discussed, as are several botanicals. Data indicate that, in general, evidence for benefit gained from taking prenatal MVMM supplements is not well established except for women who smoke, abuse alcohol or drugs, are anemic, or have poor quality diets. Because of folate's well-established effect on decreasing risk for neural tube defects, it is recommended that all women of childbearing age consume supplemental folic acid daily (0.4 mg/day) or obtain that amount from fortified foods. Similarly, it is recommended that all pregnant women be provided with iron supplementation (30–60 mg/day), and a recent policy statement by the American Thyroid Association suggests that all pregnant women living in the United States or Canada consume 150 mcg/day supplemental iodine to prevent iodine deficiency disorders. Currently, there is insufficient evidence to advise population-wide use of other dietary supplements, although zinc may be warranted for women consuming a vegan diet. Use of all botanical products should be carefully monitored and evaluated during pregnancy, especially those (e.g., chamomile and blue cohosh) that are contraindicated during this time. Clinicians are advised to periodically review current recommendations concerning these products, as research in this area is ongoing.

Keywords: Pregnancy, Dietary supplements, Nutrition, Prenatal vitamins, Women, Health

From: *Nutrition and Health: Handbook of Nutrition and Pregnancy*
Edited by: C.J. Lammi-Keefe, S.C. Couch, E.H. Philipson © Humana Press, Totowa, NJ

14.1 INTRODUCTION

During pregnancy, nutrient requirements increase to support fetal growth and development as well as maternal metabolism and tissue development specific to this period in the lifespan [1]. Although meeting these increased nutrient requirements can and perhaps should be achieved by the consumption of appropriate amounts of foods in a balanced and varied diet, the use of dietary supplements may be beneficial in some situations [2]. Indeed, prenatal vitamin and mineral supplementation is common among women living in the United States.

Using the 1988 National Maternal and Infant Health Survey, Yu and colleagues concluded that 97% of US women are advised to take multivitamin multimineral (MVMM) supplements as part of their routine prenatal care, and 67 and 84% of African-American and Caucasian women, respectively, comply with this recommendation [3]. These authors also reported that women who do not choose to consume prenatal MVMM preparations tend to be nonsmokers, less educated, younger, and/or unmarried. Data from another study suggested that prenatal supplements were taken by 86% of a culturally diverse group of low-income pregnant women [4]. In addition to MVMM supplements, some pregnant women use botanical preparations to treat common symptoms such as nausea.

The fundamental purpose of this chapter is to review the strength of the evidence that use of MVMM, single-nutrient, and botanical supplements during pregnancy can safely enhance pregnancy outcome and pregnancy-related symptoms. In particular, we will focus mainly on (1) if and when they may be indicated, and (2) whether they are efficacious and safe when taken by otherwise apparently healthy women living in economically developed countries. It is noteworthy that the topic of nutritional supplement intake during pregnancy is also covered within other chapters of this publication. For example, the relationship between folate (folic acid) intake and birth defects is discussed in detail in Chap. 17, "Folate: a Key to Optimal Pregnancy Outcome," and the potential impact of iron supplementation on birth outcome is covered in Chap. 16, "Iron Requirements and Adverse Outcomes." These nutrients, therefore, will be discussed here only briefly, and the reader is directed to these more detailed chapters as appropriate. This chapter will also provide basic information regarding functions and food sources of each of the nutrients covered as well as published guidelines for dietary supplement usage during pregnancy.

14.2 BASIC DEFINITIONS AND CONCEPTS RELATED TO SUPPLEMENT USE

14.2.1 Definition and Regulation of Dietary Supplements

The Dietary Supplement Health and Education Act (DSHEA) of 1994 issued by the US Food and Drug Administration (FDA) defines the term "dietary supplement" as a product that collectively meets the following requirements [5]:

- A product (other than tobacco) intended to supplement the diet or contain one or more of the following: vitamin, mineral, herb or other plant-derived substance (e.g., ginseng, garlic), amino acid, concentrate, metabolite, constituent, or extract.

- A product intended for ingestion in pill, capsule, tablet, or liquid form
- A product not represented for use as a conventional food or as the sole item of a meal or diet

The FDA requires that all dietary supplements be labeled as such. However, unlike drugs, dietary supplements do not need approval before they are marketed. The manufacturers and distributors of supplements are responsible for ensuring their safety and making sure that label claims are accurate and truthful. For more information concerning the regulation of dietary supplements marketed within the United States, the reader is referred to the FDA's Center for Food Safety and Applied Nutrition help-line (1-888-723-3366) or their website (http://www.cfsan.fda.gov/list.html).

14.2.2 Recommended Nutrient Intakes and Dietary Supplement Use

Although during pregnancy a number of metabolic adaptations are orchestrated to support both increased maternal and fetal needs for many nutrients, the body's requirements for some nutrients cannot be met without increased dietary intake. Indeed, available evidence indicates that dietary requirements for 14 of the 21 essential micronutrients increase during pregnancy. These nutrients comprise seven vitamins, five minerals, and choline [6]. As such, it is important to increase one's intake of these nutrients to prevent deficiencies. It is also important during this period of the lifespan to not consume *too much* of each nutrient to reduce risk for levels of intake that may be harmful.

The Institute of Medicine's (IOM) Dietary Reference Intakes (DRIs) are considered to be the gold standard in recommendations for nutrient intake, and having a basic knowledge of this set of dietary reference standards is important for understanding nutrient requirements and potential impacts of dietary supplements during pregnancy. The DRIs comprise several sets of nutrient intake standards, perhaps the most clinically important being the Recommended Dietary Allowances (RDA), the Adequate Intake Levels (AI), and the Tolerable Upper Intake Levels (UL). The RDA values represent evidence-based estimates of nutrient intakes that meet the need of approximately 97% of the population, and the AI values are derived when less solid scientific data are available. Both RDA and AI values are often used by practitioners as recommended intakes. Conversely, UL values represent the amount above which one should not consume a particular nutrient, and many of these values are based on findings of toxicity due to dietary supplement usage. As will be described throughout this chapter, dietary supplement intake should never exceed UL values. The relationships among the RDA, AI, and UL values are illustrated in Fig. 14.1.

14.3 MULTIVITAMIN MULTIMINERAL SUPPLEMENTATION DURING PREGNANCY

14.3.1 Introduction

As previously mentioned, use of MVMM during pregnancy is very common in the United States. In addition, programs to distribute MVMM in developing countries are thought by some to be desirable for the purposes of increasing birth weight and decreasing perinatal mortality. In this section, we summarize the evidence for benefit of the consumption of these multinutrient supplements during pregnancy.

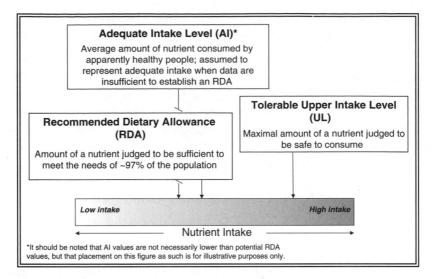

Fig. 14.1. Simplified illustration of the relationship among nutrient intake levels, Recommended Dietary Allowances (*RDA*), Adequate Intake Levels (*AI*), and Tolerable Upper Intake Levels (*UL*). The Institute of Medicine recommends that dietary intake (including that from supplements) for most healthy individuals fall between the RDA or AI and UL values

14.3.2 Evidence that MVMM Supplementation during Pregnancy is Beneficial

Although many healthcare providers recommend that pregnant women consume a standard prenatal MVMM supplement as an "insurance policy" against inadequate micronutrient intake, evidence to support a benefit from this practice for most women in developed countries is weak. Nonetheless, there are a handful of reports suggesting that MVMM use is related to decreased risk of fetal malformations. For example, Botto and coworkers conducted a retrospective, case-control study in which they compared incidence of conotruncal heart defects in 2 groups of women living in the Atlanta, Georgia area: those giving birth to infants with conotruncal defects (*n* = 158) and unaffected women (*n* = 3,026) [7]. "Multivitamin use" was defined as "reported regular use from 3 months before conception through the third month of pregnancy." Data from this study suggest that periconceptional multivitamin use is associated with a reduced risk for this form of malformation. However, because this study was not an intervention trial, there was no standardization of the type of prenatal supplement taken by the women, and it is quite possible confounding factors may have influenced the findings.

Another study that utilized a more rigorous study design provided additional, but limited, evidence that periconceptional multivitamin supplementation may be beneficial [8, 9]. This study was conducted as a randomized, double-blind, "placebo-like" controlled study with 4,682 subjects. Data suggest that multivitamin supplementation can reduce the risk of urinary tract abnormalities, cardiovascular malformations, and neural tube defects. Interestingly, these authors also found an

Table 14.1
Suggested Composition of Daily Multivitamin Multimineral Supplement for Use During Pregnancy and Comparison with the Dietary Reference Intake (DRI) Values

Nutrient	Amount[a]	RDA/AI	UL
Iron	30 mg	27 mg	45 mg
Zinc	15 mg	11 mg	40 mg
Copper	2 mg	1 mg	10 mg
Calcium	250 mg	1,000 mg	2,500 mg
Vitamin B6	2 mg	1.9 mg	100 mg
Folate	300 mcg	600 mcg	1,000 mcg
Vitamin C	50 mg	85 mg	2,000 mg
Vitamin D	5 mcg	5 mcg	50 mcg

From [2, 31]

[a]Note that this supplement formulation was suggested prior to the establishment of (1) the DRI values, (2) the American Thyroid Association recommendation that all pregnant women receive iodine supplements (150 mcg/day), and (3) the current recommendation that all women of childbearing age consume 0.4 mg/day (400 mcg/day) folic acid in the form of supplements or fortified foods

effect on fertility and multiple births, such that the women receiving the multivitamin supplement had a 5% shorter time in the achievement of subsequent pregnancy and experienced higher rates of monozygotic twinning [10, 11]. There was no effect of multivitamin consumption on rate of fetal deaths, low birth weight, or preterm birth in singletons. However, care should be taken when drawing conclusions from this study, because the placebo-like group actually received a supplement containing several trace elements. Other observational studies support possible effects of prenatal MVMM use on risk of preeclampsia, preterm birth, cleft palate, and other fetal malformations [12–15]. However, more controlled, prospective studies are needed to confirm these effects.

14.3.3 Recommendations

The American Dietetic Association and the IOM recommend that all pregnant women who smoke or abuse alcohol or drugs take MVMM supplements as should those with iron deficiency anemia or poor quality diets [2, 16]. This recommendation also applies to vegans and women carrying two or more fetuses. The recommended amounts of micronutrients in prenatal MVMM preparations are provided in Table 14.1. Care should always be taken that intake from these supplements not result in consumption of nutrient levels above the ULs set for pregnancy.

14.4 SINGLE-VITAMIN SUPPLEMENTS DURING PREGNANCY

14.4.1 Introduction

Aside from MVMM supplements, there are also single-vitamin preparations that are considered by some to provide benefits to some pregnant women. Some of these (e.g., folate) are generally prescribed to all pregnant women, whereas others (e.g., vitamin B6) may be prescribed to treat particular pregnancy-related issues. This section will review the evidence that several of the most commonly used single-vitamin supplements are effective and safe to use during this period of the lifecycle.

14.4.2 Folate/Folic Acid

The term "folate" refers to a group of chemically-similar, water-soluble compounds required for one-carbon transfer reactions. These chemical reactions are vital for life, being required for the synthesis, interconversion, and modifications of nucleotides (e.g., DNA), amino acids, and a myriad of other cellular components [17]. The term "folic acid" is used to describe the synthetic form found in nutritional supplements and fortified foods. It should be noted that, beginning in 1998 the FDA mandated that all enriched cereal-grain products marketed in the United States be fortified with folic acid. Aside from enriched cereal grains, good sources of folate include many legumes, leafy vegetables, and orange juice [18].

Among all of the micronutrients that are frequently prescribed or recommended during pregnancy, it is folate for which there is the most convincing evidence of a beneficial effect [19]. Indeed, it has long been recognized that folate supplementation during pregnancy could decrease the risk of megaloblastic anemia, prompting the US Food and Nutrition Board in 1970 to recommend folic acid supplementation (200–400 mcg/day) for all pregnant women [20, 21]. Since that time, hundreds of papers have been published concerning the impact of maternal folate nutriture on pregnancy complications, fetal growth and development, and fetal malformations, and the reader is directed to Chap. 17 within this book for a more complete discussion of these topics. Of great relevance to this chapter is the effect of periconceptional folate supplementation on neural tube defects (NTD).

The relationship between folate deficiency and increased risk of NTD was first suggested by Smithells and colleagues in 1976 when they reported that folate deficiency was *related to* increased risk of a woman having a baby with this type of malformation [22]. Following this first report, others published additional findings that periconceptional consumption of folic acid alone or in combination with other vitamins was related to decreased risk for NTD; these data, however, were often criticized for their lack of scientific stringency [23–25].

In 1991, Britain's Medical Research Council published a well-controlled, randomized intervention trial in which women planning to conceive and who had histories of having infants with NTD consumed either a placebo or a daily periconceptional folic acid supplement (4 mg/day). These women were considered "high risk" for delivering a subsequent baby with a NTD. Data showed that <1% (5 of 593) of the women consuming folic acid had a child with a NTD, compared with >3% (21 of 602) of the women receiving the placebo. Several other studies followed, all providing additional evidence that folic acid supplementation during the periconceptional period decreased risk for both first-time occurrence and recurrence of NTD [26, 27]. For example, Czeizel and Dudás conducted a large-scale ($n = 4,704$) intervention trial in which they found that periconceptional folic acid supplementation at a much lower dose (0.8 mg/day) could decrease the incidence of NTD in women with no history of delivering infants with these types of malformations. Thus, the relationship between periconceptional folic acid supplementation and risk of NTD in some women is now well established.

14.4.2.1 RECOMMENDATIONS

Due to the convincing evidence that periconceptional folic acid supplementation can decrease NTD in some women, many health organizations recommend routine folic acid supplementation during this period. For example, the US Centers for Disease Control and Prevention (CDC) began in 1991 to recommend that women at high risk of having a baby with NTD should plan subsequent pregnancies, and consume 4 mg/day of folic acid from the time they begin trying to become pregnant through the first trimester of pregnancy [28]. In 1992, they expanded their recommendation by stating that all women of childbearing age who are capable of becoming pregnant should consume 0.4 mg/day of folic acid to reduce their risk of having a pregnancy affected with spina bifida or other NTDs. This recommendation has been adopted by several clinical practice associations, such as the American Academy of Pediatrics and the National Healthy Mothers, Healthy Babies Coalition [29, 30]. Similarly, the IOM recommends that "to reduce the risk of NTD, women able to become pregnant should take 0.4 mg of folic acid daily from fortified foods, supplements, or both, in addition to consuming food folate from a varied diet" [31]. It should be noted that data from the 2005 March of Dimes Gallup survey indicated that only 33% of US women of childbearing age reported taking supplemental folic acid daily [32].

Because folic acid supplementation can make the diagnosis of a coexisting vitamin B_{12} deficiency difficult, the IOM has established a UL for folic acid of 1 mg/day for women aged 19 years and older. This UL does not apply to food folate, but only to synthetic forms obtained from supplements, fortified foods, or a combination of the two. The CDC also recommends that care be taken to keep total folate consumption at less than 1 mg/day, except under the supervision of a physician [33].

14.4.3 Vitamin B6

The term "vitamin B6" describes three related compounds: pyridoxine, pyridoxal, and pyridoxamine. These vitamins act as coenzymes involved in the synthesis and metabolism of proteins and amino acids. Vitamin B6 is also important for a variety of other physiologic functions such as carbohydrate metabolism and steroid hormone regulation. Important food sources of this vitamin include meat, poultry, fish, milk, nuts, whole and fortified grain products, potatoes, and bananas [34].

Vitamin B6 has long been thought to decrease nausea and vomiting during pregnancy, and it is still widely used for this common condition [35, 36]. Vitamin B6 was also a component in the once widely used medication doxylamine–pyridoxine (marketed as Debendox® and Bendectin®). This drug was removed from the market in 1983, when its use was suspected to cause limb defects [37]. It is noteworthy, however, that this putative detrimental effect of doxylamine–pyridoxine on birth defects was not confirmed in subsequent research, and the combination of doxylamine and pyridoxine is currently available for treating nausea and vomiting in pregnancy under the trade name Diclectin® in Canada [38, 39].

Indeed, some studies provide evidence that vitamin B6 supplementation during pregnancy can decrease the severity of nausea and vomiting without dangerous side-effects. For example, Vutyavanich and colleagues conducted a randomized, double-blind, placebo-controlled trial in which pregnant women received either oral pyridoxine hydrochloride (30 mg/day) or a placebo [40]. Women receiving the pyridoxine reported less severe nausea ($P = 0.0008$), and had a trend toward having fewer vomiting episodes ($P = 0.06$). Other investigators have found similar results, especially in women experiencing severe nausea [41]. However, there are very few reports of well-controlled trials in this area, and conclusions drawn by systematic reviews of the literature are mixed. For example, whereas Jewell and Young reported in 2003 that evidence supports an effect of vitamin B6 on decreasing severity of nausea, Thaver and colleagues reported in 2006 that there is not enough evidence to conclude that vitamin B6 supplementation in pregnancy has clinical benefits [42, 43]. Without a doubt, the conduct of additional clinical trials is warranted on this topic, as vomiting and nausea represent serious complications for many pregnant women.

14.4.3.1　RECOMMENDATIONS

The American College of Gynecology in 2004 in issuing its most recent guidance on treatment of morning sickness during pregnancy stated that, "taking vitamin B6 … is safe and effective and should be considered a first-line treatment." The IOM has established the UL for this vitamin to be 100 mg/day during pregnancy. It should be noted that, although no detrimental effects have been associated with high intakes of vitamin B6 from foods, very large oral doses (2,000 mg/day or more) of supplemental pyridoxine are associated with the development of sensory neuropathies and dermatological lesions. Thus, this level of supplementation should always be avoided.

14.4.4　Vitamin A

The first to be recognized as a vitamin per se, the term "vitamin A" actually refers to a group of lipid-soluble compounds (called retinoids) all having biological activity similar to all-*trans*-retinol. Vitamin A is essential for vision—both in low- and high-intensity light [44, 45]. It also plays a role in intercellular communication as a transcription factor in a variety of signaling pathways, including those important for cell cycle progression [46–48]. Because of its diverse functions, vitamin A deficiency can influence almost every physiologic system including the immune system and reproduction. Foods supplying the majority of vitamin A to the American diet include meat (especially liver), whole milk, butter, and dark-green or deep-yellow fruits, and vegetables (via provitamin A carotenoids).

During pregnancy, additional vitamin A is needed to support maternal metabolism and tissue growth as well as fetal growth and development. It is generally accepted that this increased need is relatively small and can be best met through diet (not supplements). In fact, there is likely more concern about vitamin A *toxicity* during pregnancy in healthy populations, as excess vitamin A from supplements has been shown to be teratogenic. For example, Rothman and colleagues found that consumption of >3,000 mcg/day vitamin A in the form of supplements during the first trimester was related to a fourfold increase in the risk for neural crest defects [49].

However, in areas of the world with endemic vitamin A deficiency, supplementation with vitamin A during pregnancy may be beneficial for both maternal and infant health. For example, vitamin A supplementation may enhance night vision, maternal immune function, and weight gain during pregnancy, especially in women with comorbidities [50–52]. Recently, the relationship between vitamin A status and transplacental transmission of HIV from mother to child has also been studied. However, in a systematic review of this literature, Wiysonge and colleagues concluded that, overall, there is no evidence that vitamin A supplementation decreases maternal-to-infant HIV transmission [53]. There is limited evidence, in fact, that vitamin A supplementation may *increase* transmission. Not surprisingly, this topic is one of great current intensity, and the reader is directed to Chap. 12, "HIV/AIDS in Pregnancy," which provides a more thorough discussion.

14.4.4.1 RECOMMENDATIONS

In summary, vitamin A supplementation is likely not warranted or desired in otherwise healthy women. However, in areas of the world with chronic and endemic vitamin A deficiency, routine low-dose daily supplementation or weekly higher-dose supplementation may be recommended [54]. Because of the risk of fetal malformations, the IOM has established the UL for this nutrient to be 3,000 mcg/day during pregnancy [55].

14.4.5 Vitamin D

There are two main forms of vitamin D in foods: ergocalciferol (vitamin D_2) which is found in plant foods, and cholecalciferol (vitamin D_3), which is found in animal foods. In the body, these compounds are converted to the active form of vitamin D ($1,25[OH]_2D_3$; calcitriol) by metabolism in the liver and kidneys. It is important to note that calcitriol can also be synthesized endogenously via exposure of the skin to ultraviolet light, thus sometimes bypassing the need for exogenous (dietary) sources. The functions of vitamin D are diverse, although its roles in facilitating calcium absorption and bone formation are best described [56, 57]. Because fish oils and some mollusks constitute by far the most vitamin D–dense sources of this micronutrient to the diet, populations consuming little or no marine products are at elevated risk for low vitamin D intakes [58]. To help prevent vitamin D deficiency, most liquid milk products are fortified with this nutrient in the United States.

Recent reports of rickets among some US infants—especially those who are dark-skinned and exclusively breastfed—have prompted renewed concern about maternal vitamin D status [59, 60]. Indeed, there is controversy concerning the optimal vitamin D intake during pregnancy, and some suggest that it may be higher than the currently recommended amount (5 mcg/day; 200 IU/day) especially in dark-skinned women and/or those exposed to very little sunlight [61]. Clearly, this area deserves additional focused research.

Several randomized trials of vitamin D supplementation during pregnancy have been conducted, most of which show a positive effect of supplementation on neonatal calcium handling [62, 63]. However, these finding tend to be more robust when investigators primarily consider women at high risk for vitamin D deficiency because of low vitamin D intake or decreased endogenous vitamin D synthesis due to lack of sunlight [64]. Some studies have also found a beneficial effect of vitamin D supplementation on birth weight, although these findings are inconsistent and require further confirmation [65, 66].

14.4.5.1 RECOMMENDATIONS

Most experts agree that there is little evidence supporting a benefit of prenatal vitamin D supplements above amounts routinely required to prevent classical vitamin D deficiency [67, 68]. With regard to this topic, the IOM states the following: "Women, whether pregnant or not, who receive regular exposure to sunlight do not need vitamin D supplementation. However, an intake of 10 mcg (400 IU)/day, which is supplied by prenatal vitamin supplements, would not be excessive" [69].

14.5 SINGLE-MINERAL SUPPLEMENTS DURING PREGNANCY

14.5.1 Introduction

Like vitamins, there are several minerals that are generally recommended to all pregnant women (e.g., iron) and others that are sometimes recommended in specific situations (e.g., magnesium). In this section we describe several commonly used single-mineral supplements in terms of their need, safety, and efficacy during pregnancy.

14.5.2 Iron

Iron facilitates the movement of oxygen from the environment to the body's aerobic tissues via hemoglobin and is also intimately associated with the electron transport chain, making this mineral essential for ATP production. Iron is also a component of several other enzymatic and nonenzymatic proteins [70, 71]. Iron deficiency results in a wide variety of signs and symptoms including anemia, poor immune function, compromised work performance, and altered behavior and cognitive function. Iron bioavailability varies greatly and is highest from animal products. Good dietary sources include meat (and organ meats), enriched cereal products, legumes, nuts, and seeds.

Because of the increased need for iron during pregnancy, the IOM recommends that iron intake during this period increase by 9 mg/day to a total of 27 mg/day. Because this level of intake is difficult for most women to achieve, supplementation is often recommended. Nonetheless, iron deficiency remains a significant public health problem in both developed and developing countries, and there is a plethora of well-designed dietary intervention studies suggesting that routine oral iron supplementation during pregnancy is beneficial [72]. For example, Cogswell and colleagues studied the effects of iron supplementation in a group of low-income yet healthy American women ($n = 275$) with hemoglobin concentrations ≥ 110 g/l and ferritin concentrations ≥ 20 mcg/l [73]. Women consumed either an iron supplement (30 mg/day ferrous sulfate) or a placebo until 28 weeks of gestation, at which time those with frank iron deficiency (serum ferritin <12 mcg/l) were provided with 60 mg/day iron; women with depleted iron stores (serum ferritin between 12 and 20 mcg/l) received 30 mg/day iron, regardless of initial assignment. Although iron supplementation during the first 28 weeks of gestation did not influence prevalence of anemia or the incidence of preterm births, it did lead to higher birth weight, lower incidence of low-birth-weight infants, and lower incidence of preterm low-birth-weight infants. Many other randomized, placebo-controlled trials have produced similar results, and the reader is directed to Chap. 16, which provides an extensive review of this topic.

14.5.2.1 RECOMMENDATIONS

Because of the recognized benefits of additional iron during pregnancy, the WHO recommends daily iron supplementation (60 mg/day) for all pregnant women for 6 months or, if 6 months of treatment cannot be achieved during the pregnancy, either continuation of supplementation during the postpartum period or increased dosage of 120 mg/day iron during pregnancy [74, 75]. Other recommendations include that of the CDC, which is that oral low-dose (30 mg/day) supplements of iron be provided to all pregnant women at the first prenatal visit [76].

The IOM has set the RDA and UL for this nutrient to be 27 and 45 mg/day, respectively, from all sources. It should be noted that, because of the potential negative impacts of iron toxicity, the use of prophylactic iron supplements in nonanemic pregnant women continues to be an issue of debate within the research and health communities [77]. Thus, as is the case for all dietary supplements, it is recommended that clinicians keep informed of evolving professional positions concerning iron supplementation during pregnancy.

14.5.3 Zinc

Much has been learned about the importance of zinc to human health during the past decade. These physiologic functions include its role as a cofactor for over 50 metalloenzymes as well as a structural element for numerous other proteins such as zinc fingers [78]. As many of these proteins modulate cellular differentiation, proliferation, and adhesion, zinc is a critical element during fetal growth. Adequate zinc status has also been associated with optimal immune function, protection against oxidative damage, and regulation of cellular death (apoptosis) [79–81]. Zinc also appears to be vital in promoting adequate growth and cognitive function in children [82, 83]. In general, good dietary sources of this mineral include meat, fish, whole milk, and whole-grain products although bioavailability of animal products is greater than plant foods.

Researchers have long known that zinc deficiency during pregnancy can cause poor fetal growth and development as well as congenital malformations in animals [84, 85]. However, controlled intervention trials designed to examine this relationship in humans have provided mixed results [86, 87]. In fact, most randomized, placebo-controlled zinc supplementation trials have found no effect of enhanced maternal zinc intake on infant weight, length, head circumference, or reduction in small-for-gestational age infants [88–90]. One exception to this is a trial conducted by Goldenberg and colleagues in which they provided low-income, zinc-deficient pregnant women ($n = 580$) with either a zinc supplement (25 mg/day) or a placebo from 19 weeks of gestation to delivery [91]. Their results suggest a beneficial effect of zinc supplementation on birth weight, incidence of low birth weight, and head circumference. There is also limited evidence that zinc supplementation may reduce congenital malformations, but additional studies of more sufficient sample size will need to be conducted to further investigate this putative effect [92, 93]. In addition, recent reports by Marialdi and colleagues provide evidence that zinc supplementation in poorly nourished women living in Peru positively influenced fetal femur diaphysis length and neurobehavioral development [94, 95]. Thus, although more studies are needed, it is possible that there are beneficial effects of zinc supplementation during pregnancy, but only in subgroups of the population that are poorly nourished.

14.5.3.1 RECOMMENDATIONS

In conclusion, most experts agree that in light of the currently available information, routine zinc supplementation should not be advocated to improve pregnancy outcome in most women. However, it is important that pregnant women consume adequate zinc, and the RDA for this nutrient has been set at 11 mg/day. Because cereals are the primary source of dietary zinc for vegetarians—and zinc bioavailability from cereals is low—these individuals may need to consume up to 50% more zinc in order to meet their requirements. In these cases, zinc supplementation may be prudent as long as it does not result in zinc consumption above the UL value (40 mg/day). In addition, because iron may interfere with the absorption and utilization of zinc, the IOM recommends supplementation with approximately 15 mg of zinc when therapeutic levels of iron (>30 mg/day) are given to treat anemia.

14.5.4 Magnesium

Like most minerals, magnesium plays a major role as a cofactor for numerous metalloenzymes in the body, many of which are important for DNA, RNA, and protein synthesis; cellular growth; reproduction; and ATP production [96]. Magnesium is, therefore, critical for maternal and fetal growth and development during pregnancy. It is also involved in regulation of cardiovascular function, apparently playing a major role in modulating blood pressure in some populations [97–99]. Some studies also suggest that magnesium can help regulate blood glucose concentration in people with diabetes [100]. The best dietary sources of this mineral include leafy vegetables, whole grains, and nuts.

Because of magnesium's possible role in helping regulate blood pressure, there has been interest in determining whether magnesium supplementation during pregnancy might decrease the risk of pre-eclampsia and its complications [101]. Although there is relatively good evidence from clinical trials that magnesium administered intravenously can dramatically decrease preeclampsia, the data from oral supplementation trials are not convincing [102, 103]. It should be noted, however, that most studies conducted to date have not employed adequate study design and proper control groups to test the effects of oral magnesium administration on pregnancy outcome variables. Thus, further research is warranted.

14.5.4.1 RECOMMENDATIONS

Magnesium supplementation is not generally indicated during pregnancy. It is noteworthy that excessive magnesium intake from nonfood sources can cause gastrointestinal distress, and the IOM's UL for this mineral (350 mg/day) refers to intake only from supplements and medications such as milk of magnesia, not from food sources.

14.5.5 Iodine

Severe maternal iodine deficiency during pregnancy has long been known to increase risks for stillbirths, abortions, and congenital abnormalities [104]. In its more serious condition, prenatal iodine deficiency causes cretinism, which is characterized by stunting, difficulty in hearing and speaking, and sometimes-profound mental retardation [105]. In fact, iodine deficiency is considered the world's most frequent cause of preventable mental retardation. In addition to detrimental effects of iodine deficiency on infants and children, maternal postpartum thyroid dysfunction is relatively common and related to chronic iodine deficiency as well [106]. The complications of both maternal and infant iodine deficiency—collectively known as iodine deficiency disorders—are complex, and the etiology of these deficiency characteristics is owed to iodine's critical role as a component of the quaternary structures of the thyroid hormones thyroxine (T_4) and triiodothyronine (T_3).

Aside from marine plants and animals, which obtain large amounts of iodine from the sea, most foods are not good sources of this mineral. Milk products, however, do contribute important amounts of iodine due to the fact that iodine-containing products are used to disinfect milk collection vessels. Because iodine deficiency is endemic in some areas, many nations support iodinization programs such as that of salt in the United States.

Despite iodinization programs, iodine intake even in some developed countries remains low and quite variable [107]. Hollowell and colleagues, using data from the National Health and Nutrition Examination Surveys (NHANES), reported that 6.7% of pregnant US women had evidence of iodine deficiency (urinary iodine < 5 mcg/dl) between the years of 1988 and 1994 [108]. Importantly, this represented an almost sevenfold increase in prevalence of iodine deficiency in this subpopulation since 1971–1974.

In response to these and other data, several investigators have conducted iodine intervention trials during pregnancy to assess whether this might be advantageous to the maternal-infant dyad. In one such study, pregnant, Belgian women ($n = 180$) with excessive thyroid stimulation but not abnormal serum thyroid stimulating hormones (TSH) levels or thyroid autoantibodies were randomized to receive either a placebo, 100 mcg/day iodine (as KI), or a combination of 100 mcg/day iodine and T_4 [109]. Women assigned to the placebo group exhibited a 30% increase in thyroid volume during pregnancy, and 16% developed a goiter. Furthermore, compared with the treatment groups, their newborns had significantly larger thyroid volumes at birth. Conversely, measures of thyroid function in both groups of women receiving an active treatment were improved, although the effects were clearly more rapid and marked in the group receiving the combination treatment. Other studies have found similar results [110, 111].

14.5.5.1 RECOMMENDATIONS

The American Thyroid Association in 2006 issued its first recommendation that all pregnant women living in the United States or Canada consume 150 mcg of supplemental iodine daily. As the UL for iodine is set at 1,100 mcg/day from all sources, it is unlikely that consumption of this level of iodine in supplemental form will lead to excessive intake of this nutrient, and it is generally agreed that the benefits of correcting iodine deficiency far outweigh the risks of iodine supplementation [112]. Further research is needed to determine if this recommendation should indeed be applied to all pregnant women or just those at elevated risk for iodine deficiency.

14.6 USE OF BOTANICAL SUPPLEMENTS DURING PREGNANCY

14.6.1 Introduction

The National Institutes of Health's (NIH) Office of Dietary Supplements (ODS) defines a *botanical* as "a plant or plant part valued for its medicinal or therapeutic properties, flavor, and/or scent" and an *herb* as a type of botanical [113]. As such, depending on their purpose, products made from botanicals may be called herbal products, botanical products, or phytomedicines. Although it is commonly believed that products such as botanicals labeled as "natural" are necessarily safe and in fact healthful, this contention is not necessarily true. In fact, the actions and efficacy of botanicals range from nonexistent, to mild, to potent.

The prevalence of botanical preparation use during pregnancy has been documented in several settings. Using a cross-sectional survey design, Forster and colleagues reported that 36% of women living in Australia used at least one herbal supplement during pregnancy [114]. The most common supplements taken were raspberry leaf, ginger, and chamomile. Women were more likely to take herbal supplements if they were older, well educated, English speaking, nonsmokers, and primiparous. Studies conducted in the United States and Norway suggest that 7–36% of pregnant women use these products with echinacea, iron-rich herbs, ginger, and chamomile being the most commonly used [115–117].

It is important to emphasize that there is a relative dearth of high-quality, rigorous research relating the use of most botanicals to health and well-being in any phase of the lifecycle. Thus, it is essential that clinicians keep up to date on current evidence that might support or refute benefits, or perhaps contraindications, of these products. Along with the ODS (http://ods.od.nih.gov/), another good source of reliable information concerning botanical supplements is the National Center for Complementary and Alternative Medicine (http://nccam.nih.gov/). Abstracts of research conducted with botanical supplements can be directly accessed at the website "CAM on PubMed" (http://nccam.nih.gov/camonpubmed/).

In this section, we highlight two botanicals (ginger and echinacea), which are commonly taken by pregnant women. In addition, we will review the available evidence concerning two botanicals (chamomile and blue cohosh) that are contraindicated during pregnancy.

14.6.2 *Ginger*

The herb commonly known as ginger (*Zingiber officinale*) is a root used both in the cuisines of many cultures as well as for "medicinal" purposes. Ginger is often used in the form of teas, pills, tablets, capsules, or liquid extracts (tinctures) to treat nausea, vomiting, diarrhea, and motion sickness [118]. Indeed, one intervention trial suggested that ginger is as effective as vitamin B6 for treatment of nausea and morning sickness [119]. In this study, women ($n = 291$) took 1.05 g of ginger or 75 mg of vitamin B6 for 3 weeks. Another study found that consumption of either 0.5 g of ginger or 10 mg of vitamin B6 three times daily decreased vomiting and nausea [120]. However, in neither of these studies was there a true control group; thus, it is possible that the placebo effect might explain these results. Nonetheless, after conducting a systematic review of the literature, Borrelli and coworkers concluded that ginger may be an effective treatment for nausea and vomiting in pregnancy, but that larger studies must be conducted to confirm this [121].

14.6.2.1 RECOMMENDATIONS

There is much to learn about this botanical preparation, and like any other medication, one should use it with caution [122]. Nonetheless, the American College of Gynecology in 2004 in issuing its most recent guidance on treatment of morning sickness during pregnancy stated, "…ginger has shown beneficial effects and can be considered a non-pharmacologic option" [123].

14.6.3 *Echinacea*

Echinacea (also known as purple coneflower) is a perennial herb found in the eastern and central United States and southern Canada. Although there are at least nine species of this plant, three of them (*Echinacea angustifolia*, *E. pallida*, and *E. purpurea*) are most commonly used for medicinal purposes. Preparations of echinacea have been long used for a plethora of conditions including treatment of the common cold, upper respiratory infections, wound healing, toothaches, joint pain, and insect bites. Thus, use of echinacea during pregnancy is typically not related to pregnancy-associated issues, but instead as an alternative to other over-the-counter or prescription medications to treat these ailments.

Limited research suggests that use of echinacea for the prevention and treatment of upper respiratory infections may be somewhat effective, although these findings are not always consistent [124–126]. Conversely, most well-controlled trials do not support an effect of echinacea in preventing or treating the common cold [127, 128]. Nonetheless, many women turn to echinacea to treat colds and upper respiratory infections during pregnancy, because they do not wish to take other medications. As with all supplements, research concerning safety is imperative when deciding whether it should be taken during pregnancy. In the case of echinacea, one case-control study has been reported. Gallo and colleagues studied 206 women who used echinacea during their first trimester of pregnancy and a matched control group ($n = 206$) and found no evidence that gestational use of echinacea was associated with poor pregnancy outcome [129].

14.6.3.1 Recommendations

Although efficacy is not well established, use of echinacea during pregnancy is generally considered safe [130]. It should be noted, however, that the American Academy of Pediatrics recommends that pregnant women limit consumption of all herbal teas to two 8-oz. servings per day, and that they choose only those in filtered tea bags [131].

14.7 CONTRAINDICATED BOTANICALS

Although there is insufficient research conducted on the vast majority of botanicals to make conclusions concerning efficacy and safety, there are several preparations which are commonly classified as being potentially harmful during pregnancy. These include sage (*Salvia officinalis*), St. John's wort (*Hypericum perforatum*), lemon balm (*Melissa officinalis*), chamomile (*Marticaria recutita*), ginkgo (*Ginkgo biloba*), horse chestnut (*Aesculus hippocastanum*), raspberry leaf (*Rubus idaeus folio*), bearberry (*Arctostaphylos uva-ursi*) and black (*Actaea racemosa* or *Cimicifuga racemosa*) or blue (*Caulophyllum thalictroides*) cohosh [132]. It is noteworthy that many of these botanicals are not used by pregnant women for pregnancy-related issues, but instead for treating other conditions for which they do not want to use prescription or other more traditional over-the-counter drugs. It is also important to note that there have been no controlled human studies conducted on most of these compounds during pregnancy. Instead, they are generally contraindicated during this period of the lifecycle due to expert opinion, observation, and clinical evidence gleaned from their effects in men or nonpregnant women. Here we will discuss two contraindicated herbs, chamomile and blue/black cohosh.

14.7.1 Chamomile

Chamomile is commonly used as a tea to cause mild sedation and treat gastrointestinal distress. However, there exists one case report of a maternal anaphylactic response associated with the use of a chamomile-containing enema during labor, which resulted in fetal death [133]. In this unfortunate situation, researchers were able to determine that a homologue of the birch pollen allergen Bet v 1 found in the product was likely the causative antigen, highlighting the fact that we know very little about the biologically active chemicals found in most botanical preparations. Thus, use of chamomile during pregnancy and labor may be potentially hazardous and is not recommended.

14.7.2 Blue and Black Cohosh

Another herbal preparation that has been shown clinically to be associated with detrimental effects in pregnant women is blue cohosh, which has long been recommended by midwives to induce labor [134]. Use of blue cohosh, however, may be associated with neonatal congestive heart failure and ischemic infarct. Jones and Lawson in 1998 reported a case study describing a woman who had taken blue cohosh to promote uterine contractions [135]. This 36-year-old woman was otherwise healthy aside from being euthyroid, and was advised to take 1 tablet of blue cohosh beginning 1 month before delivery by her midwife; however, she elected to take 3 tablets per day. No other naturopathic remedies were used. After a precipitous labor (1 h), a normal weight (3.66 kg) baby was delivered. However, within 20 min, the infant required intubation and mechanical ventilatory support and was later diagnosed as having acute anterolateral myocardial infarction. Of course, it is impossible to draw a true cause-and-effect conclusion from such a case study.

Another more recent report described a normal weight, term infant born to a healthy 20-year-old woman whose obstetrician advised her to drink a tea made from blue cohosh to facilitate labor [136]. At approximately 26 h of age, the newborn infant began to have seizures and was found to have an evolving infarct in the distribution of the left middle cerebral artery. Urine and meconium were positive for the cocaine metabolite benzoylecgonine, and testing of the contents of the blue cohosh ingested by the mother produced similar results. The authors concluded that either benzoylecgonine is a metabolite of both cocaine and blue cohosh, or the blue cohosh itself was contaminated with cocaine. Regardless, use of this herb during pregnancy is contraindicated. Researchers also caution against the use of black cohosh, which is similarly used to induce labor, although clinical evidence of detrimental effects is lacking [137].

14.8 SUMMARY AND CONCLUSION

In summary, although most pregnant women take one or more dietary supplement, there is strong evidence to support the efficacy of only three of these products during this period of the lifespan, especially for otherwise healthy women. These include folic acid, iodine, and iron, and published recommendations concerning these nutrients are provided in Table 14.2. Additional supplements may be useful in specific circumstances. For example supplemental zinc may be necessary for vegans; and women who smoke, have poor quality diets, or are carrying more than one fetus should consider taking a MVMM supplement. Other nutrient supplements such as vitamin A may be warranted for poorly nourished women, especially those with comorbidities. There is very little high quality research on the efficacy and safety of botanical supplements during pregnancy, and extreme care should be taken when recommending their use during this time.

It is important to recognize that as scientists learn more about the mechanisms by which nutrients interact with genetic or epigenetic predisposition, future studies will undoubtedly identify subpopulations of individuals who might benefit from additional supplementation during pregnancy. Clinicians are strongly urged to stay abreast of the current research concerning the use of all types of dietary supplements during pregnancy, and a list of reputable resources of such information can be found in Table 14.3.

Table 14.2
Recommendations for Dietary Supplement Use in Pregnancy

Nutrient(s)	Guidelines
Multivitamin-multimineral (MVMM)	• The American Dietetic Association and the Institute of Medicine recommend that all pregnant women who consume poor-quality or vegan diets, have iron deficiency anemia, smoke, abuse alcohol or drugs, or are carrying more than 1 fetus take MVMM supplements
Folic acid	• US Centers for Disease Control and Prevention, American Academy of Pediatrics, and the National Healthy Mothers, Healthy Babies Coalition recommend that all women of childbearing age who are capable of becoming pregnant should consume 0.4 mg/day folic acid, keeping total folate consumption to less than 1 mg/day • The Institute of Medicine recommends that women able to become pregnant should take 0.4 mg/day folic acid daily from fortified foods, supplements, or both, in addition to consuming food folate from a varied diet
Vitamin D	• The Institute of Medicine recommends that women—whether pregnant or not—who receive regular exposure to sunlight *do not* need vitamin D supplementation. However, they also state that an intake of 10 mcg/day (400 IU/day) supplied by prenatal vitamin supplements would not be excessive
Iron	• The World Health Organization recommends daily iron supplementation of 60 mg/day for all pregnant women for 6 months or, if 6 months of treatment cannot be achieved, either continuation of supplementation during the postpartum period or increased dosage of 120 mg/day iron during pregnancy • The US Centers for Disease Control and Prevention recommend that oral, low-dose (30 mg/day) supplements of iron be provided to all pregnant women beginning at the first prenatal visit
Zinc	• The Institute of Medicine recommends that, because iron may interfere with the absorption and utilization of zinc, individuals should be supplemented with 15 mg/day zinc when therapeutic levels of iron (>30 mg/day) are given to treat anemia
Iodine	• The American Thyroid Association recommends that all pregnant women living in the United States or Canada consume 150 mcg/day of supplemental iodine.
Botanicals	• See McGuffin M, Hobbs C, Upton R, Goldberg A (eds) (1997) Botanical safety handbook: guidelines for the safe use and labeling for herbs in commerce. CRC Press, Boca Raton, Fla.

Table 14.3

Missions and Contact Information for Selected Reputable Organizations Providing Information Concerning the Efficacy and Safety of Dietary Supplements

Organization	Mission	Website
National Academy of Science's Institute of Medicine (IOM)	To serve as adviser to the nation to improve health by providing unbiased, evidence-based, and authoritative information and advice concerning health and science policy to policy-makers, professionals, leaders in every sector of society, and the public at large	http://www.iom.edu/
National Institutes of Health's Office of Dietary Supplements (ODS)	To strengthen knowledge and understanding of dietary supplements by evaluating scientific information, stimulating and supporting research, disseminating research results, and educating the public to foster an enhanced quality of life and health for the U.S. population	http://ods.od.nih.gov/
National Institutes of Health's National Center for Complementary and Alternative Medicine (NCCAM)	To explore complementary and alternative healing practices in the context of rigorous science; train complementary and alternative medicine researchers; and disseminate authoritative information to the public and professionals	http://nccam.nih.gov/
US Department of Health and Human Services' Centers for Disease Control and Prevention (CDC)	To promote health and quality of life by preventing and controlling disease, injury, and disability	http://www.cdc.gov/
US Food and Drug Administration's Center for Food Safety and Applied Nutrition (CFSAN)	To promote and protect the public's health by ensuring that the nation's food supply is safe, sanitary, wholesome, and honestly labeled, and that cosmetic products are safe and properly labeled	http://www.cfsan.fda.gov/list.html
World Health Organization (WHO)	To facilitate the attainment by all peoples of the highest possible level of health	http://www.who.int/en/

ACKNOWLEDGMENTS

We are indebted to Susan Pilch, Ph.D., M.L.S., for her assistance in comprehensively searching the literature for this review.

REFERENCES

1. Picciano MF (2003) Pregnancy and lactation: physiological adjustments, nutritional requirements and the role of dietary supplements. J Nutr 1997S–2002S
2. Subcommittee on Dietary Intake and Nutrient Supplements During Pregnancy (1990) In: Nutrition during pregnancy. National Academy Press, Washington, D.C.
3. Yu SM, Keppel KG, Singh GK, Kessel W (1996) Preconceptional and prenatal multivitamin-mineral supplement use in the 1988 National Maternal Infant Health Survey. Am J Public Health 86:240–242
4. Suitor CW, Gardner, JD (1990) Supplement use among a culturally diverse group of low-income pregnant women. J Am Diet Assoc 90:268–271
5. Center for Food Safety and Applied Nutrition (2001) Overview of dietary supplements. US Food and Drug Administration, Washington, D.C. Available via http://www.cfsan.fda.gov/~dms/ds-oview.html
6. Allen LH. Pregnancy and lactation (2006) In: Bowman BA, Russell RM (eds) Present knowledge in nutrition, 9th edn., vol. 2. International Life Sciences Institute, Washington, DC, pp 529–543
7. Botto LD, Khoury MJ, Mulinare J, Erickson JD (1996) Periconceptional multivitamin use and the occurrence of conotruncal heart defects: results from a population-based, case-control study. Pediatrics 98:911–917
8. Czeizel AE (1996) Reduction of urinary tract and cardiovascular defects by periconceptional multivitamin supplementation. Am J Med Genet 62:179–183
9. Czeizel AE, Dudas I (1992) Prevention of the first occurrence of neural-tube defects by periconceptional vitamin supplementation. N Engl J Med 327:1832–1835
10. Czeizel AE, Metneki J, Dudas I (1996) The effect of preconceptional multivitamin supplementation on fertility. Int J Vitam Nutr Res 66:55–58
11. Czeizel AE, Metneki J, Dudas I (1994) The higher rates of multiple births after periconceptional multivitamin supplementation: an analysis of causes. Acta Genet Med Gemellol (Roma) 43:175–184
12. Bodnar LM, Tang G, Ness RB, Harger G, Roberts JM (2006) Periconceptional multivitamin use reduces the risk of preeclampsia. Am J Epidemiol 164:470–477
13. Lammer EJ, Shaw GM, Iovannisci DM, Finnell RH (2004) Periconceptional multivitamin intake during early pregnancy, genetic variation of acetyl-N-transferase 1 (NAT1), and risk for orofacial clefts. Birth Defects Res A Clin Mol Teratol 70:846–852
14. Correa A, Botto L, Liu Y, Mulinare J, Erickson JD (2003) Do multivitamin supplements attenuate the risk for diabetes-associated birth defects? Pediatrics 111:1146–1151
15. Vahratian A, Siega-Riz AM, Savitz DA, Thorp JM Jr (2004) Multivitamin use and the risk of preterm birth. Am J Epidemiol 160:886–892
16. American Dietetic Association (2002) Position of the American Dietetic Association: nutrition and lifestyle for a healthy pregnancy outcome. J Am Diet Assoc 102:1479–1490
17. Bailey LB, Gregory JF III (2006) Folate. In: Bowman BA, Russell RM (eds) Present knowledge in nutrition, 9th edn., vol. 2. International Life Sciences Institute, Washington, D.C., pp 278–301
18. Center for Nutrition Policy and Promotion (1999) Nutrient content of the US food supply, 1909–a summary report. United States Department of Agriculture, Washington, D.C. Available via http://www.usda.gov/
19. Tamura T and Picciano MF (2006) Folate and human reproduction. Am J Clin Nutr. 83:993–1016
20. Wills L (1931)Treatment of "pernicious anaemia of pregnancy" and "tropical anaemia" with special reference to yeast extract as a curative agent. Br Med J 1:1059–1064
21. Food and Nutrition Board, National Research Council (1970) Maternal nutrition and the course of pregnancy. National Academy Press, Washington, D.C.
22. Smithells RW, Sheppard S, Schorah CJ (1976) Vitamin deficiencies and neural tube defects. Arch Dis Child 51:944–950

23. Smithells RW, Nevin NC, Seller MJ, Sheppard S, Harris R, Read AP, Fielding DW, Walker S, Schorah CJ, Wild J (1983) Further experience of vitamin supplementation for prevention of neural tube defect recurrences. Lancet 1:1027–1031

24. Laurence KM, James N, Miller MH, Tennant GB, Campbell H (1981) Double-blind randomized controlled trial of folate treatment before conception to prevent recurrence of neural-tube defects. Br J Med 282:1509–1511

25. Seller MJ, Nevin NC (1984) Periconceptional vitamin supplementation and the prevention of neural tube defects in south-east England and Northern Ireland. J Med Genet 21:325–330

26. Czeizel AE, Dudás I (1992) Prevention of the first occurrence of neural-tube defects by periconceptional vitamin supplementation. N Engl J Med 327:1832–1835

27. Kirke PN, Daly LE, Elwood JH (1992) A randomized trial of low dose folic acid to prevent neural tube defects. Arch Dis Child 67:1442–1446

28. US Centers for Disease Control and Prevention (1991) Effectiveness in disease and injury prevention: use of folic acid for prevention of spina bifida and other neural tube defects, 1983–1991. MMWR 40:513–616

29. Committee on Genetics, American Academy of Pediatrics (1999) Folic acid for the prevention of neural tube defects. Pediatrics 104:325–327

30. National Healthy Mothers, Healthy Babies Coalition. Folic acid position statement. Available via http://www.hmhb.org/ps_folicacid.html

31. Institute of Medicine (2006) Otten JJ, Hellwig JP, Meyers LD (eds) Dietary reference intakes; the essential guide to nutrient requirements. National Academies Press, Washington, D.C.

32. Centers for Disease Control and Prevention (2005) Use of dietary supplements containing folic acid among women of childbearing age—United States, 2005. MMWR 54:955–958

33. US Centers for Disease Control and Prevention (1992) Recommendations for the use of folic acid to reduce the number of cases of spina bifida and other neural tube defects. MMWR 41 (no. RR-14)

34. McCormick DB (2006) Vitamin B$_6$. In: Bowman BA, Russell RM (eds) Present knowledge in nutrition, 9th edn., vol. 2. International Life Sciences Institute, Washington, D.C. pp 269–277

35. Tiran D (2002) Nausea and vomiting in pregnancy: safety and efficacy of self-administered complementary therapies. Complement Ther Nurs Midwifery 8:191–196

36. Zeidenstein L (1998) Alternative therapies for nausea and vomiting of pregnancy. J Nurse Midwifery 43:392–393

37. Aselton P, Jick H, Chentow SF, Perera DR, Hunter JR, Rothman KJ (1984) Pyloric stenosis and maternal benectin exposure. Am J Epidemiol 120:251–256

38. McKeigue PM, Lamm SH, Linn S, Kutcher JS (1994) Benectin and birth defects: I. A meta-analysis of the epidemiologic studies. Teratology 50:27–37

39. Kutcher JS, Engle A, Firth J, Lamm SH (2003) Bendectin and birth defects. II: Ecological analyses. Birth Defects Res A Clin Mol Teratol 67:88–97

40. Vutyavanich T, Wongtra-ngan S, Ruingsri R (1995) Pyridoxine for nausea and vomiting of pregnancy: a randomized, double-blind, placebo-controlled trial. Am J Obstet Gynecol 173:881–884

41. Sahakian V, Rouse D, Sipes S, Rose N, Niebyl J (1991) Vitamin B$_6$ is effective therapy for nausea and vomiting of pregnancy: a randomized, double-blind, placebo-controlled study. Obstet Gynecol 78:33–36

42. Thaver D, Saeed MA, Bhutta Za (2006) Pyridoxine (vitamin B6) supplementation in pregnancy. Cochrane Database Syst Rev 2:CD000179

43. Jewell D, Young G (2003) Interventions for nausea and vomiting in early pregnancy. Cochrane Database of Syst Rev 4:CD000145

44. Wald G (1968) Molecular basis of visual excitation. Science 162:230–239

45. Wolf G (2004) The visual cycle of the cone photoreceptors of the retina. Nutr Rev 62:283–286

46. Pulukari S, Sitaramayya A (2004) Retinaldehyde, a potent inhibitor of gap junctional intercellular communication. Cell Commun Adhes 11:25–33

47. Bohnsack BL, Hirschi KK (2004) Nutrient regulation of cell cycle progression. Annu Rev Nutr 24:433–453

48. Marill J, Idres N, Capron CC, Nguyen E, Chabot GG (2003) Retinoic acid metabolism and mechanism of action: a review. Curr Drug Metab 4:1–10

49. Rothman KJ, Moore LL, Singer MR, Nguygen UDT, Mannino S, Milunsky B (1995) Teratogenicity of high vitamin A intake. N Engl J Med 333:1369–1373
50. Villamor E, Msamanga G, Spiegelman D, Antelman G, Peterson KE, Hunter DJ, Fawzi WW (2002) Effect of multivitamin and vitamin A supplements on weight gain during pregnancy among HIV-1-infected women. Am J Clin Nutr 76:1082–1090
51. Haskell MJ, Pandey P, Graham JM, Peerson JM, Shrestha RK, Brown KH (2005) Recovery from impaired dark adaptation in nightblind pregnant Nepali women who receive small daily doses of vitamin A as amaranth leaves, carrots, goat liver, vitamin A–fortified rice, or retinyl palmitate. Am J Clin Nutr 81:461–471
52. Cox SE, Arthur P, Kirkwood BR, Yeboah-Antwi K, Riley EM (2006) Vitamin A supplementation increases ratios of proinflammatory to anti-inflammatory cytokine responses in pregnancy and lactation. Clin Exp Immunol 144:392–400
53. Wiysonge CS, Shey MS, Sterne JA, Brocklehurt P (2005) Vitamin A supplementation for reducing the risk of mother-to-child transmission of HIV infection. Cochrane Database Syst Rev(4) CD003648
54. Underwood BA and the IVACG Steering Committee. IVACG statement; safe doses of vitamin A during pregnancy and lactation. 1998. Available via http://www.who.int/reproductive-health/docs/vitamina4p.pdf
55. Standing Committee on the Scientific Evaluation of Dietary Reference Intakes (2001) Dietary reference intakes for vitamin A, vitamin K, arsenic, boron, chromium, copper, iodine, iron, manganese, molybdenum, nickel, silicon, vanadium, and zinc. National Academy Press, Washington, D.C.
56. Norman AW, Henry HH. (2006) Vitamin D. In: Bowman BA, Russell RM (eds) Present knowledge in nutrition, 9th edn., vol. 2. International Life Sciences Institute, Washington, D.C. pp 198–210
57. Feldman D, Pike JW, Glorieux FH (eds) (2005) Vitamin D. Elsevier San Diego, Calif.
58. Collins ED, Normal AW (1990) Vitamin D. In: Machlin LJ (eds) Handbook of vitamins. Dekker, New York, N.Y., pp 59–98
59. Kreiter SR, Schwartz RP, Kirkman HN Jr (2000), Charlton PA, Calikoglu AS, Davenport ML. Nutritional rickets in African American breast-fed infants. J Pediatr 137:153–157
60. Gessner BD, de Schweinitz E, Petersen KM, Lewandowski C (1997) Nutritional rickets among breast-fed black and Alaska Native children. Alaska Med 39:72–74, 87
61. Hollis BW, Wagner CL (2004) Assessment of dietary vitamin D requirements during pregnancy and lactation. Am J Clin Nutr 79:717–726
62. Delvin EE, Salle BL, Glorieux FH, Adeleine P, David LS (1986) Vitamin D supplementation during pregnancy: effect on neonatal calcium homeostasis. J Pediatr 109:328–334
63. Brooke OG, Brown IR, Bone CD, Carter ND, Cleeve HJ, Maxwell JD, Robinson VP, Winder SM (1980) Vitamin D supplements in pregnant Asian women; effects on calcium status and fetal growth. Br Med J 280:751–754
64. Specker B (2004) Vitamin D requirements during pregnancy. Am J Clin Nutr 80(Suppl):1740S–1747S
65. Marya RK, Rathee S, Lata V, Mudgil S (1981) Effects of vitamin D supplementation in pregnancy. Gynecol Obstet Invest 12:155–161
66. Mallett E, Gugi B, Brunelle P, Henocq A, Basuyau JP, Lemeur H (1986) Vitamin D supplementation in pregnancy: a controlled trial of two methods. Obstet Gynecol 68:300–304
67. Ward LM (2005) Vitamin deficiency in the 21st century: a persistent problem among Canadian infants and mothers. Can Med Assoc J 172:769–770
68. Mahomed K, Gulmezoglu AM. Vitamin D supplementation in pregnancy. Cochrane Database Syst Rev 2000:CD000228
69. Standing Committee on the Scientific Evaluation of Dietary Reference Intakes (1997) Dietary reference intakes for calcium, phosphorus, magnesium, vitamin D, and fluoride. National Academy Press, Washington, D.C., p 276
70. Beard J (2006) Iron. In: Bowman BA, Russell RM (eds) Present knowledge in nutrition, 9th edn., vol. 2. International Life Sciences Institute, Washington, D.C. pp 430–444
71. Dallman PR (1986) Biochemical basis for the manifestations of iron deficiency. Annu Rev Nutr 6:13–40
72. Scholl TO (2005) Iron status during pregnancy: setting the stage for mother and infant. Am J Clin Nutr 81:1218S–1222S

73. Cogswell ME, Parvanta I, Ickes L, Yip R, Brittenham (2003) Iron supplementation during pregnancy, anemia, and birth weight: a randomized controlled trial. Am J Clin Nutr 78:773–781
74. World Health Organization (2007) Iron and folate supplementation. Available via http://www.who.int/making_pregnancy_safer /publications/Standards1.8N.pdf
75. Stolzfus RJ, Dreyfuss ML (1998) Guidelines for the use of iron supplements to prevent and treat iron deficiency anemia. International Life Sciences Institute Press, Washington, D.C. Available via http://inacg.ilsi.org/file/b2_VUHUQ8AK.pdf
76. US Centers for Disease Control and Prevention (1998) Recommendations to prevent and control iron deficiency in the United States (1998) MMWR 47:1–36
77. Cogswell, ME, Kettel-Khan L, Ramakrishnan U (2003) Iron supplement use among women in the United States: science, policy and practice. J Nutr 1974S–1977S
78. Lu D, Searles MA, Klug A (2003) Crystal structure of a zinc-finger-RNA complex reveals two modes of molecular recognition. Nature 426:96–100
79. Zangger K, Oz G, Haslinger E, Kunert O, Armitage IM (2001) Nitric oxide selectively releases metals from the amino terminal domain of metallothioneins: potential role at inflammatory sites. FASEB J 15:1303–1305
80. Cousins RJ, Blanchard RK, Popp MP, Liu L, Cao J, Moore JB, Green CL (2003) A global view of the selectivity of zinc deprivation and excess on genes expressed in human THP-1 mononuclear cells. Proc Natl Acad Sci USA 100:6952–6957
81. Fong LY, Zhang L, Jiang Y, Farber JL (2005) Dietary zinc modulation of COX-2 expression and lingual and esophageal carcinogenesis in rats. J Natl Cancer Inst 97:40–50
82. Walker CF, Black RE (2004) Zinc and the risk for infectious disease. Annu Rev Nutr 24:255–275
83. Black MM (2003) The evidence linking zinc deficiency with children's cognitive and motor functioning. J Nutr 133:1473S–1476S
84. Apgar J (1968) Effect of zinc deficiency on parturition in the rat. Am J Physiol 215:160–163
85. Hurley LS, Swenerton H (1966) Congenital malformations resulting from zinc deficiency in rats. Proc Soc Exp Biol Med 123:692–696
86. Shah D, Sachdev HPS (2006) Zinc deficiency in pregnancy and fetal outcome. Nutr Rev 64:15–30
87. Castillo-Duran C, Weisstaub G (2003) Zinc supplementation and growth of the fetus and low birth weight infant. J Nutr 133:1494S–1497S
88. Hunt IF, Murphy NJ, Cleaver AE, Faraji B, Swendseid ME, Coulson AH, Clark VA, Browdy BL, Cabalum T, Smith JC Jr (1984) Zinc supplementation during pregnancy: effects on selected blood constituents and on progress and outcome of pregnancy in low-income women of Mexican descent. Am J Clin Nutr 40:508–521
89. Jonsson B, Hauge B, Larsen MF, Hald F (1996) Zinc supplementation during pregnancy: a double blind randomized controlled trial. Acta Obstet Gynecol Scand 75:725–729
90. Caulfield LE, Zavaleta N, Figueroa A, Leon Z (1999) Maternal zinc supplementation does not affect size at birth or pregnancy duration in Peru. J Nutr 129:1563–1568
91. Goldenberg RL, Tamura T, Neggers Y, Copper RL, Johnston KE, DuBard MB, Hauth JC (1995) The effect of zinc supplementation on pregnancy outcome. JAMA 274:463–468
92. Mahamed K (2000) Zinc supplementation in pregnancy. Cochrane Database Syst Rev 2:CD000230
93. Osendarp SF, West CE, Black RE (2003) The need for maternal zinc supplementation in developing countries: an unresolved issue. J Nutr 133:817S–827S
94. Merialdi M, Caulfield LE, Zavaleta N, Figueroa A, Costigan KA, Dominici F, Dipietro JA (2004) Randomized controlled trial of prenatal zinc supplementation and fetal bone growth. Am J Clin Nutr 79:826–830
95. Merialdi M, Caulfield LE, Zavaleta N, Figueroa A, Costigan KA, Dominici F, Dipietro JA (2004) Randomized controlled trial of prenatal zinc supplementation and fetal heart rate. Am J Obstet Gynecol 190:1106–1112
96. Elin RJ (1994) Magnesium: the fifth but forgotten electrolyte. Am J Clin Pathol 102:616–622
97. Rosenlund M, Berglind N, Hallqvist J, Bellander T, Bluhm G (2005) Daily intake of magnesium and calcium from drinking water in relation to myocardial infarction. Epidemiology 16:570–576

98. Ascherio A, Rimm EB, Giovannucci EL, Colditz GA, Rosner B, Willett WC, Sacks F, Stampfer MJ (2002) A prospective study of nutritional factors and hypertension among US men. Circulation 86:1475–1484

99. Ma J, Folsom AR, Melnick SL, Eckfeldt JH, Sharrett AR, Nabulsi AA, Hutchinson RG, Metcalf PA (1995) Associations of serum and dietary magnesium with cardiovascular disease, hypertension, diabetes, insulin, and carotid arterial wall thickness: The ARIC study. Atherosclerosis Risk in Community Study. J Clin Epidemiol 48:927–940

100. Paolisso G, Sgambato S, Gambardella A, Pizza G, Tesauro P, Varricchio M, D'Onofrio F (1992) Daily magnesium supplements improve glucose handling in elderly subjects. Am J Clin Nutr 55:1161–1167

101. Durlach J (2004) New data on the importance of gestational Mg deficiency. J Am Coll Nutr 694S–700S

102. The Magpie Trial Collaborative Group (2002) Do women with preeclampsia, and their babies, benefit from magnesium sulphate? The Magpie Trial: a randomized placebo-controlled trial. Lancet 359:1877–1890

103. Makrides M, Crowther CA (2001) Magnesium supplementation in pregnancy. Cochrane Database Syst Rev (4):CD000937

104. Zimmermann MB (2006) Iodine and the iodine deficiency disorders. In: Bowman BA, Russell RM (eds) Present knowledge in nutrition, 9th edn., vol. 2. International Life Sciences Institute, Washington, D.C., pp 471–479

105. Delange J, Hetzel B (1998) Chap. 20. The iodine deficiency disorders. In: DeGroot LE, Hannemann G (eds) The thyroid and its diseases. Available via http://www.thyroidmanager.org/thyroidbook.htm

106. Nohr SB, Jorgensen A, Pedersen KM, Laurberg P (2000) Postpartum thyroid dysfunction in pregnant thyroid peroxidase antibody-positive women living in an area with mild to moderate iodine deficiency: is iodine supplementation safe? J Clin Endocrinol Metab 85:3191–3198

107. Pearce EN, Pino S, He X, Bazrafshan HR, Lee SL, Braverman LE (2004) Sources of dietary iodine: bread, cows' milk, and infant formula in the Boston area. J Clin Endocrinol Metab 89:3421–2324

108. Hollowell JG, Staehling NW, Hannon WH, Flanders DW, Gunter EW, Maberly GJ, Braverman LE, Pino S, Miller DT, Garbe PL, DeLozier DM, Jackson RJ (1998) Iodine nutrition in the United States. Trends and public health implications: Iodine excretion data from National Health and Nutrition Examination Surveys I and II (1971–1974 and 1988–1994). J Clin Endocrinol Metab 83:3401–3408

109. Glinoer D, de Nayer Ph, Delange F, Lemone M, Toppet V, Spehl M, Grun J-P, Kinthaert J, Lejeune B (1995) A randomized trial for the treatment of mild iodine deficiency during pregnancy: maternal and neonatal effects. J Clin Endocrinol Metab 80:258–269

110. Glinoer D (1998) Iodine supplementation during pregnancy: importance and biochemical assessment. Exp Clin Endocrinol Diabetes 106:S21

111. Pedersen KM, Laurberg P, Iversen E, Knudsen PR, Gregersen HE, Rasmussen OS, Larsen KR, Eriksen GM, Johannesen PL (1993) Amelioration of some pregnancy-associated variations in thyroid function by iodine supplementation. J Clin Endocrinol Metab 77:1078–1083

112. Delange J, Lecomte P (2000) Iodine supplementation: benefits outweigh risks. Drug Safety 22:89–95

113. Office of Dietary Supplements. Dietary supplement fact sheet. Botanical dietary supplements: background information. Available via http://ods.od.nih.gov/factsheets/BotanicalBackground_pf.asp

114. Forster DA, Denning A, Wills G, Bolger M, and McCarthy E (2006) Herbal medicine use during pregnancy in a group of Australian women. BMC Pregnancy Childbirth 6:21–29

115. Hepner DL, Harnett M, Segal S, Camann W, Bader AM, Tsen LC (2002) Herbal medicine use in parturients. Anesth Analg 94:690–693

116. Gsui B, Dennehy C, Tsourounis C (2001) A survey of dietary supplement use during pregnancy at an academic medical center. Am J Obstet Gynecol 185:433–437

117. Nordeng H, Havnen GC (2004) Use of herbal drugs in pregnancy: a survey among 400 Norwegian women. Pharmacopeia and Drug Safety 13:371–380

118. Coates P, Blackman M, Cragg GM, Levine MA, Moss J, White JD (eds) (2005) Ginger (*Zingiber officinale*). In: Encyclopedia of dietary supplements. Dekker, New York, N.Y., pp 241–248

119. Smith C, Browther C, Willson K, Hotham N, McMillian V (2004) A randomized controlled trial of ginger to treat nausea and vomiting in pregnancy. Obstet Gynecol 103:639–645

120. Sripramote M, Lekhyananda N (2003) A randomized comparison of ginger and vitamin B_6 in the treatment of nausea and vomiting of pregnancy. J Med Assoc Thai 86:846–853

121. Borelli F, Capasso R, Aviello G, Pittler MN, Izzo AA (2005) Effectiveness and safety of ginger in the treatment of pregnancy-induced nausea and vomiting. Obstet Gynecol 105:849–856
122. Marcus DM, Snodgrass WR (2005) Do no harm: avoidance of herbal medicines during pregnancy. Obstet Gynecol 105:1119–1122
123. American College of Obstetrics and Gynecology (2004) Bulletin #52: nausea and vomiting of pregnancy. Obstet Gynecol 103:803
124. Fugh-Berman A (2003) Echinacea for the prevention and treatment of upper respiratory infections. Sem Integr Med 1:106–111
125. Grimm W, Müller H-H (1999) A randomized controlled trial of the effect of fluid extract of Echinacea purpurea on the incidence and severity of colds and respiratory infections. Am J Med 106:138–143
126. Narimanian M, Badalyan M, Panosyan V, Gabrielyan E, Panossian A, Wikman G, Wagner H (2005) Randomized trial of a fixed combination (KanJang) of herbal extracts containing *Adhatoda vasica*, *Echinacea purpurea* and *Eleutherococcus senticosus* in patients with upper respiratory tract infections. Phytomedicine 12:539–547
127. Taylor JA, Weber W, Standish L, Quinn H, Goesling J, McGann M, Calabrese C (2003) Efficacy and safety of Echinacea in treating upper respiratory tract infections in children: a randomized controlled trial. JAMA 290:2824–2830
128. Barrett BP, Brown RL, Locken K, Maberry R, Bobula JA, D'Alessio D (2002) Treatment of the common cold with unrefined echinacea: a randomized, double-blind, placebo-controlled trial. Ann Intern Med 137:939–946
129. Gallo M, Sarkar M, Au W, Pietrzak K, Comas B, Smith M, Jaeger TV, Einarson A, Koren G (2000) Pregnancy outcome following gestational exposure to echinacea. Arch Intern Med 160:3141–3143
130. Perri D, Dugoua J-J, Mills E, Koren G (2006) Safety and efficacy of echinacea (*Echinacea Angustifolia*, *E. Purpurea* and *E. Pallida*) during pregnancy and lactation. Can J Clin Pharmacol 13:e262–e267
131. American Academy of Pediatrics, Committee on Nutrition (1998) Pediatric nutrition handbook. American Academy of Pediatrics, Elk Grove Village, Ill.
132. Newell CA, Anderson LA, Phillipson JD (1996) Herbal medicines. A guide for health-care professionals. The Pharmaceutical Press, Cambridge, UK.
133. Jensen-Jarolim E, Reider N, Fritsch R, Breiteneder H (1998) Fatal outcome of anaphylaxis to chamomile-containing enema during labor: a case study. J Allergy Clin Immunol 102:1041–1042
134. Castlemen M (2001) The new healing herbs: the classic guide to nature's best medicines featuring the top 100 time-tested herbs. Rodale, Emmaus, Pa.
135. Jones TK, Lawson BM (1998) Profound neonatal congestive heart failure caused by maternal consumption of blue cohosh herbal medication. J Pediatr 132:550–552
136. Rinkel RS, Zarlengo KM (2004) Blue cohosh and perinatal stroke. N Engl J Med 351:302–303
137. Dugoua J-J, Seely D, Perri D, Koren G, Mills E (2006) Safety and efficacy of black cohosh (*Cimicifuga racemosa*) during pregnancy and lactation. Can J Clin Pharmacol 13:e257–e26145

15 Vegetarian Diets in Pregnancy

Ann Reed Mangels

Summary A vegetarian diet, defined as an eating style that avoids meat, fish, and poultry, can be healthful and nutritionally adequate for a pregnant woman. Some vegetarians, called vegans, avoid dairy products and eggs as well as meat, fish, and poultry. Vegan diets can also be healthful and nutritionally adequate for pregnancy. Vegetarian diets can provide numerous long-term health benefits including a lower risk of cardiovascular disease, some forms of cancer, and hypertension. Key nutrients for vegetarian pregnancy include protein, iron, zinc, calcium, vitamin D, vitamin B_{12}, iodine, and omega-3 fatty acids. Vegetarian women should also be counseled to follow standard weight gain recommendations. A vegetarian or vegan diet can meet requirements for all of these nutrients although in some instances, fortified foods or supplements can be especially useful in meeting recommendations. The nutrient content of supplements targeted to pregnant vegetarians should be evaluated to make sure nutrient needs are being met. Dietetics professionals play important roles in counseling pregnant vegetarians and may be called upon to address a variety of issues including family concerns and pressure, making the change to a vegetarian diet during pregnancy, foods and food preparation, meal planning, and coping with common concerns of pregnancy such as nausea and constipation. Practitioners should be able to provide current, accurate information and resources about vegetarian diets.

Keywords: Vegetarian, Vegan, Pregnancy, Iron, Cobalamin, Phytate, Zinc, Omega-3 fatty acids, Calcium, Vitamin D, Iodine

15.1 INTRODUCTION

15.1.1 Definition and Types of Vegetarians

A vegetarian is a person who does not eat meat, fish, poultry, or products containing these foods [1]. Within the broad category, there are numerous subcategories. The most common are lacto-ovo vegetarians, lacto vegetarians, and vegans. Lacto-ovo vegetarians are vegetarians who eat eggs and dairy products. Lacto vegetarians use dairy products but not eggs, and vegans (pronounced VEE-guns) avoid all animal products including dairy products, eggs, honey, and gelatin.

From: *Nutrition and Health: Handbook of Nutrition and Pregnancy*
Edited by: C.J. Lammi-Keefe, S.C. Couch, E.H. Philipson © Humana Press, Totowa, NJ

Other types of vegetarian (or near vegetarian) diets include macrobiotic, raw foods, and fruitarian diets [2]. Macrobiotic diets consist mainly of grains, vegetables, especially sea vegetables, beans, fruits, nuts, soy products, and possibly fish. As the name suggests, those choosing a raw foods diet mainly or exclusively consume uncooked and unprocessed foods. Foods used include fruits, vegetables, nuts, seeds, and sprouted grains and beans; unpasteurized dairy products and even raw meat and fish may be used [3]. Fruitarian diets are based on fruits, nuts, and seeds and often include vegetables that are botanically fruits like avocado and tomatoes; other vegetables, grains, beans, and animal products are excluded [2].

Many people, who do not strictly avoid meat, fish, or poultry, describe themselves as vegetarian [4–6]. This can have a significant impact on food choices and nutrient intake so individual assessment of the diets of self-identified vegetarian clients is essential.

15.2   REASONS FOR VEGETARIANISM

Reasons for vegetarianism are highly individual and include health considerations, environmental concerns, and animal welfare issues [7]. Other factors include religion, economics, ethical issues, and a desire to reduce world hunger [5].

15.3   HOW MANY VEGETARIANS ARE THERE?

No information is available on the number of pregnant vegetarians in the United States. We do know, based on a poll conducted in 2006, that 2.3% of the adult population consistently follows a vegetarian diet and 1.4% follows a vegan diet [8]. In most countries, as in the United States, only a small percentage of the population is vegetarian. In India, as much as 35% of the population follows a vegetarian diet [9].

15.4   ADEQUACY OF VEGETARIAN DIETS IN PREGNANCY

The American Dietetic Association and Dietitians of Canada have reviewed current information on vegetarian diets and concluded that, "Well-planned vegan and other types of vegetarian diets are appropriate for all stages of the life cycle including during pregnancy [and] lactation…" [1].

15.5   HEALTH ADVANTAGES OF VEGETARIAN DIETS

Numerous health advantages are associated with use of a vegetarian diet. These include [1]:

- A lower body mass index (BMI)
- Reduced rates of cardiovascular disease and of risk of death from ischemic heart disease
- Lower blood pressure and markedly lower rates of hypertension
- Reduced risk of type 2 diabetes
- Lower risk of colorectal cancer

These are long-term advantages. For the pregnant women, there is less information about specific, immediate health advantages. The higher fiber content [10, 11] of many vegetarian diets can help to alleviate the constipation that commonly occurs in pregnancy. Another positive aspect of vegetarian diets is that pregnant vegetarians tend to have higher intakes of both folate and magnesium than do nonvegetarians [12, 13]. One small study has shown a marked reduction in risk of preeclampsia in vegans compared to the general population [14]; however, another study did not find this reduced risk [15].

15.6 WEIGHT GAIN AND BIRTH WEIGHT IN VEGETARIAN PREGNANCY

Vegetarians as a group tend to be leaner than do nonvegetarians, with vegans tending to have a lower BMI than other vegetarians [16, 17]. This suggests that vegetarian women tend to begin pregnancy with a lower BMI than do nonvegetarians. Standard weight gain recommendations should be used for vegetarians [18]. Weight gain of pregnant lacto-ovo vegetarians and vegans is generally adequate [14, 19, 20]. Birth weights of infants of vegetarian women have been frequently shown to be similar to those of infants born to nonvegetarian women and to birth weight norms [19–22].

Low birth weights have been reported in some macrobiotic populations [23, 24]. These low birth weights appear to be due to low maternal weight gain secondary to inadequate energy intake [23, 24].

Suggestions for vegetarian women who have difficulty gaining weight in pregnancy include:

- Use small, frequent meals and snacks
- Emphasize concentrated sources of energy and nutrients such as nuts and nut butters, full-fat soy products, dried fruits, and bean spreads
- Use some refined foods (i.e., enriched grains, fruit juices) if dietary fiber intake is high
- Increase use of unsaturated oils in cooking
- Make beverages count – instead of drinking tea, coffee, seltzer, or diet drinks, try smoothies (made with fruit, juice, and milk) or milkshakes

15.7 NUTRITIONAL CONSIDERATIONS

Recommendations for most nutrients do not differ based on vegetarian status, although the main sources for some nutrients may vary. These include protein, vitamin B_{12}, and omega-3 fatty acids, and calcium and vitamin D (for vegans). Iron, and possibly zinc recommendations are higher for vegetarians than for nonvegetarians. Other nutrients, including vitamin C and vitamin A (as beta-carotene) are generally adequate in the diets of vegetarians eating a wide variety of foods.

15.7.1 Protein

Protein is rarely below recommendations in the diets of vegetarian women [2, 10, 11]. If, as women increase their energy intake in pregnancy, their protein intake also increases, then they will achieve the higher protein intake recommended for pregnancy. One study has found a mean protein intake by pregnant lacto-ovo vegetarians of 78 g/day [5], close to the current RDA of 71 g/d in the second and third trimesters [25]. Choosing foods that are good sources of protein (Table 15.1) along with adequate energy can insure that protein needs during pregnancy are met. The Institute of Medicine has concluded that the protein requirement for vegetarians consuming a variety of plant proteins is not different from that of nonvegetarians [25]. Conscious combining of proteins within a meal is not necessary when a variety of plant foods is eaten over the day [1].

Table 15.1
Protein Content of Foods Commonly Eaten by Vegetarians

Food, serving size	Protein (g)
Seitan (wheat gluten), 3 oz.	31
Soybeans, 1 cup	29
Tofu, firm, 5 oz.	24
Lentils, cooked, 1 cup	18
Tempeh, 1/2 cup	15
Veggie burger, 1	13
Dried beans, cooked, 1 cup	10–15
Yogurt, 8 oz.	9
Peanut butter, 2 T	8
Veggie dog, 1	8
Milk, 8 oz.	8
Soymilk, 8 oz.	7
Cheese, 1 oz.	7
Egg, 1 large	6
Grains, cooked, 1 cup	4–8
Soy yogurt, 6 oz.	4
Nuts, 2 T	2–4

Source: USDA Nutrient Database for Standard Reference, Release 19, 2006, and manufacturer's information

15.7.2 Iron

Iron in vegetarian diets is in the form of non-heme iron. Non-heme iron is much more sensitive than heme iron is to factors affecting absorption. Absorption of non-heme iron is increased markedly in iron-deficient individuals [26]. Non-heme iron absorption is also affected by factors in foods. Phytate, a phosphorus-containing compound found in whole grains and legumes, inhibits non-heme iron absorption as do coffee, calcium, and tannic acids in tea and some spices [2]. Vitamin C and other organic acids enhance non-heme iron absorption, and will partially counteract the inhibitory effects of phytate. Because bioavailability of iron from plant-based diets is lower than that from animal-based diets, the Institute of Medicine has established a separate higher dietary reference intake (DRI) for iron for vegetarians [27]. For pregnant vegetarians, the recommended dietary allowance (RDA) for iron is 48.6 mg/day, a level that is difficult to achieve without the use of iron supplements. See Chap. 16, ("Iron Requirements and Adverse Outcomes") for further discussion on meeting iron needs.

Vegetarian women in Western countries are no more likely to have iron deficiency anemia than have non-vegetarian women [28]. Vegetarians, however, are more likely to have lower iron stores, as indicated by serum ferritin [28].

Dietary iron intakes of vegetarian women of childbearing age vary with mean iron intakes of 11–15 mg/day reported for lacto-ovo vegetarians and 14–23 mg/day for vegans [2]. Mean dietary iron intakes of 13.8 mg/day and 17 mg/day have been reported in pregnant lacto-ovo vegetarians; supplemental iron increased total mean iron intake to 57 mg/day and 37 mg/day, respectively [5, 21].

Iron sources for vegetarians include dried beans, whole and enriched grains, soy foods, enriched meat analogs, pumpkin and squash seeds, dried fruits, and baked potatoes.

Iron supplementation, if indicated based on iron status, should be started early in pregnancy so that maternal iron status is adequate throughout both pregnancy and the postpartum period [29, 30].

15.7.3   Zinc

While zinc intakes of vegetarian women are often similar to those of nonvegetarians [2], both groups frequently have intakes below recommendations. In addition, factors in vegetarian diets including phytate and fiber can interfere with zinc absorption. The Institute of Medicine has not specified a zinc RDA for vegetarians, but suggests that the dietary requirement for zinc may be as much as 50% higher for vegetarians, especially for those relying mainly on high-phytate grains and legumes [27]. Thus, zinc recommendations for pregnant vegetarians may be as high as 16.5 mg/day. Zinc sources include dried beans, wheat germ, fortified cereals, and nuts and seeds. Food preparation techniques such as leavening bread and soaking and sprouting beans can increase zinc bioavailability [31]. A zinc supplement or a prenatal supplement containing zinc may be necessary, especially if a woman's diet is high in phytate [32].

15.7.4   Iodine

Iodine intakes in the United States have declined over the past 30 years [33], partly because of changes in the production of bread and milk. In addition, reliance on processed food has increased and food processors frequently use non-iodized salt [34]. The iodine content of most fruits, nuts, and vegetables is low, but can vary depending on soil iodine content, irrigation, and fertilization practices [35].

Vegetarians who do not use iodized salt may be at increased risk of developing iodine deficiency because, in general, plant-based diets are relatively low in iodine [36–39]. This is of special concern in pregnancy because of the effects of iodine deficiency on the developing brain [27].

Use of iodized salt (0.75 teaspoon) in cooking and at the table will provide enough iodine to meet the iodine RDA for pregnancy of 220 mcg/day. Other alternatives include iodine supplements and sea vegetables. Some, but not all prenatal supplements contain iodine. Sea vegetables like nori and hiziki can provide iodine but their iodine content is quite variable [40]. Excessive maternal iodine (2,300–3,200 mcg/day) from sea vegetables has been linked to hypothyroidism in newborn infants in Japan [41] and to postpartum thyroiditis in China [42].

15.7.5   Calcium and Vitamin D

Although calcium absorption increases in pregnancy [43], low calcium intakes can be problematic. Pregnant women with a habitually low calcium intake (<500 mg/day) experience calcium losses from bone that may adversely affect maternal bone status [44]. In addition, low calcium intakes in pregnancy have been associated with a lower bone mineral content in newborns [45]. Many lacto-ovo vegetarian women have intakes of calcium that meet current recommendations [5, 11]. Vegan women tend to have lower calcium intakes [11, 46, 47] and may benefit from information about non-dairy calcium sources.

Non-dairy sources of calcium include low-oxalate green vegetables (bok choy, collards, Chinese cabbage, kale, broccoli, turnip greens, sweet potato greens, and okra), almonds, figs, soybeans, calcium-set tofu, and fortified foods (soy and rice milks, fruit juice, and breakfast cereals) [1, 48–50].

Table 15.2 groups vegetarian calcium sources by their calcium content to assist in menu planning. For example, a pregnant vegan might choose 16 oz. of calcium-fortified soymilk, 1 cup of collards, and 1 cup of vegetarian baked beans to meet the Adequate Intake (AI) of 1,000 mg/day of calcium.

Sources of vitamin D for vegetarians include fortified foods (milk, soymilk, rice milk, breakfast cereals, margarine), egg yolks, and vitamin D supplements [1]. In addition, vitamin D requirements can be met by sun exposure. Factors such as season, skin pigmentation, location, and sunscreen use can affect cutaneous vitamin D synthesis so dietary and supplemental vitamin D are often used to meet needs.

15.7.6   Vitamin B_{12}

A regular source of vitamin B_{12} is essential in pregnancy since vitamin B_{12} from maternal stores does not appear to cross the placenta [51, 52]. Low maternal vitamin B_{12} levels in pregnancy increase risk for neural tube defects and preeclampsia [53, 54]. Infants may be born with low vitamin B_{12} stores if maternal status during pregnancy is marginal [55]. Lacto-ovo vegetarians can obtain vitamin B_{12} from dairy products and eggs. Vegan women must use foods fortified with vitamin B_{12} or a supplement containing vitamin

Table 15.2
Calcium Sources for Vegetarians

Foods, serving size	Approximate calcium content (mg)
Collards, cooked, 1 cup	300
Milk, 8 oz.	
Orange juice, fortified, 8 oz.	
Soymilk, fortified, 8 oz.	
Yogurt, 8 oz.	
Blackstrap molasses, 1 T	200
Kale, cooked, 1 cup	
Okra, cooked, 1 cup	
Tofu, calcium-set, 1/4 cup	
Turnip greens, cooked, 1 cup	
Bok choy, cooked, 1 cup	150
Cheese, 3/4 oz.	
Mustard greens, cooked, 1 cup	
Soybeans, cooked, 1 cup	
Soy yogurt, 6 oz.	
Almonds, 1/4 cup	100
Almond butter, 2 T	
Broccoli, cooked, 1 cup	
Dried beans, cooked, 1 cup	
Figs, dried, 5	
Tahini, 2 T	
Tempeh, 1/2 cup	

Source: USDA Nutrient Database for Standard Reference, Release 19, 2006, and manufacturer's information

Table 15.3
Vitamin B$_{12}$ Content in Vegetarian Foods

Food, serving size	Vitamin B$_{12}$ (mcg)
Cereals, fortified, 1 oz.	0.6–6
Cow's milk, 8 oz.	1
Egg, large, 1	0.6
Meat analogs, fortified, 1 oz.	0.5–1.2
Soymilk, fortified, 8 oz.	0.8–3.2
Vegetarian Support Formula Nutritional Yeast, miniflakes, 1 T	1.5

Source: USDA Nutrient Database for Standard Reference, Release 19, 2006 and manufacturer's information

B$_{12}$. Fortified foods include some brands of soymilk, breakfast cereals, nutritional yeast, and meat analogs (Table 15.3). Foods such as sea vegetables, tempeh, and miso cannot be counted on as reliable sources of vitamin B$_{12}$ [1].

15.7.7 Omega-3 Fatty Acids

Omega-3 fatty acids include alpha-linolenic acid (ALA), eicosapentaenoic acid (EPA), and docosahexaenoic acid (DHA). DHA is a component of neural and retinal membranes and accumulates in the brain and retina, especially in the last trimester and the early postnatal period [56]. DHA is transferred to the fetus through the placenta [57] and provided to the breast-fed infant in breast milk [58, 59].

Evidence suggests that higher maternal intakes of DHA during pregnancy can have beneficial effects on gestational length, infant visual function, and neurodevelopment [60–63]. A recent meta-analysis of supplementation of DHA or DHA and EPA throughout pregnancy found a small but significant increase in pregnancy duration and head circumference [64]. Maternal DHA supplementation in pregnancy has also been found to improve the visual acuity of infants at 4 months of age although not at 6 months [65]. Women with higher plasma DHA levels during pregnancy (suggestive of higher maternal DHA intakes) were shown to have infants with more mature sleep patterns [66]. This, along with other evidence [67], suggests that an improved maternal DHA status in pregnancy can play a role in the development of the infant's central nervous system.

Unsupplemented vegetarian diets contain little and vegan diets contain virtually no DHA or EPA; these omega-3 fatty acids are mainly found in oily fish. Vegetarians' limited intakes of EPA and DHA are reflected in blood and breast milk concentrations. Lacto-ovo vegetarians and vegans, including pregnant women [22] have lower blood concentrations of EPA and DHA than have nonvegetarians [68–70]. Breast milk concentrations of EPA and DHA reflect the amounts present in the mother's diet and are lower in breast milk of vegetarian and vegan women [71, 72].

EPA and DHA can be synthesized from ALA through a series of desaturation and elongation reactions [61]. Table 15.4 provides a list of vegetarian sources of ALA. Limiting use of oils high in linoleic acid and *trans*-fats can enhance the conversion of ALA to EPA [61, 73]. Figure 15.1 provides some suggestions for ways to limit consumption of foods containing *trans*-fats. The rate of conversion of ALA to EPA and

Table 15.4
Alpha-Linolenic Acid Sources for Vegetarians

Food, serving size	Alpha-linolenic acid (g/serving)	Comments
Flaxseed oil, 1 T	7.2	Should not be heated
Ground flaxseed, 2 T	3.2	Grinding may increase bioavailability of alpha-linolenic acid. Should be refrigerated or frozen
Walnuts, 1 oz.	2.6	
Walnut oil, 1 T	1.4	
Canola oil, 1 T	1.3	
Soybeans, 1 cup cooked	1.0	
Soybean oil, 1 T	0.9	
Tofu, firm, 1/2 cup	0.7	
Soy nuts, 1/4 cup	0.6	

Source: USDA Nutrient Database for Standard Reference, Release 19, 2006, and [65]

- Read the ingredient listing on packaged foods and choose foods that do not contain hydrogenated or partially hydrogenated oils, Crackers, cookies, margarine, pie crusts, chocolate coatings, shortening, cakes, frosting, chips, and popped or microwave popcorn frequently are sources of *trans*-fats although many companies are reformulating their recipes to eliminate *trans*-fats.
- Check the Nutrition Facts Panel for the amount of *trans*-fat in a food. Choose foods containing zero grams of *trans*-fats. Be aware, however, that if a food contains half a gram or less of *trans*-fats per serving, the label can still say that it has zero grams of *trans*-fats in a serving. It's a good idea to also check the ingredient list for partially hydrogenated oils as well.
- When eating out, avoid foods that have been fried or grilled in hydrogenated or partially hydrogenated oils. This would include French fries, fried onion rings, doughnuts, and other fried foods as well as grilled sandwiches. Ask your server what kind of oil will be used to prepare your food.
- Avoid commercial pastries made with hydrogenated or partially hydrogenated oils.

Fig. 15.1. Suggestions for decreasing consumption of foods with *trans*-fats

DHA is very limited, however, although it may be somewhat higher in pregnant women than in non-pregnant women or men [61]. ALA supplementation in pregnancy did not increase maternal or infant DHA levels although EPA concentrations were higher [74]. Similarly, ALA supplementation in lactation led to an increase in maternal plasma and breast milk ALA concentrations but had no effect on breast milk DHA levels [75]. Although further study has been recommended, it is likely that ALA cannot substitute for DHA [76].

DHA-rich microalgae provides a direct, non-animal-derived source of DHA [69, 73, 77]. Microalgae-derived DHA, when used by lactating women, has effectively increased plasma phospholipid DHA concentrations of breast-fed infants [78]. No reports have been published of its use in pregnancy. Some commercial supplements contain microalgae-derived DHA either in liquid or vegan gelatin capsules. Eggs from hens fed DHA-rich microalgae are another potential source of DHA and have been effectively used to increase the DHA intake of pregnant women [79, 80]. Other foods that have been fortified with microalgae-derived DHA include soymilk, energy bars, yogurt, and veggie burgers [81]. One expert panel has recommended a DHA intake of 300 mg/day in pregnancy and lactation [82].

15.8 PRACTICAL CONSIDERATIONS

15.8.1 Family Concerns

Pregnant vegetarians may face pressure from nonvegetarian family members who are concerned about the adequacy of the vegetarian diet during pregnancy. Dietetics professionals can provide support for the vegetarian client by providing accurate current information about vegetarian nutrition [1]. An unbiased evaluation of the pregnant woman's diet and nutritional status with recommendations for dietary modifications if necessary may reassure family members, support the client's dietary choices, and improve nutritional status.

15.8.2 Becoming Vegetarian during Pregnancy

Some women may decide to adopt a vegetarian diet during pregnancy. Dietetics professionals can assist the new vegetarian by assessing nutritional status, providing specific guidelines for planning meals during pregnancy, and providing information about meeting needs for key nutrients.

15.8.3 Foods That Are Frequently Included in Vegetarian Diets

Vegetarians typically eat a variety of fruits, vegetables, breads, cereals, beans, and possibly dairy products and eggs. A number of products, while not unique to vegetarians, may be less familiar to practitioners. Fig. 15.2 provides information about some foods that are commonly used by vegetarians.

15.8.4 Use of Soy Products in Pregnancy

Soy products, including soymilk, tofu, textured vegetable protein, and meat analogs, are often used by vegetarians to replace animal products, for convenience, and to add dietary variety. Isoflavones, phytoestrogens found in soy, appear to be transferred to the fetus [83]. Fetal exposure to isoflavones does not appear to cause adverse developmental or reproductive effects [84–86].

15.8.5 Nausea

Vegetarian women, like nonvegetarian women, may experience nausea and food aversions, especially in early pregnancy. Suggestions for coping with nausea and food aversions can be modified to meet the needs of vegetarians. Bland, starchy foods like rice, pasta, cereal, potatoes, and crackers are frequently better tolerated than sweet or fatty foods. For many women, the salads and raw vegetables that are dietary mainstays may not be appealing in early pregnancy. Vegetables incorporated in soups or mixed with mashed potatoes or rice may be better tolerated. Vegetable juice is another option.

Deli slices	Meat analogs made to resemble foods commonly used in sandwiches like bologna, salami, and ham
Egg replacer	A commercial powder containing potato starch and tapioca flour used to replace eggs in baked products
Hummus	A dip or sandwich spread made from pureed chickpeas, tahini, and spices
Meat analogs	A general term for imitation meats, usually made from soy but sometimes made from wheat gluten, grains, beans, and/or nuts
Nori	A sea vegetable available in flat sheets and often used to make sushi
Nutritional yeast	A yeast that is grown on a nutrient-rich media; good source of vitamins and minerals (specific nutrients and amounts vary by brand)
Rice milk	A rice-based beverage that can be used in place of cow's milk or soymilk although it is lower in protein. Often available in fortified form
Seitan	A meat substitute made from wheat gluten.
Soymilk	A beverage made from soybeans that can be used in place of cow's milk. Often available in fortified form and in a variety of flavors
Tahini	A spread, similar in consistency to peanut butter, made from ground sesame seeds
Tempeh	A product made from fermented soybeans, and sometimes grains, that are pressed into a solid cake
Textured soy protein (TSP or TVP)	A product made from soy flour, available as granules or chunks and used in place of ground beef or other meat products in chili, stews, soups, and other dishes
Tofu	A mild-tasting product made from coagulated soymilk. Tofu prepared using calcium sulfate as a coagulating agent is higher in calcium than tofu coagulated with nigari (magnesium chloride). Tofu can be purchased in a variety of textures (soft, silken, firm, extra firm)
Tofu hot dogs	Meat analogs that resemble hot dogs and are made from soy
Veggie burgers	Vegetarian burgers made from a variety of products including tofu, TSP, soybeans, other beans, grains, wheat gluten, and nuts. Commercially available frozen, refrigerated, and as a powdered mix

Fig. 15.2. Glossary of vegetarian foods

15.9   MEAL PLANS

A number of food guides have been developed for pregnant vegetarians [2, 87–89]. These can serve as a general guide but will need modification depending on individual energy needs, preexisting conditions, and food preferences. Fig. 15.3 shows a sample menu for pregnancy that features three meals and three snacks, a pattern that can be useful for women who prefer frequent small meals.

15.10   SUPPLEMENTS

Pregnant vegetarians who consume an adequate diet do not routinely require a daily multivitamin-mineral supplement although supplements of individual nutrients such as iron or vitamin B_{12} may be indicated. The Institute of Medicine recommends that all

BREAKFAST
1/2 cup oatmeal with maple syrup
1 slice whole-wheat toast with fruit spread
1 cup fortified soymilk
1/2 cup calcium and vitamin
D-fortified orange juice

SNACK
1/2 whole-wheat bagel with margarine
Banana

LUNCH
Veggie burger on whole-wheat bun
with mustard and catsup

1 cup steamed collard greens
1 medium apple
1 cup fortified soymilk

SNACK
3/4 cup ready-to-eat cereal with 1/2 cup
blueberries
1 cup fortified soy milk

DINNER
3/4 cup tofu stir-fried with 1 cup
vegetables
1 cup brown rice
Medium orange

SNACK
Whole-grain crackers with 2 Tbsp peanut
butter
4 oz apple juice

Nutritional analysis of sample menu

RDA/AI

2,240 calories
100 g protein (18% of calories)   71 g
55 g fat (22% of calories)
336 g carbohydrate (60% of calories)
1,688 mg calcium   1,000 mg
32.5 mg iron   49 mg (supplemental iron may be needed to
                                                         meet the iron RDA for vegetarians)
11.2 mg zinc   11 mg
2.1 mg thiamin   1.4 mg
1.4 mg riboflavin   1.4 mg
23.1 mg niacin   18 mg
9 mcg vitamin B_{12} 2.6 mcg
4.2 mcg vitamin D   5 mcg (supplement/sun exposure indicated)
850 mcg folate   600 mcg

Fig. 15.3. Sample menu plan for pregnant vegans. (Adapted with permission from Wasserman D, Mangels AR (2006) Simply Vegan, 4th edn. The Vegetarian Resource Group, Baltimore, Md.)

women capable of becoming pregnant consume 400 micrograms of folate daily from supplements, fortified food, or a combination of fortified food and supplements. Multivitamin-mineral supplements specifically identified as "vegetarian" or "suitable for vegetarians" are available for women whose diets may not be adequate or who have increased needs.

15.10.1   Vegetarian Multivitamin–Mineral Supplements

Multivitamin–mineral supplements marketed to vegetarians differ widely in nutrient content. Some contain herbs, amino acids, bioflavonoids, fiber, DHA, and other substances.

Table 15.5
Levels of Selected Nutrients in Several Brands of Vegetarian Prenatal Supplements

Supplement[a]	Serving size	Vitamin D (mcg)	Folate (mcg)	Vitamin B₆ (mg)	Vitamin B₁₂ (mcg)	Calcium (mg)	Iron (mg)	Iodine (mcg)	Zinc (mg)	Other nutrients at levels >100% of pregnancy DRI	Comments
Country Life Maxi PreNatal (a)	6 caps	5	1,000	100	40	800	18	225	15	C, E, Thia, Ribo, Nia, Bio, Pan, Mg, Mn, Cr	Also contains DHA, ginger, bioflavonoids, and other substances
Freeda Daily Prenatal (b)	1 tab	10	800	3	10	200	27	–	15	C, Thia, Ribo, Nia, Bio, Pan, Cu	
Freeda KPN Prenatal (b)	3 tabs	10	800	3	6	1,000	27	–	22.5	C, Nia, Pan	
MegaFood Baby & Me Daily-Foods (c)	6 tabs	10	800	10	50	200	18	150	15	C, E, Thia, Ribo, Nia, Bio, Pan, Mn, Cr	Also contains bioflavonoids, inositol, alfalfa, and other substances
Perfect Prenatal (d)	3 tabs	10	800	2	6	30	18	150	7.5	E, Thia, Ribo, Nia, Bio, Pan	Also contains mixed carotenoids, herbal extracts, sprouted seeds, soy, and other substances
Solgar Prenatal Nutrients (e)	4 tabs	10	800	2.5	8	1,300	27	150	15	C, E, Thia, Ribo, Nia, Bio, Pan, Mg, Cu	Also contains inositol, soy protein, carotenoid mix, and other substances
SuperNutrition PreNatal Blend (f)	6 tabs	25	1,000	110	200	1,200	40	150	30	C, E, Thia, Ribo, Nia, Bio, Pan, Mg, Se, Cu, Mn, Cr, Mo	Also contains bioflavonoids, Fiber, taurine, and other substances
DRI, pregnancy, 19–50 years		5	600	1.9	2.6	1,000	27	220	11		

Thia thiamin, *Ribo* riboflavin, *Nia* niacin, *Pan* pantothenic acid, *Bio* biotin, *Mg* magnesium, *Se* selenium, *Cu* copper, *Mn* manganese, *Cr* chromium, *Mo* molybdenum
All supplements were identified on the product label or website as being vegetarian. Some may contain dairy or other substances unacceptable to some vegetarians. Product formulations frequently change so it is important to update the information in this table regularly
[a]Source of product information: (a) www.country-life.com; (b) www.freedavitamins.com; (c) www.megafoodonline.com; (d) www.newchapter.info; (e) www.solgar.com; (f) www.supernutritionusa.com

Some provide large amounts of vitamins, especially thiamin, riboflavin, niacin, vitamin B_6, and vitamin B_{12} that are much higher than Dietary Reference Intakes.

Other prenatal supplements contain no iodine or are low in calcium, two nutrients that may be low in diets of some vegetarians. If dietary calcium is low and a woman's prenatal supplement is also low in calcium, then a calcium supplement is indicated. Calcium supplements can interfere with iron and zinc absorption, so are best used between meals.

Table 15.5 compares several brands of vegetarian prenatal supplements. Individual assessment is needed when determining whether or not a supplement is needed and the amount and type of supplementation. Since supplement content and availability may vary, clients should be encouraged to bring in their supplement label and practitioners should be aware of commonly used products.

15.11   CONCLUSION

Appropriately planned vegetarians diets can be healthful and nutritionally adequate and are appropriate for use in pregnancy. Key nutrients for pregnant vegetarians include protein, iron, zinc, vitamin B_{12}, omega-3 fatty acids, and calcium and vitamin D (for vegans). Practitioners should be aware of good sources of these nutrients and be able to assess the need for supplements. Practitioners should also be able to provide current, accurate information about vegetarian nutrition and foods (see Tables 15.1–15.4).

REFERENCES

1. Mangels AR, Messina V, Melina V (2003) Position of The American Dietetic Association and Dietitians of Canada: Vegetarian diets. J Am Diet Assoc 103:748–765
2. Messina V, Mangels R, Messina M (2004) The Dietitian's Guide to Vegetarian Diets: Issues and Applications, 2nd edn. Jones and Bartlett Publishers, Sudbury, Mass.
3. Koebnick C, Garcia AL, Dagnelie PC, Strassner C, Lindemans J, Katz N, Leitzmann C, Hoffmann I (2005) Long-term Consumption of a Raw Food Diet is Associated with Favorable Serum LDL Cholesterol and Triglycerides but also with Elevated Plasma Homocysteine and Low Serum HDL Cholesterol in Humans. J Nutr 135:2372–2378
4. Barr SI, Chapman GE (2002) Perceptions and practices of self-defined current vegetarian, former vegetarian, and non-vegetarian women. J Am Diet Assoc 102:354–360
5. Finley DA, Dewey KG, Lonnerdal B, Grivetti LE (1985) Food choices of vegetarians and nonvegetarians during pregnancy and lactation. J Am Diet Assoc 85:676–685
6. Haddad EH, Tanzman JS (2003) What do vegetarians in the United States eat? Am J Clin Nutr 78(Suppl):626S–632S
7. Lindeman M, Sirelius M (2001) Food choice ideologies: the modern manifestations of normative and humanist views of the world. Appetite 37:175–184
8. The Vegetarian Resource Group. How many adults are vegetarians? Available via http://www.vrg. org/journal/vj2006issue4/vj2006issue4poll.htm
9. Key TJ, Appleby PN, Rosell MS (2006) Health effects of vegetarian and vegan diets. Proc Nutr Soc. 65:35–41
10. Haddad EH, Berk LS, Kettering JD, Hubbard RW, Peters WR (1999) Dietary intake and biochemical, hematologic, and immune status of vegans compared with nonvegetarians. Am J Clin Nutr 70(Suppl): 586S–593S
11. Davey GK, Spencer EA, Appleby PN, Allen NE, Knox KH, Key TJ (2003) EPIC-Oxford lifestyle characteristics and nutrient intakes in a cohort of 33883 meat-eaters and 31546 non meat-eaters in the UK. Public Health Nutr 6:259–68

12. Koebnick C, Heins UA, Hoffmann I, Dagnelie PC, Leitzmann C (2001) Folate status during pregnancy in women is improved by long-term high vegetable intake compared with the average Western diet. J Nutr 131:733–739

13. Koebnick C, Leitzmann R, Garcia AL, Heins UA, Heuer T, Golf S, Katz N, Hoffmann I, Leitzmann C (2005) Long-term effect of a plant-based diet on magnesium status during pregnancy. Eur J Clin Nutr 59:219–225

14. Carter JP, Furman T, Hutcheson HR (1987) Preeclampsia and reproductive performance in a community of vegans. South Med J. 80:692–697

15. Thomas J, Ellis FR (1977) The health of vegans during pregnancy. Proc Nutr Soc. 36:46A

16. Fraser GE (2003) Diet, life expectancy, and chronic disease. Studies of Seventh-Day Adventists and Other Vegetarians. Oxford University Press, New York, N.Y.

17. Spencer EA, Appleby PN, Davey GK, Key TJ (2003) Diet and body mass index in 38,000 EPIC-Oxford meat-eaters, fish-eaters, vegetarians and vegans. Int J Obesity 27:728–734

18. Institute of Medicine (1990) Nutrition during pregnancy. National Academy Press, Washington, D.C.

19. King JC, Stein T, Doyle M (1981) Effect of Vegetarianism on the Zinc Status of Pregnant Women. Am J Clin Nutr 34:1049–1055

20. Ward RJ, Abraham R, McFadyen IR, Haines AD, North WR, Patel M, Bhatt RV (1988) Assessment of trace metal intake and status in a Gujerati pregnant Asian population and their influence of the outcome of pregnancy. Br J Obstet Gynecol 95:676–682

21. Drake R, Reddy S, Davies J (1998) Nutrient intake during pregnancy and pregnancy outcome of lacto-ovo-vegetarians, fish-eaters and non-vegetarians. Veg Nutr 2:45–52

22. Lakin V, Haggarty P, Abramovich DR (1998) Dietary intake and tissue concentrations of fatty acids in omnivore, vegetarian, and diabetic pregnancy. Prost Leuk Ess Fatty Acids 58:209–220

23. Dagnelie PC, van Staveren WA, van Klaveren JD, Burema J (1988) Do children on macrobiotic diets show catch-up growth? Eur J Clin Nutr 42:1007–1016

24. Dagnelie PC, van Staveren WA, Vergote FJ, Burema J, van't Hof MA, van Klaveren JD, Hautvast JG (1989) Nutritional status of infants aged 4 to 18 months on macrobiotic diets and matched omnivorous control infants: A population-based mixed-longitudinal study. II. Growth and psychomotor development. Eur J Clin Nutr 43:325–338

25. Food and Nutrition Board, Institute of Medicine (2002) Dietary Reference Intakes for Energy, Carbohydrate, Fiber, Fat, Fatty Acids, Cholesterol, Protein, and Amino Acids. National Academies Press, Washington, D.C.

26. Cook JD (1990) Adaptation in iron metabolism. Am J Clin Nutr 51:301–308

27. Food and Nutrition Board, Institute of Medicine (2000) Dietary Reference Intakes for vitamin A, vitamin K, arsenic, boron, chromium, copper, iodine, iron, manganese, molybdenum, nickel, silicon, vanadium, and zinc. National Academies Press, Washington, D.C.

28. Hunt JR (2003) Bioavailability of Iron, Zinc, and Other Trace Minerals from Vegetarian Diets. Am J Clin Nutr 78(Suppl):633S–639S

29. Allen LH (2005) Multiple Micronutrients in Pregnancy and Lactation: An Overview. Am J Clin Nutr 81(Suppl):1206S–1212S

30. Scholl TO (2005) Iron status during pregnancy: setting the stage for mother and infant. Am J Clin Nutr 81(Suppl):L1218S–1222S

31. Gibson RS, Hotz C (2001) Dietary diversification/modification strategies to enhance micronutrient content and bioavailability of diets in developing countries. Br J Nutr 85(Suppl 2):S159–S166

32. King JC (2000) Determinants of maternal zinc status during pregnancy. Am J Clin Nutr 71(Suppl):1334S–1343S

33. Caldwell KL, Jones R, Hollowell JG (2005) Urinary iodine concentration: United States National Health and Nutrition Examination Survey 2001–2002. Thyroid 15:692–699

34. Pearce EN, Pino S, He X, Bazrafshan HR, Lee SL, Braverman LE (2004) Sources of dietary iodine: bread, cows' milk, and infant formula in the Boston area. J Clin Endocrinol Metab 89:3421–3424

35. Pennington JAT, Schoen SA, Salmon GD, Young B, Johnson RD, Marts RW (1995) Composition of core foods of the U.S. food supply, 1982–1991. III. Copper, manganese, selenium, iodine. J Food Comp Anal 8:171–217

36. Remer T, Neubert A, Manz F (1999) Increased risk of iodine deficiency with vegetarian nutrition. Br J Nutr 81:45–49
37. Lighttowler HJ, Davis GJ (1998) The effect of self-selected dietary supplements on micronutrient intakes in vegans Proc Nutr Soc 58:35A
38. Krajcovicova M, Buckova K, Klimes I, Sebokova E (2003) Iodine deficiency in vegetarians and vegans. Ann Nutr Metab 47:183–185
39. Key TJA, Thorogood M, Keenant J, Long A (1992) Raised thyroid stimulating hormone associated with kelp intake in British vegan men. J Human Nutr Diet 5:323–326
40. Teas J, Pino S, Critchley A, Braverman LE (2004) Variability of iodine content in common commercially available edible seaweeds. Thyroid 14:836–841
41. Nishiyama S, Mikeda T, Okada T, Nakamura K, Kotani T, Hishinuma A (2004) Transient hypothyroidism or persistent hyperthyrotropinemia in neonates born to mothers with excessive iodine intake. Thyroid 14:1077–1083
42. Guan H, Li C, Li Y, Fan C, Teng Y, Shan Z, Teng W (2005) High iodine intake is a risk factor of postpartum thyroiditis: result of a survey from Shenyang, China. J Endocrinol Invest 28:876–81
43. Prentice A (2000) Maternal calcium metabolism and bone mineral status. Am J Clin Nutr 71(Suppl):1312S–1316S
44. O'Brien KO, Donangelo CM, Vargas Zapata CL, Abrams SA, Spencer EM, King JC (2006) Bone calcium turnover during pregnancy and lactation in women with low calcium diets is associated with calcium intake and circulating insulin-like growth factor 1 concentrations. Am J Clin Nutr 83:317–323
45. Koo WW, Walters JC, Esterlitz J, Levine RJ, Bush AJ, Sibai B (1999) Maternal calcium supplementation and fetal bone mineralization. Obstet Gynecol 94:577–582
46. Donaldson MS (2001) Food and nutrient intake of Hallelujah vegetarians. Nutr Food Sci 31:293–303
47. Waldmann A, Koschizke JW, Leitzmann C, Hahn A (2003) Dietary intake and lifestyle factors of a vegan population in Germany: results from the German Vegan Study. Eur J Clin Nutr 57:947–955
48. Heaney R, Dowell M, Rafferty K, Bierman J (2000) Bioavailability of the calcium in fortified soy imitation milk, with some observations on method. Am J Clin Nutr 71:1166–1169
49. Weaver C, Plawecki K (1994) Dietary calcium: adequacy of a vegetarian diet. Am J Clin Nutr 59:1238S–1241S
50. Weaver C, Proulx W, Heaney R (1999) Choices for achieving adequate dietary calcium with a vegetarian diet. Am J Clin Nutr 70:543S–548S
51. Luhby AL, Cooperman JM, Donnenfeld AM, Herman JM, Teller DN, Week JB (1958) Observations on transfer of vitamin B_{12} from mother to fetus and newborn. Am J Dis Child 96:532–533
52. Allen LH (1994) Vitamin B-12 metabolism and status during pregnancy, lactation, and infancy. Adv Exp Med Biol. 352:173–186
53. Groenen PM, van Rooij IA, Peer PG, Gooskens RH, Zielhuis GA, Steegers-Theunissen RP (2004) Marginal maternal vitamin B_{12} status increases the risk of offspring with spina bifida. Am J Obstet Gynecol 191:11–17
54. Sanchez, SE, Zhang C, Rene-Mallinow M, Ware-Jauregui S, Larrabure G, Williams MA (2001) Plasma folate, vitamin B_{12} and homocysteine concentrations in preeclamptic and normotensive Peruvian women. Am J Epidemiol 153:474–840
55. Bjørke Monsen AL, Ueland PM, Vollset SE, Guttormsen AB, Markestad T, Solheim E, Refsum H. (2001) Determinants of cobalamin status in newborns. Pediatrics 108:624–630
56. Martinez M (1992) Tissue levels of polyunsaturated fatty acids during early human development. J Pediatr 120:S129–S38
57. Koletzko B, Larque E, Demmelmair H (2007) Placental transfer of long-chain polyunsaturated fatty acids (LC-PUFA) (1997) J Perinat Med 35(Suppl):S5–S11
58. Gibson RA, Neumann MA, Makrides M (1997) Effect of increasing breast milk docosahexaenoic acid on plasma and erythrocyte phospholipid fatty acids and neural indices of exclusively breast fed infants. Eur J Clin Nutr. 51:578–84
59. Jensen CL, Maude M, Anderson RE, Heird WC (2000) Effect of docosahexaenoic acid supplementation of lactating women on the fatty acid composition of breast milk lipids and maternal and infant plasma phospholipids. Am J Clin Nutr 71:292S–299S

60. McCann JC, Ames BN (2005) Is docosahexaenoic acid, an *n*-3 long-chain polyunsaturated fatty acid, required for development of normal brain function? An overview of evidence from cognitive and behavioral tests in humans and animals. Am J Clin Nutr 82:281–295

61. Williams CM, Burdge G (2006) Long-chain *n*-3 PUFA: plant v. marine sources. Proc Nutr Soc. 65:42–50

62. Jensen CL (2006) Effects of *n*-3 fatty acids during pregnancy and lactation. Am J Clin Nutr 83(Suppl):1452S–1457S

63. Cheatham CL, Colombo J, Carlson SE (2006) *n*-3 fatty acids and cognitive and visual acuity development: methodologic and conceptual considerations. Am J Clin Nutr 83(Suppl):1458S–1466S

64. Szajewska H, Horvath A, Koletzko B (2006) Effect of *n*-3 long-chain polyunsaturated fatty acid supplementation of women with low-risk pregnancies on pregnancy outcomes and growth measures at birth: a meta-analysis of randomized controlled trials. Am J Clin Nutr 83:1337–3144

65. Judge MP, Harel O, Lammi-Keefe CJ (2007) A docosahexaenoic acid-functional food during pregnancy benefits infant visual acuity at four but not six months of age. Lipids 42:117–122

66. Cheruku SR, Montgomery-Downs HE, Farkas SL, Thoman EB, Lammi-Keefe CJ (2002) Higher maternal docosahexaenoic acid during pregnancy is associated with more mature neonatal sleep-state patterning. Am J Clin Nutr 76:608–613

67. Colombo J, Kannass KN, Shaddy DJ, Kundurthi S, Maikranz Jm, Anderson CJ, Blega OM, Carlson SE (2004) Maternal DHA and the development of attention in infancy and toddlerhood. Child Develop 75:1254–67

68. Krajcovicova-Kudlackova M, Simoncic R, Babinska K, Bederova A (1995) Levels of lipid peroxidation and antioxidants in vegetarians. Eur J Epidemiol 111:207–211

69. Geppert J, Kraft V, Demmelmair H, Koletzko B (2005) Docosahexaenoic acid supplementation in vegetarians effectively increases omega-3 index: a randomized trial. Lipids 40:807–814

70. Rosell MS, Lloyd-Wright Z, Appleby PN, Sanders TA, Allen NE, Key TJ (2005) Long-chain *n*-3 polyunsaturated fatty acids in plasma in British meat-eating, vegetarian, and vegan men. Am J Clin Nutr 82:327–334

71. Sanders TAB, Reddy S (1992) The influence of a vegetarian diet on the fatty acid composition of human milk and the essential fatty acid status of the infant. J Pediatr 120:S71–S77

72. Uauy R, Peirano P, Hoffman D, Mena P, Birch D, Birch E (1996) Role of essential fatty acids in the function of the developing nervous system. Lipids 31:S167–S176

73. Davis B, Kris-Etherton P (2003) Achieving optimal essential fatty acid status in vegetarians: Current knowledge and practical implications. Am J Clin Nutr 78(Suppl):640S–646S

74. DeGroot RH, Hornstra G, van Houwelingen AC, Roumen F (2004) Effect of alpha-linolenic acid supplementation during pregnancy on maternal and neonatal polyunsaturated fatty acid status and pregnancy outcome. Am J Clin Nutr 79:251–260

75. Francois CA, Connor SL, Bolewicz LC, Connor WE (2003) Supplementing lactating women with flaxseed oil does not increase docosahexaenoic acid in their milk. Am J Clin Nutr 77:226–233

76. Akabas SR, Deckelbaum RJ (2006) Summary of a workshop on n-3 fatty acids: current status of recommendations and future directions. Am J Clin Nutr 93(Suppl):1536S–1538S

77. Conquer JA, Holub BJ (1996) Supplementation with an algae source of docosahexaenoic acid increases (*n*-3) fatty acid status and alters selected risk factors for heart disease in vegetarian subjects. J Nutr 126:3032–3039

78. Jensen CL, Voigt RG, Prager TC, Zou YL, Fraley JK, Rozelle JC, Turcich MR, Llorente AM, Anderson RE, Heird WC (2005) Effects of maternal docosahexaenoic acid on visual function and neurodevelopment in breastfed term infants. Am J Clin Nutr 82:125–132

79. Smuts CM, Huang M, Mundy D, Plasse T, Major S, Carlson SE (2003) A randomized trial of docosahexaenoic acid supplementation during the third trimester of pregnancy. Obstet Gynecol 101:469–479

80. Smuts CM, Borod E, Peeples JM, Carlson SE (2003) High-DHA eggs: feasibility as a means to enhance circulating DHA in mother and infant. Lipids 38:407–414

81. Martek Biosciences Corporation. Finding life's DHA. Available via http://consumer.martek.com/findinglifesdha/

82. Simopoulos AP, Leaf A, Salem N (1999) Conference report: workshop on the essentiality of and recommended dietary intakes for omega-6 and omega-3 fatty acids. J Am Coll Nutr 18:487–489

83. Adlercreutz H, Yamada T, Wahala K, Watanabe S (1999) Maternal and neonatal phytoestrogens in Japanese women during birth. Am J Obstet Gynecol 180:737–743

84. Munro IC, Harwood M, Hlywka JJ, Stephen AM, Doull J, Flamm WG, Adlercreutz H (2003) Soy isoflavones: a safety review. Nutr Rev 61:1–33

85. National Toxicology Program, Center for the Evaluation of Risks to Human Reproduction (2006) Report on the reproductive and developmental toxicity of Genistein. Avilable via http://cerhr.niehs.nih.gov/chemicals/genistein-soy/genistein/Genistein_Report_final.pdf

86. National Toxicology Program, Center for the Evaluation of Risks to Human Reproduction (2006) Report on the reproductive and developmental toxicity of soy formula. Available via http://cerhr.niehs.nih.gov/chemicals/genistein-soy/soyformula/Soy-report-final.pdf

87. Messina V, Melina V, Mangels AR (2003) A new food guide for North American vegetarians. J Am Diet Assoc 103:771–775

88. Davis B, Melina V (2000) Becoming Vegan. Book Publishing, Summertown, Tenn.

89. Melina V, Davis B (2003) The New Becoming Vegetarian. Book Publishing, Summertown, Tenn.

16 Iron Requirements and Adverse Outcomes

John Beard

Summary Iron deficiency continues to be one of the most prevalent nutritional deficiency diseases in the world and has a particularly high prevalence in pregnancy. The incidence and severity are greater in developing countries but even in developed countries, the prevalence may reach 30–40% in the third trimester. The assessment of iron status in pregnancy can be challenging due to the rapid expansion of the maternal blood volume and then rapid fetal and placental growth. The recommendation for iron intervention is based on a multibiomarker approach that includes serum ferritin, soluble transferrin receptor, transferrin saturation, and hemoglobin. There are significant negative outcomes to iron deficiency in pregnancy; these include maternal and infant mortality in severe cases, but also shortened gestation, prematurity, and poorer infant development in less severe cases. A substantial scientific and medical literature shows a substantial adverse outcome to iron deficiency in the first trimester, with additive risk if the iron deficiency persists throughout pregnancy. Infants born to iron-deficient mothers are more likely to become iron deficient themselves in early postnatal life; this in turn, appears to be causally related to delayed neuron maturation. The reversibility of the cognitive and behavioral deficits that occur due to iron deficiency between 6–12 months of postnatal life is questionable and is the subject of several current research projects.

Keywords: Iron deficiency, Ferritin, Transferrin receptor, Neurodevelopment

16.1 INTRODUCTION

The estimated prevalence of iron deficiency in pregnancy varies from 30% in industrialized countries to >60% in other less developed parts of the world. The disparity likely represents myriad effects of access to health care, dietary quality and quantity, and reproductive frequency. Iron fortification and supplementation programs have reduced the prevalence of iron deficiency and anemia in pregnancy in developed countries, but a residual prevalence of around 10–20% in the United States suggests there is still room for more progress. An inspection of the NHANES III national survey data shows the lower quartile for median iron intakes of reproductive age women is only between 8.4 and 9.9 mg/day, far below the habitual intakes that could sustain the very large increase

From: *Nutrition and Health: Handbook of Nutrition and Pregnancy*
Edited by: C.J. Lammi-Keefe, S.C. Couch, E.H. Philipson © Humana Press, Totowa, NJ

in requirements with pregnancy [1]. Several recent reviews have thoroughly examined the relationship between maternal iron status and infant outcomes with some unsurprising results [2, 3]. Poor iron status, even in the first trimester, is associated with attenuated intrauterine growth and development, while emerging experimental and clinical data point toward persistent consequences for infant neurodevelopment and functioning. The high prevalence of iron deficiency and the possibility of significant negative outcomes as a result of an iron deprived intrauterine environment thus requires a clear understanding by the clinical and public health world of issues regarding iron homeostasis during pregnancy.

16.2 IRON BALANCE IN PREGNANCY

16.2.1 Iron Needs for Mother and Fetus

Iron (Fe) requirements increase quite dramatically during pregnancy for expansion of the maternal blood volume, placental growth, and fetal growth. Quantitatively, these requirements change from <1 mg Fe/day in a reproductive age female to a median requirement of 4.6 mg Fe/day and a 90th percentile requirement of nearly 6.75 mg Fe/day by the third trimester [1]. A factorial model has been used by the Recommended Dietary Intakes (RDI) committee to estimate iron needs during pregnancy (Table 16.1). The components throughout the entire pregnancy include an expansion of the red cell mass (450 mg), needs for fetal and placental iron (370 mg), and blood losses during and after delivery (150–250 mg). Thus, the total estimated additional needs are between 1,040 and 1,240 mg of iron. Of course, these requirements are not equal in all trimesters: in the first trimester iron for the fetus (25 mg) and the umbilicus and placenta (5 mg) total 30 mg of iron. But in the second trimester, this increases dramatically to 75 mg for fetal growth and 25 mg for the placenta. In the third trimester, there is another large increase in requirements to 145 mg Fe for the fetus and >45 mg for the umbilicus and placenta for a total requirement of >220 mg. These uneven demands for iron during pregnancy are related to the differences in median hemoglobin (Hb) concentration during pregnancy even in iron supplemented healthy women (Fig. 16.1). Much of the expansion of the red cell mass occurs in the second trimester, while most of the fetal deposition of iron occurs in the third trimester. The corresponding cutoff levels for the diagnosis of anemia in pregnancy vary accordingly as do the apparent consequences of iron deficiency and anemia.

16.2.2 Iron for the Maternal Red Cell Mass and Anemia

The red cell mass in pregnancy is not a static number and can be affected by the amount of iron supplementation that has occurred during the pregnancy [4]. For example,

Table 16.1
Estimated Median Iron Requirements (mg) During Pregnancy

	1st trimester	2nd trimester	3rd trimester	Total
Fetus	25 mg	75 mg	145 mg	245 mg
Placenta and umbilicus	5 mg	25 mg	45 mg	75 mg
Red cell mass	5–10 mg	225 mg	225 mg	450 mg
Total	35–40 mg	325 mg	415 mg	

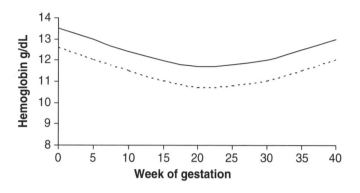

Fig. 16.1. Hemoglobin concentration in healthy women in developed countries. The *solid curve* is the median Hb concentration, while the *dashed line* is the 5th percentile

when supplemental iron was provided to a group of women, the expansion of the red cell mass was approximately 570 mg of Fe, whereas when no supplementation was provided, the expansion was only 260 mg of Fe. It has been suggested that for every 10 g/l increase in maternal Hb desired, there is a need for an additional 175 mg of absorbed iron [5]. The amount of additional iron needed for expansion of the red cell mass is also dependent on the numbers of fetuses in the womb. If twins are expected, then the expansion is estimated to be 680 ml, while triplets increase the blood volume to around 900 ml. The World Health Organization recommends iron supplements of between 30–60 mg/day if the woman has iron stores (e.g., ferritin >30 mcg/l). The recommendation is quite close to the Institute of Medicine recommendation of 30 mg/day for the second and third trimesters if stores are also present at the first clinic visit. If stores are absent, then a much more aggressive approach is usually taken with intakes of 120–240 mg/day advocated. It is easy to assume that more iron and higher Hb is better, but there are data that actually demonstrate a negative outcome to an overly elevated Hb concentration [6]. Consumption of large doses of iron supplements have been related to oxidative damage, and the gastrointestinal side effects may be related to the poor compliance in many populations of pregnant women [7, 8]. The alternative approach of non-daily low-dose iron supplementation appears to be effective in situations of only modest iron deficits.

16.2.3 Iron Supplementation and Maternal Red Cell Responses

The decline in Hb in the first trimester is now seen as a normal physiologic event and is the result of expansion of the plasma volume. Overzealous supplementation to prevent this physiological anemia has been associated with risk of poor fetal outcomes in at least one study [9]. The normal nadir of Hb is between 24 and 32 weeks of gestation, after which the Hb concentration again rises to levels similar to that seen in the first trimester. The extent of this Hb readjustment may be affected by iron reserves as the large expansion of the red cell mass in the second trimester and early third trimester usually depletes all iron reserves, and physiologic anemia may now be replaced with nutritional anemia.

Maternal Hb concentration and infant outcomes have a U-shaped curve, with an increased risk for poor outcomes at each end of the distribution [10]. High Hb likely reflects an improper expansion of the plasma volume, as in preeclampsia, with increased infant mortality and morbidity [11]. The variation in the amount of hemodilution is considerable, and makes the relatively simple Hb measurement quite unreliable with regard to diagnosis of iron deficiency anemia. Current target Hb concentrations in each trimester are based on supplementation trials, which suggest that Hb > 110 g/l in first and third trimesters and 105 g/l in the second trimester represent reasonable clinical expectations of lower normal levels [1].

16.2.4 Iron Absorption

Maternal iron stores are usually limited, and the capacity to increase the efficiency of absorption of dietary iron appears to maximize at around 40–60% for non-heme iron in the second trimester [12]. The efficiency of iron absorption is strongly associated with the iron status of the woman; women with a serum ferritin >30 mcg/l are much less efficient than women with a ferritin level < 30 mcg/l [13, 14]. Given the fact that there are limited stores and an upper limit to dietary intake, iron balance is very difficult to maintain. When there is insufficient dietary and reserve iron to meet the demands, essential body iron from maternal pools are sacrificed with a resulting maternal iron deficiency. The adaptive responses in placental transfer of iron with severe iron deficiency include upregulation of placental ferritin receptors and presumably an increase in placental-fetal transfer of iron [15, 16]. These attempts to maximize iron delivery to the fetus often has limited success as there is still a smaller-than-normal endowment of iron for the newborn and subsequent postnatal infant iron deficiency results (see sections below) [17].

16.3   ASSESSMENT OF IRON STATUS

The assessment of iron status cannot rely solely on the concentration of Hb or a hematocrit (Hct) value, as there is usually a large overlap in pregnancy between subjects that have iron deficiency and those that do not [18, 19]. The various categories of iron status, adequate, low or depleted iron stores, iron deficient erythropoiesis, and iron deficiency anemia, are all characterized by a range of values of a number of biomarkers that are sensitive to iron storage, iron transport, or tissue iron deficiency (Table 16.2).

Table 16.2
Cutoff Values to Define Iron Status during Pregnancy

	Iron replete	Iron depleted	Iron-deficient erythropoiesis	Iron-deficient
Serum ferritin	>30 mcg/l	<30 mcg/l	<15 mcg/l	<15 mcg/l
Soluble TfR[a]	<2–4 mg/dl	<2–4 mg/dl	4–5 mg/dl	>6 mg/dl
Hemoglobin[b]	>110 g/l (1st)	>110 g/l (1st)	>110 g/l (1st)	<110 g/l (1st)
	>105 g/l (2nd)	>105 g/l (2nd)	>105 g/l (2nd)	<105 g/l (2nd)
	>110 g/l (3rd)	>110 g/l (3rd)	>110 g/l (3rd)	<110 g/l (3rd)
Transferrin saturation	>20–25%	>20–25%	<15%	<15%
Serum Fe	>115 mcg/dl	115 mcg/dl	<60 mcg/dl	< 40 mcg/dl

[a]Soluble transferrin receptor levels depend on the clinical assay method utilized. These values correspond to values developed from the Ramco ELISA kit.

[b]Hb and hematocrit cut-off levels will vary by trimester; these values were derived from the IOM report [1].

A ferritin of <30 mcg/l indicates low iron stores, while a level <12–15 mcg/l is used to indicate depleted stores. The soluble form of the transferrin receptor, TfR, is sensitive to inadequate delivery of iron to cells and thus becomes a sensitive biomarker of iron inadequacy once the serum ferritin is <15 mcg/l. TfR is heavily expressed on the surface of developing red cells and is thus quite responsive to iron deficiency erythropoiesis. There is a strong inverse correlation between ferritin and sTfR such that combinations of these two markers can be used to generate composite indicators that are sensitive to changes in iron status across a wide range [19].

One of the important characteristics of the TfR measurement is that this biomarker is not sensitive to infection and inflammation and in contrast to serum ferritin, a strongly reactive acute phase protein. While these indicators have been validated in a number of settings, their utilization in studies in pregnancy has been quite limited with many clinical trials of supplementation relying only on Hb response to intervention [20]. Some have attempted to come up with proxy prediction rules for iron deficiency, based only on the measurement of Hb and red cell distribution width (RDW) [21], but the need to validate all of the measures within a particular clinical environment makes this approach problematic. Zinc protoporphyrin measurements are a useful indicator of iron status, but these are frequently not available as part of the routine clinical laboratory workup. The utility of these biomarkers to diagnose iron status in pregnancy clearly varies with the expected changes in iron status given the high iron requirements in pregnancy. For example, ferritin becomes insensitive as an indicator in many women by the second trimester because iron stores are already low. At this time, the sTfR or the composite indicator, body Fe, would be useful for indicating further changes in iron status before there is evidence of overt anemia [22].

16.4   CONSEQUENCES OF THIS NEGATIVE IRON BALANCE

The consequences of depletion of the essential body pools of iron include anemia, altered hormone metabolism, altered energy metabolism, depressed immune functioning, and changes in behavior and cognition [18, 23]. The impact of each of these consequences on maternal and fetal survival, fetal growth, and postnatal development are still being examined. The possible causal routes include direct and indirect effects of anemic hypoxia, placental delivery of iron, and alterations in hormonal control of pregnancy due to alterations in the stress: hypothalamic–pituitary–adrenal axis system [17]. Maternal anemia has been related to maternal mortality, fetal mortality, fetal growth retardation, pregnancy complications, and a small amount on infant growth [2, 3, 24, 25]. A vast majority of the studies on anemia and pregnancy outcome have not delineated effects of iron deficiency from effects of anemia.

16.5   ANEMIA AND BIRTH WEIGHT, GESTATIONAL AGE, AND INFANT MORTALITY

Most reviewers of the scientific literature will agree that there is a U-shaped curve relationship between the maternal hemoglobin concentration and the proportion of LBW infants [2]. The cause of the elevation in prevalence of LBW infants at the upper end of the distribution of Hb is believed to be improper expansion of the maternal plasma volume [26], while insufficient erythropoiesis and poor volume expansion may be

Table 16.3
Common Iron Supplement Prescribed during Pregnancy

Chemical form	Trade names	Oral forms
Ferrous sulfate	Feosol, Feratab, Fer-gen-sol, Ferospace, Ferralyn, Lanacaps Ferra-TD, Slow Fe	Capsules, extended release, oral liquid, syrup, tablets
Ferrous gluconate	Fergon, Ferralet, Simron	Capsules, tablets, extended release tablets
Ferrous fumarate	Femiron, Feostat, Ferretts, Fumasorb, Fumerin, Hemocyte, Ircon, Nephro-Fer, Span-FF	Extended release, oral solution, oral suspension, tablets, chewable tablets
Iron-polysaccharide	Hytinic, Niferex, Nu-Iron	Capsules, oral solution, tablets

associated with the low Hb concentrations at the other end of the distribution curve. The optimal maternal Hb for minimal incidence of LBW in the published literature varies (Table 16.3). The hemoglobin concentration and the definition of anemia are trimester dependent, with a clear nadir of concentration in mid-gestation. Since many of these studies did not use finite times of sampling, the variations may well reflect the timing of sampling and not true discrepancies in the relationship of data to outcomes [27].

Severity of anemia is an additional factor associated with an increased risk of LBW and prematurity with severe anemia (Hb < 80 g/l) [27]. There is a median relative risk (RR) of 4.9 for severe anemia with moderate anemia having a median risk of approximately 2. The causes of anemia were not known in most studies; thus, the contribution of iron deficiency anemia cannot be evaluated. In a study of pregnancy outcome where malaria is endemic, Verhoeff et al. [28] reported a RR of 1.6 for intrauterine growth retardation if maternal Hb was <80 g/l at the first clinic visit compared to a RR of 1.4 (not significant) if moderate anemia was present at delivery. Interestingly, the prevalence of anemia decreased from 23.6 to 11.4% between the first trimester and delivery. In this study of 1,423 live-born singleton births in rural Malawi, there was *no benefit* to intrauterine growth retardation of iron-folate administration during pregnancy. The authors did observe, however, a significant reduction in the prevalence of prematurity with micronutrient supplementation. In contrast, malaria intervention, even in mid gestation was effective in promoting fetal growth.

16.6   MATERNAL ANEMIA AND MORTALITY

Maternal mortality is correlated with the severity of anemia in pregnancy [29]. In his review of reports from 1950 to 1999, Rush examined the relationship of maternal anemia, usually at delivery, with both antepartum and postpartum death. He arrived at the conclusion that severe anemia (Hb < 6–7 g/l or Hct < 0.14) is associated with an increased rate of maternal death. In very severe anemia, the death rate may be as high as 20%, greater than the comparison group of minimum mortality. Transfusion is an accepted clinical practice in developed countries but is often unavailable in the Third World. When the Hb is this low, compensatory mechanisms begin to fail, lactic acid levels rise, and cardiac failure may occur. While no direct causal relationships between iron deficiency anemia and mortality is usually demonstrated, it is reasonable to expect that it is contributory to death rates.

16.7   IRON DEFICIENCY ANEMIA AND PREGNANCY OUTCOMES

Several research groups have computed relative risks for the impact of iron deficiency anemia on pregnancy complications while controlling for other causes of anemia [2, 30]. In a study in Camden, N.J., Scholl and colleagues [31] showed that iron-deficiency anemia in the first trimester was more strongly related to prematurity and LBW (RR = 3.1) than anemia of any cause later in pregnancy. They concluded from this that iron deficiency in the first trimester was important, but anemia at other times had little effect. In a study in Malawi, Verhoeff et al. [28] collected samples early in pregnancy and at mid-gestation. In their evaluation within the context of coexistence of anemia associated with malaria, they observed a RR of 6 for LBW of iron deficiency early in pregnancy but not in late pregnancy. In a third study, investigators measured iron status in Chinese mothers in early pregnancy (<8 weeks) and observed that moderate iron deficiency anemia conferred a RR of 2.96 for prematurity and LBW [32]. These three studies, taken together, suggest that iron deficiency anemia has an impact on fetal growth and development similar to anemia in general. Lower maternal iron status is associated with lower cord blood iron, prematurity, and lower Apgar scores [33]. The authors did not measure ferritin in either cord or maternal blood so attribution to ferritin status cannot be concluded.

The utility of ferritin as an indicator of maternal iron stores loses sensitivity by the middle of the second trimester and as a result has little sensitivity as a predictor of poor fetal outcomes [34]. Based on several studies, there is a relationship of elevated ferritin with preterm birth, LBW, and preeclampsia [31, 35]. Higher ferritin concentrations may be more an indication of upper genital tract infection and a subsequent development of spontaneous preterm delivery than an indication that higher iron status is bad for fetal growth and development. For example, Lao et al. [36] reported on an analysis of birth outcomes for 488 nonanemic women. They observed a significant *inverse relationship* between maternal ferritin quartiles and infant birth weight, with an increased risk of prematurity and neonatal asphyxia in those mothers with the highest quartile of ferritin. In an analysis of the Preterm Prediction Study of the National Institute of Child Health and Human Development Maternal–Fetal Medicine Units network conducted from 1992 to 1996, there was a similar relationship [35, 37]. Regardless of the gestational age at sampling (19, 26, and 36 weeks), ferritin in the highest quartile was associated with the lower mean birth weights than those in the other three quartiles of ferritin. The adjusted odds ratio was significant, however, only at the 26-week sample with an odds ratio of 2 for premature delivery and 2.7 for small birth weight (<1,500 g). In the 2002 follow-up analysis, utilizing cervical ferritin concentrations at 22–24 weeks of gestation, the adjusted odds ratio of very premature delivery (<32 weeks) was as high as 6.3. There was also a strong correlation with other markers of inflammation from cervical fluid. These studies suggest that elevations of ferritin in mid-gestation increase the risk for pregnancy complications. Iron supplementation trials can answer the question of whether prevention of the decrease in iron status can improve birth outcomes. Prophylactic iron supplementation during the first trimester of pregnancy in poor women improved birth weight, lowered the incidence of prematurity, but did not alter the incidence of SGA deliveries [38]. The timing and dose of supplementation, as well as the frequency of supplementation, are important considerations in interventions during pregnancy [7, 39]. Large and frequent doses of iron as usually prescribed in the United States (Table 16.3) are frequently associated with complaints of constipation, dark stool, and gastrointestinal (GI) upset. Alternative strategies include nondaily supplementation at a much lower dose, delayed release supplements, and forms of iron salts that are less irritating to the GI tract. Since it is unlikely that dietary intakes alone will meet the iron requirements during pregnancy, it is imperative that cost-effective supplementation vehicles be developed that avoid these side effects while maintaining a positive iron balance.

A number of side effects occur with nearly all of these iron supplements and thus reduce compliance. The most common side effects include black stools, constipation, GI upset, and vomiting. This however is common only when high doses of >150 mg of iron a day are being prescribed to rapidly treat significant anemia. It is relatively uncommon for women taking the smaller doses of 30 mg per day to have any noticeable side effects. When there are side effects, switching to a slow release variety or a different formulation will often times relieve the symptoms of GI distress.

Alternative strategies to minimize the need for oral iron supplements include a diet that is high in iron-containing foods; most notably this is red meat. Fish, lamb, and other meats also contain significant amounts of iron as the highly bioavailable heme-iron. A reduction in inhibitors of iron absorption (tea, coffee, high-phytate-containing grains and breads) may provide someal addition benefit in terms of bioavailable iron. It is unlikely however, that diet alone will be sufficient to meet the very large requirements for pregnancy unless the woman enters pregnancy with substantial iron stores (serum ferritin >50–60 mcg/l). Iron stores less than this imply that dietary sources need to compensate, however there is a finite limit to efficiency of absorption.

16.8 MATERNAL IRON STATUS AND OTHER FETAL OUTCOMES

16.8.1 Infant Development

Despite the concept that the fetus is an effective parasite for iron, the previously discussed information indicates that fetal development is compromised when maternal iron status is compromised (Table 16.4). One dimension now receiving attention vis-à-vis iron status is neurodevelopment of the infant [40]. In an important study several years ago, Tamura et al. [41] noted a relationship between newborn cord ferritin levels and cognition and behavior at 5 years of age. The children were compared by their cord blood ferritin in the two median ferritin quartiles: Those in the lowest quartile scored lower on a number of tests including language ability, fine motor skills, and tractability. Since cord blood ferritin is correlated with maternal iron status, these data suggest that poor iron status at birth is related to later infant development. The intervention study of Presozio et al. [42] reached a similar conclusion regarding the benefit of iron supplementation in pregnancy on infant scores in tests of motor and mental development at 12 months of age. More recently, a study in South Africa [43, 44] showed that infants of iron-deficient anemic mothers had lower developmental scores, assessed with the Griffith scale at 9 months of age, than had infants of mothers who were not anemic. All

Table 16.4
Studies of Maternal Anemia and Fetal Outcomes

Study and authors	Hb range (g/l), optimal birth weight	Minimal prematurity	Minimal infant mortality	Notes
National Collaborative Perinatal Project	105–125	115–125	95–105	Variable dates of samplings
National Collaborative Perinatal Project African-Americans	85–95	105–115	85–95	Variable dates of samplings
United Kingdom	86–95	96–105	N/a	Variable dates of samplings
Cardiff Birth Studies	104–132		104–132	Regardless of samples <13 weeks,13–19 weeks, or >20 weeks gestation
Chinese mothers	110–119	110–119		Before 8th week of gestation
American mothers		110–119		Control for length of gestation

infants in this study were of full gestational age and weight, thus intrauterine growth failure and severe maternal anemia (<85 g/l) were excluded. These anemic mothers had increased amounts of depression and altered mother–child interactions compared with iron-supplemented mothers. Indeed, maternal postpartum depression related to Hb concentration in the months after delivery of the infant may contribute to changes in infant development. While it is an area of maternal nutrition not frequently considered, it is important to consider that maternal functioning in the postpartum period can be heavily influenced by her nutritional status [43], and Chap. 19, "Postpartum Depression and the Role of Nutritional Factors". This in turn, has a strong influence on infant development. This is not to suggest that iron supplementation will lead to smarter children. A recent study showed that very modest iron supplementation (20 mg/day) of mothers in New Zealand had no effect on the IQ of the infants at 4 years of age despite a reduction in prevalence of IDA from 11 to 1% during pregnancy [45]. The authors did show that behaviors of the infants were affected by the iron intervention in pregnancy, but they could not separate direct biological effects of the iron on fetal growth and development from the indirect effect that would be expected through the improved iron status of the mother during and after the pregnancy. Mother–child interactions can be quite sensitive to the nutritional status of the infant and the mother.

16.8.2 Infant Iron Status

The relationship between moderately anemic mothers and infant hematology (Table 16.4) has been reviewed [3, 46]. There is a general correlation between maternal Hb in the third trimester with the infant Hb at 9 months of age [47]. However, this relationship cannot be observed when the overall prevalence rates of anemia are so high as to remove the possibility that there are "normal" Hb concentrations in both mothers and infants [48]. In a number of others studies reviewed by Scholl, Allen, and Milman, there is a reoccurring theme that maternal anemia may be related to infant anemia in early life on some occasions, but commonly the relationship is more strongly expressed at 9–12 months of age when infant iron stores have been exhausted [2, 3, 46]. Thus, there is the concept that iron-depleted and moderately anemic mothers provide sufficient delivery of iron for infant growth and erythropoiesis in utero, but fail to provide sufficient iron for normal growth and development over the next 12 months of life.

16.9   CONCLUSION

Iron deficiency and anemia during pregnancy have functional outcomes for both the mother and the developing infant. The strong epidemiological data show a strong impact of anemia during the first trimester on the short-term outcomes of pregnancy such as gestational age and birth weight. The severity of iron deficiency and anemia over the course of pregnancy appears to be a determinant of postnatal development of infants and their neurodevelopment in the first and second years of life. As it is very difficult to begin oral iron treatment before the 8–10th week of pregnancy, due to failure of many women to seek clinical care before this time, it might be prudent to adopt the approach of the folic acid supplementation recommendations and suggest that women who plan to become pregnant be certain their iron status is good. This means the serum ferritin should be higher than 40–50 mcg/l prior to pregnancy and the woman should be quite faithful in her consumption of modest doses of iron supplements [6].

REFERENCES

1. IOM (2001) Dietary Reference Intakes for Micronutrients. Food and Nutrition Board Reports. National Academy of Science Press, Washington, D.C.
2. Scholl TO (2005) Iron status during pregnancy: setting the stage for mother and infant. Am J Clin Nutr 81:1218S–1222S
3. Allen LH (2005) Multiple micronutrients in pregnancy and lactation: an overview. Am J Clin Nutr 81:1206S–1212S
4. De Leeuw NK, Lowenstein L, Hsieh YS (1966) Iron deficiency and hydremia in normal pregnancy. Medicine 45:291–315
5. Beaton GH (2000) Iron needs during pregnancy: do we need to rethink our targets? Am J Clin Nutr 72(1 Suppl):265S–271S
6. Milman N (2006) Iron prophylaxis in pregnancy-general or individual and in which dose? Ann Hematol 85:821–828
7. Casanueva E, Viteri FE, Mares-Galindo M, Meza-Camacho C, Loria A, Schnaas L, Valdes-Ramos R (2006) Weekly iron as a safe alternative to daily supplementation for nonanemic pregnant women. Arch Med Res 37:674–682
8. Pena-Rosas JP, Nesheim MC, Garcia-Casal MN, Crompton DW, Sanjur D, Viteri FE, Frongillo EA, Lorenzana P (2004) Intermittent iron supplementation regimens are able to maintain safe maternal hemoglobin concentrations during pregnancy in Venezuela. J Nutr 134:1099–104
9. Allen LH (1997) Pregnancy and iron deficiency: unresolved issues. Nutr Rev 55:91–101
10. Murphy JF, O'Riordan J, Newcombe RG, Coles EC, Pearson JF (1986) Relation of haemoglobin levels in first and second trimesters to outcome of pregnancy. Lancet 1:992–995
11. Steer P, Alam MA, Wadsworth J, Welch A (1995) Relation between maternal haemoglobin concentration and birth weight in different ethnic groups. BMJ 310:489–491
12. Barrett JF, Whittaker PG, Williams JG. & Lind T (1994) Absorption of non-haem iron from food during normal pregnancy. Brit Med J 309:79–82
13. O'Brien KO (1999) Regulation of mineral metabolism from fetus to infant: metabolic studies. Acta Paediatr 88:88–91
14. O'Brien KO, Zavaleta N, Caulfield LE, Yang DX, Abrams SA (1999) Influence of prenatal iron and zinc supplements on supplemental iron absorption, red blood cell iron incorporation, and iron status in pregnant Peruvian women. Am J Clin Nutr 69:509–515

15. Gambling L, Charania Z, Hannah L, Antipatis C, Lea RG, McArdle HJ. (2002) Effect of iron deficiency on placental cytokine expression and fetal growth in the pregnant rat. Biol Reprod 66:516–523

16. Hindmarsh PC, Geary MP, Rodeck CH, Jackson MR. & Kingdom JC (2000) Effect of early maternal iron stores on placental weight and structure. Lancet 356:719–723

17. Allen LH (2001) Biological mechanisms that might underlie iron's effects on fetal growth and preterm birth. J Nutr 131:581S–589S

18. Beard J (2006) Iron. In: Present Knowledge in Nutrition, 9 edn. International Life Sciences Press, Emmaus, pp. 127–145

19. Mei Z, Cogswell ME, Parvanta I, Lynch S, Beard JL, Stoltzfus RJ. & Grummer-Strawn LM (2005) Hemoglobin and ferritin are currently the most efficient indicators of population response to iron interventions: an analysis of nine randomized controlled trials. J Nutr 135:1974–1980

20. Milman N et al (2006) Body iron and individual iron prophylaxis in pregnancy-should the iron dose be adjusted according to serum ferritin? Ann Hematol 85:567–573

21. Casanova BF, Sammel MD, Macones GA (2005) Development of a clinical prediction rule for iron deficiency anemia in pregnancy. Am J Obstet Gynecol 193:460–466

22. Cook JD, Flowers CH, Skikne BS (2003) The quantitative assessment of body iron. Blood 101:3359–3364

23. Beard J (2003) Iron deficiency alters brain development and functioning. J Nutr 133(Suppl):1468S–1472S

24. Allen LH (2000) Anemia and iron deficiency: effects on pregnancy outcome. Am J Clin Nutr 71(Suppl):1280S–1284S

25. Brabin BJ, Hakimi M, Pelletier D (2001) An analysis of anemia and pregnancy-related maternal mortality. J Nutr 131:604S–614S; discussion 614S–615S

26. Yip R (2000) Significance of an abnormally low or high hemoglobin concentration during pregnancy: special consideration of iron nutrition. Am J Clin Nutr 72(Suppl):272S–279S

27. Rasmussen K (2001) Is there a causal relationship between iron deficiency or iron-deficiency anemia and weight at birth, length of gestation and perinatal mortality? J Nutr 131(2S-2):590S–601S; discussion 601S–603S

28. Verhoeff FH et al (2001) An analysis of intra-uterine growth retardation in rural Malawi. Eur J Clin Nutr 55:682–689

29. Rush D (2000) Nutrition and maternal mortality in the developing world. Am J Clin Nutr 72(Suppl):212S–240S

30. Brabin BJ, Premji Z, Verhoeff F (2001) An analysis of anemia and child mortality. J Nutr 131(2S-2): 636S–645S; discussion 646S–648S

31. Scholl TO, Reilly T (2000) Anemia, iron and pregnancy outcome. J Nutr 130(Suppl):443S–447S

32. Zhou LM et al (1998) Relation of hemoglobin measured at different times in pregnancy to preterm birth and low birth weight in Shanghai, China. Am J Epidemiol 148:998–1006

33. Lee HS et al (2006) Iron status and its association with pregnancy outcome in Korean pregnant women. Eur J Clin Nutr 60:1130–1135

34. Goldenberg RL, Tamura T (1996) Prepregnancy weight and pregnancy outcome. J Am Med Assoc 275:1127–1128

35. Tamura T et al (1996) Serum ferritin: a predictor of early spontaneous preterm delivery. Obstet Gynecol 87:360–365

36. Lao TT, Tam KF, Chan LY (2000) Third trimester iron status and pregnancy outcome in non-anaemic women; pregnancy unfavourably affected by maternal iron excess. Hum Reprod 15:1843–1848

37. Ramsey PS et al (2002) The preterm prediction study: elevated cervical ferritin levels at 22 to 24 weeks of gestation are associated with spontaneous preterm delivery in asymptomatic women. Am J Obstet Gynecol 186:458–463

38. Siega-Riz AM et al (2006) The effects of prophylactic iron given in prenatal supplements on iron status and birth outcomes: a randomized controlled trial. Am J Obstet Gynecol 194:512–519

39. Milman N et al (2006) Side effects of oral iron prophylaxis in pregnancy–myth or reality? Acta Haematol 115:53–57

40. Lozoff B et al (2006) Long-lasting neural and behavioral effects of iron deficiency in infancy. Nutr Rev 64:S34–S43; discussion S72–S91

41. Tamura Y et al (2002) Cord ferritin concentrations and psychomotor development of children at five years of age. J. Pediatr. 140:165–170
42. Preziosi P et al (1997) Effect of iron supplementation on the iron status of pregnant women: consequences for newborns. Am J Clin Nutr 66:1178–1182
43. Beard JL et al (2005) Maternal iron deficiency anemia affects postpartum emotions and cognition. J Nutr 135:267–272
44. Perez EM et al (2005) Mother-infant interactions and infant development are altered by maternal iron deficiency anemia. J Nutr 135:850–855
45. Zhou SJ et al (2006) Effect of iron supplementation during pregnancy on the intelligence quotient and behavior of children at 4 y of age: long-term follow-up of a randomized controlled trial. Am J Clin Nutr 83:1112–1117
46. Milman N (2006) Iron and pregnancy-a delicate balance. Ann Hematol 85:559–565
47. Savoie N, Rioux FM (2002) Impact of maternal anemia on the infant's iron status at 9 months of age. Can J Public Health 93:203–207
48. Kilbride J et al (1999) Anaemia during pregnancy as a risk factor for iron-deficiency anaemia in infancy: a case-control study in Jordan. Int J Epidemiol 28:461–468

17 Folate: A Key to Optimal Pregnancy Outcome

Beth Thomas Falls and Lynn B. Bailey

Summary Folate is a water-soluble vitamin required for cell division and normal growth. Studies have definitively shown that when the synthetic form of the vitamin, folic acid, is taken during the periconceptional period, there is a significant reduction in risk for neural tube defects (NTDs) and findings have been translated into public health policy to increase intake through supplementation and fortification. Few women adhere to the recommendations to take folic acid supplements primarily because they have not been advised to do so by their health care provider. Restricted folate intake during pregnancy has also been associated with poor pregnancy outcomes including preterm delivery, low infant birth weight, and fetal growth retardation. Clinicians and health care practitioners have a unique opportunity to improve pregnancy outcome by providing advice to their patients who may become pregnant to take daily folic acid supplements, consume folic acid-fortified foods including ready-to-eat breakfast cereals, and to consume concentrated sources of natural dietary folate including dark green leafy vegetables, orange juice, nuts, and dried peas. It is imperative that all health care providers that work with women of childbearing age (including pregnant women) become knowledgeable about folate's key role in reproduction and deliver the message that folate is a unique key to optimizing reproductive outcome.

Keywords: Folate, Requirements, Neural tube defects, Fortification

17.1 INTRODUCTION

Folate is a water-soluble vitamin occurring either naturally in food or as folic acid, which is the synthetic form in supplements or fortified foods [1]. Folate must be consumed in adequate amounts prior to and during pregnancy to ensure an optimal pregnancy outcome as recently reviewed [2]. DNA synthesis is dependent on folate and when intake is limited, cell division slows down at a time when the developing embryo has the greatest need. One of the most significant public health discoveries of this century was the finding that periconceptional folic acid significantly reduces the risk of neural tube defects (NTDs). This scientific fact has been translated into public health policy throughout the world, including widespread recommendations from professional

From: *Nutrition and Health: Handbook of Nutrition and Pregnancy*
Edited by: C.J. Lammi-Keefe, S.C. Couch, E.H. Philipson © Humana Press, Totowa, NJ

organizations for periconceptional folic acid supplementation in addition to mandated and voluntary folic acid fortification policies.

In addition to the role of folic acid in the development of the neural tube, which takes place during the first 28 days of gestation, folate is of vital importance throughout gestation for a positive pregnancy outcome. Folate coenzymes are also involved in one-carbon transfer reactions required for amino acid metabolism including the remethylation of homocysteine to form the essential amino acid methionine and the body's primary methylating agent, *S*- adenosylmethionine (SAM), which is utilized in over 100 different methylation reactions including DNA methylation. Folate requirements are increased in pregnancy to meet the demands for increased DNA synthesis and thus cell division [3]. The increase in cell division is associated with the rapidly growing fetus and placenta coupled with the increasing size of the maternal reproductive organs. Folate is required for the formation of red blood cells, and the expansion in the number of red blood cells for maternal and fetal circulation further increases the requirement for folate during pregnancy [4].

Restricted folate intake during pregnancy has been associated with poor pregnancy outcomes including preterm delivery, low infant birth weight, and fetal growth retardation [3]. Biochemical indicators of depleted maternal folate status have been linked to increased spontaneous abortion and pregnancy complications (e.g., abruptio placenta or placental infarction with fetal growth retardation and preeclampsia), which increase the risk of low birth weight and preterm delivery [5, 6]. The requirement for folate to support the rapid cell division and growth of pregnancy is clear, and evidence for an increased folate requirement during pregnancy is well documented. This chapter will address specific recommendations for periconceptional folic acid use to reduce NTD risk in addition to recommendations for folate intake throughout pregnancy to optimize pregnancy outcome.

17.2   FOLATE REQUIREMENTS DURING PREGNANCY

Dietary reference intakes (DRIs) are recommendations for nutrient intakes used to plan and assess the adequacy of diets for healthy people [7]. The DRIs include a number of reference values including Estimated Average Requirement, Recommended Dietary Allowance (RDA), Adequate Intake, and Tolerable Upper Intake Level. For the purposes of planning and assessing a diet for a healthy individual, clinicians should generally utilize the RDA [7]. The RDA for folate for pregnant women is the average daily dietary folate intake that is sufficient to meet the nutrient requirements of nearly all healthy pregnant women (97–98%) [8]. New units (Dietary Folate Equivalents, DFEs) were derived to express the folate RDA that account for the higher bioavailability of folic acid compared to naturally occurring food folate [7]. When expressed as DFEs, folic acid intake is converted to an amount that is equivalent to food folate. It was estimated that folic acid in fortified foods is ~85% bioavailable and food folate is ~50% bioavailable, which means that folic acid in food is 1.7 times more bioavailable than is folate occurring naturally in food. If a mixture of synthetic folic acid plus food folate is consumed, then the DFE is calculated as follows: DFE = food folate + (1.7 × synthetic folic acid). For example, a meal that contains 100 micrograms (mcg) of folate from orange juice and 100 mcg of folic acid from fortified cereal would contain a total of 270 mcg DFE (100 mcg from orange juice + [100 mcg folic acid × 1.7]).

The current folate DRIs for pregnant women were based on data from a controlled metabolic study and a series of population-based studies in which dietary folate intake was reported [7].

17.3 NEURAL TUBE DEFECTS AND PERICONCEPTIONAL FOLATE REQUIREMENT

Neural tube defects are a group of birth defects that affect the developing embryonic brain or spine and occur when the developing neural tube fails to close during the first 28 days of gestation [9]. The two most common NTDs are spina bifida and anencephaly, which can cause lifelong disability or death. Birth records collected through birth defect surveillance by the Center for Disease Control and Prevention (CDC) suggest that approximately 2,500 babies with NTDs, or 1 to 2 per 1,000, are born each year in the United States [10, 11]. The rate of NTD affected pregnancies is approximately 40% higher in women of Hispanic descent compared with Caucasian women [12], while the rate in African American women is approximately 30% lower than in Caucasian women [11].

A large body of epidemiological evidence preceded the definitive controlled intervention studies that established the scientific fact that periconceptional folic acid will significantly reduce the risk of NTDs [9]. Public health policies have been established in countries around the globe based on this scientific evidence, with the primary approach being periconceptional folic acid supplement use [13]. However, the most successful public health approach has been the implementation of mandatory folic acid fortification in countries including the United States, Canada, and Chile. A number of researchers have examined population-based data for various US birth defects registries and found that compared to prefortification, the prevalence of NTDs at birth decreased significantly after fortification [10, 11]. However, the percent decline reported is well below the estimated 70% decrease in incidence of NTDs if 100% compliance with public health recommendations was met [14]. This observation emphasizes the need for renewed efforts by practitioners to advise all women capable of becoming pregnant to take a daily folic acid supplement even if they are not planning a pregnancy, since the majority of pregnancies are unplanned. The neural tube develops within one month of conception, often before women know they are pregnant which is the basis for the recommendation that folic acid be taken prior to and during the very early phase of embryogenesis.

In 1998, the Food and Nutrition Board of the Institute of Medicine (IOM) made a specific recommendation that all women capable of becoming pregnant should take 400 mcg of synthetic folic acid per day from supplements or fortified foods, in addition to dietary folate, to reduce NTD risk [7]. The IOM recommendation is not the same as the RDA, a common misconception, since the recommendation specifies that the supplemental form of the vitamin, folic acid (400 mcg/day), be taken (or consumed as fortified food) in addition to folate in a varied diet [7]. One reason for this separate recommendation is that for prevention of NTDs, research evidence supports the fact that supplemental folic acid must be taken during the periconceptional period—that is, beginning several months prior to conception and continuing through the end of the embryogenesis period of pregnancy (approximately 8 weeks postconception) in addition to diet.

Many professional organizations have folic acid NTD-related position statements. For example, the American College of Medical Genetics' (ACMG) position statement states, "All women capable of becoming pregnant should take 400 mcg of folic acid daily, in the form of supplement, multivitamin, and/or through fortified foods, in addition to eating a healthy diet. This is particularly important before conception and through the first trimester of pregnancy" [15].

With regard to women who have had a previous NTD-affected pregnancy, the CDC and a number of professional organizations recommend taking 4,000 mcg of folic acid starting at least 1 month before conception and continuing throughout the first 3 months of pregnancy to reduce the risk for recurrence [13, 16]. The evidence that the 4,000-mcg dose is the effective dose for recurrence is based on data from one large-scale intervention trial in which this dose was found to reduce the NTD risk by 72% relative to a placebo [17]. Since lower doses were not evaluated, it is not possible to rule out the possibility that lower doses (e.g., 400 mcg) would be as effective in preventing NTD recurrence. The ACMG, along with other professional and government agencies, recommends that women who have had a prior NTD-affected pregnancy, who have a first-degree relative with a NTD, or who are themselves affected should consult with their physician before becoming pregnant. They should obtain genetic counseling concerning their occurrence or recurrence risks, pregnancy management, and the appropriate folic acid intake for them [15].

The CDC, in collaboration with more than 35 federal, public, and private partners, recently released national recommendations designed to encourage women to take steps toward good health before becoming pregnant. One of these steps is the recommendation to take 400 mcg of folic acid to help prevent NTDs. This new 2006 folic acid NTD recommendation reaffirms the earlier public health recommendation released in 1992 and the recommendations of many professional organizations, which emphasizes the fact that this message has not been translated into a behavioral change (i.e., preconceptional folic acid use) [18].

To achieve compliance with the public health recommendations, the March of Dimes, the CDC, and several other organizations of health care professionals have conducted public and professional campaigns urging women of childbearing age to consume folic acid daily beginning before conception and continuing into the early months of pregnancy [14]. However, in spite of these efforts, folic acid supplement use among women of childbearing age remains relatively unchanged [19]. Data from US, Puerto Rican, Dutch, and Australian surveys all indicate that folic acid supplementation during the periconceptional period remains low (~30%) [20].

The factors leading to low supplement use are not all clear. The fact that the majority of pregnancies are unplanned is one of the major reasons that women of productive age do not take daily folic acid supplements to reduce the risk of having an NTD-affected pregnancy. Cultural bias against taking supplements may exist among Hispanic women due to the misconception that vitamin supplements are associated with weight gain [21]. Information received from physicians significantly impacts use of folic acid supplements by women. Although only approximately one third of American women of childbearing age take folic acid daily, many women report they would take folic acid if their physicians advised them to do so. A recent survey found that 68% of women reported taking a folic acid supplement after they received brief folic acid counseling, compared with 20% of women from a group that did not receive counseling [22]. This finding points out the vital role health care providers play in reducing the incidence of NTDs. Continued efforts to educate health care providers, as well as to identify factors that inhibit clinicians and nutritionists from talking to their patients and clients about the role of folic acid in risk reduction of NTDs is a critical need.

17.4   FOLATE AND NON–NEURAL TUBE DEFECT BIRTH DEFECTS

In addition to the well-established relationship between folate and NTDs, the etiology of other birth defects may also be related to impaired folate status. A key example are the research findings that suggest that periconceptional use of multivitamins containing folic acid is associated with a reduction in the occurrence of congenital heart defects [23]. Although the data are less convincing, periconceptional folate may be related to a reduced occurrence of orofacial clefts as well [23].

17.5   SOURCES OF FOLATE

17.5.1   *Food Folate*

Pregnant women consume folate as naturally occurring food folate, folic acid in fortified foods, and supplements that contain folic acid. Naturally occurring dietary folate is concentrated in certain foods, including orange juice, dark green leafy vegetables, and dried beans such as black beans and kidney beans [24] (Table 17.1). With the exception of liver, meat is generally not a good source of folate.

Table 17.1
Best Sources of Food Folate

	Weight (g)	Measure	Folate content (mcg)
Fruits			
Orange juice, ready-to-drink	249	1 cup	80
Orange	131	1 medium	49
Strawberries, fresh	151	8 medium	80
Vegetables			
Asparagus, cooked	75	5 spears	100
Avacado	87	1/2 cup	55
Broccoli, cooked	92	1/2 medium	50
Brussels sprouts, cooked	78	1/2 cup	80
Corn, on the cob	123	1 large	55
Mustard greens, cooked	75	1/2 cup	90
Okra, cooked	92	1/2 cup	135
Spinach, raw	56	1 cup	110
Spinach, cooked	95	1/2 cup	100
Tomato juice	243	1 cup	50
Turnip greens, cooked	75	1/2 cup	85
Legumes			
Beans, black, cooked	86	1/2 cup	130
Beans, kidney, cooked	91	1/2 cup	115
Beans, navy, cooked	91	1/2 cup	125
Beans, pinto, cooked	86	1/2 cup	145
Black-eyes peas, cooked	83	1/2 cup	105
Chickpeas, cooked	82	1/2 cup	140
Lentils, cooked	99	1/2 cup	180

From [24]

17.5.2   Fortified Foods

As of January 1, 1998, all cereal grain products in the United States labeled as "enriched" (e.g., bread, pasta, flour, breakfast cereal and rice) and mixed food items containing these grains were mandated by the Food and Drug Administration to be fortified with folic acid. It is required that all enriched products contain 140 mcg of folic acid per 100 g of product [25]. It is estimated that several thousand food items in the US food supply now contain folic acid derived from enriched cereal grain ingredients [26]. Significant increases in blood folate concentration in response to folic acid fortification in the United States have been documented in a number of studies including the National Health and Nutrition Examination Survey [27]. The observed increase in blood folate postfortification is greater than expected and, based on the analysis of a large number of fortified foods, has been attributed to over-ages by the food industry [28]. It is estimated that the increase in folic acid intake due to fortification may be as high as two times greater than originally predicted [29].

Ready-to-eat (RTE) breakfast cereals contribute significantly to folate intake in the United States since the majority of RTE breakfast cereals in the US marketplace contain approximately 100 mcg/serving of folic acid, and a smaller number contain 400 mcg/serving. Folic acid is an added ingredient in a large number of other commonly consumed RTE products including breakfast bars, nutritional bars, and snack foods.

17.5.3   Supplemental Folic Acid

Supplemental forms of folic acid are available as folic acid only or as a component of a multivitamin. The majority of over-the-counter folic acid supplements or multivitamins with folic acid contain 400 mcg, which is the dose recommended to reduce NTD risk, making it easy for women of reproductive age to have access to a supplement with the recommended dose. All of the commonly prescribed prenatal vitamin supplements contain 1 mg folic acid (1,000 mcg × 1.7 = 1700 mcg DFE) and over-the-counter prenatal supplements generally have 800 mcg folic acid (800 mcg × 1.7 = 1360 mcg DFE) both of which provide significantly more than the RDA of 600 mcg/d DFE for pregnant women.

17.5.4   Potential Adverse Effects of Folic Acid

Folate is not associated with any toxicity symptoms [7]. However, because of a potential concern that high doses of supplemental folic acid may interfere with the diagnosis of a vitamin B_{12} deficiency (referred to as "masking"), the IOM recommends the total daily intake of folic acid should not exceed 1,000 mcg (1 mg) unless prescribed by a physician. Recent evidence indicated that folic acid fortification has not lead to an increase in masking of vitamin B_{12} deficiency [30]. However, health care practitioners prescribing prenatal vitamins containing 1 mg folic acid should monitor vitamin B_{12} status, especially if folic acid fortified foods are consumed in addition to the supplement. Although vitamin B_{12} deficiency is not a common finding among women of childbearing age, women who avoid animal-based foods, the sole source of vitamin B_{12}, should be advised to take supplemental vitamin B_{12}.

During pregnancy, over-the-counter (OTC) multivitamin supplements that contain lower levels of folic acid intended for nonpregnant women should not be taken without the knowledge and approval of health care providers. Many OTC supplements contain retinol that could be harmful to a developing embryo [31].

Several reports have suggested the possibility of a significant increase in the occurrence of miscarriage or multiple births associated with folic acid supplementation. As recently reviewed, research evidence does not support these suggestions [23].

Health care providers can be confident that by encouraging nonpregnant women of childbearing age to take folic acid supplements and pregnant women to take prenatal vitamins with folic acid that they are significantly improving pregnancy outcomes without increasing risks.

17.6 ASSESSMENT OF FOLATE STATUS

Serum folate concentration is considered a sensitive indicator of recent dietary folate intake [7] in contrast to red blood cell folate concentration, which is considered an indicator of long-term status. Folate is not taken up by the mature red blood cell in the circulation; hence, red blood cell (RBC) folate concentration represents folate taken up in the developing reticulocyte early in the approximately 120-day erythrocyte lifespan [7]. This is especially relevant during pregnancy, when production of red cells increases by approximately 33% [32]. Based on associations with liver folate concentrations determined by biopsy, red blood cell folate concentration is considered representative of tissue folate stores [7]. The most commonly used cutoffs for defining low serum and red blood cell folate levels are 6.8 and 317 nmol/l, respectively [27].

17.7 DRUG AND ALCOHOL IMPACT ON FOLATE STATUS

A large number of drugs can affect the absorption and metabolism of folate. Folate antagonists, including methotrexate, have played a key role in cancer treatment for half a century [33]. Methotrexate has also been frequently used to treat diseases such as rheumatoid arthritis, psoriasis, asthma, and inflammatory bowel disease [34]. An increased risk of birth defects has been associated with doses of methotrexate as low as 10 mg weekly, and may be associated with the antifolate drug action [35]. Women of childbearing age treated with methotrexate for one of these non-neoplastic diseases should be thoroughly educated about the risk of their medication to a developing fetus. These women should be counseled about the importance of avoiding an unplanned pregnancy.

There are numerous reports of impaired folate status associated with chronic use of anticonvulsants (e.g., carbamazepine, phenobarbital, primidone, valproic acid) [7]. Although these drugs appear to be folate antagonists, the precise mechanism for the drug-nutrient interaction in not known [36]. Maternal periconceptional exposure to these folate antagonists appears to increase the risk of NTDs [37]. It is unclear whether folic acid supplementation protects against the effect of these drugs, but the recommendation for folic acid supplementation for women with epilepsy is the same as for other women of childbearing age. Even with supplemental folic acid, women taking antiepileptic drugs should undergo perinatal diagnostic ultrasound to rule out NTDs [38].

Alcohol consumed chronically in large amounts has been shown to contribute to folate deficiency by interfering with folate absorption, decreasing hepatic folate uptake, and increasing urinary excretion [39]. Women of childbearing age should be educated about the effect of alcohol on their folate status, and thus the risk of abnormal fetal development associated with impaired folate status during pregnancy. If chronic alcohol use is suspected during the preconception period, then women should be counseled that alcohol consumption during the periconceptional period as well as postconception should be avoided.

17.8 APPLICATION

The potential for optimizing folate intake to positively impact pregnancy outcome has become a major public health initiative in the United States. Ensuring that a diet provides the RDA of 600 mcg/day DFE during pregnancy is essential for normal cell division and growth during pregnancy. In 1998, the Food and Nutrition Board of the IOM made a specific recommendation that all women capable of becoming pregnant should take 400 mcg of folic acid daily from supplements and fortified foods, in addition to a varied diet. This recommendation was reaffirmed in 2006 by the CDC in collaboration with more than 35 federal, public, and private partners [18]. Compliance with this recommendation will lead to a reduction in the occurrence of NTDs and emerging evidence suggests that it may reduce the incidence of non-neural tube birth defects as well.

Health care providers (e.g., MDs, RNs, RDs) have the unique opportunity to influence the behavior of individual women. It is important that all health care providers that work with women of childbearing age (including pregnant women) become knowledgeable about folate's roles in reproduction and the relationship between periconceptional folic acid intake and NTD risk. Clinicians should assess folate intake and status, and provide education and referrals as needed. Many women may be confused about the separate recommendations for dietary folate intake during pregnancy (RDA of 600 mcg/day DFE) and the supplemental folic acid recommendation for NTD prevention (400 mcg/day periconceptionally). Also, most women may not yet understand the use of DFEs. Health care providers can clarify these issues to make it easier for women to comply with the appropriate recommendations.

A number of prescription medications, as well as alcohol, are folate antagonists and use of these substances may lead to folate deficiency with increased risk of pregnancy complications. Women of childbearing age should be informed about these concerns and counseled to avoid an unplanned pregnancy while taking a folate-depleting medication.

In summary, health care providers should make concerted efforts to deliver the message to all women of reproductive age that daily folic acid supplements should be taken to reduce the risk of NTDs in future pregnancies that may be unplanned events. In addition, efforts should be made to ensure that the diets of non-pregnant women provide the RDA, which is 400 mcg/day DFE. Once pregnancy is confirmed, women should be advised to make sure that their daily folate intake meets the RDA, which is 600 mcg/day DFE. Although currently formulated prenatal supplements prescribed for pregnant women contain more than the RDA, OTC multivitamin supplements with lower amounts of folic acid are not advised due to the presence of retinol in these supplements. High doses of folic acid have not been associated with any toxic effects or any negative side effects; therefore, currently available prenatal supplements should be considered safe for pregnant women. Vitamin B_{12} status should be evaluated when prescribing high doses of folic acid since masking of hematological abnormalities associated with the vitamin B_{12} deficiency may occur. Special attention should be focused on maintaining optimal folate status in women who are taking antifolate medications or consume alcohol excessively.

In conclusion, folic acid is a unique key to optimizing reproductive outcome. Health care providers have access to that key and have the ability to positively influence birth outcome by delivering the "take folic acid supplements daily" message to all women capable of becoming pregnant. It is indeed an exciting opportunity for clinicians and practitioners to make a life-changing difference with a minimal amount of effort in the health care setting.

REFERENCES

1. Bailey L, Gregory J (2006) Folate. In: Bowman B, Russell R (eds) Present knowledge in nutrition, 9th edn. ILSI Press, Washington, D.C. pp 278–301
2. Tamura T, Picciano MF (2006) Folate and human reproduction. Am J Clin Nutr 83:993–1016
3. Scholl TO, Johnson WG (2000) Folic acid: influence on the outcome of pregnancy. Am J Clin Nutr 71(Suppl):1295S–303S
4. Physiology in pregnancy (1993) In: Cunningham F, MacDonald P, Fant N, Leveno K, Gilstrap L (eds) Williams obstetrics. Appleton & Lange, Norwalk, Conn., pp 209–247
5. Put NJM van der, Heil SG, Eskes TKAB, Blom HJ (2000) A common mutation in the 5,10-methylene-tetrahydrofolate reductase gene as a new risk factor for placental vasculopathy. Am J Obstet Gynecol 182:1258–1263
6. Emblem BM, Tverdal A, Gjessing HK, Monsen ALB, Ueland PM (2000) Plasma total homocysteine, pregnancy complications, and adverse pregnancy outcomes: the Hordaland Homocysteine Study. Am J Clin Nutr 71:962–968
7. Institute of Medicine (1998) Folate. In: Dietary reference intakes for thiamin, riboflavin, niacin, vitamin B_6, folate, vitamin B_{12}, pantothenic acid, biotin, and choline. Washington, DC: National Academy Press, Washington, D.C., pp 196–305
8. Bailey LB (2000) New standard for dietary folate intake in pregnant women. Am J Clin Nutr 71(Suppl):1304S–1307S
9. Botto LD, Moore CA, Khoury MJ, Erickson JD (1999) Neural-tube defects. N Engl J Med 341:1509–1519
10. Khoury MJ, Kirby RS, Shaw GM, Velie EM, Merz RD, Forrester MB, Williamson RA, Krishnamurt DS, Stevenson RE, Dean JH (1995) Surveillance for anencephaly and spina bifida and the impact of prenatal diagnosis—United States, 1985–1994. MMWR CDC Surveill Summ 44:1–13
11. Williams LJ, Mai CT, Kirby RS, Pearson K, Devine O, Mulinare J (2005) Changes in the birth prevalence of selected birth defects after grain fortification with folic acid in the United States: findings from a multi-state population-based study. Birth Defects Res A Clin Mol Teratol 73:679–689
12. Feuchtbaum LB, Currier RJ, Riggle S, Roberson M, Lorey FW, Cunningham GC (1999) Neural tube defect prevalence in California (1990–1994): eliciting patterns by type of defect and maternal race/ethnicity. Genet Test 3:265–272
13. Centers for Disease Control Epidemiology Program Office (1992) Recommendations for the use of folic acid to reduce the number of cases of spina bifida and other neural tube defects. In: MMWR Recomm Rep 1–7
14. Green NS (2002) Folic acid supplementation and prevention of birth defects. J Nutr 132:2356S–23560
15. Toriello HV (2005) Folic acid and neural tube defects. Genet Med 7:283–284
16. Committee on Genetics (1999) Folic acid for the prevention of neural tube defects. Pediatrics 104:325–327
17. Wald N, Sneddon J, Densem J, Frost C, Stone R (1991) Prevention of neural tube defects: results of the Medical Research Council Vitamin Study. MRC Vitamin Study Research Group. Lancet 338:131–137
18. Cordero JF, Atrash HK, Parker CS, Boulet S and Curtis MG (2006) Recommendations to improve preconception health and health care—United States. A report of the CDC/ATSDR Preconception Care Work Group and the Select Panel on Preconception Care. MMWR Recomm Rep 55(RR-6):1–23
19. March of Dimes Birth Defects Foundation (2005) Folic acid and the prevention of birth defects. A national survey of pre-pregnancy awareness and behavior among women of childbearing age 1995–2005. The Gallup Organization, Princeton, N.J.
20. Bailey LB, Rampersaud GC, Kauwell GP (2003) Folic acid supplements and fortification affect the risk for neural tube defects, vascular disease and cancer: evolving science. J Nutr 133:1961S–1968S

21. Higgins PG, Learn CD (1999) Health practices of adult Hispanic women. J Adv Nurs 29:1105–11012
22. Robbins JM, Cleves MA, Collins HB, Andrews N, Smith LN, Hobbs CA (2005) Randomized trial of a physician-based intervention to increase the use of folic acid supplements among women. Am J Obstet Gynecol 192:1126–1132
23. Bailey LB, Berry RJ (2005) Folic acid supplementation and the occurrence of congenital heart defects, orofacial clefts, multiple births, and miscarriage. Am J Clin Nutr 81:1213S–S1217
24. Suitor CW, Bailey LB (2000) Dietary folate equivalents: interpretation and application. J Am Diet Assoc 100:88–94
25. US Food and Drug Administration (1996) Food standards: amendment of standards of identity for enriched grain products to require addition of folic acid. Final rule. 21 CFR Parts 136, 137, and 139. Fed Reg 8781–8789
26. Lewis CJ, Crane NT, Wilson DB, Yetley EA (1999) Estimated folate intakes: data updated to reflect food fortification, increased bioavailability, and dietary supplement use. Am J Clin Nutr 70:198–207
27. Pfeiffer CM, Caudill SP, Gunter EW, Osterloh J, Sampson EJ (2005) Biochemical indicators of B vitamin status in the US population after folic acid fortification: results from the National Health and Nutrition Examination Survey 1999–2000. Am J Clin Nutr 82:442–450
28. Rader JI, Weaver CM, Angyal G (2000) Total folate in enriched cereal-grain products in the United States following fortification. Food Chem 70:275–289
29. Choumenkovitch SF, Selhub J, Wilson PWF, Rader JI, Rosenberg IH, Jacques PF (2002) Folic acid intake from fortification in United States exceeds predictions. J Nutr 132:2792–2798
30. Zeller JA, Cox C, Williamson RE, Dufour DR (2003) Low vitamin B-12 concentrations in patients without anemia: the effect of folic acid fortification of grain. Am J Clin Nutr 77:1474–1477
31. Azais-Braesco V, Pascal G (2000) Vitamin A in pregnancy: requirements and safety limits. Am J Clin Nutr 71:1325S–S1333
32. Blackburn S, Loper D (eds) (1992) The hematologic and hemostatic systems. In: Maternal, fetal, and neonatal physiology: a clinical perspective. Philadelphia, Saunders, pp 159–200
33. Priest DG, Bunni MA (1995) Folates and folate antagonists in cancer chemotherapy. In: Bailey LB (ed) Folate in health and disease. Marcel Dekker, New York, N.Y., pp 379–404
34. Morgan SL, Baggot JE (1995) Folate antagonists in nonneoplastic disease: proposed mechanism of efficacy and toxicity. In: Bailey LB (ed) Folate in health and disease. Marcel Dekker, New York, N.Y., pp 435–62
35. Lloyd ME, Carr M, McElhatton P, Hall GM, Hughes RA (1999) The effects of methotrexate on pregnancy, fertility and lactation. QJM 92:551–63
36. Young SN, Ghadirian AM (1989) Folic acid and psychopathology. Prog Neuropsychopharmacol Biol Psychiatry 13:841–863
37. Hernandez-Diaz S, Werler MM, Walker AM, Mitchell AA (2001) Neural tube defects in relation to use of folic acid antagonists during pregnancy. Am J Epidemiol 153:961–968
38. Yerby MS (2003) Management issues for women with epilepsy: neural tube defects and folic acid supplementation. Neurology 61:23S–26S
39. Halsted CH, Villanueva JA, Devlin AM, Chandler CJ (2002) Metabolic interactions of alcohol and folate. J Nutr 132:2367S–2372S

IV THE POSTPARTUM PERIOD

Nutrition Issues During Lactation

Deborah L. O'Connor, Lisa A. Houghton, and Kelly L. Sherwood

Summary Breastfeeding is the gold standard and strongly recommended method of feeding infants. The World Health Organization recommends human milk as the exclusive nutrient source for the first 6 months of life, with introduction of solids at this time, and continued breastfeeding until at least 12 months postpartum. It will come as a surprise to many readers that the energy and nutrient needs of lactating women adhering to these optimal infant feeding guidelines will exceed those of pregnancy. During the first 4–6 months of life, an infant will double its birth weight accumulated during the entire 9 months of pregnancy. The nutrient output via breast milk to support this growth is tremendous. Early postpartum, weight is an issue for many women, as they are anxious to return to their prepregnancy body size. For many, weight management will be difficult given personal circumstances and multiple demands on their time. Given the elevated nutrient requirements of lactation, women will need to plan meals with care to maximum nutrient intake while limiting energy dense foods. The purpose of this chapter is to provide an overview of the energy demands of lactation as well as a select list of nutrients known to be sometimes provided in short supply for reproductive age women in developed countries. The specific nutrients to be examined include calcium, vitamin D, folate, vitamin B_{12}, and iron. As energy balance is a current area of concern for many lactating women, and their health care providers, we will also review the literature in relation to dieting and exercise during lactation. Finally, we will close by talking about long-chain polyunsaturated fatty acids (LC-PUFAs), variability in breast milk content, and the implications of maternal LC-PUFAs supplementation on infant outcomes.

Keywords: Breastfeeding, lactation, postpartum, nursing

18.1 INTRODUCTION

Breastfeeding is the gold standard and strongly recommended method of feeding infants. The World Health Organization and the American Academy of Pediatrics recommend human milk as the exclusive nutrient source for the first 6 months of life, and indicates that breastfeeding be continued at least through the first 12 months of life, and

From: *Nutrition and Health: Handbook of Nutrition and Pregnancy*
Edited by: C.J. Lammi-Keefe, S.C. Couch, E.H. Philipson © Humana Press, Totowa, NJ

thereafter as long as mother and baby mutually desire [1, 2]. The scientific rationale for recommending breastfeeding as the preferred feeding choice for infants stems from its acknowledged benefits to infant nutrition; gastrointestinal function; host defense; neurodevelopment; and psychological, economic, and environmental well-being [2, 3]. The American Academy of Pediatrics Policy Statement entitled, "Breastfeeding and the Use of Human Milk" is an excellent resource, which includes a succinct discussion of the specific benefits of breastfeeding [2]. Briefly, the policy acknowledges that research provides good evidence that breastfeeding decreases the rate of postneonatal infant mortality (~21%), and reduces the incidence of a wide range of infectious diseases including bacterial meningitis, bacteremia, diarrhea, respiratory tract infection, necrotizing enterocolitis, otitis media, urinary tract infection, and late-onset sepsis rates in preterm infants. Breast-feeding is also associated with slight improvements in cognitive development in both term-born and prematurely born infants, although the benefits appear to be greatest for the latter group of infants [4].

The nutritional needs of lactating women adhering to these optimal infant feeding guidelines will exceed those of pregnancy. During the first 4–6 months of life, an infant will double its birth weight accumulated during the entire 9 months of pregnancy [5]. The energy content of breast milk secreted in the first 4 months postpartum alone well exceeds the energy demands of an entire pregnancy. While the nutrient demands of lactation are high, few data exist to support recommended nutrient intakes, how well women are doing in meeting these recommendations, and the consequences of suboptimal nutritional status. For the most part, nutrient requirements for the lactating woman are based on those of nonpregnant, nonlac-tating women with an incremental amount added to account for the amount of the nutrient secreted into breast milk. Lactation success has traditionally been defined by infant outcomes, such as growth, volume, and nutrient content of milk consumed and optimal nutritional status [5]. Few studies, to date, have directly examined the nutritional status of lactating women or the short- or long-term consequences of suboptimal maternal nutrition on their own health.

The purpose of this chapter is to examine the energy demands of lactation as well as a select list of nutrients known to be sometimes provided in short supply for reproductive age women in developed countries. The specific nutrients to be examined include calcium, vitamin D, folate, vitamin B_{12}, and iron. As energy balance is a current area of concern for many lactating women and their health care providers, we also review the literature in relation to dieting and exercise during lactation. Finally, we will close by talking about long-chain polyunsaturated fatty acids (LC-PUFAs), variability in breast milk content, and the implications of maternal LC-PUFAs supplementation on maternal and infant outcomes.

18.2   ENERGY

18.2.1   *Estimated Energy Requirements*

The incremental energy cost of lactation is determined by the amount of milk produced (exclusivity and duration), the energy density of the milk secreted, and the energy cost of milk synthesis [6]. The Estimated Energy Requirements (EERs) for lactation, or the average daily energy intake predicted to maintain energy balance in a healthy lactating woman, of a given age, weight, height, and level of physical activity can be estimated by a factorial approach from the sum of the (1) EER of a nonpregnant, nonlactating woman (of a given age, weight and activity level), plus (2) estimated milk energy output, plus (3) energy mobilization from tissue stores (i.e., weight loss) [7]:

1. The EER of a nonpregnant nonlactating woman can be calculated using information provided in Table 18.1. The current age, weight and relative physical activity level must be known.
2. Milk energy output is tabulated by multiplying the volume of milk produced by its energy density (Table 18.2). The figure used by the Institute of Medicine to estimate the daily volume of milk produced from birth to six months is 0.78 l/day [7]. From 7 to 12 months, mean milk production is estimated to be 0.6 l/day, reduced with the introduction of solid foods. While the daily volume of breast milk produced among exclusively breastfeeding mothers is remarkably consistent from woman to woman and country to country, it varies considerably, of course, if a woman is partially or totally breastfeeding [6]. The US Institute of Medicine reviewed studies where human milk energy density was measured by bomb calorimetry and found an average value of 0.67 kcal/g.
3. Energy mobilization from tissue stores is the energy derived from the weight lost in the first six months postpartum (Table 18.2). For the purposes of calculating the EERs for lactation this value was set at 0.8 kg/month [7].

Research shows that women meet most of the incremental energy requirements of lactation by eating more calories, decreasing physical activity early postpartum, and mobilizing fat stores laid down during pregnancy [6]. Mobilization of fat stores laid down during pregnancy is not obligatory, and the extent to which they are used to support lactation depends on the nutritional status of the lactating mother and the amount of weight gained during pregnancy. A well-nourished woman will mobilize approximately 0.72 MJ/day (~170 kcal/day) of fat stores to help support breastfeeding in the first 6 months postpartum. Physical activity tends to be lower in the early postpartum period among women in the developed world, as daily activities change in response to caring for a newborn. While the expectation of many women is that they will lose weight rapidly by breastfeeding, weight changes postpartum are highly variable, though they are greater and more consistent for women who breastfeed exclusively. Some women may actually gain weight postpartum. Generally, well-nourished women will lose on average 0.8 kg/month (1.8 pounds/month) for the first 6 months postpartum; undernourished women can expect to lose 0.1 kg/month.

The reported energy intakes of lactating women in the literature are generally lower than that recommended by the Institute of Medicine [7]. Under-reporting may be a reason for these low energy intakes or the EER is set too high. Alternatively, mobilization of fat stores may play a greater role in energy balance or energy expenditure is lower than expected. There is little evidence to suggest energy conservation in the lactating woman, i.e., more efficient metabolism due to a change in the hormonal milieu, for example. Most research data do not suggest that an individual's basal

Table 18.1

Calculating the Estimated Energy Requirement (EER) for a Nonpregnant, Nonlactating Woman 30 Years of Age[a] [7]

Height m (in)	PAL[b]	Weight for BMI of 18.5 kg/m² kg (lb)	Weight for BMI of 24.99 kg/m² kg (lb)	EER, women (kcal/day)[c]	
				BMI of 18.5 kg/m²	BMI of 24.99 kg/m²
1.45 (57)	Sedentary	38.9 (86)	52.5 (116)	1,563	1,691
	Low active			1,733	1,877
	Active			1,946	2,108
	Very active			2,201	2,386
1.50 (59)	Sedentary	41.6 (92)	56.2 (124)	1,625	1,762
	Low active			1,803	1,956
	Active			2,025	2,198
	Very active			2,291	2,489
1.55 (61)	Sedentary	44.4 (98)	60.0 (132)	1,688	1,834
	Low active			1,873	2,036
	Active			2,104	2,290
	Very active			2,382	2,593
1.60 (63)	Sedentary	47.4 (104)	64.0 (141)	1,752	1,907
	Low active			1,944	2,118
	Active			2,185	2,383
	Very active			2,474	2,699
1.65 (65)	Sedentary	50.4 (111)	68.0 (150)	1,816	1,981
	Low active			2,016	2,202
	Active			2,267	2,477
	Very active			2,567	2,807
1.70 (67)	Sedentary	53.5 (118)	72.2 (159)	1,881	2,057
	Low active			2,090	2,286
	Active			2,350	2,573
	Very active			2,662	2,916
1.75 (69)	Sedentary	56.7 (125)	76.5 (168)	1,948	2,134
	Low active			2,164	2,372
	Active			2,434	2,670
	Very active			2,758	3,028
1.80 (71)	Sedentary	59.9 (132)	81.0 (178)	2,015	2,211
	Low active			2,239	2,459
	Active			2,519	2,769
	Very active			2,855	3,140
1.85 (73)	Sedentary	63.3 (139)	85.5 (188)	2,082	2,290
	Low active			2,315	2,548
	Active			2,605	2,869
	Very active			2,954	3,255
1.90 (75)	Sedentary	66.8 (147)	90.2 (198)	2,151	2,371
	Low active			2,392	2,637
	Active			2,692	2,971
	Very active			3,053	3,371
1.95 (77)	Sedentary	70.3 (155)	95.0 (209)	2,221	2,452
	Low active			2,470	2,728
	Active			2,781	3,074
	Very active			3,154	3,489

[a]For each year below 30, add 7 kcal/day. For each year above 30, subtract 7 kcal/day.

[b]*PAL* physical activity level.

[c]EER for women can be calculated as follows: EER = 354 − (6.91 × age [years] + PA × (9.36 × weight [kg] + 726 × height [m]), where PA is the physical activity coefficient of 1 for sedentary PAL, 1.12 for low active PAL, 1.27 for active PAL, and 1.45 for very active PAL.

Table 18.2
Calculating the Estimated Energy Requirement (EER) for a Lactating Woman [7]

= adult EER	+ milk energy output	− weight loss
See Table 1(7)	*500 kcal (1ˢᵗ six months)*	*170 kcal (1ˢᵗ six months)*
	400 kcal (2ⁿᵈ six months)	*0 kcal (2ⁿᵈ six months)*
	WHY?	**WHY?**
calculated based on a non-pregnant woman's weight, age, and physical activity level	= milk production × energy density 1ˢᵗ 6 months: 0.78 L/d × 0.67 kcal/g rounded to 500 kcal/d 2ⁿᵈ 6 months: 0.6 L/d × 0.67 kcal/g rounded, to 400 kcal/d	Weight loss seen in 1ˢᵗ six months with an average loss of 0.8 kg/month equivalent to 170-kcal/d deficit

metabolic rate is lower during lactation than prepregnancy or that more energy is used to complete a physical task. Basal metabolic rate is the rate of energy expenditure in an individual resting comfortably, awake, and motionless, 12–14 h after last consuming food.

18.2.1.1 A Sample Calculation of the Estimated Energy Requirement for Lactation

Using Tables 18.1 and 18.2, the EER for lactating women may be calculated. Using the example of a "low active," exclusively, for a lactating woman 4 months postpartum who is 35 years of age, and weighs 60 kg and is 1.6 m tall, you would first refer to Table 18.1, and calculate her EER as if she were neither pregnant nor lactating. Alternatively, you could calculate her estimated energy requirement as if she was nonpregnant, nonlactating using footnote a of Table 18.1. Using the table itself, our sample lactating woman would have a nonpregnant, nonlactating estimated energy requirement of 2,083 kcal. To this value, you would add 500 kcal to account for the amount of energy required to produce breast milk, and subtract 170 kcal for the contribution from fat stores laid down in pregnancy (information found in Table 18.2). Hence, our sample lactating woman's EER would be approximately 2,413 kcal (2,083 + 330). It is important to stress this is an *estimate* of the energy requirements of an *average* woman only, and follow-up is required to ascertain its appropriateness at an individual level, i.e., some women can gain or lose weight using this recommendation for energy intake.

18.3 POSTPARTUM WEIGHT RETENTION

Research suggests that more than a third of pregnant women gain more weight during pregnancy than is recommended, and this is particularly a problem among women who are already overweight or obese [8]. Given the increasing prevalence of obesity among women in both developed and developing nations raises the question of whether retention of this excess weight gain postpartum and lifestyle changes in the postpartum period are likely contributors to obesity among women [9]. While many women express a desire to lose weight postpartum and return to their prepregnancy weight, weight loss among women postpartum is highly variable. In general, most women will retain between 0.5 and 3 kg (1.1–6.6 pounds) of weight from their previous pregnancy over the longer term [9, 10]. At 18 months postpartum, 20% of women will be more than 5 kg (11 lb) heavier than they were before pregnancy. Nutrition advice for lactating women has historically been to avoid dieting while breastfeeding to ensure appropriate nutritional status for both infant and mother. Given the global epidemic of obesity and associated health consequences this advice needs to be reevaluated.

A woman who is lactating has the same physiologic requirements for regulating body weight as one that is not, except that she is producing a continuous supply of milk creating a much higher energy output. As noted above, the total energy cost to a woman who is exclusively breastfeeding an infant 0 to 6 months is estimated to be 500 kcal/day; theoretically, this output of energy could result in 0.5 kg/week (1.1 pound/week) of weight loss, provided energy intake and physical activity remain unchanged. While the woman-to-woman variability is tremendous, this rate of postpartum weight loss is seldom achieved as energy intake and/or a decrease in physical activity in the early postpartum period compensates, at least in part, for the energy costs of lactation. Higher energy intakes in lactating women versus nonlactating women may be attributed to enhanced appetite due to increased prolactin levels and higher energy demands. Prolactin is a hormone released by the anterior pituitary gland, which stimulates breast development and milk production in women.

The most consistent and strongest determinant of weight loss during lactation is pregnancy weight gain [11, 12]. Other factors that have been shown to influence postpartum weight loss, albeit inconsistently, include prepregnancy weight, age, parity, race, smoking, exercise, return to work outside the home, and lactation. While the impact is modest, the portfolio of evidence suggests that breastfeeding results in a faster rate of postpartum weight loss than formula feeding [13]. The average difference in weight loss by 12 months postpartum between lactating and nonlactating women is about 0.6–2.0 kg (1.3–4.4 lb) [13].

Body composition also changes throughout pregnancy and lactation. Due to high levels of estrogen in pregnancy, a pregnant body favors the gynoid shape. Specifically, body fat distributes to the thigh area and to a lesser extent the suprailiac, subscapular, costal, biceps and triceps areas [11]. Changes in body composition with breastfeeding are then reversed and fat is mobilized from the trunk and thigh areas.

18.4 EXERCISE AND LACTATION

Physical activity at any stage of the life cycle is associated with a decreased prevalence of cardiovascular disease, colon cancer, type 2 diabetes, and overweight, and it decreases mortality rates from all causes. Specifically in lactation, regular activity improves cardiovascular fitness, plasma lipid levels, and insulin response [14]. Regular activity also has the potential to benefit psychosocial well-being in lactation, such as improving self-esteem and reducing depression and anxiety. Other potential benefits include promotion of body weight regulation and optimizing bone health. Engagement in regular activity by the mother may also encourage the same in her offspring, promoting a healthy lifestyle and body weight management for the entire family.

Women can actively engage in moderate exercise during lactation without affecting milk production, milk composition, or infant growth [15, 16]. Lovelady et al. demonstrated that overweight sedentary lactating women randomized to a regimen of reduced energy intake (−500 kcal/day) and aerobic exercise (45 min/4 days each week) had babies that grew similarly to those of women who were not on an energy restricted diet and exercised once or never per week [15]. Aerobic exercise in this study consisted of walking, jogging, and dancing at 65–80% of maximum heart rate. The duration of exercise was initially 15 min, increased by at least 2 min each day until the women were exercising for 45 min at the target heart rate. There is some evidence that exercise in the absence of energy restriction will not promote weight loss postpartum, and diet restriction alone results in a greater percentage of lean body mass loss compared to exercise in combination with energy restriction [16, 17].

The American College of Obstetricians and Gynecologists has developed guidelines for exercising in pregnancy and the postpartum period [18]. For the postpartum period, these guidelines include resuming physical activity gradually, and only when a woman's body has healed substantially from pregnancy and delivery (usually 4–6 weeks postpartum). They encourage that women obtain clearance from their primary care physician to resume physical activity. In addition, Larsen-Meyer recommends that women avoid becoming excessively fatigued, remain well hydrated, and watch for abnormal bleeding or pain [14].

18.5 ACHIEVING A BALANCE OF DIET AND EXERCISE FOR MOM AND BABY

Some basic guidelines regarding energy control for lactating women are summarized in Table 18.3. Maintaining a healthy diet during lactation is not only essential as a weight loss strategy, but it also ensures that macro- and micronutrient intake is adequate to support optimal maternal health and breastfeeding success.

18.6 CALCIUM

18.6.1 Background

Calcium is important for the normal development and maintenance of the skeleton, with over 99% of total body calcium found in bones and teeth [21]. The remainder of total body calcium is tightly regulated in blood, extracellular fluid, and muscle, where it plays a role in blood pressure regulation, muscle contraction, nerve transmission, and hormone secretion. Calcium homeostasis is maintained by parathyroid hormone, which increases blood calcium, and calcitonin, which lowers blood calcium. If blood calcium levels fall, parathyroid hormone is secreted, stimulating the release of calcium from bone. Chronic calcium deficiency, due to inadequate intake, will result in progressive loss of skeletal mass and osteoporosis.

During lactation, secretion of calcium into breast milk averages about 200 mg/day to accommodate the whole-body mineral accretion rate of the infant [22]. Although renal calcium excretion rates are lowered to meet the elevated calcium demands of lactation, the primary source of calcium secreted in breast milk appears to be from increased maternal bone resorption. The concentration of calcium in breast milk decreases after 3–6 months and thus, the greatest loss of bone mineral content occurs within the first

Table 18.3
Guidelines for Energy Control During Lactation

Diet

- Eat a well balanced diet. Compare your typical daily food choices against the US Department of Agriculture Dietary Guidelines and MyPyramid, or dietary guidance from your country of origin, and make appropriate modifications [19, 20]. In the event that you need help making modifications, see your primary care physician or a clinical dietitian. A vitamin and/or mineral supplement may be necessary
- Restricting dietary intake by 500 kcal/day is safe as is moderate weight loss
- Reduce consumption of foods high in fat and simple sugars (e.g., sucrose, fructose)
- Emphasize fruit and vegetable consumption
- Emphasize foods high in calcium and vitamin D
- If capable of becoming pregnant, then consume 400 mcg/day of folic acid for neural tube defect prevention

Exercise

- Prepregnancy activities may be resumed gradually after medical clearance (usually around 4–6 weeks postpartum)
- Gradually work up to 30 min of moderate exercise each day for most days of the week
- An exercise regimen consisting of 45 min of moderate aerobic exercise 4 days/week (60–80% maximum heart rate) in combination with a 500 kcal/day diet restriction has been shown to promote postpartum weight loss and does not negatively affect breastfeeding
- Avoid excessive fatigue and keep hydrated
- Wear a bra that is supportive to your activity

few months postpartum [22]. Serial measurements of bone density after 2–6 months of lactation have shown a decrease of 3–10% in bone mineral content of trabecular bone (sponge-like interior) in the lumbar spine, hip, femur, and distal radius, with smaller losses occurring with cortical bone (exterior shell) [22–24]. Loss of calcium from the maternal skeleton is not prevented by increased dietary calcium, even among women with low baseline calcium intakes [25–27]. Upon return of menses, and restoration of estrogen, maternal bone lost during lactation is restored within 3–6 months of cessation of breastfeeding [23, 24, 26, 28, 29]. Based on the majority of epidemiological studies, there is no adverse effect of lactation history on peak bone mass, bone density, or hip fracture risk [30]. Thus, the evidence suggests that the bone mineral changes that occur during and following lactation are a normal physiological response, and an increased requirement for calcium is not needed.

18.6.2 Recommended Dietary Intake for Calcium

The Adequate Intake (AI) for calcium during lactation is set at 1,000 mg/day for women who are 19–50 years of age [21]. It is recommended that breastfeeding women less than 19 years of age consume 1,300 mg calcium/day due to the increased need to support ongoing bone growth of the teen herself [21]. By definition an "adequate intake level," as defined by the US Institute of Medicine, is the average daily nutrient intake level of apparently healthy people who are assumed to have adequate nutritional state. The adequate intake level is expected to meet or exceed the needs of most individuals.

18.6.3 Calcium Intakes of Women

The lack of effect of calcium supplementation on maternal bone metabolism during lactation does not lessen the importance of consuming foods rich in calcium. Available data from the US National Health and Nutrition Examination Surveys (NHANES) suggest that women, regardless of reproductive stage, are not meeting recommended intakes for calcium [31]. Daily median calcium intake by adult women aged 20–39 years is about 684 mg/day [31]. In African-American and other ethnic minority groups, calcium intake is particularly low [32]. While the calcium intakes of lactating women, specifically, have not been extensively studied in North America, data from a small sample of lactating women ($n = 16$) participating in the 1994 Continuing Food Survey of Food Intake by Individuals (CSFII), suggest a median calcium intake of 1,050 mg/day [33]. Similarly, in another study of lactating women ($n = 52$), average calcium intakes were approximately 1,218 mg and 1,128 mg/day at 3 and 6 months postpartum, respectively [34].

18.6.4 Sources of Calcium in the Diet

Approximately two thirds of dietary calcium intake in the United States is from fluid milk and other dairy products [35]. Nondairy sources include calcium-fortified orange juice, and rice or soy beverages. Salmon with bones and some green leafy vegetables such as broccoli may also contribute to the intake of calcium; however, in general these sources contain less calcium per serving than do milk and dairy products (Table 18.4). The calcium bioavailability of nondairy foods is variable [36, 37]. For most solid foods, the bioavailability of calcium is inversely associated with its oxalate content. For example, the calcium bioavailability from foods high in oxalates such as spinach and rhubarb is low, whereas it is high in foods with low concentrations of oxalates such as kale, broccoli, and bok choy [38]. Supplemental sources of calcium come in a variety of preparations, both liquid and solid. Calcium from carbonate and citrate are the most common forms of calcium supplements [39]. Ingestion of a meal with dietary and supplemental calcium results in a 20–25% improvement in absorption relative to absorption obtained when a source is ingested on an empty stomach [39, 40]. The absorption of supplemental calcium is greatest when calcium is taken in doses of 500 mg or less [21].

18.7 VITAMIN D

18.7.1 Background

The most well appreciated function of vitamin D is to maintain normal blood calcium and phosphorus concentrations thereby promoting bone health. Vitamin D can be obtained from food, or synthesized in the skin by exposure to ultraviolet light. Solar ultraviolet-B (UVB) photons are absorbed by 7-dehydrocholesterol in the skin, transformed to previtamin D, and then rapidly converted to vitamin D. Total body exposure to 10–15 min peak sunlight during the summer months in a Caucasian is equivalent to approximately 20,000 IU of vitamin D [41, 42]. Seasonal changes, time of day, latitude,

Table 18.4
Dietary Sources of Calcium

Food	Portion size	Calcium content per serving (mg)
Milk, whole, 2%, 1%, skim	1 cup	300
Yogurt, low fat, plain	3/4 cup	300
Calcium-enriched orange juice	1 cup	300
Fortified rice or soy beverage	1 cup	300
Yogurt, fruit bottom	3/4 cup	250
Cheese, hard	1 oz.	240
Sardines	4 medium, 1 3/4 oz.	185
Salmon, canned with bones	3 oz.	180
Tofu, firm, made with calcium sulfate	3 1/2 oz.	125
Cottage cheese 1% milk fat	3/4 cup	120
White beans	1/2 cup	100
Almonds, dry roast	1/4 cup	95
Turnip greens	1/2 cup	95
Ice cream, vanilla	1/2 cup	85
Navy beans	1/2 cup	60
Oysters, canned	1/2 cup	60
Orange	1 medium	55
Dried figs	2 medium	54
Kale	1/2 cup	50
Chickpeas	1/2 cup	40
Broccoli	1/2 cup	35
Regular soy beverage	1 cup	20

Adapted from British Columbia Ministry of Health (2005) BC health file no. 69e, Nutrition Series. Available at: http://www.bchealthguide.org/healthfiles/hfile68e.stm

aging, sunscreen use, and skin pigmentation can influence the cutaneous production of vitamin D. Above 37°N latitude during the months of November to February, there is marked reduction in the UVB radiation reaching the earth's surface [43]. To give the reader an approximate idea of the location of the 37°N latitude, Richmond, Virginia, and Oakland, California, are located here. Likewise, most of Europe lies above 37°N. Therefore, very little, if any, vitamin D is produced in the skin in the winter north of Richmond or Oakland or in Europe. Once produced in the skin, or absorbed, vitamin D is metabolized in the liver to the major circulating form, 25-hydroxyvitamin D. The circulating concentration of 25-hydroxyvitamin D is a good indicator of the cumulative effects of sunlight exposure and dietary vitamin D intake. Following production, 25-hydroxyvitamin D is converted in the kidney to its biologically active form, 1,25-dihydroxyvitamin D, and transported to major target tissues. 1,25-Dihydroxyvitamin D is responsible for an increase in intestinal calcium transport and mobilization of calcium from the bone.

Human milk contains low amounts of vitamin D, ranging from 4 to 40 IU/l [44]. Infant formula is routinely fortified with 400 IU vitamin D per liter, while the breastfed infant is primarily dependent upon endogenous synthesis or supplemental sources of vitamin D.

Currently the American Academy of Pediatrics recommends that infants <6 months of age not be exposed to direct sunlight [45]; hence, the opportunity for cutaneous exposure is limited. Although the vitamin D content of human milk is related to maternal dietary intake [21], maternal consumption of less than 600–700 IU vitamin D per day will not provide sufficient vitamin D in breast milk to meet the infants' vitamin D requirements [46]. Thus, breastfed infants are recommended to be given a 400 IU vitamin D supplement each day [47, 48]. High-dose maternal vitamin D supplementation (>2,000 IU/day) has been shown to improve both the vitamin D status of lactating women and their infants [49]. A maternal intake of 4,000 IU/day increased the vitamin D activity of milk by 100 IU/l. Dietary intakes of vitamin D up to 10 times the DRI for 3 months in this small group of lactating women ($n = 18$) resulted in no adverse events as demonstrated by normal serum calcium concentrations and no observation of hypercalcuria. Ala-Houhala and colleagues also supplemented healthy mothers with 2,000 IU, 1,000 IU, or no vitamin D for a period of 8 weeks [50, 51]. Circulating levels of 25-hydroxyvitamin D of infants who were breast fed from women receiving 2,000 IU/day of vitamin D were similar to those of infants directly supplemented with 400 IU/day. Further studies are needed to assess the safety of high-dose supplementation over prolonged periods.

18.7.2   Recommended Dietary Intake for Vitamin D

Due to the very small and insignificant amounts of vitamin D secreted in human milk, it has historically been concluded that there is no evidence that lactation increases maternal requirements for vitamin D. Therefore, the current recommended adequate intake remains similar to nonlactating adults and is set at 200 IU/day [21]. Since the establishment of this recommended dietary intake of vitamin D in 1997, concerns about the wide spread prevalence of vitamin D deficiency have surfaced in the medical and scientific literature. Furthermore, the basis of these recommendations was made prior to the use of circulating 25-hydroxyvitamin D as an indicator of vitamin D status. To date, there is no scientific literature available pertaining to the minimum vitamin D intake needed to maintain normal concentrations of maternal circulating 25-hydroxyvitamin D. The appropriate dose of vitamin D during lactation appears to be greater than the current dietary reference intake of 200 IU/day. Supplemental intake of 400 IU vitamin D per day has only a moderate effect on maternal blood concentrations of 25-hydroxyvitamin D [52]. Many experts agree that a desirable 25-hydroxyvitamin D concentration is ≥75 nmol/l (30 ng/ml) [52], and attainment of these levels requires an additional intake of approximately 1,700 IU/day [53].

Currently, the US Institute of Medicine considers an intake of 2,000 IU/day for lactating women to be the tolerable upper intake level. The upper tolerable level, as defined by the US Institute of Medicine, is the highest level of continuing daily nutrient intake that is likely to pose no risk of adverse health effects in almost all individuals. Hathcock and colleagues [54] recently focused on the risk of hypercalcemia and demonstrated that the margin of safety for vitamin D consumption for adults is likely greater than ten times any current recommended level. These authors conclude that the tolerable upper limit for vitamin D consumption by adults should be set at 10,000 IU/day [54].

Furthermore, vitamin D is a fat-soluble vitamin and is stored in body fat. As a result, several studies have linked obesity with poorer vitamin D status, as demonstrated by lower circulating 25-hydroxyvitamin D concentrations [55–58]. A study conducted by Wortsman and colleagues [58] confirmed that obese patients had lower basal 25-hydroxyvitamin D and higher serum parathyroid hormone concentrations than nonobese patients. Following exposure to an identical amount of UVB radiation, the blood concentration of vitamin D was 57% less in obese than in nonobese subjects. It was proposed that the lower serum 25-hydroxyvitamin D levels seen among obese subjects were the result of increased sequestering of vitamin D in fat tissue. Likewise, body mass index (BMI) was inversely correlated with peak blood vitamin D concentrations after oral dosing. In conclusion, obese subjects may have a greater requirement for vitamin D than their nonobese counterparts do.

18.7.3   Dietary Intake of Vitamin D

Since the primary source of vitamin D is synthesis in the skin, very little survey data are available regarding dietary vitamin D intake. As the widespread use of sunscreens and public health recommendations to avoid sun exposure limits this endogenous source of vitamin D, most people necessarily rely on vitamin D from either dietary or supplemental sources. Although dietary sources may provide an amount to meet the currently published 1997 recommendations for vitamin D, they fall short of meeting the suggested requirement proposed in recent studies [49, 53]. A supplemental source of vitamin D is likely required to meet these latter proposed recommendations, at least in the winter months when sun exposure is limited.

18.7.4   Sources of Vitamin D in the Diet

Only a few foods are natural sources of vitamin D. These include liver, fatty fish such as salmon, and eggs yolks. Cod liver oil is an excellent source of vitamin D, containing approximately 1,360 IU/tablespoon. The major dietary sources of vitamin D, however, are vitamin D fortified foods including milk (100 IU per 8-oz. serving), some orange juices (100 IU per 8-oz. serving), and some margarines (60 IU/tablespoon). Breakfast cereals, breads, crackers, cereal grain bars and other foods may be fortified with 10–15% of the recommended daily value for vitamin D. Supplemental vitamin D is available in two distinct forms, vitamin D_2 and vitamin D_3. Vitamin D_3, however, has proven to be a more potent form, with a 70% greater increase in 25-hydroxyvitamin D concentrations [59].

18.8   FOLATE

18.8.1   Background

Folate is a generic term used to describe a number of related compounds that are involved in the metabolism of nucleic and amino acids, and therefore the synthesis of DNA, RNA, and proteins. Folate plays a role in the conversion of homocysteine to methionine. Folic acid is a synthetic form of the vitamin, used in vitamin supplements and food fortification. It exhibits a high degree of stability, and is more bioavailable than naturally occurring folate from food. Unlike folic acid, naturally occurring food folates are usually reduced, and contain a polyglutamate tail consisting of one to several glutamate molecules. Prior to active transport across the small intestine, this polyglutamate tail must be hydrolyzed to produce the monoglutamate form. Traditionally synthetic folic acid was thought to be completely reduced at the gut and to enter portal circulation primarily in the form of 5-methyltetrahydrofolate; however, synthetic folic acid (>200 mcg), even in very modest doses, can be absorbed by a nonsaturable mechanism involving passive diffusion. Thus, small amounts of unmetabolised folic acid (1–5%) have been shown to be present in circulation [60, 61].

The average amount of folate secreted into human milk is estimated to be 85 mcg/liter/day [62]. With the exception of severe maternal folate deficiency (i.e., megaloblastic anemia), the content of folate in human milk remains stable and appears to be conserved at the expense of the mother's folate stores [63].

18.8.2   Recommended Dietary Intake for Folate

The bioavailability of naturally occurring folates in food and synthetic forms of the vitamin is thought to differ considerably. A folic acid supplement taken on an empty stomach is thought to be 100% bioavailable compared to about 50% for naturally occurring food folate (Table 18.5) [62]. In an effort to take into account the different bioavailability of folate from natural versus synthetic sources, folate requirements are now expressed as dietary folate equivalents (micrograms of DFE): micrograms of food folate + (1.7 × mcg of folic acid).

The recommended dietary allowance (RDA) for folate published by the US Institute of Medicine for breastfeeding women aged 14–50 years is 500 mcg DFEs per day. The scientific evidence necessary to establish an RDA is more robust than that for an "adequate intake level." An RDA is the average daily dietary intake level that is sufficient to meet the nutrient requirement of nearly all (97–98%) healthy individuals. The RDA of 500 mcg DFEs per day is the amount of folate estimated to replace the folate secreted daily in human milk plus the amount of folate required by the nonlactating woman to maintain healthy folate status, but does not factor in the metabolic cost of milk synthesis [62]. Lactating women who are planning a subsequent pregnancy, or who are not taking effective precautions to prevent one, should be encouraged to consume 400 mcg folic acid supplement daily for at least 4 weeks before and 12 weeks after conception to reduce the risk of having a subsequent child with a neural tube defect.

The tolerable upper limit for folate for lactating women aged 14–18 years and aged 19 years and older is set at 800 mcg and 1,000 mcg of folic acid from fortified foods or supplements [62]. It should be noted that the upper limit for folate does not include naturally occurring food folate, as no adverse effects have been linked with the consumption of excess food folate. Overzealous use of folic acid supplements is not risk free. For example, very high intakes of folic acid could mask a vitamin B_{12} deficiency by correcting its characteristic symptom, megaloblastic anemia. Delayed diagnosis

Table 18.5

Relative Bioavailability of Naturally Occurring and Synthetic Folate

- 1 mcg of folate from food provides 1 mcg of dietary folate equivalent (DFE)
- 1 mcg of folic acid supplement taken on an empty stomach provides 1.7 mcg of DFE
- 1 mcg of folic acid supplement taken with meals or from fortified food provides 1.7 mcg of DFE

of vitamin B_{12} deficiency can result in increased risk of irreversible neurological damage.

There are situations in which a larger folic acid supplement may prove worthwhile during lactation and these should be discussed with the patient. For example, a woman, who has lactated for a long duration, has not taken supplemental folic acid, and who has difficulty in remembering to take it every day may be a good candidate for a higher level of folic acid supplementation. Likewise, a woman who has had a previous pregnancy affected by a neural tube defect may quite rightfully be recommended by their physician to consume higher amounts of supplemental folic acid if she is capable of becoming pregnant. In the event that a high folic acid supplement is recommended (i.e., >1 mg/day), it is advisable that the first 1 mg be consumed with a B_{12}-containing supplement and any folate above 1 mg/day be consumed as a folic-acid only supplement to ensure that fat soluble vitamin intakes (particularly vitamin A) do not reach unsafe levels.

18.8.3 Dietary Intake of Folate

Prior to folic acid fortification of the food supply in North America in 1998, a reduction in maternal folate stores during lactation was observed and was likely due to poor dietary folate intakes [64–67]. Since implementation of the fortification program, significant improvements in blood folate status of reproductive age women, including pregnant and lactating women, have been described [68, 69]. Dietary folate intakes from unfortified foods during lactation, however, remain suboptimal for approximately one third of women as demonstrated in a sample of well-nourished lactating Canadian women [70]. On average in this study, natural food folate provided 283 ± 71 mcg/day folate, while folic acid from fortified foods supplied approximately 125 ± 35 mcg/day folic acid. The investigators concluded that without mandatory folic acid fortification, 98% of lactating women would not have met their requirements for folate from diet alone [70].

18.8.4 Sources of Folate in the Diet

Natural rich sources of folate are green leafy vegetables as well as citrus fruit juices, liver, and legumes. After folic acid fortification of the food supply, the category "bread, rolls, and crackers" became the single largest contributor of total folate in the American diet, contributing 16% of total intake, surpassing natural vegetable folate sources [71]. Table 18.6 presents data on the major dietary contributors of folate in the diets of a sample of pregnant and lactating Canadian women [70]. Orange juice was the largest source of total dietary folate (11.1%), while enriched pasta products were the second largest contributor (8.8%). Based on the US Department of Agriculture's (USDA's) Dietary Guidelines and MyPyramid, or Canada's Food Guide for Healthy Eating, the grains food group provided 41% of total dietary folate [19, 20]. Thus, women avoiding white bread and enriched pasta to lose weight may be at particular risk of low folate intake.

The principal form of supplemental folate used in the world today is folic acid; however, supplemental 5-methyltetrahydrofolate is now available in some vitamin and mineral supplements, including prenatal supplements. At the time of writing they are available for use in the United States but not Canada, for example.

Table 18.6
Major Folate Contributors to the Daily Diet of Lactating Women Post-Folic Acid Fortification [70]

Food	Serving size	Folate (mcg DFE)	Contribution to total folate intake (%)
Orange juice	1 cup	72	11
Pasta, dry	1 cup	391	8.8
Green salad	1 1/2 cup	77	5.2
Bagels	1 medium	226	5
Whole wheat, rye, and other dark breads	1 slice	14	4.8
Cold cereals	1 cup	166	4.5
White bread	1 slice	171	4.2
Cream, milk, eggnog	1 cup	12	3.9
Rice, cooked	1 cup	215	3.8
Cake, cookies, donuts, pies	1 medium[a]	115	3.3

[a]Example: donut

18.9   VITAMIN B$_{12}$

18.9.1   Background

Vitamin B$_{12}$, often referred to as cobalamin, is required for the formation of red blood cells and normal neurological function [62]. Similar to folate, vitamin B$_{12}$ is involved in DNA synthesis. If vitamin B$_{12}$ deficiency occurs, then DNA production is disrupted, producing megaloblastic changes in blood cells (macrocytosis). Neurological complications occur in 75–90% of individuals with clinically defined vitamin B$_{12}$ deficiency [62]. When a deficiency occurs in developed countries, it is frequently associated with inadequate absorption rather than a dietary deficiency. Inadequate absorption could be caused by chronic antacid use, atrophic gastritis, hypochlorhydria, or pernicious anemia—most frequently found in individuals >50 years of age.

High doses of synthetic folic acid (greater than 1,000 mcg) can mask vitamin B$_{12}$ deficiency by reversing megaloblastic anemia [63, 72]. Megaloblastic anemia is the clinical indicator that often leads a clinician to suspect that vitamin B$_{12}$ deficiency may be an issue. Vitamin B$_{12}$ is excreted in the bile and effectively reabsorbed such that it can take up to 20 years for a vitamin B$_{12}$ deficiency to develop due to low vitamin B$_{12}$ intake. In contrast, a deficiency due to poor absorption can take only a few years to develop.

During lactation, the concentration of vitamin B$_{12}$ in human milk varies widely, and reflects maternal vitamin B$_{12}$ intake and status [62]. Low maternal intake or poor absorption of vitamin B$_{12}$ rapidly leads to a low level of vitamin B$_{12}$ in human milk [73]. Severe deficiency can occur after approximately 4 months of age in exclusively breast-fed infants of mothers with inadequate intake [74]. It is postulated the rapid postnatal development of vitamin B$_{12}$ deficiency in the infant is due, in part, to poor in utero transfer of vitamin B$_{12}$ from mother to child. Simply put, if a mother's vitamin B$_{12}$ status is suboptimal in lactation, it could very well have been in pregnancy as well. Symptoms of infantile vitamin B$_{12}$ deficiency include irritability, abnormal reflexes, feeding difficulties, reduced level of alertness or consciousness leading to coma, and permanent development disabilities if diagnosis is delayed [75].

Despite woman-to-woman variation in milk vitamin B_{12} content, the concentration of vitamin B_{12} in human milk changes very little after the first month postpartum [76]. The average reported concentration of vitamin B_{12} secreted in the milk of well-nourished mothers is approximately 0.33 mcg/day during the first 6 months of lactation, and 0.25 mcg/day during the second 6 months [76]. In a group of women receiving vitamin B_{12} containing supplements, the average B_{12} content of milk was 0.91 mcg/l [77], while the B_{12} content of milk from unsupplemented vegetarian mothers was lower, averaging 0.31 mcg/l [73].

18.9.2   Recommended Dietary Intake for Vitamin B_{12}

The RDA for lactating women age 14–50 years is 2.8 mcg/day. This value is higher than the RDA for nonpregnant, nonlactating women (2.4 mcg/day) to account for the amount of vitamin B_{12} secreted into breast milk. No adverse effects have been linked with excess vitamin B_{12} from supplements and/or food, and thus no Upper Limit (UL) has been set by the US Institute of Medicine.

18.9.3   Dietary Intake of Vitamin B_{12}

Low dietary vitamin B_{12} intakes during lactation typically occur when either the mother is a strict vegetarian or in a developing country where the usual consumption of animal products is low. Since the frequent consumption of animal foods is common in North America, median vitamin B_{12} intake from food in the general adult population in the United States of 3–4 mcg/day and Canada of 4–7 mcg/day are well above recommended levels [62]. Nonetheless, there are data to suggest the prevalence of suboptimal vitamin B_{12} deficiency may be higher than previously appreciated in reproductive age females. For example, House et al. [78] reported that 44% of a large sample of pregnant women in the province of Newfoundland in Canada (n = 1,424) had serum vitamin B_{12} concentrations during the first trimester of pregnancy below a commonly used cut-off value indicative of below-normal or deficient vitamin B_{12} status (<130 pmol/l). Koebnick et al. [79] reported a 22% prevalence of low serum vitamin B_{12}, and elevated homocysteine concentrations, a functional index of folate, vitamin B_{12}, or vitamin B_6 deficiency, among pregnant lacto-ovo vegetarians in Germany.

18.9.4   Sources of Vitamin B_{12} in the Diet

Vitamin B_{12} is synthesized by bacteria and found primarily in meat, eggs, fish (including shellfish), and to a lesser extent dairy products. Fortified breakfast cereals provide a significant source of vitamin B_{12} (6.0 mcg/3/4 cup), particularly for vegetarians. Plant sources, such as spirulina (algae) and nori (seaweed), contain vitamin B_{12} analogues, which can compete with vitamin B_{12} and inhibit metabolism. Lactating vegetarians may need to also be advised that milk and milk products are a good source of vitamin B_{12} (0.9 mcg/250 ml), while vegans are recommended to consume a supplement (~2.8 mcg/ day) and/or ensure their diet includes foods fortified with vitamin B_{12} such as textured vegetable protein and soy milk.

The form of vitamin B_{12} most frequently used in supplements and/or fortified foods is cyanocobalamin, which is readily converted in the body to its utilizable forms of methylcobalamin and 5-deoxyadenosylcobalamin [80]. Other supplemental forms include methylcobalamin and adenosylcobalamin.

18.10   IRON

18.10.1   Background

Iron is an essential component of numerous proteins and enzymes in the human body. Over 60% of iron in the body is found in hemoglobin, the oxygen-carrying pigment of the red blood cell that transports oxygen from the lungs to tissues for use in metabolism. About 4% of iron is found in myoglobin, the oxygen binding storage protein of muscle, and trace amounts are associated with electron transport and iron-dependent enzymes. A large portion of the remaining iron in the body is found stored in the form of ferritin, primarily in the liver but also in bone marrow and the spleen. With the exception of pregnancy and menstruation where there is a net outward flux of iron, the iron content of the body is highly conserved. The secretion of iron into breast milk is low, with the average milk iron being in the order of 0.35 mg/l [81]. Maternal dietary iron intake appears to have very little effect on milk iron levels.

18.10.2   Recommended Dietary Intake for Iron

Iron requirements during lactation are considerably lower than those for nonpregnant, nonlactating women based on the assumption that *exclusively* breastfeeding women will not resume menses for a period of 6 months postpartum. The RDA for iron for nonpregnant, nonlactating women is 18 mg/day, and for lactating women aged 19 to 50 years it is 9 mg/day [81]. The RDA for iron for lactating adolescents is slightly higher at 10 mg/day to provide additional iron to support the young mother's ongoing growth and development. The UL for all breastfeeding women is 45 mg of iron per day [81].

Iron-deficiency anemia during pregnancy, particularly in the third trimester, is common in both developed and developing countries, and is well described in the literature [5, 81–88]. While less well characterized, due to the net maternal iron deficit accrued during pregnancy (RDA = 27 mg/day), available evidence suggests a high prevalence of maternal iron deficiency early postpartum, despite women meeting dietary recommendations for lactation. The recovery of iron stores and alleviation of iron deficiency during this period is important, as low maternal iron status is related to fatigue, depression, decreased work capacity, and decreased ability of the mother to care for her newborn infant [89–91].

18.10.3   Dietary Intake of Iron

Data from nationally representative surveys in the United States suggest that the median iron intake of nonpregnant nonlactating women is ~12 mg/day, and that of pregnant women is 15 mg/day [81]. Inclusion of iron supplement use did not significantly influence these national estimates and underscores why many women will complete their pregnancy at a net iron deficit. While the samples of lactating women in these national surveys are small, the iron intakes of lactating women are generally higher than that reported for other premenopausal women, including pregnant women. This may reflect a combination of factors including the small sample size, the health consciousness and socioeconomic status of lactating versus nonlactating women, and treatment for early postpartum iron deficiency anemia. Bodnar [92] reported, using NHANES III national data from the United States, that approximately 10% of postpartum women have iron deficiency anemia. Among postpartum women of low household income, >20% were iron deficient (with or without anemia) compared with 7.5% of postpartum women not defined as low income.

18.10.4   Sources of Iron in the Diet

Two types of iron are present in the diet: heme and nonheme iron. Heme iron is obtained from animal sources such as meat, poultry, and fish, and is about 20–30% absorbed. Non-heme iron, present in plant foods, iron fortificants, and iron supplements, is less bioavailable with absorption of 5–10% [81]. Dietary factors such as vitamin C and the presence of meat, fish or poultry can enhance the absorption of non-heme iron, while phytates found in legumes, grains and rice, polyphenols (in tea, coffee, and red wine) and vegetable proteins, such as those in soybeans, can inhibit non-heme iron absorption. Iron sources obtained from a typical Western diet consisting of abundant animal foods and sufficient sources of vitamin C were estimated to be approximately 18% bioavailable; the bioavailability of iron from a vegetarian diet is approximately 10% [81]. As a result, the requirement for iron is 1.8 times greater for vegetarians. The average iron content of fruit, vegetables, breads, and pasta ranges from 0.1 to 1.4 mg per serving. Some iron-fortified cereals contain up to 24 mg of iron per 1-cup serving.

18.11   LONG-CHAIN POLYUNSATURATED FATTY ACIDS

18.11.1   Background

Long-chain polyunsaturated fatty acids (LC-PUFAs) are fatty acids with a backbone of greater than 20 carbons, and are of either of the omega-3 (e.g., docosahexaenoic acid or DHA, 22:6n-3 and eicosapentaenoic acid or EPA, 20:5n-3) or omega-6 series (e.g., arachadonic acid or ARA, 20:4n-6). Humans are able to synthesis these LC-PUFAs from fatty acid precursors via a series of elongation and desaturation steps at all stages of the life cycle. DHA and EPA, for example, are synthesized from the shorter, less unsaturated omega-3 fatty acid, alpha-linolenic acid (ALA, 18:3n-3), and ARA is synthesized from linoleic acid. LC-PUCFAs are essential for the development and maturation of the fetal and neonatal brain as well as eicosanoid metabolism, fluidity in membranes, and gene expression. Whether pregnant and lactating women and infants can convert enough ALA to DHA and EPA to meet physiological requirements is uncertain and future research in this area is urgently required. Further, the 18-carbon fatty acids, linoleic acid (omega-6 series), and ALA (omega-3 series) compete for the same enzymatic machinery to synthesize ARA and DHA. The trend toward higher dietary intakes of the 18-carbon omega-6 versus the omega-3 series of fatty acids may likewise contribute to inappropriately low levels of LC-PUFAS of the omega-3 series. As has been shown in studies using stable isotopes, even infants have the enzymatic machinery to convert ALA acid to DHA and linoleic acid to ARA [93–97]. These studies alone, however, provide insufficient data to assess whether sufficient quantities of DHA and ARA are synthesized to meet the infant's requirements. Infants fed formulas without DHA and ARA, but containing adequate levels of alpha-linolenic and linoleic acid, have lower levels of DHA and ARA in their blood compared with either breastfed infants or infants fed formulas supplemented with these fatty acids [96]. As with DHA and ARA in breast milk, the profile and concentration of these fatty acids in the blood will reflect dietary intake and do not provide sufficient data to assess whether endogenous biosynthesis of

DHA and ARA is adequate to meet requirements. Results from clinical trials with term-born infants designed to evaluate whether preformed DHA and ARA need to be added to infant formula in addition to the precursor fatty acids (ALA and linoleic acid) are mixed with some showing at least a short-term benefit [98–105] on either visual or cognitive development and others showing no benefit at all [106–111].

The US Institute of Medicine assumes that the fatty acid composition of breast milk meets the requirements of most infants. However, the concentration of DHA in breast milk globally ranges widely from 0.1 to 1.4% of total fatty acids due to the fat composition of the mother's diet [112, 113]. Furthermore, Innis et al. [114] have reported a 50% decline in human milk concentrations of DHA since the late 1980s in Canada and Australia. As anticipated, maternal supplementation with DHA appears to increase breast milk DHA content in a dose-dependent manner [115, 116]. While maternal supplementation with ALA tends to increase ALA content of human milk, it appears to have little effect on milk DHA concentrations [117]. At present, there is insufficient evidence to determine whether the variation in DHA content of human milk has clinical implications for the breast-fed infant including visual function or neurodevelopment [118]; however, there are a number of interesting studies to suggest there may be, and hence, further research in this area is important [93–95, 119–122]. As is the case with other nutrients, it will be difficult to untangle the possible relative impact of DHA consumption during pregnancy versus lactation on infant development. New evidence does suggest that supplementation of women prenatally with DHA may affect maturation of the visual system of infants and their ability to problem solve [123, 124]. There is some evidence to suggest a potential role for omega-3 fatty acids in the prevention of depression during the postpartum period, but again more research needs to take place to confirm this relationship [118] and see Chap. 19, "Postpartum Depression and the Role of Nutritional Factors".

18.11.2 Recommended Dietary Intake for LC-PUFAs

Currently, there are no specific recommendations for DHA, EPA, or ARA intake in North America [7]. There are, however, very specific recommendations for ALA and linoleic acid. For nonpregnant nonlactating women, the US Institute of Medicine recommends an adequate intake level of 1.1 g/day ALA or an acceptable macronutrient distribution range of 0.6–1.2% energy. For pregnant and lactating women, they recommend 1.4 g/day. They do make the recommendation that up to 10% of this range can be consumed as DHA and/or EPA. At a workshop on the "Essentiality of and Recommended Dietary Intakes (RDIs) for Omega-6 and Omega-3 Fatty Acids" held by the National Institutes of Health (NIH) in 1999, attendees recommended that pregnant and lactating women consume 300 mg/day of DHA [125]. For nonpregnant nonlactating women, the US Institute of Medicine recommends an adequate intake level of 12 g/day linoleic acid or an acceptable macronutrient distribution range of 5–10% energy. For pregnant and lactating women, they recommend 13 g/day.

18.11.3 Dietary Intake of LC-PUFAs

The current Western diet is thought to be low in omega-3 fatty acids (e.g., ALA, EPA, DHA) and high in the omega-6 series, particularly linoleic acid. The 1994–1996 USDA Continuing Survey of Food Intakes by Individuals, a nationally representative analysis of consumption, provided data on the major polyunsaturated fatty acids in the

Table 18.7
Guidance for Fish Consumption during Pregnancy and Lactation [126]

Females who are or may become pregnant or who are breastfeeding:

- May benefit from consuming seafood, especially those with relatively higher concentrations of EPA and DHA—i.e., hake, herring, pollock, salmon, rainbow trout, king or snow crab, shrimp, clams, mussels.
- Can reasonably consume two 3-oz. (cooked) serving but can safely consume 12 oz./week.
- Can consume up to 6 oz. of white (albacore) tuna per week.
- Should avoid large predatory fish such as shark, swordfish, or king mackerel.

food supply in a subset of 112 pregnant or lactating women. Median daily intakes of this sample of women in this report indicated DHA intakes of ~44 mg/day [7], well below the 300 mg/day recommended by expert consensus at the NIH workshop [125].

18.11.4   Sources of LC-PUFAs

Meat and eggs are rich sources of ARA, while EPA and DHA are derived mainly from fatty fish such as mackerel, salmon, herring, trout, and sardines. Several foods are available that have added omega-3 fats including eggs, milk, yogurt, cheese, pasta, and bread. There is considerable concern about increasing the omega-3 series LC-PUFAs via fish consumption during pregnancy and lactation because of the possible risk of contaminants. This issue, for the most part, centers on the methylmercury and PCB (polychlorinated biphenyls) contamination of fish. (Refer to Table 18.7 below for guidance surrounding fish consumption in pregnancy and lactation.) Recommendations do differ from country to country. For those interested in more specific details regarding the species-to-species methylmercury and PCB content of different fish, a handy "Fish List" can be found at www.seachoice (download "Canada's Seafood Guide"), or http://www.cfsan.fda.gov/~frf/sea-mehg.html, which provides information provided by the US Department of Health and Human Services and the US Environmental Protection Agency.

18.12   CONCLUSION

Breastfeeding is the gold standard and strongly recommended method of feeding infants. The World Health Organization recommends human milk as the exclusive nutrient source for the first 6 months of life, with introduction of solids at this time, and continued breastfeeding until at least the first 12 months postpartum. Early postpartum weight is an issue for many women as they are anxious to return to their prepregnancy body size. For many, weight management will be difficult given personal circumstances and multiple demands on their time. Given the elevated nutrient requirements of lactation, women will need to plan meals with care to maximum nutrient intake while limiting energy dense foods. Women are encouraged to use the USDA Dietary Guidelines and My Pyramid or Canada's Food Guide or guides from their country of origin to select food choices—including number and portion size. Following this dietary guidance should facilitate adequate intakes of key nutrients including calcium, vitamin D, folate, vitamin B_{12}, and iron. In the event that a woman is unable to select food choices in the amounts and portions described in the food guides, a referral to a dietitian will be worthwhile as well as prescribing an appropriate vitamin or mineral supplement. Exercise postpartum is strongly encouraged after medical clearance, usually after 4–6 weeks postpartum. Women are encouraged to gradually work up to 30 min of moderate exercise each day for most days of the week. Women who wish to tackle their weight more aggressively should be reassured that an exercise regimen consisting of 45 min of moderate aerobic exercise 4 days/week day (60–80% maximum heart rate) in combination with a 500 kcal/day diet restriction has been shown to promote postpartum weight loss and does not negatively affect breastfeeding.

ACKNOWLEDGMENT

The authors wish to acknowledge the expert assistance of Aneta Plaga in helping consolidate the written work of the three authors and manuscript preparation.

REFERENCES

1. World Health Organization (2001) Global Strategies for Infant and Young Child Feeding. Resolution Passes at: Fifty-fourth World Health Assembly 9 May 2001
2. American Academy of Pediatrics (2005) Breastfeeding and the use of human milk. Pediatrics 115:496–506
3. American Academy of Pediatrics (2004) Breastfeeding. In: Kleinman RE (ed) Pediatric nutrition handbook, 5th edn. Elk Grove, Ill., pp 55–86
4. Anderson JW, Johnstone BM, Remley DT (1999) Breast-feeding and cognitive development: a meta-analysis. Am J Clin Nutr 70:525–35
5. Picciano MF (2003) Pregnancy and lactation: physiological adjustments, nutritional requirements and the role of dietary supplements. J Nutr 133:1997S–2002S
6. Butte NF, King JC (2005) Energy requirements during pregnancy and lactation. Public Health Nutr 8:1010–1027
7. Institute of Medicine (2002) Dietary Reference Intakes for energy, carbohydrate, fiber, fat, fatty acids, cholesterol, protein, and amino acids. National Academy Press, Washington, D.C.
8. Lederman SA (2004) Influence of lactation on body weight regulation. Nutr Rev 62:S112–S119
9. Lovelady CA, Stephenson KG, Kuppler KM, Williams JP (2006) The effects of dieting on food and nutrient intake of lactating women. J Am Diet Assoc 106:908–912
10. Rossner S, Ohlin A (1995) Pregnancy as a risk factor for obesity: lessons from the Stockholm Pregnancy and Weight Development Study. Obes Res 3 Suppl 2:267s–275s
11. Butte NF, Hopkinson JM (1998) Body composition changes during lactation are highly variable among women. J Nutr 128:381S–385S
12. Chou TW, Chan GM, Moyer-Mileur L (1999) Postpartum body composition changes in lactating and nonlactating primiparas. Nutrition 15:481–484
13. Dewey KG (2004) Impact of breastfeeding on maternal nutritional status. Adv Exp Med Biol 554:91–100
14. Larson-Meyer DE (2002) Effect of postpartum exercise on mothers and their offspring: a review of the literature. Obes Res 10:841–853
15. Lovelady CA, Garner KE, Moreno KL, Williams JP (2000) The effect of weight loss in overweight, lactating women on the growth of their infants. N Engl J Med 342:4494–53
16. McCrory MA, Nommsen-Rivers LA, Mole PA, Lonnerdal B, Dewey KG (1999) Randomized trial of the short-term effects of dieting compared with dieting plus aerobic exercise on lactation performance. Am J Clin Nutr 69:959–967
17. Dewey KG, McCrory MA (1994) Effects of dieting and physical activity on pregnancy and lactation. Am J Clin Nutr 59:446S-452S; discussion 452S–453S
18. ACOG Committee opinion (2002) Number 267, January 2002: Exercise during pregnancy and the postpartum period. Obstet Gynecol 99:171–173
19. United States Department of Agriculture (2007) MyPyramid. Available via http://www.mypyramid.gov/index.html

20. Health Canada (2007) Canada's Food Guide. Available at: http://www.hc-sc.gc.ca/fn-an/food-guide-aliment/index_e.html
21. Institute of Medicine (1997) Dietary Reference Intakes for calcium, phosphorus, magnesium, vitamin D and fluoride. National Academy Press, Washington, D.C.
22. Prentice A (2003) Micronutrients and the bone mineral content of the mother, fetus and newborn. J Nutr 133:1693S–1699S
23. Kovacs CS, Kronenberg HM (1997) Maternal-fetal calcium and bone metabolism during pregnancy, puerperium, and lactation. Endocr Rev 18:832–872
24. Sowers M (1996) Pregnancy and lactation as risk factors for subsequent bone loss and osteoporosis. J Bone Miner Res 11:1052–1060
25. Cross NA, Hillman LS, Allen SH, Krause GF (1995) Changes in bone mineral density and markers of bone remodeling during lactation and postweaning in women consuming high amounts of calcium. J Bone Miner Res 10:1312–1320
26. Kalkwarf HJ, Specker BL, Bianchi DC, Ranz J, Ho M (1997) The effect of calcium supplementation on bone density during lactation and after weaning. N Engl J Med 337:523–528
27. Prentice A, Jarjou LM, Cole TJ, Stirling DM, Dibba B, Fairweather-Tait S (1995) Calcium requirements of lactating Gambian mothers: effects of a calcium supplement on breast-milk calcium concentration, maternal bone mineral content, and urinary calcium excretion. Am J Clin Nutr 62:58–67
28. Polatti F, Capuzzo E, Viazzo F, Colleoni R, Klersy C (1999) Bone mineral changes during and after lactation. Obstet Gynecol 94:52–56
29. Ritchie LD, Fung EB, Halloran BP, Turnlund JR, Van Loan MD, Cann CE, King JC (1998) A longitudinal study of calcium homeostasis during human pregnancy and lactation and after resumption of menses. Am J Clin Nutr 67:693–701
30. Kovacs CS (2001) Calcium and bone metabolism in pregnancy and lactation. J Clin Endocrinol Metab 86:2344–2348
31. Ervin RB, Wang CY, Wright JD, Kennedy-Stephenson J (2004) Dietary intake of selected minerals for the United States population: 1999–2000. Adv Data 341:1–5
32. Fulgoni V III, Nicholls J, Reed A, Buckley R, Kafer K, Huth P, DiRienzo D, Miller GD (2007) Dairy consumption and related nutrient intake in African-American adults and children in the United States: continuing survey of food intakes by individuals 1994–1996, 1998, and the National Health And Nutrition Examination Survey 1999–2000. J Am Diet Assoc 107:256–264
33. Nusser SM, Carriquiry AL, Dodd KW, Fuller WA. (1996) A semiparametric transformation approach to estimating usual daily intake distributions. J Am Stat Assoc 91:1440–1449
34. Mackey AD, Picciano MF, Mitchell DC, Smiciklas-Wright H (1998) Self-selected diets of lactating women often fail to meet dietary recommendations. J Am Diet Assoc 98:297–302
35. Nicklas TA (2003) Calcium intake trends and health consequences from childhood through adulthood. J Am Coll Nutr 22:340–356
36. Heaney RP (2005) Measuring calcium absorption Am J Clin Nutr 81: author reply 1451–1452
37. Heaney RP, Rafferty K, Bierman J (2005) Not all calcium-fortified beverages are equal. Nutr Today 40:39–44
38. Miller GD, Jarvis JK, McBean LD (2001) The importance of meeting calcium needs with foods. J Am Coll Nutr 20:168S–185S
39. Straub DA (2007) Calcium supplementation in clinical practice: a review of forms, doses, and indications. Nutr Clin Pract 22:286–296
40. Heaney RP (1991) Calcium supplements: practical considerations. Osteoporos Int 1:65–71
41. Haddad JG, Matsuoka LY, Hollis BW, Hu YZ, Wortsman J (1993) Human plasma transport of vitamin D after its endogenous synthesis. J Clin Invest 91:2552–2555
42. Matsuoka LY, Wortsman J, Haddad JG, Hollis BW (1989) In vivo threshold for cutaneous synthesis of vitamin D_3. J Lab Clin Med 114:301–305
43. Holick MF (2004) Sunlight and vitamin D for bone health and prevention of autoimmune diseases, cancers, and cardiovascular disease. Am J Clin Nutr 80:1678S–1688S
44. Lammi-Keefe CJ (1995) Vitamin D and E in human milk. In: Jensen RG, ed. Handbook of milk composition. Academic Press, San Diego, Calif., pp 706–717

45. American Academy of Pediatrics (1999) Committee on Environmental Health. Ultraviolet light: a hazard to children. Pediatrics 104:328–333

46. Kalkwarf HJ, Specker BL (2004) Vitamin D Metabolism in Pregnancy and Lactation. In: Pike JW, Glorieux F, Feldman D (eds) Vitamin D. Academic, San Diego, Calif., pp 839–850

47. Health Canada (2004) Vitamin D Supplementation for Breastfed Infants. 2004 Health Canada Recommendation. Available via www.healthcanada.ca/nutrition

48. Gartner LM, Greer FR (2003) Prevention of rickets and vitamin D deficiency: new guidelines for vitamin D intake. Pediatrics 111:908–910

49. Hollis BW, Wagner CL (2004) Vitamin D requirements during lactation: high-dose maternal supplementation as therapy to prevent hypovitaminosis D for both the mother and the nursing infant. Am J Clin Nutr 80:1752S–1758S

50. Ala-Houhala M (1985) 25-Hydroxyvitamin D levels during breast-feeding with or without maternal or infantile supplementation of vitamin D. J Pediatr Gastroenterol Nutr 4:220–226

51. Ala-Houhala M, Koskinen T, Terho A, Koivula T, Visakorpi J (1986) Maternal compared with infant vitamin D supplementation. Arch Dis Child 61:1159–1163

52. Vieth R, Bischoff-Ferrari H, Boucher BJ et al (2007) The urgent need to recommend an intake of vitamin D that is effective. Am J Clin Nutr 85:649–650

53. Barger-Lux MJ, Heaney RP, Dowell S, Chen TC, Holick MF (1998) Vitamin D and its major metabolites: serum levels after graded oral dosing in healthy men. Osteoporos Int 8:222–230

54. Hathcock JN, Shao A, Vieth R, Heaney R (2007) Risk assessment for vitamin D. Am J Clin Nutr 85:6–18

55. Arunabh S, Pollack S, Yeh J, Aloia JF (2003) Body fat content and 25-hydroxyvitamin D levels in healthy women. J Clin Endocrinol Metab 88:157–161

56. Buffington C, Walker B, Cowan GS Jr, Scruggs D (1993) Vitamin D Deficiency in the Morbidly Obese. Obes Surg 3:421–424

57. Liel Y, Ulmer E, Shary J, Hollis BW, Bell NH (1988) Low circulating vitamin D in obesity. Calcif Tissue Int 43:199–201

58. Wortsman J, Matsuoka LY, Chen TC, Lu Z, Holick MF (2000) Decreased bioavailability of vitamin D in obesity. Am J Clin Nutr 72:690–693

59. Trang HM, Cole DE, Rubin LA, Pierratos A, Siu S, Vieth R (1998) Evidence that vitamin D3 increases serum 25-hydroxyvitamin D more efficiently than does vitamin D2. Am J Clin Nutr 68:854–858

60. Kelly P, McPartlin J, Goggins M, Weir DG, Scott JM (1997) Unmetabolized folic acid in serum: acute studies in subjects consuming fortified food and supplements. Am J Clin Nutr 65:1790–1795

61. Troen AM, Mitchell B, Sorensen B et al (2006) Unmetabolized folic acid in plasma is associated with reduced natural killer cell cytotoxicity among postmenopausal women. J Nutr 136:189–194

62. Institute of Medicine (1998) Dietary Reference Intakes for thiamin, riboflavin, niacin, vitamin B_6, folate, vitamin B_{12}, pantothenic acid, biotin and choline. National Academy Press, Washington, D.C.

63. O'Connor DL, Green T, Picciano MF (1997) Maternal folate status and lactation. J Mammary Gland Biol Neoplasia 2:279–289

64. Butte NF, Calloway DH, Van Duzen JL (1981) Nutritional assessment of pregnant and lactating Navajo women. Am J Clin Nutr 34:2216–2228

65. Keizer SE, Gibson RS, O'Connor DL (1995) Postpartum folic acid supplementation of adolescents: impact on maternal folate and zinc status and milk composition. Am J Clin Nutr 62:377–384

66. Sneed SM, Zane C, Thomas MR (1981) The effects of ascorbic acid, vitamin B6, vitamin B12, and folic acid supplementation on the breast milk and maternal nutritional status of low socioeconomic lactating women. Am J Clin Nutr 34:1338–1346

67. Tamura T, Yoshimura Y, Arakawa T (1980) Human milk folate and folate status in lactating mothers and their infants. Am J Clin Nutr 33:193–197

68. Houghton LA, Sherwood KL, Pawlosky R, Ito S, O'Connor DL (2006) [6S]-5-Methyltetrahydrofolate is at least as effective as folic acid in preventing a decline in blood folate concentrations during lactation. Am J Clin Nutr 83:842–850

69. Ray JG, Vermeulen MJ, Boss SC, Cole DE (2002) Increased red cell folate concentrations in women of reproductive age after Canadian folic acid food fortification. Epidemiology 13:238–240

70. Sherwood KL, Houghton LA, Tarasuk V, O'Connor DL (2006) One-third of pregnant and lactating women may not be meeting their folate requirements from diet alone based on mandated levels of folic acid fortification. J Nutr 136:2820–2826

71. Dietrich M, Brown CJ, Block G (2005) The effect of folate fortification of cereal-grain products on blood folate status, dietary folate intake, and dietary folate sources among adult non-supplement users in the United States. J Am Coll Nutr 24:266–274

72. Morris MS, Jacques PF, Rosenberg IH, Selhub J (2007) Folate and vitamin B-12 status in relation to anemia, macrocytosis, and cognitive impairment in older Americans in the age of folic acid fortification. Am J Clin Nutr 85:193–200

73. Specker BL, Black A, Allen L, Morrow F (1990) Vitamin B-12: low milk concentrations are related to low serum concentrations in vegetarian women and to methylmalonic aciduria in their infants. Am J Clin Nutr 52:1073–106

74. Jones KM, Ramirez-Zea M, Zuleta C, Allen LH (2007) Prevalent vitamin B-12 deficiency in twelve-month-old Guatemalan infants is predicted by maternal B-12 deficiency and infant diet. J Nutr 137:1307–133

75. Stabler SP, Allen RH (2004) Vitamin B_{12} deficiency as a worldwide problem. Annu Rev Nutr 24: 299–326

76. Trugo NM, Sardinha F (1994) Cobalamin and cobalamin-binding capacity in human milk. Nutr Res 14:22–33

77. Donangelo CM, Trugo NM, Koury JC, Barreto Silva MI, Freitas LA, Feldheim W, Barth C (1989) Iron, zinc, folate and vitamin B12 nutritional status and milk composition of low-income Brazilian mothers. Eur J Clin Nutr 43:253–266

78. House JD, March SB, Ratnam S, Ives E, Brosnan JT, Friel JK (2000) Folate and vitamin B12 status of women in Newfoundland at their first prenatal visit. CMAJ 162:1557–1559

79. Koebnick C, Hoffmann I, Dagnelie PC, Heins UA, Wickramasinghe SN, Ratnayaka ID, Gruendel S, Lindemans J, Leitzmann C (2004) Long-term ovo-lacto vegetarian diet impairs vitamin B-12 status in pregnant women. J Nutr 134:3319–3326

80. Brody T (1999) Nutritional biochemistry, 2nd edn. Academic, San Diego, Calif.

81. Institute of Medicine (2001) Dietary Reference Intakes for vitamin A, vitamin K, arsenic, boron, chromium, copper, iodine, iron, manganese, molybdenum, nickel, silicon, vanadium and zinc. National Academy Press, Washington, D.C.

82. Akesson A, Bjellerup P, Berglund M, Bremme K, Vahter M (2002) Soluble transferrin receptor: longitudinal assessment from pregnancy to postlactation. Obstet Gynecol 99:260–266

83. Beard JL, Dawson H, Pinero DJ (1996) Iron metabolism: a comprehensive review. Nutr Rev 54:295–317

84. Dijkhuizen MA, Wieringa FT, West CE, Muherdiyantiningsih, Muhilal (2001) Concurrent micronutrient deficiencies in lactating mothers and their infants in Indonesia. Am J Clin Nutr 73:786–791

85. Ettyang GA, van Marken Lichtenbelt WD, Oloo A, Saris WH (2003) Serum retinol, iron status and body composition of lactating women in Nandi, Kenya. Ann Nutr Metab 47:276–283

86. Haidar J, Muroki NM, Omwega AM, Ayana G (2003) Malnutrition and iron deficiency in lactating women in urban slum communities from Addis Ababa, Ethiopia. East Afr Med J 80:191–194

87. Takimoto H, Yoshiike N, Katagiri A, Ishida H, Abe S (2003) Nutritional status of pregnant and lactating women in Japan: a comparison with nonpregnant/nonlactating controls in the National Nutrition Survey. J Obstet Gynaecol Res 29:96–103

88. Villalpando S, Latulippe ME, Rosas G, Irurita MJ, Picciano MF, O'Connor DL (2003) Milk folate but not milk iron concentrations may be inadequate for some infants in a rural farming community in San Mateo, Capulhuac, Mexico. Am J Clin Nutr 78:782–789

89. Beard JL, Hendricks MK, Perez EM, Murray-Kolb LE, Berg A, Vernon-Feagans L, Irlam J, Isaacs W, Sive A, Tomlinson M (2005) Maternal iron deficiency anemia affects postpartum emotions and cognition. J Nutr 135:267–272

90. Corwin EJ, Murray-Kolb LE, Beard JL (2003) Low hemoglobin level is a risk factor for postpartum depression. J Nutr 133:4139–4142

91. Perez EM, Hendricks MK, Beard JL, Murray-Kolb LE, Berg A, Tomlinson M, Irlam J, Isaacs W, Njengele T, Sive A, Vernon-Feagans L (2005) Mother-infant interactions and infant development are altered by maternal iron deficiency anemia. J Nutr 135:850–855

92. Bodnar LM, Cogswell ME, Scanlon KS (2002) Low income postpartum women are at risk of iron deficiency. J Nutr 132:2298–2302
93. Colombo J, Kannass KN, Shaddy DJ et al (2004) Maternal DHA and the development of attention in infancy and toddlerhood. Child Dev 75:1254–1267
94. Jensen CL, Voigt RG, Prager TC et al (2005) Effects of maternal docosahexaenoic acid intake on visual function and neurodevelopment in breastfed term infants. Am J Clin Nutr 82:125–132
95. Lauritzen L, Jorgensen MH, Mikkelsen TB et al (2004) Maternal fish oil supplementation in lactation: effect on visual acuity and n-3 fatty acid content of infant erythrocytes. Lipids 39:195–206
96. Raiten DJ, Talbot JM, Waters JH (1998) Assessment of nutrient requirements for infant formulas. J Nutr 128:i–iv, 2059S–2293S
97. Salem N Jr (1996), Wegher B, Mena P, Uauy R. Arachidonic and docosahexaenoic acids are biosynthesized from their 18-carbon precursors in human infants. Proc Natl Acad Sci U S A 93:49–54
98. Agostoni C, Trojan S, Bellu R, Riva E, Bruzzese MG, Giovannini M (1997) Developmental quotient at 24 months and fatty acid composition of diet in early infancy: a follow up study. Arch Dis Child 76:421–424
99. Birch EE, Garfield S, Castaneda Y, Hughbanks-Wheaton D, Uauy R, Hoffman D (2007) Visual acuity and cognitive outcomes at 4 years of age in a double-blind, randomized trial of long-chain polyunsaturated fatty acid-supplemented infant formula. Early Hum Dev 83:279–284
100. Birch EE, Garfield S, Hoffman DR, Uauy R, Birch DG (2000) A randomized controlled trial of early dietary supply of long-chain polyunsaturated fatty acids and mental development in term infants. Dev Med Child Neurol 42:174–181
101. Birch EE, Hoffman DR, Castaneda YS, Fawcett SL, Birch DG, Uauy RD (2002) A randomized controlled trial of long-chain polyunsaturated fatty acid supplementation of formula in term infants after weaning at 6 wk of age. Am J Clin Nutr 75:570–580
102. Birch EE, Hoffman DR, Uauy R, Birch DG, Prestidge C (1998) Visual acuity and the essentiality of docosahexaenoic acid and arachidonic acid in the diet of term infants. Pediatr Res 44:201–209
103. Carlson SE, Ford AJ, Werkman SH, Peeples JM, Koo WW (1996) Visual acuity and fatty acid status of term infants fed human milk and formulas with and without docosahexaenoate and arachidonate from egg yolk lecithin. Pediatr Res 39:882–888
104. Makrides M, Neumann M, Simmer K, Pater J, Gibson R (1995) Are long-chain polyunsaturated fatty acids essential nutrients in infancy? Lancet 345:1463–1468
105. Willatts P, Forsyth JS, DiModugno MK, Varma S, Colvin M (1998) Effect of long-chain polyunsaturated fatty acids in infant formula on problem solving at 10 months of age. Lancet 352:688–691
106. Auestad N, Halter R, Hall RT et al (2001) Growth and development in term infants fed long-chain polyunsaturated fatty acids: a double-masked, randomized, parallel, prospective, multivariate study. Pediatrics 108:372–381
107. Auestad N, Montalto MB, Hall RT et al (1997) Visual acuity, erythrocyte fatty acid composition, and growth in term infants fed formulas with long chain polyunsaturated fatty acids for one year. Ross Pediatric Lipid Study. Pediatr Res 41:1–10
108. Auestad N, Scott DT, Janowsky JS et al (2003) Visual, cognitive, and language assessments at 39 months: a follow-up study of children fed formulas containing long-chain polyunsaturated fatty acids to 1 year of age. Pediatrics 112:e177–e83
109. Lucas A, Stafford M, Morley R et al (1999) Efficacy and safety of long-chain polyunsaturated fatty acid supplementation of infant-formula milk: a randomised trial. Lancet 354:1948–1954
110. Makrides M, Neumann MA, Simmer K, Gibson RA (2000) A critical appraisal of the role of dietary long-chain polyunsaturated fatty acids on neural indices of term infants: a randomized, controlled trial. Pediatrics 105:32–38
111. Scott DT, Janowsky JS, Carroll RE, Taylor JA, Auestad N, Montalto MB (1998) Formula supplementation with long-chain polyunsaturated fatty acids: are there developmental benefits? Pediatrics 102:E59
112. Brenna JT, Varamini B, Jensen RG, Diersen-Schade DA, Boettcher JA, Arterburn LM (2007) Docosahexaenoic and arachidonic acid concentrations in human breast milk worldwide. Am J Clin Nutr 85:1457–1464
113. Innis SM (1992) Human milk and formula fatty acids. J Pediatr 120:S56–61
114. Innis SM, Elias SL (2003) Intakes of essential n-6 and n-3 polyunsaturated fatty acids among pregnant Canadian women. Am J Clin Nutr 77:473–478

115. Henderson RA, Jensen RG, Lammi-Keefe CJ, Ferris AM, Dardick KR (1992) Effect of fish oil on the fatty acid composition of human milk and maternal and infant erythrocytes. Lipids 27:863–869

116. Jensen CL, Maude M, Anderson RE, Heird WC (2000) Effect of docosahexaenoic acid supplementation of lactating women on the fatty acid composition of breast milk lipids and maternal and infant plasma phospholipids. Am J Clin Nutr 71:292S–299S

117. Francois CA, Connor SL, Bolewicz LC, Connor WE (2003) Supplementing lactating women with flaxseed oil does not increase docosahexaenoic acid in their milk. Am J Clin Nutr 77:226–233

118. Jensen CL (2006) Effects of n-3 fatty acids during pregnancy and lactation. Am J Clin Nutr 83:1452S–1457S

119. Helland IB, Smith L, Saarem K, Saugstad OD, Drevon CA (2003) Maternal supplementation with very-long-chain n-3 fatty acids during pregnancy and lactation augments children's IQ at 4 years of age. Pediatrics 111:e39–e44

120. Innis SM (2003) Perinatal biochemistry and physiology of long-chain polyunsaturated fatty acids. J Pediatr 143:S1–S8

121. Innis SM, Gilley J, Werker J (2001) Are human milk long-chain polyunsaturated fatty acids related to visual and neural development in breast-fed term infants? J Pediatr 139:532–538

122. Williams C, Birch EE, Emmett PM, Northstone K (2001) Stereoacuity at age 3.5 y in children born full-term is associated with prenatal and postnatal dietary factors: a report from a population-based cohort study. Am J Clin Nutr 73:316–322

123. Judge MP, Harel O, Lammi-Keefe CJ (2007) A docosahexaenoic acid-functional food during pregnancy benefits infant visual acuity at four but not six months of age. Lipids 42:117–122

124. Judge MP, Harel O, Lammi-Keefe CJ (2007) Maternal consumption of a docosahexaenoic acid-containing functional food during pregnancy: benefit for infant performance on problem-solving but not on recognition memory tasks at age 9 mo. Am J Clin Nutr 85:1572–1577

125. Simopoulos AP, Leaf A, Salem N Jr (1999) Workshop on the Essentiality of and Recommended Dietary Intakes for Omega-6 and Omega-3 Fatty Acids. J Am Coll Nutr 18:487–489

126. Institute of Medicine (2007) Seafood choices: balancing benefits and Risks. National Academy Press, Washington, D.C.

19 Postpartum Depression and the Role of Nutritional Factors

Michelle Price Judge and Cheryl Tatano Beck

Summary Postpartum depression is the number one complication of childbirth [1], and healthcare providers need to have a keen understanding of the disorder in order to provide support and advice. In the first portion of this chapter, the prevalence and onset of postpartum depression is discussed, with a consideration for risk factors that have been associated with the disorder. Within this context, there is a discussion of how postpartum depression affects the mother, the mother–infant relationship, and infant development. Detection of postpartum depression is key to treatment, making repeated screening throughout the first year postpartum highly important.

Nutrition plays an integral and complex role in the brain. Nutrients provide structural substrates and serve as cofactors in many biological reactions. There are wide varieties of nutrients that are attributed to having a role in normal function. To name a few, the macronutrients, B vitamins and some trace minerals have been noted as factors integral to central nervous system (CNS) function. In order to describe the important role of nutrients in mental health it is important to first discuss general principles of CNS anatomy and physiology including nerve impulse conduction, neuroanatomy relating to mood and emotions, and the role of neurotransmitters in the brain. The latter section of this chapter outlines the potential role of key nutrients in postpartum depression with practical nutritional strategies for the postpartum period. This chapter closes with a discussion of the current literature related to breastfeeding and postpartum depression.

Keywords: Postpartum depression, screening scale

19.1 INTRODUCTION

Postpartum depression is the most common complication of childbirth and it is a major public health problem [1]. Up to as many as 50% of all cases of this mood disorder go undetected [2]. One of the major challenges for clinicians in dealing with postpartum depression is early recognition. A striking characteristic of this devastating mood disorder is how covertly it is suffered by mothers. Another obstacle to recognizing postpartum depression is the failure of clinicians to question women about related symptoms after delivery [3]. Postpartum depression has been a term applied to a wide range of postpartum

From: *Nutrition and Health: Handbook of Nutrition and Pregnancy*
Edited by: C.J. Lammi-Keefe, S.C. Couch, E.H. Philipson © Humana Press, Totowa, NJ

emotional disorders. As such, this catchall phrase has resulted in women often being misdiagnosed.

Postpartum depression is a major depressive episode, which has a duration of at least 2 weeks. Women experience either depressed mood or a loss of interest or pleasure in activities. Women must experience at least four other symptoms from the following list: "changes in appetite or weight, sleep and psychomotor activity; decreased energy; feelings of worthlessness or guilt; difficulty thinking, concentrating, or making decisions; or recurrent thoughts of death or suicidal ideation, plans, or attempts" [4]. The *Diagnostic and Statistical Manual of Mental Disorders, 4th edition, Text Revision* (DSM-IV-TR) includes a postpartum onset specifier, which states that the onset of this disorder must occur within the first 4 weeks after delivery. Clinicians and researchers both attest to this time criterion as being much too limited. Postpartum depression can occur any time during the first 12 months after the birth of an infant.

19.2 PREVALENCE/INCIDENCE

The Agency for Healthcare Research and Quality (AHRQ) recently conducted a systematic review of studies on the prevalence and incidence of postpartum depression during the first 12 months after delivery [5]. During the postpartum period, the point prevalence of major and minor depressive episodes starts rising and is at its highest in the third month at 12.9%. During the fourth month through the seventh month postpartum, the prevalence decreases slightly to between 9.9 and 10.6%. [5]. When looking at the point prevalence for major depression alone, major depressive episodes peak at 2 months (5.7%) and 6 months (5.6%) after delivery.

Regarding period prevalence, the AHRQ report revealed that after delivery up to 19.2% of mothers have either major or minor depressive episodes during the first 3 months, with 7.1% having a major depressive episode. Incidence of a new episode of major or minor depression during the period of the first 3 months postpartum can be up to 14.5% of mothers, with 6.5% of these women experiencing major depressive episodes.

In a large population based study in Denmark, Munk-Olsen and colleagues [6] investigated first lifetime onset of psychiatric illness in 1,171 mothers over the first 12 months after their baby's birth. Prevalence of severe mental disorders through the first 3 months after delivery was reported to be 1.03 per 1000 births. Primiparous mothers had an elevated risk of hospital admission with any mental disorder through the first 3 months after birth, with the highest risk 10–19 days after delivery.

In *Listening to Mothers II*, a report of the second National US survey of women's childbearing experiences [7], up to 63% of new mothers reported experiencing some depressive symptomatology on the Postpartum Depression Screening Scale [8].

19.3 ONSET

Based on Kendall's et al. classic research [9], for the majority of women the onset of postpartum depression starts within the first 3 months after birth. In this epidemiological study, there was a definite peak in psychiatric admission rate in the first months after delivery.

Stowe and colleagues [10] reviewed 209 consecutive referrals to a mental health program for mothers with major postpartum depression. Sixty-six percent of the sample reported an early onset for this mood disorder (mean = 2.2 weeks) and 22% reported late onset (mean =13.3 weeks after delivery).

19.4 CAUSES

Postpartum depression is a complex phenomenon that includes interaction between biochemical, genetic, psychosocial, and situational life-stress factors. No clear consensus on the cause of postpartum depression currently exists. For example, conflicting reports have not supported any one hormonal etiology. A disruption of neurotransmitters in the brain has also been proposed as one biochemical cause of postpartum depression [11].

19.5 CULTURAL PERSPECTIVES OF POSTPARTUM DEPRESSION

Evidence is mounting that postpartum depression is a universal phenomenon. Oates et al. [12], for example, investigated women's experiences of postpartum depression in the following 11 countries: United States, United Kingdom, France, Italy, Sweden, Ireland, Japan, Australia, Switzerland, Portugal, and Uganda. This postpartum mood disorder, called "morbid unhappiness" in some countries, was reported to be a common experience after delivery. In some countries, it was not recognized as a mental illness with the specific label of postpartum depression. The symptoms of postpartum depression in all countries closely approximated the Western concept of the signs and symptoms of this mental illness.

Current international studies that included a formal diagnosis of postpartum depression have reported prevalence rates of 18.7% in Morocco [13], 34.7% in South Africa [14] 27.6% in Japan [15], and 40% in Costa Rica [16].

Examples of reported international rates of postpartum depressive symptomatology include 17.3% in China [17], 17.6% in Portugal [18], 20.7% in Malaysia [19], 22% in the United Arab Emirates [20], and 25.6% in Turkey [21].

Horowitz and colleagues [22] conducted focus groups with mothers between 2 and 4 months after delivery in nine different countries: the United States, Australia, Finland, Guyana, India, Italy, Korea, Sweden, and Taiwan. How mothers described their postpartum depressive symptoms was remarkably similar across these countries. Common cognitive symptoms reported included poor concentration, worry, and indecisiveness. The most frequently cited emotional symptoms were anger, irritability, depression, sadness, guilt, anxiety, loneliness, fear, inadequacy, and tearfulness.

19.6 POSTPARTUM DEPRESSION RISK FACTORS

Results of individual studies in which predictors of this crippling mood disorder were investigated have been summarized in four meta-analyses [23–25]. In Beck's [23] meta-analyses the following risk factors for postpartum depression were significant: prenatal depression, self-esteem, child care stress, life stress, social support, prenatal anxiety, maternity blues, marital satisfaction, history of previous depression, infant temperament, marital status, socioeconomic status, and unplanned/unwanted pregnancy. Prenatal depression was one of the strongest risk factors. O'Hara and Swain's [24] and Robertson's et al. [25] meta-analyses corroborated the predictors identified by Beck [23]. The strongest predictors of postpartum depression reported by O'Hara and Swain [24] were psychopathology history and psychologic disturbance during the prenatal period, poor marital relationship, low social support, and life stressors. In the most recent meta-analysis, Robertson et al. [25] also reported that the strongest risk factors for developing postpartum depression were prenatal depression, prenatal anxiety, stressful life events, low levels of social support, and a previous history of depression. In a recent study of 4,332 postpartum women, income level, occupational prestige, marital status, and number of children were significant risk factors for postpartum depression [26]. The strongest of these risk factors was income level. Financially poor women were at higher risk for postpartum depression than financially affluent women were.

Research is also revealing that other women at risk for postpartum depression are mothers who have preterm infants, multiple infants, or infants in the neonatal intensive care units [27, 28].

19.7 PHENOMENOLOGY OF POSTPARTUM DEPRESSION

Loss of control is the basic problem women grapple with when suffering from postpartum depression [29]. Mothers try to resolve this loss of control in a four-stage process as outlined in Fig. 19.1 [29, 30]. In the first stage, mothers are bombarded with horrifying anxiety, relentless obsessive thoughts, and difficulty concentrating. In the second stage, women feel that their normal selves are "gone." The women describe feeling "unreal," like they were just robots going through the motions caring for their infants. In this stage, women often isolate themselves and may begin to contemplate harming themselves. The third stage involves women strategizing ways to survive postpartum depression, including battling the health care system to get appropriate mental health treatment, prayer, and seeking solace in postpartum depression support groups. In the final stage, women finally regain control of their thoughts and emotions as their depression lifts. During this transition period, mothers describe having "good days" and "bad days"; however, when they wake up in the morning they never know what kind of a day it will be. As the postpartum depression lifts, mothers go through a mourning period

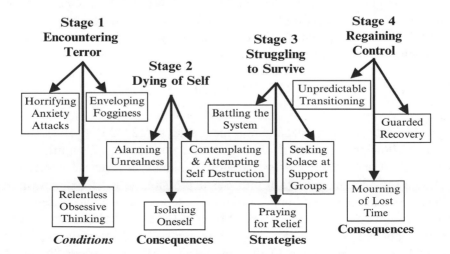

Fig. 19.1. The four-stage process of "teetering on the edge." (Reprinted with permission from [29])

where they grieve over their lost time with their infants, which had been stolen from them by their depression. When mothers finally recover, they feel fragile and vulnerable and so they call their recovery a "guarded recovery."

19.8   EFFECTS OF POSTPARTUM DEPRESSION ON MOTHER–INFANT INTERACTION

Research is confirming that postpartum depression negatively affects mother–infant interactions during the first year of life. Field's program of research has repeatedly reported that postpartum depression negatively affects maternal–infant interaction. Field [31], for example, reported a dysregulation profile for infants of mothers suffering from postpartum depression. Infants of postpartum depressed mothers have lower responsivity on the Brazelton scale, higher levels of indeterminate sleep, and elevated levels of norepinephrine and cortisol, activation of right frontal electroencephalogram, decreased responsivity to facial expressions, lower vagal tone, neurological delays, decreased play, decreased Bayley mental and motor scale scores, and lower weight percentiles [32]. Field [31] reported that postpartum depressed mothers displayed two predominant styles of interaction, withdrawn or intrusive. Mother–infant dyads matched negative behavior states more often and positive states less frequently than nondepressed mother–infant dyads [33].

Postpartum depressed mothers displayed significantly lower contingent responsiveness and higher negative contingent responsiveness to their infants [34]. Recently Paulson et al. [35] found that mothers depressed at 9 months after birth were 1.5 times more likely to engage in less positive enrichment activity with their child such as reading, singing songs, and telling stories. Forman et al. [36] reported that at 6 months postpartum depressed mothers were less responsive to the infants, experienced higher levels of parenting stress, and perceived their infants more negatively than nondepressed mothers. Since a mother constitutes the infant's primary social environment during the first months of life, the effects of postpartum depression on the rapidly developing baby is of great concern and merits closer scrutiny and study.

19.9   EFFECTS OF POSTPARTUM DEPRESSION ON CHILD DEVELOPMENT

Longitudinal research is revealing that there are long-term sequelae for children whose mothers suffered from postpartum depression. Toddlers of postpartum depressed mothers displayed more insecure attachment to their mothers than toddlers of nondepressed mothers [37].

Eighteen-month-olds whose mothers suffered from postpartum depression performed significantly poorer on Bayley scales for object concepts tasks [37] as compared with children the same age of nondepressed mothers. In a different study of 10-month-old boys, children of postpartum depressed mothers performed significantly poorer on Bayley Scales of Infant Development than did boys of nondepressed mothers [38].

Teachers rated kindergarteners of postpartum depressed mothers as displaying more internalizing problems, i.e., overanxious and depressed, plus more externalizing problems, i.e., defiant and aggressive, than children of nondepressed mothers [39].

In a longitudinal study, Hay and colleagues [40] found that 11-year-old children of postpartum depressed mothers displayed more violent behaviors than did children whose mothers had not experienced postpartum depression. Halligan et al. [41] recently published findings from their longitudinal study. Thirteen-year-olds whose mothers had suffered from postpartum depression were at an increased risk for depression if their mothers had later episodes of depression following the postpartum period. Anxiety disorders in these adolescents were increased in the group whose mothers had been postpartum depressed regardless of whether their mothers had suffered from subsequent depressive episodes.

19.10   SCREENING

Postpartum depression is treatable, but the women suffering must first be identified. Mothers may not seek help for postpartum depression due to any number of reasons including a lack of knowledge regarding this devastating illness and/or because of the tremendous stigma attached to mental illness. Also, women may fear that if they are diagnosed with postpartum depression, child welfare authorities may take their infants.

Women should be screened for postpartum depression periodically during the first year after delivery. The standard practice of screening just one time during the early postpartum period (i.e., at 6 weeks postpartum) may not detect postpartum depression that develops later. Because a woman is adjusting well during the early postpartum period does not mean she will not develop postpartum depression sometime later during the first 12 months after birth. Without repeated screenings, a mother may fall through the cracks in the health care system. Prior to screening women for postpartum depression, health care providers need to dispel the idealized myths of motherhood and provide a trusting environment in which women can feel free to discuss any negative feelings or thoughts they may be experiencing.

The Postpartum Depression Screening Scale (PDSS) is a survey available to clinicians for screening [8]. This self-report scale consists of 35 items that assess the presence, severity, and type of postpartum depressive symptoms. It has a five-point Likert response format in which women are asked to respond to statements about how they have been feeling since delivery. The response options range from 1 = strongly disagree, to 5 = strongly agree (Table 19.1). Agreement with a statement indicates the mother is experiencing that depressive symptom. The PDSS consists of seven symptoms content scales: Sleeping/Eating Disturbances, Loss of Self, Anxiety/Insecurity, Guilt/Shame, Emotional Lability, Mental Confusion, and Suicidal Thoughts. The range of possible scores is 35–175. A cutoff score of 80 or above indicates a positive screen for postpartum depression and the need to refer the mother for a formal diagnostic evaluation by a mental health clinician. Using this cutoff score of 80, Beck and Gable [8] reported the PDSS had a sensitivity of 94% and a specificity of 98%.

The Edinburgh Postnatal Depression Scale (EPDS) is a second instrument that has been developed to screen for depression [42]. It consists of ten items in a Likert format that assess the following common depressive symptoms: inability to laugh or look forward to things with enjoyment, feeling scared or panicky, feeling like "things have gotten on top of me," difficulty sleeping, and feeling sad. Using that cutoff, reported sensitivity (86%) and specificity (78%) have been reported by Cox et al. [42].

The Edinburgh Postnatal Depression Scale's 10 items assess depression in general. None of the items is written in the context of new motherhood. With a sample of 150

Table 19.1
Postpartum Depression Screening Scale: Selected Items by Dimension*

During the past 2 weeks I...
Sleeping/eating disturbances
    No. 1: I had trouble sleeping even when my baby was asleep
    No. 8: I lost my appetite
Loss of self
    No. 19: I did not know who I was anymore
    No. 5: I was afraid that I would never by my normal self again
Anxiety/insecurity
    No. 23: I felt all alone
    No. : I felt really overwhelmed
Guilt/shame
    No. 20: I felt guilty because I could not feel as much love for my baby
    as I should
    No. 27: I felt like I had to hide what I was thinking or feeling toward the
    baby
Emotional lability
    No. 3: I felt like my emotions were on a roller coaster
    No. 31: I felt full of anger ready to explode
Mental confusion
    No. 11: I could not concentrate on anything
    No. 4: I felt like I was losing my mind
Suicidal thoughts
    No. 14: I started thinking I would be better off dead
    No. 28: I felt that my baby would be better off without me

new mothers, when using the published recommended cutoff points for major depression, the PDSS achieved a sensitivity of 94% and specificity of 98% while the Edinburgh Postpartum Depression Scale's sensitivity was 78% and specificity was 99% [116].

Formal diagnosis of postpartum depression can be made by conducting a Structured Clinical Interview for DSM-IV Axis 1 Disorders (SCID) [43].

19.11 BRIEF OVERVIEW OF CENTRAL NERVOUS SYSTEM ANATOMY AND PHYSIOLOGY IN RELATION TO POSTPARTUM DEPRESSION

As discussed earlier in this chapter, the etiology of postpartum depression is currently unknown and likely to be multifactorial. With a focus on the brain specifically, it is important to consider how nutrients can affect the basic structure of the nerve cell and the surrounding neurochemical environment. Nerve cells or neurons are arranged in a highly organized fashion and collections of neurons comprise different functional areas of the brain including vision, hearing, memory, speech, emotions, and mood. A nutrient deficit can negatively influence the communication between nerve cells (i.e., neurotransmission) and have a collective negative effect on different functional areas of the brain including the regulation of mood and emotional response. The following sections outline the basic principles of neuroscience and review evidence of the role of key nutrients that may play a role in neurophysiological processes related to postpartum depression.

19.11.1   Nerve Impulse Conduction

The nerve cell or neuron is comprised of the cell body, dendrites, and an axon. The dendrites receive communications from other cells and have extensive branched projections that serve to maximize cellular signaling. Information received by the dendrites is sent to the body of the cell, where the nucleus of the cell passes information on to the axon. The axon of the cell also has branching at the terminal end, and it is at this end that the one nerve cell communicates with the next (Fig. 19.2). The point of communication between

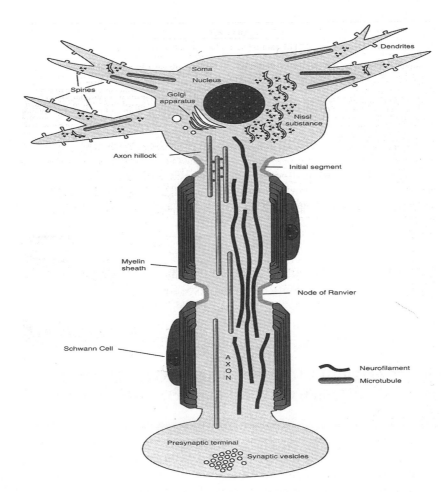

Fig. 19.2. Anatomy of a neuron [48]

the terminal end of one nerve cell and the dendrite projection of another is referred to as the synapse. Neuronal cell excitation and inhibition typically occur by direct or indirect mechanisms that involve the activation of ion channels that are regulated by membrane bound protein channels. For example, the activation of an ion channel, like chloride, results in an intercellular influx of ions with a negative charge resulting in inhibition of the neuron. Conversely, ion channels that allow the flow of positive ions typically result in excitation of the neuron [44]. Inhibition and excitation of neurons is stimulated via the release of chemicals called neurotransmitters (produced in the axon terminal) and it is these that have a major influence on brain activity. Impulse transmission is expedited via a phospholipid-based sheath that covers the neurons, the myelin sheath. The myelin acts as a barrier for signal transmission and causes a "hopping" of transmission signals between myelinated areas. This "hopping" effect, or saltatory conduction, speeds the rate of impulse transmission [44].

Nutrient deficits interfering with the release of important cellular signals or the formation of myelin could have a profoundly negative impact on cellular communication.

19.11.2   Neuroanatomy Relating to Mood and Emotions

The cerebral cortex is comprised of two distinct tissue layers, the gray and white matter. The cortex surrounds the entire perimeter of the brain, comprising 80% of the volume of the human brain [45]. In the gray matter, blood vessels are present as well as neuronal cell bodies. The white matter of the brain is composed of the nerve axons connected to the cell bodies in the gray matter. Different areas of the cortex are responsible for receiving and processing stimuli. The four functional areas of the cerebral cortex include motor, sensory, visual, and auditory (Fig. 19.3). The primary functional areas of the cerebral cortex process initial sensory information and the secondary and tertiary areas are responsible for association and processing. For example, visual information is

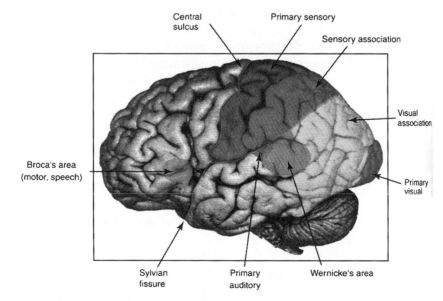

Fig. 19.3. Functional areas of the cortex [48]

sent to the primary visual cortex and processing related to assessment of shape, color and categorization occur in secondary and tertiary visual areas.

Newer evidence suggests that memory occurs in various regions of the cortex and is dependent upon the type of sensory stimulus. These regions of visual, motor, sensory, or visual cortex then communicate with central brain structures.

The central brain structures are generally referred to as the brainstem which is categorized into three main components: diencephalon, midbrain, and hindbrain. The diencephalon is comprised of three thalamic regions: epithalamus, thalamus, and hypothalamus. The thalamus is a very important relay station for the brain's cerebral cortex. The majority of sensory information is first sent to the thalamus before the information is relayed to the appropriate cortical area. The hypothalamus is involved in all aspects of motivated behavior or function including feeding, sexual, sleeping, emotional, temperature regulation, endocrine, and movement [45].

The midbrain consists of many structures. As examples, two important structures, the superior and inferior colliculi, are involved in visual and auditory processing. The superior colliculi receive projections from the retina and are involved in regulating behaviors related to visual stimuli. The inferior colliculi receive auditory information and regulate auditory-related behaviors [45]. For example, the sound of a phone ringing stimulates the auditory cortex, and it is the inferior colliculus that orients the individual to the phone and prompts the behavior to walk toward the phone.

Last, the forebrain extends from the brainstem and is made up largely of the limbic system and the basal ganglia. Studies using functional magnetic resonance imaging have demonstrated that the forebrain is highly specialized for working memory involved in new learning [46]. The central structures of the forebrain are commonly referred to as the limbic system composed of the hippocampus, septum, and cingulate gyrus. Memory, new learning, and emotions are functions attributed to the limbic system. Current research also points to the limbic system's central role in spatial behavior [45]. The basal ganglia include the putamen, globus pallidus, and caudate nucleus, and are involved in motor responses that require sequencing and smooth execution. These structures are also involved in habit learning and storage of older learned information.

From an anatomical standpoint, alterations in the function of the limbic system are most likely to cause disturbances in emotion and mood associated with postpartum depression.

19.11.3 Neurotransmitters and Brain Function Including Mood Regulation

The proper balance of neurotransmitters is important to maintaining normal brain chemistry and function. Behavior, mood, learning, and memory are just a few of the important functions requiring proper neurotransmission. Different cells have varying levels of the enzymes involved in the production of the neurotransmitters. The availability of different enzymes in the presynaptic membrane controls the type and quantity of neurotransmitter produced. Numerous neurotransmitters influence the complex chemical communicating systems in the brain. The monoamines, acetylcholine, amino acids, and peptides comprise the majority of the neurotransmitters [47].

Dopamine, norepinephrine, epinephrine, and serotonin are referred to as monoamines due to the existence of one amine group in their molecular structure. Dopamine, norepinephrine, and epinephrine comprise the catecholamine group of monoamines. Dopamine, norepinephrine, and epinephrine are produced via the same biochemical pathway referred to as the dopaminergic pathway [47]. In this pathway, tyrosine is converted to L-dopa, an intermediate in the biochemical pathway, via the enzyme tyrosine hydroxylase [47].

Serotonin is classified as an indolamine produced by the serotonergic pathway. In a manner similar to the dopaminergic pathway, the serotonergic pathway begins with an amino acid, in this case tryptophan [47]. The conversion of tryptophan to the next intermediary requires tryptophan hydroxylase. In both the dopaminergic and serotonergic pathways, tyrosine hydroxylase and tryptophan hydroxylase, respectively, are present in limited amounts. Therefore, it is the enzymes and not the availability of tyrosine or tryptophan that are the limiting factors for the production of these monoamines. Working memory involved in new learning has been attributed to dopamine in the prefrontal cortex (mesensyphalic region) [48]. Norepinephrine is thought to play an important role in stimulating processes involved in attention. Low dopamine and serotonin have been associated with depression and psychomotor slowing, which affects motivational processes. Drugs that alter the reuptake of these monoamines have been shown to elevate mood. The drug class called monoamine reuptake inhibitors (MAOI) are typically the first line of treatment in patients experiencing depression and postpartum depression [49, 50].

Acetylcholine is produced from acetyl CoA and choline and it is synthesized through the action of choline acetyl transferase in the cholinergic pathway. Deficits in the production of acetylcholine are associated with learning and long-term memory deficits [51]. Acetylcholine also appears to have a role in maintaining normal motor activity [51].

The amino acids involved in neurotransmission can have either excitatory or inhibitory actions on nerve signal transmission. Glutamate and aspartate are the main excitatory amino acids and gamma amino butyric acid (GABA) and glycine are main inhibitory amino acids. Long-term potentiation in the hippocampus related to memory has been shown to require a reduction in GABA [52].

19.12   ROLE OF NUTRITION IN POSTPARTUM DEPRESSION

19.12.1   Carbohydrate

Carbohydrates (i.e., bread, cereal, rice, potatoes, pasta, beans) play a vital role in delivering energy to the body and can influence mood. Carbohydrates are the brain's primary source of energy, making adequate dietary intake important to postpartum mental health. The delicate balance between carbohydrate and insulin can also affect mood. Balanced and consistent carbohydrate intake throughout the day can help ensure this balance between carbohydrate and insulin. Insulin increases markedly throughout the course of a normal pregnancy, and levels fall dramatically after delivery. Although the mechanism requires elucidation, it has been hypothesized that this drop in insulin levels following delivery may induce depression through a reduction in serotonin production [53]. Crowther et al. [54] demonstrated in a large randomized clinical trial that women with gestational diabetes mellitus (GDM) who received individualized dietary advice had lower rates of postpartum depression compared to those women with GDM receiving standard care.

19.12.2   Protein

Protein (meat, poultry, fish eggs, cheese, nuts, legumes) can play a role in mood regulation. During the postpartum period, women should ensure that they are consuming an adequate amounts of protein, especially as a full complement of the essential amino acids. While protein intake in pregnancies of women who are consuming a full range of foods is generally not a concern (see Chap. 1, "Nutrient Recommendations and Dietary Guidelines for Pregnant Women"), women who are depressed may not be eating normally. Women who are breastfeeding should pay particular attention to their protein intake, as protein needs are the same as during pregnancy. A full complement of all the essential amino acids will help ensure the synthesis of neurotransmitters. As reviewed in the section on neurotransmitters, the amino acids can have a direct impact on neurotransmission. Glutamate, aspartate, GABA, and glycine can be excitatory or inhibitory with respect to neurotransmission depending upon the amino acid of interest. The essential amino acid tryptophan stimulates the production of serotonin, which plays an important role in the regulation of anger, aggression, body temperature, mood, sleep, sexuality and appetite.

19.12.3   Fat/Omega-3 Fatty Acids: Docosahexaenoic Acid

Docosahexaenoic acid (DHA, 22:6n-3) has a central role in regulating the biophysical properties of neural membranes [1]. Based upon animal studies, specific regions of the brain, including the cerebral cortex, synapses, and retinal rod photoreceptors, have a particularly high DHA concentration [55–57]. Studies conducted in animals provide evidence for disturbances in brain development of offspring relating to DHA deficiency induced during the gestational period [52, 58–61]. In the United States and Canada, maternal intake of DHA, found in cold-water marine fish, is far below the current recommended level of 300 mg/day during pregnancy [62–66], which raises concern for maternal health and infant neurodevelopment as developmental advantages have been reported for infants of mothers who consumed DHA during pregnancy [67–69].

19.12.4   The Role of DHA in Neurotransmission

DHA deficiency during gestation in rats decreases dopamine production [61, 70] in the brain, which in turn induces behavioral disturbances resulting in decreased learning ability in their offspring. These findings are of particular interest because they link DHA deficiency, subsequent altered brain development, and impaired functional status. Conversely, the offspring of rats deficient in DHA during the gestational period exhibit an increase in acetylcholine [60] and GABA [52] production. Adequate levels of dopamine are necessary for mood elevation and for learning processes. Alternatively, acetylcholine and GABA appear to be increased in DHA deficiency. Learning, long-term memory and motor activity have been linked to acetylcholine, and it is unknown currently how such alterations affect mood and emotions. All of the observed alterations of the brain's neurochemical profile provide a compelling basis for further exploration regarding DHA deficiency and the CNS.

Although multiple reports have provided evidence of an inverse relationship between DHA intake and depression [reviewed in 71], few investigations have focused directly on postpartum depression specifically [72–75]. Hibbeln [72] conducted a meta-analysis of 23 countries that used the EPDS to screen for depression and reported that the DHA content of mother's milk and seafood consumption rates were associated with lower prevalence rates of postpartum depression.

Otto and colleagues [73] investigated plasma phospholipid DHA in 112 women at delivery and at 32 weeks postpartum. The EPDS was given to the women at the 32-week time point to assess postpartum depression. There was an inverse relationship between DHA status and depressive symptoms.

De Vriese et al. [74] conducted a similar investigation of maternal DHA status immediately following delivery. DHA and total n-3 fatty acids were significantly lower in the women who developed postpartum depression compared to the women who did not.

In a study of 865 Japanese women Miyake and colleagues [75] investigated risk of postpartum depression related to dietary fatty acid intake. Again, the EPDS was used to evaluate postpartum depression and diet history questionnaires were self-administered to measure dietary fatty acid intake. There were no significant relationships between dietary fish consumption or n-3 fatty acid intake and postpartum depression. Likewise, Browne et al. [76] investigated maternal fish consumption and plasma DHA status after birth in relation to postpartum depression diagnosed using the Composite International Diagnostic Interview. There were no associations between maternal fish consumption during pregnancy or maternal DHA status following delivery and depressive symptoms in the postpartum period.

In conclusion, although the evidence exists to support notion that DHA status may have a protective effect in the prevention of postpartum depression [72–74], further investigations are necessary to better define this relationship in light of some of the more recent conflicting reports [75, 76], that have failed to provide evidence to support this idea. To date, although DHA appears help prevent postpartum depression it has not been found to be beneficial in the treatment of postpartum depression [77, 78]. Future investigations should focus on dose–response relationships in the treatment of postpartum depression as well as expand investigations to include measures of baseline DHA status [71].

19.13 MICRONUTRIENTS AND POSTPARTUM DEPRESSION

19.13.1 Iron

Iron is a component of hemoglobin in red blood cells (RBC) and as such, an iron deficient diet can result in iron deficiency anemia characterized by the production of RBC that do not contain a full complement of hemoglobin and are inefficient at delivering oxygen to cells. Pregnancy and childbirth place women at risk for iron deficiency anemia due to a marked blood volume expansion during pregnancy, increased maternal needs, fetal requirements, and blood loss associated with childbirth. Iron deficiency anemia is the most common nutritional deficiency in the United States and the world, affecting 7.8 million adolescent girls and women of childbearing age [79]. The iron requirements of pregnancy are thoroughly discussed in Chap. 16, ("Iron requirements and Adverse Outcomes").

Although the role of iron status remains unclear with respect to postpartum depression, current investigations point to increased risk for postpartum depression in women who have anemia [80, 81]. Corwin and colleagues [80] measured hemoglobin levels at 7, 14, and 28 days postpartum and depressive symptoms using the Center for Epidemiological Studies-Depressive Symptomatology Scale (CES-D) at 28 days postpartum. Hemoglobin levels on day 7 were negatively correlated with CES-D scores obtained on day 28 postpartum. Further, Beard et al. [81] demonstrated that iron treatment resulted in a 25% improvement in previously iron deficient mothers' depression and stress scales. Anemic mothers who did not receive iron treatment did not display improvements in depression or stress scales. Beard further discusses these findings in Chap. 16 of this volume.

19.13.2   Folic Acid

Folate deficiency has been associated with problems in nerve development and function and has classically been thought to pose more of a risk to fetuses, infants, children who are growing; however, recent reports have linked folate deficiency with depression [87–91]. Neural tube defects have been associated with inadequate folate intake prior to conception and during pregnancy are described in Chap. 17, "Folate: a Key to Optimal Pregnancy Outcome"). Excellent sources of folate include liver, yeast, asparagus, spinach, oranges, legumes, and fortified cereals/grain products [79 and see Chap. 17]. During the postpartum period, folate needs are 400 mcg/day unless the woman is breast-feeding, which increases the need to 500 mcg/day [79].

Folate deficiency has been linked to depression in several investigations [82–86]. Although the mechanism is unknown at this time, folate has been hypothesized to be related to serotonin production involving S-adenosylmethionine (SAMe), a major methyl donor formed from methionine, which is formed during regeneration of homocysteine [87]. To date, no significant associations have been reported for folate and postpartum depression [88]. The current folate fortification programs make dietary folate deficiency a unlikely culprit in current rates for postpartum depression.

19.13.3   Riboflavin/Vitamin B$_2$

Riboflavin is important in the formation of key enzymes necessary for energy production via the citric acid cycle/electron transport chain. Based upon this role, adequate riboflavin intake during pregnancy and the postpartum period could be beneficial to maternal energy level and mood. Milk is the best source of riboflavin in the North American diet [79], with liver, red meat, poultry, fish, whole grains and enriched breads and cereals, asparagus, broccoli, mushrooms, and green leafy vegetables identified as other excellent sources. During the postpartum period, riboflavin needs are similar to those prior to pregnancy 1.1 mg/day [79]. The riboflavin requirement for breastfeeding women is 1.6 mg/day, greater than that for pregnant women.

To our knowledge, there is only one report for the role of riboflavin in postpartum depression. Miyake and colleagues [88] reported that pregnant women with riboflavin consumption in the third quartile were independently related to a decreased risk for postpartum depression. It has been hypothesized that riboflavin coenzymes are involved in the regeneration of homocysteine which is involved in serotonin production [89].

19.13.4   Vitamin B$_{12}$

Vitamin B$_{12}$ is necessary for the maintenance of myelin, which insulates nerves and affects neurotransmission [79]. Although dietary B$_{12}$ deficiencies are rare due to efficient recycling, strict vegetarians and individuals with decreased appetite/anorexia should consider supplementing their diet. Neurological symptoms associated with deficiency include numbness and tingling, abnormalities in gait, memory loss, and disorientation. Vitamin B$_{12}$ is found almost exclusively in animal products. Fortified cereal and grain products provide an alternative for those individuals who do not consume animal products. Although vitamin B$_{12}$ is important to CNS functions, no associations have been reported for vitamin B$_{12}$ and depression [90] or postpartum depression [88].

19.13.5   Pyridoxine/Vitamin B$_6$

Inadequate dietary intake of vitamin B$_6$ intake leads to decreased production of pyridoxal phosphate a key compound in the metabolism of all energy nutrients. This vitamin is also important for the synthesis of nonessential amino acids, which in turn effect the production of both important neurotransmitters, and the lipids that comprise myelin in nerve tissue [79]. Deficiency causes neurological symptoms including depression, headaches, confusion, and numbness and tingling in the extremities and seizures [79]. Vitamin B$_6$ is found in a variety of foods including chicken, fish, pork, organ meats, whole-wheat products, brown rice, soybeans, sunflower seeds, bananas, broccoli, and spinach [79]. During the postpartum period, women require 1.3 mg of B$_6$ daily if not breastfeeding. Breastfeeding women need 2 mg/day of B$_6$.

Few investigations have been conducted investigating the role of B$_6$ in depression; however, an inverse association between plasma B$_6$ levels and depressive symptoms has been reported [91]. With respect to postpartum depression, only one investigation has included the assessment of the association with B$_6$ and reported no measurable association with postpartum depression [88]. Further work is necessary to determine if a definitive relationship exists between B$_6$ status and postpartum depression.

19.14   BREASTFEEDING

Breastfeeding is known to be very beneficial to both mother and infant, and recent reports suggest that it may reduce risk for postpartum depression in women by reducing stress [92–97]. Although the majority of investigations point to breastfeeding as protective in postpartum depression, results are equivocal, as other investigations have reported no relationship between depressive symptoms and breastfeeding [98–100]. As earlier described, Infants of mothers with postpartum depression are at risk for cognitive and emotional impairments [37–41], and breastfeeding can help protect infants against these negative outcomes. Breastfed infants of depressed mothers exhibited decreased depressive symptoms compared to those who were bottle fed [101].

19.15   ANTIDEPRESSANTS AND BREASTFEEDING

The clinicians of breastfeeding women diagnosed with postpartum depression must consider the different treatment options for their patients including antidepressants, hormonal therapy, or psychotherapy. In situations where the postpartum depression requires antidepressants, the safety of the nursing infant must be considered. Antidepressants taken during breastfeeding can induce adverse symptoms in the infant. The antidepressants that have been particularly problematic are nefazodone [102], citalopram [103], doxepin [104, 105], and fluoxetine [106, 107]. Given the negative infant outcomes associated with maternal antidepressant therapies, the US Food and Drug Administration (FDA) has not approved any antidepressant for use during lactation [49]. Alternatively, depression during the postpartum period can impair maternal–infant interactions [108], which in turn negatively affect infant cognitive development [109], emotional development [109], anxiety, and self-esteem [110]. In some cases, the clinician may decide that the potential benefits of maternal antidepressant therapy outweigh the risks in which case paroxetine, sertraline, and nortriptyline should be considered for use first as these have been investigated and are reportedly without adverse infant-related outcomes [49].

19.16   EFFECT OF POSTPARTUM DEPRESSION ON BREASTFEEDING SUCCESS

Given the benefits of breastfeeding for both mother and infant, breastfeeding mothers with postpartum depression may benefit from this choice of feeding. However, the additional demands of breastfeeding could also be overwhelming for women experiencing postpartum depression, and care should be taken to support mothers deciding to formula feed. Those women who decide to breastfeed will likely need additional support to foster the continuation of breastfeeding during this difficult time. Although breastfeeding may reduce depressive symptoms during the postpartum period, mothers with depressive symptoms are more likely to discontinue breastfeeding [111–115]. Referrals to area lactation consultants and breastfeeding support groups such as La Leche League can be extremely helpful to mothers with PPD who are interested in continuing breastfeeding.

19.17   CONCLUSION

Postpartum depression is a treatable mood disorder that often goes undetected. The symptoms of postpartum depression can have a profound impact on maternal–infant bonding as well as other family dynamics. Infants of mothers with postpartum depression are more likely to have delays with cognitive and emotional development. Postpartum depression can occur any time throughout the first year postpartum, with the highest prevalence occurring during the first three postpartum months. The etiology of postpartum depression is unknown currently but is thought to be a complex problem including biochemical, genetic, psychosocial, and situational life-stress factors. Postpartum depression appears to transcend cultural background as it affects women throughout the world; however, it appears to affect more women of low socio-economic status compared with women who are more affluent. Although there are multiple risk factors that place women at risk for postpartum depression, prenatal depression is in the foreground as placing women at the highest risk. Women with postpartum depression suffer from an overwhelming feeling of loss of control. Screening is key to treating postpartum depression as many women go undetected. Screenings should be repeated throughout the first postpartum year as postpartum depression can develop at any point.

Nutrients play key roles in maintaining a healthy CNS, including serving as structural components of brain tissue, altering neurochemical properties of membranes involved in neurotransmission, as precursors in the formation of neurotransmitters, and serving directly as neurotransmitters. A well-balanced diet comprised of adequate carbohydrate, protein and fat will help ensure a steady source of fuel to the brain. Key nutrients including omega-3 fatty acids, iron, folate, riboflavin, and vitamin B_6 have been implicated in depression and/or postpartum depression, and care should be taken to ensure adequate intakes of these nutrients for women during pregnancy and the postpartum period.

Last, although breastfeeding may be beneficial to both mother and infant, care should be taken to support women in making the decision that is best for them. Women suffering from postpartum depression are at an increased risk for breastfeeding cessation and require additional support for long-term breastfeeding success.

REFERENCES

1. Wisner KL, Chambers C, Sit DKY (2006) Postpartum depression: a major public health problem. J Am Med Assoc 296:2612–2618
2. Hearn G et al (1998) Postnatal depression in the community. Br J Gen Pract 48:1064–1066
3. Nonacs R, Cohen L (1998) Postpartum mood disorders: Diagnosis and treatment guidelines Psychiatry 59 (Suppl):34–40
4. American Psychiatric Association (2000) Diagnostic and statistical manual of mental disorders, 4th edn. American Psychiatric Association, Washington, D.C.
5. Gavin NI, Gaynes BN, Lohr KN, Meltzer-Brody S, Gartlehner G, Swinson T (2005) Perinatal depression: A systematic review of prevalence and incidence. Obstet Gynecol 106:1071–1083
6. Munk-Olsen T, Laursen TM, Pedersen CB, Mors O, Mortensen PB (2006) New parents and mental disorders: a population-based register study. J Am Med Assoc 296:2582–2589
7. Declercq ER, Sakala C, Corry MP, Applebaum S (2006) Listening to mothers II: report of the second national U.S. survey of women's childbearing experiences. Childbirth Connection, New York, N.Y.
8. Beck CT, Gable RK (2002) Postpartum Depression Screening Scale Manual. Western Psychological Services, Los Angeles, Calif.
9. Kendall RE, Chalmers JC, Platz C (1987) Epidemiology of puerperal psychosis. Br J Psychiatry 150:662–672
10. Stowe ZN, Hostetter AL, Newport J (2005) The onset of postpartum depression: Implications for clinical screening in obstetrical and primary care. Am J Obstet Gynecol 192:522–6
11. Sichel DA, Driscoll JW (1999) Women's moods: what every woman must know about hormones, the brain, and emotional health. William Morrow, New York, N.Y.
12. Oates MR, Cox JL, Neena S, Asten P, Glangeud-Fredenthal N, Figuerido B et al (2004) Postnatal depression across countries and cultures: a qualitative study. Br J Psychiatry 184 (Suppl):510–516
13. Agoub M, Moussaoui, D Battas O (2005) Prevalence of postpartum depression in a Moroccan sample. Arch Womens Ment Health 8:37–43
14. Tomlinson M, Cooper PJ, Setin A, Swartz L, Moltero C (2006) Post-partum depression and infant growth in a South African peri-urban settlement. Child Care Health Dev 32:81–86
15. Ueda M, Yamashita H, Yshida K (2006) Impact of infant health problems on postnatal depression: Pilot study to evaluate a health visiting system. Psychiatry Clin Neurosci 60:182–19
16. Wolf AW, DeAndraca I, Lozoff, B (2002) Maternal depression in three Latin American samples. Soc Psychiatry Psychiatr Epidemiol 37:169–176
17. Xie RH, He G, Liu A, Bradwejn J, Walker M, Wen SW (2007) Fetal gender and postpartum depression in a cohort of Chinese women. Soc Sci Med doi:1016/j.socscimed.2007.04.003
18. Figueiredo B, Pacheco A, Costa R (2007) Depression during pregnancy and the postpartum period in adolescent and adult Portuguese mothers. Arch Womens Ment Health 10:103–109
19. Azidah AK, Shaiful BI, Rusli N, Jamil MY (2006) Postnatal depression and socio-cultural practices among postnatal mothers in Kota Bahru, Kelantan, Malaysia. Med J Malaysia 61:76–83
20. Green K, Broome H, Mirabella J (2006) Postnatal depression among mothers in the United Arab Emirates: socio-cultural and physical factors. Psychol Health Medicine 11:425–431
21. Dindar I, Erdogan S (2007) Screening of Turkish women for postpartum depression within the first postpartum year: The risk profile of a community sample. Public Health Nurs 24:176–183
22. Horowitz JA, Chang SS, Das S, Hayes B (2001) Women's perceptions of postpartum depressive symptoms from an international perspective. Int Nurs Perspect 1:5–14
23. Beck CT (2001) Predictors of postpartum depression: an update. Nurs Res 50:275–285
24. O'Hara M, Swain A (1996) Rates and risk of postpartum depression: a meta-analysis Int Rev Psychiatry 8:37–45
25. Robertson E, Grace S, Wallington T, Stewart D (2004) Antenatal risk factors for postpartum depression: a synthesis of recent literature. Gen Hosp Psychiatry 26:289–295
26. Segre LS, O'Hara MW, Arndt S, Stuart S (2007) The prevalence of postpartum depression: The relative significance of three social status indices. Soc Psychiatry Psychiatr Epidemiol 42:316–321
27. Beck CT (2003) Recognizing and screening for postpartum depression in mothers of NICU infants. Adv Neonatal Care 3:37–46

28. Maloni JA, Margevicius SP, Damato EG (2006) Multiple gestation: side effects of antepartum bed rest. Biol Res Nurs 8:115–128
29. Beck CT (1993) Teetering on the edge: a substantive theory of postpartum depression. Nurs Res 42:42–8
30. Beck CT (2007) Exemplar: teetering on the edge: a continually emerging theory of postpartum depression. In: Munhall PL (ed) Nursing research: a qualitative perspective, 4th edn. Jones and Bartlett, Sudbury, Mass., pp 273–92
31. Field T (1998) Maternal depression effects on infants and early interventions. Prevent Med 27:200–203
32. Field T, Fox N, Pickens J, Nawrocki T, Soutullo P (1995) Right frontal EEG activation in 3 to-6-month old infants of "depressed" mothers. Dev Psychol 31:358–363
33. Field T, Healy B, Goldstein S, Guthertz, M (1990) Behavior state matching in mother–infant interactions of nondepressed versus depressed mother–infant dyads. Dev Psychol 26:7–14
34. Stanley C, Murray L, Stern A (2004) The effect of postnatal depression on mother–infant interaction, infant response to the still-face perturbation and performance on an instrumental learning task. Dev Psychopathol 16:1–18
35. Paulson JF, Dauber S, Leiferman JA (2006) Individual and combined effects of postpartum depression in mothers and fathers and parenting behavior. Pediatrics 118:659–668
36. Forman DR, O'Hara MW, Stuart S, Gorman LL, Larsen KE, Coy, KC (2007) Effective treatment for postpartum depression is not sufficient to improve the developing mother–child relationship. Dev Psychopathol 19:585–602
37. Righetti-Veltema M, Bousquet A, Manzano J (2003) Impact of postpartum depressive symptoms on mother and her 18-monthh old infant. Eur Child Adolesc Psychiatry 12:75–83
38. Murray L, Fiori-Cowley A, Hooper R, Cooper P (1996) The impact of postnatal depression and associated adversity on early mother–infant interaction and later infant-outcome. Child Dev 67:2512–2526
39. Essex MJ, Klein MH, Miech R, Smider NA (2001) Timing of initial exposure on maternal major depression and children's mental health symptoms in kindergarten. Br J Psychiatry 179:151–156
40. Hay DF, Pawlby S, Angold A, Harold GT Sharp D (2003) Pathways to violence in the children of mothers who were depressed postpartum. Dev Psychol 39:1083–1094
41. Halligan SL, Murray L, Martins C, Cooper PJ (2007) Maternal depression and psychiatric outcomes in adolescent offspring: a 13-year longitudinal study. J Affect Disord 97:145–154
42. Cox JL, Holden JM, Sagovsky R (1987) Detection of postnatal depression: development of the 10-item Edinburgh Postnatal Depression Scale. Br J Psychiatry 150:782–786
43. First MB, Spitzer RL, Gibbon M, Williams JB (1997) User's guide for the structured clinical interview for DSM-IV axis 1 disorders. American Psychiatric Press, Washington, D.C.
44. Kolb B, Whishaw IQ (2003) The structure and electrical activity of neurons. In: Atkinson RC, Lindzey G, Thompson RF (ed) Fundamentals of human neuropsychology, 5th edn. Worth, New York, N.Y., pp 75–98
45. Kolb B, Whishaw IQ (2003) Organization of the nervous system. In: Atkinson RC, Lindzey G, Thompson RF (ed) Fundamentals of human neuropsychology, 5th edn. Worth, New York, N.Y., pp 46–74
46. Courtney SM, Petit L, Maisog JM, Ungerleider LG, Haxby JV (1998) An area specialized for spatial working memory in the human frontal cortex. Science 279:1347–1351
47. Kolb B, Whishaw IQ (2003) Communication between neurons. In: Atkinson RC, Lindzey G, Thompson RF (ed) Fundamentals of human neuropsychology, 5th edn. Worth, New York, N.Y., pp 99–133
48. Zigmond MJ, Bloom FE, Landis SC, Roberts JL, Squire LR (1999) Fundamental neuroscience. Academic, San Diego, Calif.
49. Gjerdingen D (2003) The effectiveness of various postpartum depression treatments and the impact of antidepressant drugs on nursing infants. JABFP 15:372–382
50. Whitby DH, Smith KM (2005) The use of tricyclic antidepressants and selective serotonin reuptake inhibitors in women who are breastfeeding. Pharmacotherapy 25:411–425
51. Groff JL, Gropper SS, Hunt SM (1995) Advanced nutrition and human metabolism, 3rd edn. West, St. Paul, Minn.
52. Hamano H, Nabekura J, Nishikawa M, Ogawa T (1996) Docosahexaenoic acid reduces GABA response in substantia nigra neuron of rat. J Neurophysiol 75:1264–1270
53. Chen TH, Lan TH, Yang CY, Juang KD (2006) Postpartum mood disorders may be related to a decreased insulin level after delivery. Med Hypotheses 66:820–823

54. Crowther CA, Hiller JE, Moss JR, McPhee AJ, Jeffries WS, Robinson JS (2005) Effect of treatment of gestational diabetes mellitus on pregnancy outcomes. N Engl J Med 352:2477–2486
55. Bowen RA, Clandinin MT (2002) Dietary low linolenic acid compared with docosahexaenoic acid alter synaptic plasma membrane phospholipid fatty acid composition and sodium-potassium ATPase kinetics in developing rats. J Neurochem 83:764–774
56. Bazan NG, Scott BL (1990) Dietary omega-3 fatty acids and accumulation of docosahexaenoic acid in rod photoreceptor cells of the retina and at synapses. J Med Sci 48:97–107
57. Sarkadi-Nagy E, Wijendran V, Diau GY, Chao AC, Hsieh AT, Turpeinen A, Nathanielsz PW, Brenna JT (2003) The influence of prematurity and long chain polyunsaturated supplementation in 4-week adjusted baboon neonate brain and related tissues. Pediatr Res 54:244–252
58. Auestad N, Innis SM (2000) Dietary *n*-3 fatty acid restriction during gestation in rats: neuronal cell body and growth-cone fatty acids. Am J Clin Nutr 71:312S–314S
59. Innis SM, Owens SD (2001) Dietary fatty acid composition in pregnancy alters neurite membrane fatty acids and dopamine in newborn rat brain. J Nutr 131:118–122
60. Aid S, Vancassel S, Poumes-Ballihaut C, Chalon S, Guesnet P, Lavialle M (2003) Effect of diet-induced *n*-3 PUFA depletion on cholinergic parameters in the rat hippocampus. J Lipid Res 44:1545–1551
61. Levant B, Radal JD, Carlson SE (2004) Decreased brain docosahexaenoic acid during development alters dopamine-related behaviors in adult rats that are differentially affected by dietary remediation. Behav Brain Res 152:49–57
62. Lewis NM, Widga AC, Buck JS, Frederick AM (1995) Survey of omega-3 fatty acids in diets of Midwest low-income pregnant women. J Agromed 2:49–56
63. Judge MP, Loosemore ED, DeMare CI, Keplinger MR, Mutungi G, Cote S, Ryan M, Ibarolla B, Lammi-Keefe CJ (2003) Dietary docosahexaenoic acid (DHA) intake in pregnant women. J Am Diet Assoc 103(9-S)
64. Loosemore ED, Judge MP, Lammi-Keefe CJ (2004) Dietary intake of essential and long-chain polyunsaturated fatty acids in pregnancy. Lipids, 39(5): 421–424.
65. Innis SM, Elias SL (2003) Intakes of essential *n*-6 and *n*-3 polyunsaturated fatty acids among pregnant Canadian women. Am J Clin Nutr 77:473–478
66. Denomme J, Stark KD, Holub, BJ (2005) Directly quantitated dietary (*n*-3) fatty acid intakes of pregnant Canadian women are lower than current dietary recommendations. J Nutr 135:206–211
67. Colombo J, Kannass KN, Shaddy DJ, Kundurthi S, Maikranz JM, Anderson CJ, Blaga OM, Carlson SE (2004) Maternal DHA and the development of attention in infancy and toddlerhood. Child Dev 75:1254–1267
68. Judge MP, Harel O, Lammi-Keefe CJ (2007) A docosahexaenoic acid-functional food during pregnancy benefits infant visual acuity at four but not six months of age. Lipids 42:117–122
69. Judge MP, Harel O, Lammi-Keefe CJ (2007) Maternal consumption of a DHA-functional food during pregnancy: comparison of infant performance on problem-solving and recognition memory tasks at 9 months of age. Am J Clin Nutr 85:1572–1577
70. Takeuchi T, Futumoto Y, Harada E (2002) Influence of a dietary *n*-3 fatty acid deficiency on the cerebral catecholamine contents, EEG and learning ability in rat. Behav Brain Res 131:193–203
71. Sontrop J, Campbell MK (2006) *n*-3 polyunsaturated fatty acids and depression: a review of the evidence and a methodological critique. Prev Med 42:4–13
72. Hibbeln JR (2002) Seafood consumption, the DHA content of mothers' milk and prevalence rates of postpartum depression: a cross-national, ecological analysis. J Affect Disord 69:15–29
73. Otto SJ, de Groot RH, Hornstra G (2003) Increased risk of postpartum depressive symptoms is associated with slower normalization after pregnancy of the functional docosahexaenioc acid status. Prostaglandins Leukot Essent Fatty Acids 69:237–243
74. De Vriese SR, Christophe AB, Maes M (2003) Lowered serum *n*-3 polyunsaturated fatty acid (PUFA) levels predict the occurrence of postpartum depression: Further evidence that lowered *n*-PUFAs are related to major depression. Life Sci 73:3181–3187
75. Miyake Y, Sasaki S, Yokoyama T, Tanaka K, Ohya Y, Fukushima W, Saito K, Ohfuji S, Kiyohara C, Hirota Y (2006) Risk of postpartum depression in relation to dietary fish and fat intake in Japan: the Osaka Maternal and Child Health Study. Psychol Med 36:1727–1735

76. Browne JC, Scott KM, Silvers KM (2006) Fish consumption in pregnancy and omega-3 status after birth are not associated with postnatal depression. J Affect Disord 90:131–139

77. Peet M, Stokes C (2005) Omega-3 fatty acids in the treatment of psychiatric disorders. Drugs 65:1051–1059

78. Freeman MP, Hibbeln JR, Wisner KL, Brumbach BH, Watchman M, Gelenberg AJ (2006) Randomized dose-ranging pilot trial of omega-3 fatty acids for postpartum depression. Acta Psychiatr Scand 113:31–35

79. Smolin L, Grosvenor M (2003) Nutrition science applications, 4th edn. Wiley, New York, N.Y.

80. Corwin EJ, Murray-Kolb LE, Beard JL (2003) Low hemoglobin level is a risk factor for postpartum depression. J Nutr 133:4139–4142

81. Beard JL, Hendricks MK, Perez EM, Murray-Kolb LE, Berg A, Vernon-Feagans L, Irlam J, Isaacs W, Sive A, Tomlinson M (2005) Maternal iron deficiency anemia affects postpartum emotions and cognition. J Nutr 135:267–272

82. Ghadirian AM, Anath J, Engelsmann F (1980) Folic acid deficiency and depression. Psychosomatics 21:926–929

83. Abou-Saleh MT, Coppen A (1989) Serum and red blood cell folate in depression. Acta Psychiatr Scand 80:78–82

84. Carney MWP, Chary TKN, Laundy M, Bottiglieri T, Chanarin I, Reynolds EH, Toone B (1990) Red cell folate concentrations in psychiatric patients. J Affect Disord 19:207–213

85. Bottiglieri T, Laundy M, Crellin R, Toone BK, Carney MWP, Reynolds EH (2000) Homocysteine, folate methylation and monoamine metabolism in depression. J Neurol Neurosurg Psychiatry 69:228–232

86. Morris MS, Fava M, Jacques PF, Selhub J, Rosenberg JH (2003) Depression and folate status in the US population. Psychother Psychosom 72:80–87

87. Young SN (2007) Folate and depression- a neglected problem. J Psychiatry Neurosci 32:80–82

88. Miyake Y, Sasaki S, Tanaka K, Yokoyama T, Ohya Y, Fukushima W, Saito K, Ohfuji S, Kiyohara C, Hirota, Y (2006) Dietary folate and vitamins B_{12}, B_6, and B_2 intake and the risk of postpartum depression in Japan: The Osaka Maternal and Child Health Study. J Affect Disord 96:133–138

89. Ganji V, Kafai MR (2004) Frequent consumption of milk, yogurt, cold breakfast cereals, peppers, and cruciferous vegetables and intakes of dietary folate and riboflavin but not vitamins B-12 and B-6 are inversely associated with serum total homocysteine concentrations in the US population. Am J Clin Nutr 80:1500–1507

90. Bjelland I, Tell GS, Vollset SE (2003) Folate, vitamin B-12, homocysteine, and the MTHFR 677C T polymorphism in anxiety and depression: the Hordaland Homocysteine Study. Arch Gen Psychiatry 60:618–626

91. Hvas AM, Juul S, Bech P, Nexo E (2004) Vitamin B-6 level is associated with symptoms of depression. Psychother Psychosom 73:340–343

92. Abou-Saleh MT, Ghubash R, Karin L, Krymski M, Bhai I (1998) Hormonal aspects of postpartum depression. Psychoneuroendocrinology 23:465–475

93. Heinrichs M, Meinlschmidt G, Neumann I, Wagner S, Kirschbaum C, Ehlert U, Hellhammer DH (2001) Effects of suckling on hypothalamic-pituitary-adrenal axis responses to psychosocial stress in postpartum lactating women. J Clin Endocrinol Metab 86:4798–4804

94. Mezzacappa ES, Katkin ES (2002) Breast-feeding is associated with reduced perceived stress and negative mood in mothers. Health Psychol 21:187–193

95. Groer MV (2005) Differences between exclusive breastfeeders, formula-feeders, and controls: a study of stress, mood and endocrine variables. Biol Res Nurs 7:106–117

96. Kendall-Tackett K (2007) A new paradigm for depression in new mothers: the central role of inflammation and how breastfeeding and anti-inflammatory treatments protect maternal mental health. Int Breastfeed J 2:6

97. Breese SJ, Beal JM, Miller-Shipman SB, Payton ME, Watson GH (2006) Risk factors for post-partum depression: a retrospective investigation at 4 weeks postnatal and a review of literature. J Am Osteopath Assoc 106:193–198

98. McKee MD, Zayas LH, Jankowski KRB (2004) Breastfeeding intention and practice in an urban minority population: relationship to maternal depressive symptoms and mother–infant closeness. J Reprod Infant Psychol 22:167–181

99. Boyd RC, Zayas LH, McKee D (2006) Mother–infant interaction, life events and prenatal and post-partum depressive symptoms among urban minority women in primary care. Matern Child Health J 10:139–148

100. McCarter-Spaulding D, Horowitz JA (2007) How does postpartum depression affect breastfeeding? Am J Matern Child Nurs 32:10–17

101. Jones NA, McFall BA, Diego MA (2004) Patterns of brain electrical activity in infants of depressed mothers who breastfeed and bottle feed: the mediating role of infant temperament. Biol Psychol 67:103–124

102. Yapp P, Ilett KF, Kristensen, JH, Hackett LP, Paech MJ, Rampono J (2000) Drowsiness and poor feeding in a breast-fed infant: association with nefazodone and its metabolites. Ann Pharmacother 34:1269–1272

103. Schmidt K, Olesen OV, Jensen PN (2000) Citalopram and breast-feeding: serum concentration and side effects in the infant. Biol Psychiatry 47:164–165

104. Frey OR, Scheidt P, von Brenndorff, A (1999) Adverse effects in a newborn infant breast-fed by a mother treated with doxepin. Ann Pharmacother 33:690–693

105. Matheson I, Pande H, Alertsen AR (1985) Respiratory depression caused by N-desmethyldoxepin in breast milk. Lancet 2:1124

106. Brent NB, Wisner KL (1998) Fluoxetine and carbamazepine concentrations in a nursing mother/infant pair. Clin Pediatr 37:41–44

107. Chambers CD, Anderson PO, Thomas RG (1999) Weight gain in infants breastfed by mothers who take fluoxetine. Pediatrics 104:e61

108. Murray L, Fiori-Cowley A, Hooper R, Cooper P (1996) The impact of postnatal depression and associated adversity on early mother–infant interactions and later infant outcome. Child Dev 67:2512–2526

109. Beck CT (1998) The effects of postpartum depression on child development: a meta-analysis Arch Psychiatr Nurs 12:12–20

110. Politano PM, Stapleton LA, Correll JA (1992) Differences between children of depressed and nonde-pressed mothers: locus of control, anxiety and self-esteem: a research note. J Child Psych Psychiatr 33:451–455

111. Henderson JJ, Evans SF, Straton JA, Priest SR, Hagan R (2003) Impact of postnatal depression on breastfeeding duration. Birth 30:175–180

112. Pippins JR, Brawarsky P, Jackson RA, Fuentes-Afflick E, Haas JS (2006) Association of breastfeeding with maternal depressive symptoms. J Womens Health 15:754–762

113. Dennis C-L, McQueen K (2007) Does maternal postpartum depressive symptomatology influence infant feeding outcomes? Acta Paediatr 96:590–594

114. Hatton DC, Harrison-Hohner J, Coste S, Dorato V, Curet LB, McCarron DA (2005) Symptoms of postpartum depression and breastfeeding. J Hum Lact 21:444–449

115. McLearn KT, Minkovitz CS, Strobino DM, Mark E, Hou W (2006) Maternal depressive symptoms at 2 to 4 months post partum and early parenting problems. Arch Pediatr Adolesc Med 160:279–284

116. Beck CT, Gable RK (2001) Comparative analysis of the performance of the Postpartum Depression Screening Scale with two other depression instruments. Nurs Res 50:242–250

V   THE DEVELOPING WORLD

20 Implications of the Nutrition Transition in the Nutritional Status on Pregnant Women

Jaime Rozowski and Carmen Gloria Parodi

Summary In most developing countries, including Chile, an epidemiologic and nutrition transition has taken place, the former characterized by an increase of the population due to a reduced mortality, followed by a decrease in fertility and an increase in longevity. The nutrition transition has been characterized by an increase in the consumption of fats and simple sugars and a decrease in fruit and vegetable intake. This, together with a decrease in physical activity, has contributed to an increase in the prevalence of obesity in fertile women. Data collected from 36 developing countries showed that in 32 of them, overweight was more prevalent than underweight in urban areas, while in 53% (19/36) underweight was more prevalent in rural areas compared to urban settings. In all of those countries, the prevalence of overweight was significantly correlated with gross national income per capita. Different surveys in Chile have shown that 90% of homes have a television set, 60% of all families own at least one car, 27.3% of women aged 17–44 are obese, and 90% of them do not perform any significant physical activity. The consequence of all this is that women are getting to pregnancy heavier than they used to, resulting in an increase in complications during pregnancy including gestational diabetes mellitus, hypertension, and delivery complications, all of which can affect the newborn at birth and in later life. This chapter defines the characteristics of the nutritional transition and concentrates in a discussion of obesity during pregnancy and its consequences at birth and in later life.

Keywords: Nutrition transition, Pregnancy, Obesity, Birth outcome, Gestational diabetes

20.1 INTRODUCTION

In 1971, Omran described an epidemiological transition that was characterized by a general increase in the age of the population and a decrease in early mortality [1]. Later on, Bobadilla et al. proposed that a link can be made between this transition and the health needs of the population, defined by changes in its age structure and its causes of death [2]. In the epidemiologic transition, Omran defined an accelerated model, observed in some developed countries like Japan, and a delayed model, which (at that time), was

From: *Nutrition and Health: Handbook of Nutrition and Pregnancy*
Edited by: C.J. Lammi-Keefe, S.C. Couch, E.H. Philipson © Humana Press, Totowa, NJ

seen in most developing countries. The main difference between these two models was in the timing and the pace of change [1].

The demographic transition is characterized by an increase in urbanization and industrialization, an expansion of education, and rising incomes. From the health point of view, it translates into an improvement in medical care and public health. As a result of these improvements, in total there is a decrease in mortality due to infectious diseases [3] but a decrease in fertility reflecting urbanization, education and increased incomes resulting in decreased family size. This precipitates an epidemiologic transition in which aging of the population is seen together with the appearance of the nontransmittable chronic diseases.

Concomitant with the last stages of the epidemiological transition, a nutrition transition, has/is occurring characterized by a change in dietary patterns favoring foods richer in fats, an increased consumption of simple carbohydrates, a reduction in the consumption of fruits and vegetables, and a decrease in physical activity. The transition is seen in practically all developing countries, although the extent of its degree varies depending on the state of development of a particular country.

Many theories have been advanced to explain nutrition transition. Although the "westernization" of the diet and life habits is responsible in part, other factors like a diminution of physical activity and a generalized laissez faire regarding personal health are also responsible [4].

Probably the most remarkable feature of the nutrition transition has been a marked increase in the prevalence of obesity, a serious chronic disease whose most important consequence is an increase in mortality due to its association with chronic ailments including cardiovascular disease. As is true for many chronic diseases, obesity is preventable, but its prevalence has increased steadily in the last decades in both developed and developing countries [4]. Many factors are linked to the development of obesity, the most important being a sedentary lifestyle and nutrition.

20.2   THE NUTRITION TRANSITION

20.2.1   Changes in Food Consumption

In the last 25 years, important changes in diet and lifestyle have taken place, which have had adverse consequences on the nutritional status of the population at large. However, to establish the causes for the changes in dietary habits across cultures is difficult. The increased consumption of processed foods, particularly fast foods, has been cited as the root cause of those changes. Fast food is usually cheaper (calorie for calorie) than fruits and vegetables, so socioeconomic factors have likely impacted changing food consumption patterns. Although there are still countries with a high prevalence of undernutrition in the developing world, most of them are undergoing a change in which the prevalence of overweight and obesity are higher than the prevalence of undernutrition. This places a burden on the health care dollars of these countries since new resources have to be allocated to fight obesity, which combined with resources needed to combat malnutrition creates a dual financial burden that often is unaffordable [5].

In Latin America, one of the factors that has influenced the changes in lifestyle is the gradual improvement in socioeconomic level. In the region, traditionally the diet is composed mainly of cereals, like wheat and maize. However, the increase in the purchasing power of the average family and the low prices of calorie dense foods, e.g, fast foods and processed snack foods, have contributed to the rise in the consumption of low-cost, high-fat foods and refined sugar. For instance, in Santiago (Chile), for the equivalent of $3, it is possible to buy a meal at a fast-food restaurant that provides more than half of the daily caloric needs of an adult woman [4]. This example underlines the increase in the availability of energy, as it is shown in Fig. 20.1, which shows calorie availability in several countries from Latin America at different stages of development. In practically all of them calorie availability has increased, even in those that still deal with a serious problem of undernutrition like Guatemala. In Brazil, a country with a wide range of population from the socioeconomic point of view, average calorie availability increased from 2,072 calories in 1980 to 3,146 calories in 2003, a 52% increase [6]. Fat availability in these countries showed a similar pattern (Fig. 20.2). Meat consumption has also gone up, showing a 50% increase from 1985 to 2003. The increased consumption of meat can be explained by this food being considered a prestigious item in the groups that develop economically. The problem is that, in the case of Chile, as well as in other countries in Latin America, together with the increase in fats and meat availability, we have seen a decrease in fruit, vegetable, and fish intake [6]. The consumption of fish in Chile (with 3,000 miles of coastline) reaches less that 1% of the average total calories consumed by the population. The increase in consumption of carbohydrates has been substantial.

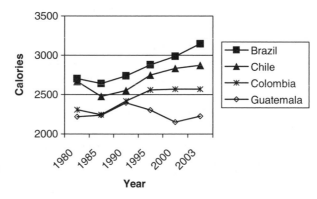

Fig. 20.1. Calorie availability in selected countries from Latin America, 1980–2003 [7]

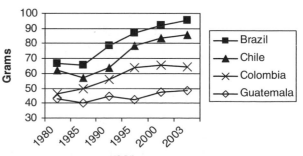

Fig. 20.2. Fat availability in selected countries from Latin America, 1980–2003 [7]

For instance, in Mexico, the diet has changed from a high-fiber, low-calorie diet, to one that is currently characterized by high consumption of fat and sugar, from 1984 to 1998, the purchase of meat, milk (and derivatives), and of fruits and vegetables decreased by 19, 27, and 29%, respectively. On the other hand, purchases of refined carbohydrates increased by 6% and that of soda by 37% during the same years [7].

20.2.2 Physical Activity

The second half of the twentieth century was characterized by remarkable technological advances both in developed and most of the developing countries, as is the case for Chile. All these advancements have had a tremendous impact on lifestyles. Fifty years ago, car availability and affordibility was fairly low in the country, and people were used to walking long distances. The census carried out in 2002 showed that 60.4% of Chilean homes own at least one car [8]. Likewise, 50 years ago, TV ownership was limited in Chile and children spent much more time than today in physical exercise. Nowadays 90% of all homes own a TV that has nonstop programming, inducing people to watch it for hours at a time, and thereby interfering with any kind of physical activity [8]. To this we have to add the hours sitting in front of a computer. Although the 1992 census did not even consider the item, 10 years later the census determined that 21% of all Chilean homes have a computer. Additionally, there is an increased concern for safety and security, so the exercise usually done on the street and parks has practically disappeared. Thus, all these factors have contributed to the fact that the population, especially women, spend a limited amount of time in exercise.

The female population presents a higher prevalence of physical inactivity than males. The National Health Survey (NHS) done in Chile in 2003 showed that 90% of the women aged 17–44 years were classified as sedentary (less than 30 min each of exercise three times a week) [9]. This was confirmed by the Quality of Life Survey of 2005 which showed that 95.9% of women interviewed had not performed any exercise outside of the work place for at least 30 min [8]. This level of inactivity has also been demonstrated in smaller surveys carried out in the country [10].

20.3 EFFECT OF THE NUTRITION TRANSITION ON PREGNANCY

Although the nutrition transition in Chile has affected people of all areas of society, the strongest affect has been in women, who show a higher prevalence of obesity and of sedentary behavior than men. Figure 20.3 shows changes in the incidence of underweight and overweight in pregnant women in Chile since 1990. As the graph shows, there is an almost perfect inverse correlation between both conditions, the prevalence of obesity reaching 32.2% in 2004 [11]. The fact is that women are arriving at pregnancy with a heavier weight than in the past, a characteristic that can also be seen in developed countries.

20.3.1 Undernutrition and Pregnancy

Pregnancy in Chile 40 years ago was characterized by a high prevalence of undernutrition. At that time, 24.9% of adult women were undernourished while 7% were obese (in the 2003 survey 4,1% were undernourished while 27.3% were obese). In 1937, the

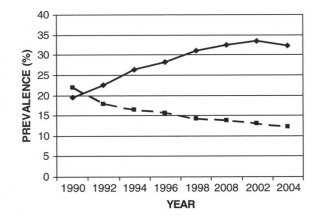

Fig. 20.3. Nutritional status of pregnant women, Chile 1990–2004 (weight/height index for gestational age). *Straight line* obesity, *dashed line* underweight. (From [12])

government implemented a supplementary feeding program for preschool children and pregnant women (National Feeding Program, PNAC), which provided milk through pregnancy to the mother [12]. Later it evolved to also providing vitamins and iron supplements and nutrition education to poor women. This program has been extremely successful and has been replicated in other countries in Latin America.

20.3.2 Obesity and Pregnancy

Chile is a good example in terms of the nutrition transition as seen in developing countries [13]. As indicated above, successful private and public programs have practically eliminated undernutrition, but the situation has gone to the other extreme, obesity being the principal problem today. The 2003 National Health Survey showed that 27.3% of all women aged 17–44 were obese, higher than the prevalence observed in men (19.2%). Figure 20.4 compares the prevalence of obesity related to age in Chilean women obtained by Berríos et al. [14–16] from observations in 1987 and 1992 in Santiago, and from the CARMEN Study in 1998 carried out in Valparaiso [10], using a body mass index (BMI) of $27.3 \, kg/m^2$ as a cutoff point. A marked increase in obesity prevalence was seen in the 25- to 34-year-old group, and prevalence consistently increased with age. A recent national survey showed that of women 17 years of age and older, 33% were overweight, and 25% were obese. In other words, less than 50% of the females of childbearing age have a BMI considered healthy. It is also important to point out that 2.3% of the women in this age group were morbidly obese, a characteristic not observed previously [9].

The increment in obesity prevalence in women is now global. For instance, in the United States, the prevalence of obesity almost doubled in a period of 20 years, from 12.7% in men and 17% in women in 1980, to 27.7 and 34% in men and women, respectively, in the year 2000 [17].

The high prevalence of female obesity is also seen in developing countries in other regions. An interesting study was done by Mendez et al. in which the authors analyzed data from demographic and health surveys (DHS; www.measuredhs.com) on underweight

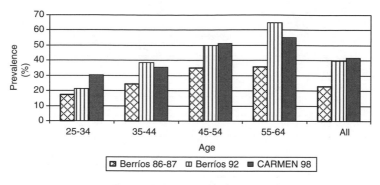

Fig. 20.4. Prevalence of obesity (BMI $\geq 27.3\,\mathrm{kg/m^2}$) in Chilean women, 1986–1998 [14–16]

(BMI $<18.5\,\mathrm{kg/m^2}$) and overweight (BMI $\geq 25\,\mathrm{kg/m^2}$) in woman from 20 to 49 years of age from developing countries in Asia, Africa, the Middle East, and Latin America [18]. The study showed an inverse relationship between the prevalence of overweight and underweight and a direct correlation between overweight and degree of urbanization. Data were collected from 36 developing countries, and in most (32/36), overweight was more prevalent than underweight in urban areas, while in 53% (19/36) underweight was more prevalent in rural areas compared to urban settings. In all areas, prevalence of overweight was significantly correlated with gross national income per capita (GNI) [18].

In the United States, the incidence of obesity during pregnancy ranges from 18.5 to 38.2%, depending on the population studied and the cutoff points used to define obesity. Overweight and obesity before pregnancy are considered risk factors for specific pregnancy complications [19].

The increase in the prevalence of obesity has been universal in Latin America without distinction of gender, race, or age, but the situation for women is more complex. Obesity in women of childbearing age translates into heavier newborns at the end of their pregnancy, and these infants have an increased chance of being obese as both children and adults, perpetuating the cycle.

20.3.2.1 COMPLICATIONS OF OBESITY DURING PREGNANCY

When obesity is present during pregnancy, it presents serious consequences for both the mother and the fetus. During this period, a series of metabolic changes take place in order to supply the fetus with all needed nutrients, and actually, in itself, pregnancy is a diabetogenic condition for the mother, as discussed below. Pregnant women have a higher risk for developing gestational diabetes mellitus (GDM) and hypertension, which can impair fetal development. These complications, in addition to problems at the time of delivery, could be fatal [20–26].

20.3.2.1.1 Gestational Diabetes Mellitus. The risk of developing GDM increases in direct relationship to BMI; that is, the higher the BMI the higher the risk. In the case of women with normal prepregnancy weight, the risk resides in weight gain during pregnancy; in this case, the higher the weight gain the higher the risk. This situation is not seen in overweight or obese women, since the weight gain is usually very controlled during pregnancy, their prepregnancy weight being the factor that contributes to GDM [24–26]. Thus, the presence of overweight or obesity before conception is extremely important, and it contributes more to the development of GDM than does the weight gained during pregnancy. This is mainly due to the presence of insulin resistance, which becomes more severe as the diabetogenic factors of pregnancy appear. Another risk factor for those that have gained too much weight during pregnancy is that it increases the chances of developing GDM in later pregnancies [24, 26].

Incidence of GDM is directly related to the risk for type 2 diabetes mellitus (DM2) later in life. Approximately 40% of the women who develop GDM will show glucose intolerance or DM2 later in life [21, 27–29], in addition to an increase in the cardiovascular (CV) risk, which is even higher when complicated with other factors like obesity [30, 31]. Our study of the incidence of GDM in Chilean women showed an incidence of 12% detected in the second trimester and 7% detected in the third trimester (unpublished data). Of the women that developed GDM, 70% of them were overweight or obese at the initiation of the study and showed a high prevalence of overweight and obesity in the pregestational stage, again confirming that prepregnancy weight is of paramount importance in the development of GDM.

Atalah and Castro, in a prospective study in 883 pregnant Chilean women showed that pregestational obesity (BMI $\geq$ 30 kg/m^2 or initial body fat mass $\geq$ 30%) increased the risk for developing GDM six times (OR 6.4, 95% CI 2.1–19.6), and it increased the risk of developing hypertension eight times (OR: 7.8, 95% CI 3–20.4) [32].

There has been relatively little research regarding physical activity before or during pregnancy. Results from the Nurses Health Study of Harvard showed that prepregnancy physical activity, including vigorous exercise and brisk walking, conferred a protection against the development of GDM [33]. In this prospective study, those women who spent 20 h/week or more watching television and who did not perform vigorous activity had a significantly higher risk of GDM than did women who spent less than 2 h/week watching television and were physically active (RR, 2.30; 95% CI, 1.06–4.97). The importance of physical activity in the maintenance of body weight has been studied for a long time, and there is a consensus on its importance before and during pregnancy, not only in controlling weight, but also in lowering fasting and postprandial glucose concentration [33].

20.3.2.1.2 Hypertension Disorders. In the United States, 5–10% of pregnant women suffer from complications of hypertension produced by their pregnancy. Although these hypertensive disorders have been known for a long time, the mechanisms involved are still unknown [34 and reviewed in Chap. 11, "Preeclampsia"]. Hypertensive disorders are classified as preeclampsia or gestational hypertension which appears during pregnancy, and preexisting hypertension or its exacerbation. The difference between preeclampsia and hypertension resides in that the latter does not show proteinuria and is more benign [34, 35]. The pathogenesis of these disorders is still unknown and could include genetic factors, immune factors, and placental abnormalities, leading to an endothelial dysfunction that is characteristic of preeclampsia [34]. Several researchers agree that insulin resistance has an important role in the development of preeclampsia, and that this condition is directly related to obesity. An elevated pregestational BMI and an excessive weight gain during pregnancy are closely correlated to the development of preeclampsia and gestational hypertension [34, 36–38]. Although insulin resistance plays an important role in this process, we also have to consider maternal hemodynamic changes, which in obese pregnant women include elevated blood pressure, hemoconcentration, and altered cardiac function [39, 40].

In the study of Atalah and Castro cited above, the authors found that obese women had almost eight times the chance of developing hypertension (OR 7.8, 95% CI 3–20.4) when compared with normal-weight women [32]. Similar results were obtained in Brazil [40].

20.3.2.1.3   Other Complications. Although found with less frequency (but not less important) in pregnancy, respiratory alterations that occur in obese pregnant women may contribute to snoring and sleep apnea, which requires a constant change of position during sleep. Other complications, like urinary or thromboembolic disorders are currently under study [19, 20, 24].

20.3.2.2   COMPLICATIONS DURING BIRTH

A relatively small number of studies have investigated the changes produced by obesity at the time of birth, and even in these, there is controversy, as to the potential impact of obesity on labor and delivery, and prematurity. Garbaciak et al. [39] comparing pregnant women with normal weight with obese pregnant women and did not find a difference in the incidence of prematurity between groups. However, Naeye showed that obesity in the mother was a significant risk for premature births [41].

During the 1990s a cohort of more than 150,000 Swedish women were followed during pregnancy. Obese women had a higher risk of complications at the time of delivery than those with normal weight. The same study showed that in obese nulliparous women, there was an increase in very premature deliveries (<32 weeks of gestation), the incidence showing an increase at the extremes, i.e., in very low weight women and those with a $BMI > 30\,kg/m^2$ [42]. Obese women also had a longer time in delivery and showed an increase in the incidence of induced deliveries compared to those women with normal weight [19, 20].

20.3.2.2.1   Cesarean Delivery. One of the main complications at the time of birth in obese woman is the increased frequency of delivery by cesarean section, showing a higher incidence than normal-weight women. This was independent of other comorbidities. Although the risk increases with a higher BMI, prepregnancy weight is even more important [43]. Brost et al. showed in a US study of 2,929 expecting women, that each increase in one unit of pregestational BMI raised the risk of having a caesarean delivery by 7% [44]. Similar results were obtained when BMI was determined in weeks 27–31, but in this case, the risk increased to 7.8%.

A study of 4,500 pregestational women in Brazil showed that 6.9% were obese and 43.1% were preobese. Caesarean delivery at end of pregnancy was performed in 52% of the obese and in 43% of the preobese women, with a relative risk of caesarean delivery for obese women of 1.8 (95% CI: 1.5–2). The rate of weight gain was also related to the frequency of caesarian section, the risk being 1.3 (95% CI: 1.2–1.5) for those with excess weight gain. A faster rate of weight gained also increased the risk of infection in all deliveries [45].

The reason for an increased incidence of caesarean delivery involves an incomplete dilation of the cervix, fetal distress, or failure of induction [23]. The risk of caesarean delivery also increases in women with GDM and/or preeclampsia, but in the case of obese women who are without these associated pathologies, it is attributed to longer gestational periods, which produce heavier newborns.

20.3.2.2.2   Anesthesia and Postpartum. For obese women having a cesarean or a vaginal delivery, the administration of anesthesia is a delicate issue. The anatomical characteristics of obese women are different from normal-weight women. The increase in subcutaneous fat may cause anatomical abnormalities, which may make it more difficult to locate physical reference points. This has more importance in morbidly obese women since certain aspects, for instance the epidural administration, are more likely to fail. General anesthesia may also cause respiratory problems due to anatomical differences (for instance, shorter and fatter necks) [24].

The postpartum period is also more complicated for obese women undergoing either vaginal or cesarean delivery vis-a-vis bleeding, infection, and urinary problems. In vaginal delivery, obese women show a higher incidence of perianal rupture due to the elevated weight of the newborn [25]. Also, they tend to stay longer in the hospital after delivery, further increasing the costs incurred [24].

20.3.2.3   LACTATION

A variety of hormonal changes occur alter delivery, and these can affect the production of breast milk. Overweight and obese women initiate lactation later than do normal-weight women due to a lower prolactin secretion in response to the infant sucking stimulation [46]. It is reasoned, though not demonstrated, that this is caused by high progesterone levels in obese women. Normally, after delivery, progesterone levels diminish, inducing an increase in the secretion of prolactin, which stimulates milk production. Since the adipocyte is an additional source of progesterone, hormone levels would be maintained, inhibiting the activation of prolactin. Fortunately, this delay in lactation is observed initially but has no relationship with the length of the period of lactation [46].

20.3.2.4   LONG-TERM EFFECTS OF MATERNAL OBESITY IN THE MOTHER AND THE INFANT

The effects of obesity are not limited to immediate effects on the mother and the newborn. Other complications appear in both later in life. Those women with GDM have a higher risk of developing DM2 later in life, and if obesity is also present, they have a higher risk of developing cardiovascular disease (CVD) [30, 47]. Children of women with GDM may present with macrosomy and have a higher risk of developing DM in adolescence and CDV in adulthood. It has been proposed that this is due to an increased proportion of body fat at birth [47, 48].

20.4   CONCLUSION

The nutrition transition and changes in lifestyles have caused a remarkable increase in obesity during pregnancy. The information described above strongly indicates that this disease is a serious problem with far-reaching consequences. Policy efforts must be directed to the prevention of obesity in young girls, as a way to break the vicious circle. An obese mother has a higher chance of having large-for-gestational-age children, who in turn may become obese in adulthood and will also have newborns with a high risk of macrosomy. We are not considering here the expenses involved in treating an individual with CVD at an early age. Thus, in our opinion, preventive measurements have to be directed to the young, thus preventing women from getting to pregnancy with overweight or obesity.

REFERENCES

1. Omran A (2005) The epidemiologic transition: A theory of the epidemiology of population change. 1971. Milbank Q 83:731–757
2. Bobadilla JL, Frank J, Lozano R, Frejka T, Stern C (1993) The epidemiologic transition and health priorities. In: Jamison DT, Mosley WH, Measham AR, Bobadilla JL (eds) Disease control priorities in developing countries, Oxford University Press, New York, N.Y., pp 51–63
3. Mosley WH, Bobadilla JL, Jamison DT (1993) The health transition: Implications for health policy in developing countries. In: Jamison DT, Mosley WH, Measham AR, Bobadilla JL (eds) Disease control priorities in developing countries, Oxford University Press, New York, N.Y., pp 673–699
4. Rozowski J, Castillo O, Moreno M (2005) Effect of Westernization of nutritional habits on obesity prevalence in Latin America. Analysis and recommendations. In: Bendich A, Deckelbaum R (eds) Preventive nutrition: the comprehensive guide for health professionals, 3rd edition. Humana, Totowa, N.J., pp 771–790
5. Doak CM, Adair, LS, Bentley Monteiro C, Popkin BM (2005) The dual burden household and the nutrition transition paradox. Int J Obesity (London) 29:129–136
6. Food and Agriculture Organization of the United Nations (2007) http://www.fao.org
7. Rivera JA, Barquera S, Gonzales Cossio T, Olaiz G, Sepúlveda J (2004) Nutrition transition in Mexico and in other Latin American Countries. Nutr Rev 62:S149–S157
8. Instituto Nacional de Estadísticas (2007) http://www.ine.cl
9. Ministry of Health. Encuesta Nacional de Salud (2003) http://www.minsal.cl
10. Jadue L, Vega J, Escobar (1999) Factores de riesgo para las enfermedades no transmisibles: Metodología y resultados globales de la encuesta de base del programa CARMEN (Conjunto de Acciones para la Reducción Multifactorial de las Enfermedades no Transmisibles). Rev Med Chile 127:1004–1013
11. Ministerio de Salud de Chile (2007) http://www.minsal.cl
12. Mardones FS, Gonzales N, Mardones FR (1986) Programa Nacional de Alimentación Complementaria en Chile en el período 1937–1982. Rev Chil Nutr 14:173–182
13. Albala C, Vio F, Kain J, Uauy R (2001) Nutrition transition in Latin America: The case of Chile. Nutr Rev 59:170–176
14. Berríos X (1990) Prevalencia de factores de riesgo en enfermedades crónicas del adulto. Rev Med Chile 118:1041–1042
15. Berríos X, Jadue L, Zenteno J, Ross M, Rodríguez H (1990) Prevalencia de factores de riesgo de enfermedades crónicas. Estudio en población general de la Región Metropolitana, 1986–1987. Rev Med Chile 118:597–604
16. Berrios X (1997) Tendencia temporal de los factores de riesgo de enfermedades crónicas ¿La antesala de una epidemia que viene? Rev Med Chile 125:1405–1407
17. Flegal KM, Carroll MD, Ogden CL, Johnson CL (2002) Prevalence and trends in obesity among US adults, 1999–2000. J Am Med Assoc 288:1723–1727
18. Mendez MA, Monteiro CA, Popkin BM (2005) Overweight exceeds underweight among women in most developing countries. Am J Clin Nutr 81:714–721
19. Galtier-Dereure F, Boegner C, Bringer J (2000) Obesity and pregnancy: complications and cost. Am J Clin Nutr 71:(Suppl):1242S–1248S
20. Zhang C, Solomon CG, Manson JE, Hu FB (2006) A prospective study of pregravid physical activity and sedentary behaviors in relation to the risk for gestational diabetes mellitus. Arch Intern Med 166:543–548
21. Castro L, Anina R (2002) Maternal obesity and pregnancy outcome. Curr Opin Obstet Gynecol 14:601–606
22. Catalano P, Kirwan JP, Hangel – DeMonzo S, King J (2003) Gestational diabetes and insulin resistance: role in short- and long-term implication for mother and fetus. J Nutr 133:1647S–1683S
23. Kabiru W, Raynor D (2004) Obstetric outcome associated with increase in BMI category during pregnancy. Am J Obstet Gynecol 191:928–932
24. Linné Y (2004) Effects of obesity on women's reproduction and complications during pregnancy. Obes Rev 5:137–143
25. Andersen KR, Andersen ML, Schantz AL (2004) Obesity and pregnancy. Acta Obstet Gynecol Scand 83:1022–1029

26. Linné Y, Dye L, Barkeling B, Rössner S (2004) Long-term weight development in women: a 15 years follow-up of effects of pregnancy. Obes Res 12:1166–1178
27. Cornier MA (2005) Obesity and diabetes. Curr Opin Endocrinol Diabetes 12:260–266
28. Jovanovic L, Pettit DJ. Gestational diabetes mellitus. J Am Med Assoc 286:2516–2518
29. Buchanan TA, Kjos SL (1999) Gestational diabetes: risk or myth? J Clin Endocrinol Metab 84:1854–1857
30. Kim C, Newton K, Knopp R (2002) Gestational diabetes and the incidence of type 2 diabetes. Diabetes Care 25:1862–1868
31. Ko GT, Chan JC, Tsang LW, Li CY, Cockram CS (1999) Glucose intolerance and others cardiovascular risk factors in Chinese women with history of gestational diabetes mellitus. Aust NZ Obstet Gynecol 39:478–483
32. Atalah E, Castro R (2004) Maternal obesity and reproductive risk. Rev Med Chile 132:923–930
33. Zhang C, Solomon CG, Manson JE, Hu FB (2006). A prospective study of pregravid physical acvtivity and sedentary behaviour in relation to the risk for gestational diabetes mellitus.
34. Seely EW, Solomon CG (2003) Insulin resistance and potential role in pregnancy-induced hypertension. J Clin Endocrinol Metab 88:2393–2398
35. Duley L (2003) Preeclampsia and hypertensive disorders of pregnancy. Br Med Bull 67:169–176
36. Villamor E, Msamanga G, Urassa W, Petrato P, Spiegelman D, Hunter DJ, Fawzi WW (2006) Trend in obesity, underweight and wasting among women attending antenatal clinics in urban Tanzania, 1995–2004. Am J Clin Nutr 83:1387–1394
37. American Diabetes Association (1999) Gestational diabetes mellitus (position statement). Diabetes Care 22(Suppl):S66
38. Saftlas A, Wang W, Risch H, Woolson R, Hsu C, Brecken M (2000) Prepregnancy body mass index and gestational weight gain as risk factors for preeclampsia and transient hypertension. Ann Epidemiol 10:475
39. Garbaciak JA, Richter M, Muller S, Barton JJ (1985) Maternal weight and pregnancy complications. Am J Obstet Gynecol 152:238–245
40. Nucci LB, Schmidt MI, Duncan BB, Fuchs SC, Fleco ET, Santos MM (2001) Nutritional status of pregnant women: prevalence and associated pregnancy outcomes. Rev Saude Publica 55:502–507
41. Naeye RL (1990) Maternal body weight and pregnancy outcome. Am J Clin Nutr 52:273–279
42. Cnattingius S, Bergstron R, Lipworth L et al. (1998). Pre-pregnancy weight and the risk of adverse pregnancy outcome. N Eng J Med, 338 (3):147–152.
43. Ehrenberg HM, Durnwald CP, Catalano P, Mercer BM (2004) The influence of obesity and diabetes on risk of caesarean delivery. Am J Obstet Gynecol 191:969–974
44. Brost BC, Goldenberg RL, Mercer BM, Iams JD, Meis PJ, Moawad AH, Newman RB, Miodovnik M, Caritis SN, Thurnan GR, Bottoms SF, Das A, Mc Nellis D (1997) The preterm prediction study: association of caesarean delivery with increases in maternal weight and body mass index. Am J Obstet Gynecol 177:333–341
45. Seligman LC, Duncan BBN, Branchtei L et al (2006) Obesity and gestational weight gain: Cesarean delivery and labor complications. Rev Saude Publica 40:457–465
46. Rasmussen KM, Kjolhede CL (2004) Pregnancy overweight and obesity diminish the prolactin response to suckling in the first week post partum. Pediatrics 113:465–471
47. Catalano P (2003) Editorial: Obesity and pregnancy—the propagation of a vicious cycle? J Clin Endocrinol Metab 88:3505–3506
48. Cetin E, Radaelli T (2005) Normal and abnormal fetal growth. In: Djelmiš J, Desoye G, Ivaniševic M (eds) Diabetology of pregnancy: front diabetes, vol. 17. Karger, Basel, Switzerland, pp 72–82

21 Nutrition and Maternal Survival in Developing Countries

Parul Christian

Summary Maternal mortality continues to be high and maternal nutrition poor in the developing world. However, the specific role of nutrition in affecting maternal health and survival remains unclear. Recent trials provide support for a specific and perhaps important place for nutrition in reducing the burden of maternal mortality in developing countries. Specific nutrition interventions have been shown to be efficacious against some causes of maternal mortality. Calcium supplementation during pregnancy in high-risk populations or populations with dietary deficiency can reduce the risk of eclampsia and severe morbidity and mortality related to hypertensive disorders of pregnancy. Magnesium sulfate is a low-technology and inexpensive means to reduce the risk of eclampsia. Maternal anemia is likely to increase the risk of maternal mortality. Antenatal iron supplementation when done adequately can bring about improvements in hemoglobin concentrations that are likely to reduce the risk of maternal mortality by about 25%. Maternal vitamin A deficiency may be associated with an increased risk of maternal deaths, but further evidence is needed. Antenatal nutritional interventions that are able to achieve high coverage may likely be an effective means for impacting maternal survival in undernourished populations of the world where the burden of maternal mortality is high.

Keywords: Maternal mortality, Causes, Pregnancy, Micronutrients, Anemia, Vitamins, Supplementation, Morbidity, Nutrition

21.1 INTRODUCTION

The World Health Organization (WHO) estimates that 529,000 maternal deaths occur worldwide each year [1]. Reducing mortality related to complications of pregnancy, labor, and delivery continues to be a priority and to receive attention globally. The *Lancet* series on maternal survival [2–6] is a call to focus attention on the high burden of maternal mortality in the developing world. Additionally, the series provides a critical overview of existing strategies that are most likely to exert an impact. The United Nations' 5th Millennium Development Goal (MDG) is a call to reduce maternal mortality by 75% between 1990 and 2015, a goal that appears to be more elusive and one that is lagging behind most [7, 8], despite ongoing efforts by the Safe Motherhood Initiative, an effort launched in 1987 to

From: *Nutrition and Health: Handbook of Nutrition and Pregnancy*
Edited by: C.J. Lammi-Keefe, S.C. Couch, E.H. Philipson © Humana Press, Totowa, NJ

reduce maternal morbidity and mortality. There appears to be a consensus on the kind of focused approaches and the necessary tools and strategies known to work to reduce maternal death. Addressing obstetric complications, be it through increasing the availability of skilled attendants or emergency obstetric care or by meeting unmet obstetric need, is a prime focus of the Safe Motherhood Initiative and the Maternal Survival Series steering group [8]. Fully recognizing the urgency and importance of effective intrapartum care [3] and similar strategies, this chapter examines the evidence for the role of nutrition in contributing to a reduction in maternal mortality in the developing world. The purpose is not to detract attention from ongoing initiatives and partnerships focused on skilled birth attendance, intrapartum, and emergency obstetric care, but to add to these a consideration of specific nutritional interventions with known established efficacies for enhancing maternal health and survival. Poor nutrition is often referred to in the same broad terms as economic development, poverty, low education, and poor access to medical care [9]—factors that are known to be underlying contributors to the burden of maternal mortality and that are likely to take much longer to address. Another well-accepted notion is that it is hard to prevent maternal health risks—because every pregnancy faces a risk. Nonetheless, evidence suggests that some nutritional interventions can reduce the burden of life threatening morbidities that result in maternal death, especially in many underserved settings of the world where health care access is poor and maternal malnutrition is high. This chapter examines such interventions and strategies and provides the biologic basis, rationale, and empirical evidence for the impact of nutrition in ameliorating the risk of maternal mortality in the developing world.

21.2   CAUSES OF MATERNAL MORTALITY AND THE LINK WITH NUTRITION

Maternal death is defined as death during pregnancy or within 42 days of the end of a pregnancy from any cause related to or brought on by the pregnancy, or its management but not from accidental and incidental causes [10]. Deaths from both direct (resulting from obstetric complications of pregnancy) and indirect causes (resulting from previous existing diseases or diseases that developed during pregnancy) are included in calculating maternal mortality. Measuring maternal mortality is complex due to the lack of adequate data on the timing and cause of deaths in many regions of the world where a large proportion of births do not occur in hospitals. A recent WHO systematic review of causes of maternal deaths revealed wide regional variations [11]. Hemorrhage was the most common cause, accounting for approximately a third of maternal deaths in both Africa and Asia. On the other hand, hypertensive disorders were a leading cause of maternal mortality in Latin America (25.7%), followed closely by hemorrhage (20.8%). Hypertensive disorders contributed about 9% of deaths in Africa and Asia, whereas deaths from sepsis/infections ranged from 9 to 12% in these countries. Anemia was an important cause of death in Asia (12.8%) but less so in Africa (3.7%), and Latin America (0.1%). Obstructed labor, e.g., a labor in which something is preventing the normal process of labor and delivery, contributed to 13.4% of maternal deaths in Latin America, 9.4% in Asia, and 4.1% in Africa. HIV/AIDS contributed to 6.2% of maternal deaths in Africa, although this may be an under representation, as deaths due to HIV/AIDS are often classified under "indirect causes" [12].

The sections that follow describe the evidence linking maternal nutritional deficiencies to maternal mortality. Specifically, the association between anemia and maternal mortality and hemorrhage is examined, including the efficacy of iron supplementation and other interventions in reducing maternal anemia. The role of calcium and antioxidants in the prevention of hypertensive disease and preeclampsia, and the efficacy of magnesium sulfate in the prevention of eclampsia is reviewed as well as the link between sepsis and infection and maternal vitamin A and zinc deficiencies. Finally, causes of obstructed labor and nutritional factors related to maternal stunting with focus on growth in childhood and adolescence are discussed.

21.2.1   Maternal Anemia and Maternal Mortality

The relationship between maternal anemia and the risk of mortality have been examined in two previous reviews [13, 14]. No randomized controlled trials to date provide data on the impact of iron supplementation on maternal mortality as the outcome. The likelihood of such trials being conducted in the future is low, mainly due to ethical and feasibility considerations. Observational studies conducted in Africa and Asia, primarily among pregnant women presenting at hospitals, provide evidence for the association between hemoglobin concentration upon admission at a hospital or clinic and maternal mortality. None of these studies provides information on iron deficiency per se or the proportion of anemia attributable to iron deficiency. Anemia, especially severe, can arise from multiple causes including malaria (mainly *Plasmodium falciparum*), hookworm, vitamin deficiencies, and chronic infections such as HIV. However, as shown by the review by Brabin et al. [14], *P. falciparum* malaria as an etiology of anemia may be less important in leading to maternal death. In holoendemic malarious settings, an estimated 9 versus 41 deaths per 100,000 are due to malaria-related severe anemia compared with nonmalarial anemia deaths among primagravidae [14].

Previously, severe anemia alone was considered to be associated with an increased risk of maternal mortality [13, 14], and the population attributable risk was strong for severe but not moderate anemia (Table 21.1) [14]. Severe anemia (normally defined as hemoglobin [Hb] < 70 g/l) can result in circulatory decompensation and increased cardiac output at rest. The added stress of labor and blood loss, whether normal or excessive, can lead to circulatory shock and death. In most settings, however, the prevalence of mild-to-moderate anemia (70–110 g/l) tends to be much higher than that of severe anemia [15]. Recently data from nine studies were examined to determine the relationship between Hb concentration and case fatality (Figs. 21.1, 21.2) (R.J. Stoltzfus and L. Mullany, unpublished data) for estimating the global burden of disease linking iron deficiency to disability and death [16]. Hb data collected in these studies when plotted by the observed proportion of maternal

Table 21.1

Relative Risk and Population Attributable Ratio (*PAR*) of Maternal Mortality for Moderate and Severe Maternal Anemia

	Relative risk	95% CI	PAR	PAR
Population prevalence			5%	20%
Moderate anemia (Hb 40–80 g/l)	1.35	0.92, 2	0.017	0
Severe anemia (Hb 27–47 g/l)	3.51	2.05, 6	0.111	0.334

From [14]

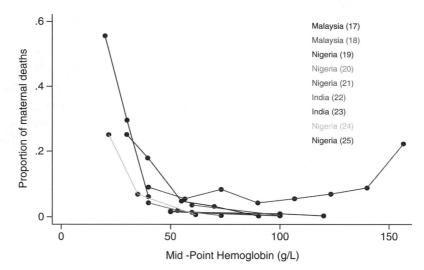

Fig. 21.1. Proportion of maternal deaths by hemoglobin concentration among pregnant women in nine studies

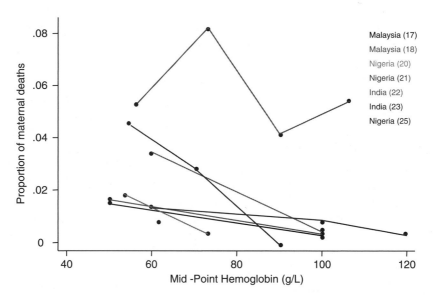

Fig. 21.2. Proportion of maternal deaths by hemoglobin concentration ranging between 50 and 120 g/l among pregnant women in nine studies

deaths revealed a threshold relationship; case fatality increased dramatically at maternal Hb concentration below 50 g/l (Fig. 21.1). Using the same data but limiting the range of Hb concentration to between 50 and 120 g/l revealed the relationship to be a linear one in most countries, suggesting an inverse, continuous relationship between the two variables within the narrower range of Hb (Fig. 21.2). This newly defined association was used to model the decrease in proportion of maternal deaths with a unit increase in Hb concentration [16]. The WHO prevalence estimates for anemia were converted to mean Hb concentrations, assuming a normal distribution. These values were further adjusted to reflect a distribution of Hb for iron deficiency anemia assuming that iron deficiency contributes to 50% of anemia, globally [26]. Using relative risk estimates from published studies, an odds ratio estimate of 0.75 (95% confidence interval [CI]: 0.62, 0.89) was calculated, indicating a decrease of 25% in the odds of maternal death with every 10g/l increase in the population in mean Hb concentration [16]. Antenatal iron supplementation at dosages ranging between 60 and 120 mg/day and duration between 10 and 12 weeks has been shown to increase Hb concentrations by about 80–140 g/l [26]. Thus, antenatal iron supplementation would be an important strategy for combating the risk of anemia-related maternal mortality in the developing world. Also, it is noteworthy that the mortality risk increases precipitously at extremely low Hb concentrations (<50 g/l) (Fig. 21.1). However, because vast majorities of pregnant women are likely to be mildly-to-moderately and not severely anemic, the relationship that is observed between Hb concentrations between 50 and 120 g/l may be more meaningful in describing the risk at a population level. Furthermore, iron deficiency is unlikely to be the cause of such severe anemia. Rather, other etiologies such as malarial infection, hookworm, or chronic diseases may be responsible [13, 14]. A note of caution with respect to the anemia-maternal mortality relationship; confounding and bias cannot be overruled since data examining this relationship were derived solely from observational studies of women who may have presented to the hospital with multiple morbidities and whose Hb was assessed at the time of booking and not in early or even mid-pregnancy.

Recent studies that are not part of the previous reviews (as mentioned above) have also linked maternal anemia to the risk of maternal mortality and morbidity. In a small case-control study in Ghana, anemic women (defined as Hb <80 g/l) experienced more deaths (5/157) compared with age- and parity-matched controls (0/152) with Hb >109 g/l [27]. Anemic cases that were treated in this study (treatment unspecified) had an increase in median Hb from 65 to 95 g/l. In a second small study from India (n = 447 pregnancies), Hb was assessed as part of an antenatal exam (gestational age unspecified) [28]. While too small to examine mortality as an outcome, this study showed that subjects with Hb < 89 g/l had a four- to sixfold higher risk of prolonged labor compared with those with Hb >110 g/l. Similarly, the risk for caesarean section and "operative" vaginal delivery was higher by about the same magnitude. Both studies were observational with low sample sizes and failed to adjust for confounding variables. The lack of adjustment for confounding is problematic, as illustrated by a recent study of women with HIV in Tanzania [29]. Data in this study were analyzed linking anemia during pregnancy to female mortality (Table 21.2). While the relative hazards for both all-cause and AIDS-specific mortality increased with increasing severity of anemia, adjustment attenuated the magnitude of the risk and reduced the differential in the excess risk between moderate and severe anemia. Although these were not maternal deaths per se, and the median follow-up period represented in the analysis was 5.9 years (interquartile range: 3.8–6.7 years), this study showed an increased risk of mortality among HIV-infected women related to anemia. HIV infection is increasingly likely to contribute to the risk of anemia especially in the context of sub Saharan Africa.

Anemia and primary postpartum hemorrhage (PPH) together contribute to 40–43% of maternal deaths in Africa and Asia, where the burden of maternal mortality is the

Table 21.2
Anemia as a Predictor of Mortality among Women with HIV in Tanzania

| | All cause mortality | | AIDS-related mortality | |
| | Relative Hazards (95% CI) | | | |
Anemia	Unadjusted	Adjusted[a]	Unadjusted	Adjusted[a]
Moderate (Hb = 85–109 g/l)	2.7 (2–3.6)	2.1 (1.5–2.8)	3 (2.1, 4.2)	2.2 (1.5–3.2)
Severe (Hb < 85 g/l)	6.2 (4.4–8.6)	3.2 (2.2–4.6)	7.2 (4.8, 10.7)	3.5 (2.2–5.3)

From [29]

[a]Adjusted for CD4 count, WHO clinical stage, age, pregnancy, treatment arm in the study, and body mass index

highest [11]. While underlying anemia is considered to exacerbate the deleterious effect of PPH and death due to this cause, on its own it contributes to 9.1% of deaths in these two regions [11]. Although it is commonly stated that severe anemia can exacerbate the risk of death due to PPH, there are no empirical data to support this. In a series of 40 PPH deaths that occurred between April 1982 and April 2002 in rural India, hospital records indicated that 47.5% had severe anemia (Hb < 70 g/l) at the time of admission, and another 45% had Hb between 70 and 90 g/l [30]. However, without controls, it is hard to predict the mortality due to PPH among nonanemic women.

It has long been considered that anemia increases the risk of PPH [31, 32], although data supporting this are scant. The main causes of PPH include uterine atony, placental retention, trauma, and coagulopathy [33].

Few studies exist that have examined the risk of PPH itself by level of anemia. The few studies that have examined the risk, indicate a weak association (Table 21.3). Among emergency room patients admitted to a hospital in Auckland, New Zealand, between January and August of 1966, of 1,743 anemic women, 159 suffered from PPH compared with 15 out of 170 nonanemic women, suggesting no difference in the risk of PPH by anemia status [34]. More recently, Geelhoed et al. [18], using a cohort design in two subdistrict hospitals of Ghana, compared the risk of PPH among 157 severely anemic (Hb < 80 g/l) and 152 nonanemic pregnant women (Hb ≥ 109 g/l) matched for age and parity and found no difference. On the other hand, Tsu [35] in a multivariate logistic regression analysis, after adjusting for other risk factors such as maternal age, parity, and antenatal hospitalization, reported that Zimbabwean women with PPH after normal vaginal deliveries were more likely to be anemic compared to those that did not experience PPH. In a study conducted in a hospital in Nigeria among 101 women who developed PPH and 107 controls, there was no difference in the prevalence of anemia measured during pregnancy [36]. Finally, in a hospital study, 374 cases of PPH were derived from 9,598 vaginal deliveries and matched to controls (ratio: 1:3) [37]. Cases had a significantly higher hematocrit (not lower) compared with controls at admission. There was no association between antenatal anemia and uterine atony, an important cause of PPH in a tertiary care hospital–based study in Pakistan [38].

In summary, study results are suggestive of neither (1) an increased risk of mortality due to PPH as a result of underlying anemia nor (2) an increased risk of PPH due to maternal anemia during pregnancy. However, as described previously, the association between maternal mortality and anemia remains consistent, albeit derived from obser-vational studies. In addition, antenatal iron-folate supplementation is likely to affect

Table 21.3

Risk of Postpartum Hemorrhage and Maternal Anemia

Reference	Population			
Risk of PPH: cohort studies		*Anemic (%)*	*Nonanemic (%)*	*P value*
25[a]	Auckland, New Zealand, booked emergency patients, 1,743 anemic vs. 170 nonanemic women	10.8	8.8	NS
18[b]	157 severely anemic and 152 nonanemic Ghanaian women	11.3	12.1	NS
Risk of anemia: case-control studies		*Cases (%)*	*Non (%)*	*P value*
26[c]	151 cases of PPH and 299 controls	39.6	19	0.05
27[d] 1997	101 cases of PPH and 107 controls	5.9	4.7	NS
28 Hct, mean±SD	374 cases and 1,122 controls	36.6±2.1	35.6±2.3	0.01

PPH postpartum hemorrhage, *NS* not significant, *Hct* hematocrit

[a]Anemia defined as Hb<105 g/l and/or hematocrit <35 % or Hb > 105 g/l but blood film showing anisocytosis, microcytosis and hypochromia

[b]Anemia defined as Hb <80 g/l and nonanemic women had Hb ≥109 g/l

[c]Hb < 120 g/l

[d]Anemia was undefined

other outcomes, reducing low birth weight and preterm delivery, and increasing infant iron stores. (These outcomes are discussed in Chap. 22, "Anemia and Iron Deficiency in Developing Countries".) There exists an international policy for iron-folate supplementation during pregnancy for women in the developing world [39]. This policy states that pregnant women should receive 60 mg iron and 400 mcg folic acid for 6 months during pregnancy in settings where the prevalence of anemia is <40%, and for 6 months during pregnancy through 3 months postpartum in settings where anemia prevalence is ≥40%. However, programs have failed to effectively reduce the prevalence of maternal anemia in many regions of the world. Anemia continues to affect 50–60% of women during pregnancy in countries in Asia and Africa. Data from Demographic and Health Surveys (DHS) reveal low rates of any antenatal iron use in developing countries, with the proportion of women taking at least 90 tablets, an amount recommended during pregnancy, being even lower in many countries (Fig. 21.3).

Iron supplementation, which usually is effective in halving anemia rates in developed countries is an insufficient strategy for combating the burden of this condition in the developing world. Malaria and hookworm control, reducing nutritional deficiencies in addition to iron and folic acid, and increasingly, treatment of HIV should be concurrent approaches for reducing maternal anemia in the developing world. The major barrier to effective supplementation programs is inadequate supply of iron tablets [41]. In a country such as Malawi, with one of the highest maternal mortality ratios (984/100,000 live births), women consider anemia to be a maternal health concern [42]. Similarly, perceptions of improvement in physical well-being and alleviation of symptoms of fatigue and poor appetite due to iron use would be helpful in efforts aimed at promoting antenatal iron supplementation in the developing world [41].

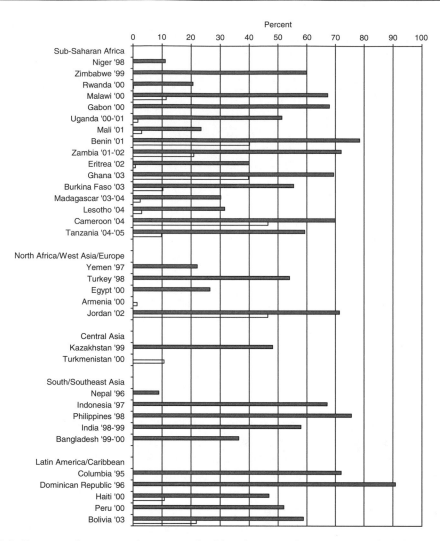

Fig. 21.3. Percent of women who reported taking iron supplements [any (dark bars) and 90+ tablets (light bars)] during a previous pregnancy in the past 3–5 years. Data on prevalence of consumption of 90+ tablets were not available for many countries. (From Demographic and Health Surveys and [40])

21.2.2 Nutrition and Hypertensive Disease of Pregnancy

It has been argued that while nutritional factors such as calcium intake may modify the risk of preeclampsia, eclampsia, and hypertension, it is unlikely that nutritional factors are important in reducing the risk of mortality due to these causes. This argument is based on data that show that mortality due to preeclampsia, eclampsia, and hypertension has not decreased or varied over time or by condition [43].The recent WHO systematic review of causes of maternal mortality shows high regional variation in preeclampsia associated mortality [11], suggesting that the above argument is refutable. Historically, observations that Mayan Indians who consumed a traditional diet of corn soaked in lime had high calcium intakes [44], and Ethiopian women whose diets contained high amounts of calcium had low rates of preeclampsia [45] support the role of calcium in reducing hypertensive disorders of pregnancy. A dietary deficit in calcium is postulated to cause an increase in blood pressure due to stimulation of the parathyroid gland and release of parathyroid hormone, resulting in movement of calcium intracellularly in vascular smooth muscles and causing an intensified reactivity and vasoconstriction [46]. A review of the numerous epidemiologic and clinical data on this topic is beyond the scope of this chapter. A recent Cochrane systematic review of 12 randomized clinical trials show reductions of 30% and 52% in the risk of high blood pressure (RR = 0.70, 95% CI: 0.57–0.86) and preeclampsia (RR = 0.48, 95% CI: 0.33, 0.69), respectively, as a result of daily 1–2 g of calcium supplementation during pregnancy [47]. The effects were evident among the high-risk women and in those with low baseline calcium intake. Based on data from four trials, a composite outcome of maternal death or serious morbidity was reduced by 20% (RR = 0.80, 95% CI: 0.65, 0.97). Maternal death was reported as an outcome in a recent multicenter, double-blinded, placebo-controlled trial of daily 1.5 g calcium, starting from before 20 weeks of gestation among women with low intake of calcium [48]. This trial, while not powered to show an impact on maternal mortality, reported 1/4151 and 6/4161 maternal deaths in the calcium and placebo arms, respectively (RR = 0.17, 95% CI: 0.03–0.76).

Previously, Bucher et al. [49] in a meta-analysis of 14 small randomized controlled trials (total n = 2459) showed statistically significant reductions in the risk of hypertension and preeclampsia in the magnitude of 60–70% because of calcium supplementation at 1–2 g per day. However, a trial (Calcium for Preeclampsia Prevention, CPEP) in the United States, which included twice the number of women as in the 14 trials combined in the meta-analysis, failed to show an effect on these outcomes with 2 g of calcium supplementation per day [50]. However, women in this trial were healthy and were consuming about 1.1 g calcium at baseline, suggesting that they were unlikely to be deficient. Nonetheless, the conflicting results from the meta-analysis and this single large trial led to equipoise with regard to programmatic and policy implications for calcium supplementation in the prevention of hypertensive disease in pregnancy. Subsequently, the WHO six-center trial of calcium supplementation among women consuming a low calcium diet (<600 mg/day) showed no affect on preeclampsia or hypertension but significantly reduced the incidence of eclampsia (alone) and severe hypertension (defined as a systolic blood pressure ≥160 or a diastolic blood pressure ≥110 mmHg) [48]. The Cochrane meta-analysis shows that calcium supplementation during pregnancy is an economical and safe way of reducing the risk of preeclampsia, especially among women with low calcium intakes or at high risk for hypertensive disorders in pregnancy [47]. However, the optimal dosage of calcium for supplementation during pregnancy to provide the most benefit remains uncertain.

The use of magnesium sulfate, an anticonvulsant, among women with symptoms of preeclampsia (hypertension and proteinuria) for the prevention of seizures in women with eclampsia is routine in the United States [51]. A Cochrane meta-analysis of data from six randomized, controlled trials showed that magnesium sulfate halved the risk of eclampsia (RR = 0.41, 95% CI: 0.29–0.58) and reduced the risk of death by 46%, although not significantly (RR = 0.54, 95% CI: 0.26, 1.10) [52]. The largest trial thus far, the Magpie Trial, contributed 10,141 women, many of whom were from developing countries, to the Cochrane analysis [53]. Flushing, a side effect that was assessed in these studies, was elevated among those who received magnesium sulfate compared with the placebo, but there were no other adverse side effects. The risk of eclampsia was reduced in women with both mild and severe preeclampsia. Magnesium sulfate treatment when tested against other anticonvulsants such as phenytoin and nimodipine, performed better, although against phenytoin it appeared to increase the risk of caesarean section deliveries [52, 54]. Thus, magnesium sulfate can be considered an inexpensive and easy means for preventing and treating eclampsia and maternal deaths [55], but policies and programs to implement its use are still not in place. There is also no screening for preeclampsia in place at prenatal clinics in many regions of the world. In fact, in many settings prenatal check-ups may not even be available to women during pregnancy. What also remains unclear is whether magnesium sulfate can prevent preeclampsia and if magnesium deficiency plays a role in the etiology of this disorder.

Oxidative stress has been proposed to have a potential role in the two-stage model of preeclampsia [54, 56, 57]. The first stage in this model is reduced placental perfusion, resulting from abnormal implantation or other pathologies. The second stage involves the maternal hypertensive/inflammatory response that may be influenced by environmental factors and oxidative stress [56]. Trophoblastic cells isolated from the placenta of preeclamptic women have increased superoxide generation and decreased superoxide dismutase activity, supporting the hypothesis that increased oxidative stress plays a role in the pathology for preeclamptic placentae [58]. In a small randomized, placebo-controlled trial, daily vitamin C (100 mg) and E (400 IU) from 16 to 22 weeks of gestation significantly reduced the risk of preeclampsia [59]. The plasminogen activator inhibitor ratio (PA1:PA2), which is elevated in preeclampsia, significantly decreased due to supplementation, suggesting a reduction in endothelial activation and placental dysfunction. Subsequently, however, a larger trial among 2,410 women at risk of preeclampsia failed to replicate these results [60]. Daily supplementation with vitamins C and E provided at the same dosages used in the previous trial failed to affect the incidence of preeclampsia (15 vs. 16% in the placebo group). Two other trials also failed to show a difference in the risk of preeclampsia between those who received vitamins C and E supplementation compared with placebo [61, 62]. Thus, at present the use of antioxidants during pregnancy does not appear to hold merit for either preventing eclampsia among high-risk women or as a prophylaxis for preventing the risk of preeclampsia during pregnancy.

21.2.3   Vitamin A, Maternal Mortality, and Infection

Maternal vitamin A deficiency affects 5.6% of pregnant women worldwide [63]. Maternal night blindness, resulting from vitamin A deficiency, annually affects 5–15% women during pregnancy in the developing world [64]. A randomized trial of vitamin A and beta-carotene supplementation to women 15–45 years of age in rural Nepal showed a 40 and 50% reduction, respectively, in all cause pregnancy-related mortality relative to a placebo, although cause-specific mortality using verbal autopsy data provided no information for type of maternal causes of death that were affected [65]. However, pregnant women who experienced night blindness in the placebo group and representing a subgroup at high risk experienced a significantly increased risk of pregnancy-related (42 days) and 2-year postpartum mortality compared to non –night-blind women or night-blind women who received either vitamin A or beta-carotene [66]. Moreover, causes of death among night-blind women were more likely (RR = 5, 95% CI: 2.2–10.6) to be infection. It is well known that vitamin A deficiency can reduce immunocompetence and resistance to severe infection, and that vitamin A maintains the integrity of epithelial and endothelial cells and preserves natural killer cell activity. While the relationship between vitamin A deficiency and childhood morbidity and mortality is well established, only two previous studies provide the link between vitamin A deficiency and puerperal infection [67]. For example, in England, a trial conducted in the 1930s showed that maternal vitamin A supplementation in later pregnancy through the first week postpartum reduced the occurrence of puerperal sepsis [68], a finding replicated more recently among Indonesian women [69]. There was a 78% reduction in puerperal infection assessed by the occurrence of fever in the first 10 days postpartum following supplementation.

Other biologically plausible mechanisms for the vitamin A effect are described by Faisel et al. [33]. The trial in Nepal reported a reduction in the length of labor (by 1.5 h among primiparas and 50 min in multiparas) associated with vitamin A supplementation [70]. The hematopoietic activity of vitamin A is well known and vitamin A may act by reducing anemia, one of the known causes of maternal deaths. Using data from Nepal, it is estimated that about 20% of all cause maternal mortality could be attributable to vitamin A deficiency [63]. Two trials in Bangladesh and Ghana are currently underway to test the efficacy of vitamin A in reducing maternal mortality. Results of these trials are awaited before any policy or program decisions are made for supplementing women with vitamin A.

The role of zinc in reproductive health is well established in animal models and in small observational studies in humans [71, 72]. Yet zinc supplementation trials have not provided evidence for an impact on pregnancy and maternal health outcomes [71, 72]. A two- to ninefold increased odds of premature rupture of membrane, preterm birth, protracted first- and second-stage labor, and inefficient uterine contractions were associated with zinc deficient diets or low plasma zinc concentrations [72]. However, supplementation trials have failed to show efficacy, perhaps due to inadequate sample size, or because the studies were conducted in populations that were not likely to be zinc deficient. In a recent review of eight randomized trials of antenatal zinc supplementation conducted in developing countries, there was no clear evidence of a benefit of maternal zinc supplementation on maternal outcomes, although pregnancy or delivery complications were not assessed in some of these studies [73]. In a study in Central Java, daily zinc supplementation failed to reduce postpartum infection [69], whereas in a study conducted in West Java, zinc supplementation was associated with significantly higher delivery complications relative to the controls ($n = 12$ vs. 3) [73]. Thus, it is not clear if zinc supplementation in pregnancy can result in a decrease in pregnancy-related infection or other adverse outcomes or perhaps even increase delivery complications.

21.2.4   Maternal Height, Nutritional Status, and Dysfunctional Labor

Unlike in developed countries, where it is no longer a cause of maternal mortality, obstructed labor is a cause of death in 10–15% of women in the developing world [11]. The inverse relationship between maternal height and the risk of dystocia (difficult labor) due to cephalopelvic disproportion (CPD) obstructed labor due to a disparity between the dimensions of the fetal head and maternal pelvis or assisted/caesarean-section deliveries is frequently described but has been weak. A meta-analysis of nine studies found low sensitivity and specificity of low maternal height (<20th percentile) in predicting the risk of dystocia [74]. Because the prevalence of caesarean section in developing countries is low (at about 2%), the predictive value of low maternal height for caesarean section as an outcome is only 5%. Further, the relationship between maternal height and dystocia appears to be relative and not absolute. Height in the analysis was expressed in percentiles and, thus, the absolute value of a 20th percentile differed in different populations despite the risk of dystocia at that cutoff being of the same magnitude. In practical terms, therefore, using a single value of height as a cut-off would not be appropriate for predicting the risk of dysfunctional labor. In an earlier WHO meta-analysis of 16 studies, the odds of nonspontaneous deliveries, including caesarean section, was 60% higher among women in the lowest height quartile compared with those in the highest quartile [75].

The size of the fetus is also a cofactor in this relationship. Harrison et al. [76] showed that the risk of operative delivery was associated with both maternal height and the size of the baby; low maternal height and high birth weight increased the risk of operative delivery (Fig. 21.4).

However, maternal height, which is correlated with uterine volume, is a strong predictor of birth weight, and thus it is unlikely that the shortest women would produce truly large babies. But in underserved settings, the risk of obstructed labor may increase even at a birth weight that would be considered "normal" in a developed country [77]. Evidence that interventions such as food supplementation during pregnancy, aimed at improving birth weight in malnourished settings, can increase the risk of CPD or obstructed labor, however, is lacking. There was no evidence of an increased risk of CPD or complications in delivery in an antenatal food supplementation trial in the Gambia in which birth weight increases ranged from 100 to 200 g, [78]. The largest increase in head circumference, where head size was more strongly correlated with CPD than birth weight per se [13], was only 3 mm. This translates to an increase of only 1 mm in diameter, which is unlikely to raise the prevalence of CPD. Perinatal mortality was reduced due to maternal supplementation, further attesting to the lack of evidence for any harmful outcome as a result of the increase in birth weight but rather a benefit [78]. Whether these findings can hold in more malnourished South Asian settings where maternal stunting is more prevalent is

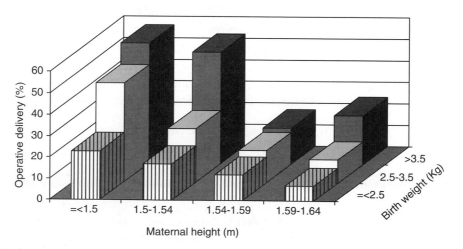

Fig. 21.4. Rate of operative delivery by maternal height and birth weight. (From [76])

unclear. In two recent randomized controlled trials of antenatal multiple micronutrient supplementation in rural Nepal, mean birth weight improved by about 60–100 g [79, 80]. However, neonatal mortality was slightly, although not significantly, elevated in both studies, an increase that was significant when the data were pooled [81]. When treatment effects were examined by percentiles of birth weight in one of the studies, it was apparent that maternal multiple micronutrient supplementation increased birth weight both in the lower and upper tails of the distribution [82]. The increase on the upper tail may explain the increased risk of birth asphyxia and mortality in these infants as a result of maternal micronutrient supplementation [83]. However, other mechanisms may have also resulted in an adverse effect of the intervention on infant outcomes. Currently antenatal micronutrient supplementation beyond iron–folic acid is being evaluated for its safety and efficacy in the developing world.

Addressing the problem of short maternal stature and stunting is important for improving reproductive health in the developing world. However, few interventions are known to influence maternal height. Food supplementation between 6 and 24 or even up to 36 months of age may promote accelerated linear growth, but interventions beyond this period do not show further benefit [84]. Interventions during school age or adolescence appear to only modulate onset of menarche [5]. Based on adoption studies, relocation from environments that give rise to stunting may promote catch up growth but only among younger children (<2 years old) [84, 13]. In older adopted children accelerated maturation (early onset of menarche) resulted in a shorter growth period that led to little overall benefit in terms of attained height. Based on data from nine European countries, mean age at menarche has decreased in Europe by 44 days (18–58) as indicated by follow-ups of 5-year birth cohorts (youngest cohort: 1915–1919, oldest cohort: 1960–1964) [85]. Simultaneously, European women have grown taller over time, from 0.42 to 0.98 cm per 5-year birth cohort. However, women grew 0.31 cm taller when menarche occurred a year later due to the later closure of the epiphyseal plates of the long bones following onset of menarche. Will a decrease in the age of onset of menarche in the developing world as a result of nutrition interventions result in yet a shorter period of growth as suggested in adoption studies? This is an intriguing question.

21.2.5 *Maternal Undernutrition and Maternal Mortality*

The risk of maternal mortality related to undernutrition or wasting malnutrition has not been examined in many studies. While it is commonly assumed that women who die each year from pregnancy-related causes are likely to be "short, thin, and anemic" [86], empirical data to support that the underlying cause for mortality is wasting malnutrition are lacking. In an analysis of risk factors collected during the first half of gestation among ~22,000 pregnant women in two supplementation studies in Nepal, maternal mid–upper arm circumference (MUAC), an indicator of thinness, was significantly and independently associated with the risk of maternal mortality in a multivariate analysis [87]. For each 1-cm increase in maternal first-trimester MUAC, the risk of maternal death decreased by 24% (adjusted OR = 0.76, 95% CI: 0.67, 0.87). There are no data to show that improving maternal nutrition via food supplementation either during or prior to pregnancy results in an improvement in survival. While on the one hand, maternal undernutrition may put women at an increased risk of mortality as evidenced by the link with MUAC, there is a concern that dystocia due to higher birth weight where maternal stunting is common, such as in South Asia and Latin America [13], may pose a risk. Thus, the benefits of food supplementation during pregnancy are unclear and data on a wide range of maternal, fetal, and infant outcomes from these regions are needed. The intergenerational cycle of maternal stunting leading to low birth weight and future risk of stunting [88], needs to be broken. Improved management of labor and delivery [12] may have the best potential for breaking this cycle, while attempts continue to improve nutritional status and linear growth at various stages of life. The developed world no longer faces the risk of pregnancy-related mortality due to obstructed labor.

For the developed world and countries in transition, a growing concern is the rising prevalence of overweight and obesity in pregnancy [89]; see Chap. 20, ("Implications of the Nutrition Transition in the Nutritional Status of Pregnant Women"). WHO estimates that 9–25% of women in developed countries are obese. Obesity in pregnancy may increase the risk of early miscarriage by 20–25% and preeclampsia and gestational diabetes mellitus by approximately five- to sixfold. It may also be associated with the increased risk for stillbirth, caesarean section, macrosomia, or preterm birth [89]. With developing countries and countries in transition now witnessing overweight and obesity, food supplementation and micronutrient interventions during pregnancy will need to be in light of this expanding global problem. In a randomized controlled trial of antenatal multiple micronutrient supplementation in periurban Mexico women who had received multiple micronutrients were more likely to retain weight postpartum compared to those in the control group [90].

21.3   CONCLUSION

Anemia contributes to 10% of maternal deaths in Asia and Africa. Antenatal iron supplementation needs to be heightened in many regions where the rates of maternal anemia are high. There is little evidence to suggest that anemia can increase the risk of PPH. Calcium supplementation in populations with low intakes of dietary calcium can reduce the risk of eclampsia and severe morbidity and mortality, although the optimal dosage remains unclear. Magnesium sulfate is an inexpensive means to prevent the risk of eclampsia among high-risk women with preeclampsia. However, in settings where home deliveries are common, it is unclear how management of preeclampsia with magnesium sulfate should be implemented. Maternal vitamin A supplementation reduced the risk of pregnancy-related mortality in one study; results of two other trials are awaited. Overall, there appears to be a role for nutrition interventions in reducing the risk of maternal mortality in the developing world, but antenatal programs with strong nutritional components that reach a high proportion of pregnancies need strengthening before a substantial impact can be achieved.

REFERENCES

1. World Health Organization (WHO), United Nations' Children's Fund, United Nations' Population Fund. (2004) Maternal mortality in 2000: estimates developed by WHO, UNICEF, UNFPA. World Health Organization, Geneva, Switzerland
2. Ronsmans C, Graham WJ, on behalf of the Lancet Maternal Survival Series steering group (2006) Maternal mortality: who, when, where, and why. Lancet 368:1189–1200
3. Campbell OMR, Graham WJ, on behalf of the Lancet Maternal Survival Series steering group. (2006) Strategies for reducing maternal mortality: getting on with what works. Lancet 368:1284–1299
4. Koblinsky M, Matthews Z, Hussein J, Mavalankar D, Mridha MK, Anwar I et al (2006) Going to scale with professional skilled care. Lancet 368:1377–1386

5. Borghi J, Ensor T, Somanathan A, Lissner C, Mills A, on behalf of the Lancet Maternal Survival Series steering group (2006) Mobilising financial resources for maternal health. Lancet 368:1457–1465

6. Filippi V, Ronsmans C, Campbell OMR, Graham WJ, Mills A, Borghi J et al (2006) Maternal health in poor countries: the broader context and a call for action. Lancet 368:1535–1541

7. Simwaka BN, Theobald S, Amekudzi YP, Tolhurst R (2005) Meeting millennium development goals 3 and 5. BMJ 331:708–709

8. Rosenfield A, Maine D, Freedman L (2006) Meeting MDG-5: an impossible dream? Lancet 368:1133–1135

9. Starrs AM (2006) Safe motherhood initiative: 20 years and counting. Lancet 368:1130–1132

10. WHO (1992) Statistical classification of diseases and related health problems, 10th revision. WHO, Geneva, Switzerland

11. Khan KS, Wojdyla D, Say L, Gulmezoglu AM, Van Look PF (2006) WHO analysis of causes of maternal death: a systematic review. Lancet 367:1066–1074

12. Ronsmans C (2001) Chap. 2: Maternal mortality in developing countries. In: Semba RD, Bloem MW (eds) Nutrition and health in developing countries. Humana, Totowa, N.J. pp. 31–56

13. Rush D (2000) Nutrition and maternal morbidity in the developing world. Am J Clin Nutr 72(Suppl):212S–240S

14. Brabin BJ, Hakimi M, Pelletier D (2001) An analysis of anemia and pregnancy-related maternal mortality. J Nutr 131:604S–615S

15. Stoltzfus RJ (1997) Rethinking anaemia surveillance. Lancet 349:1764–1766

16. Stoltzfus RJ, Mullany L, Black RE (2004) Chap. 3: Iron deficiency anaemia. In: Ezzati M et al (eds) Comparative quantification of health risks: global and regional burden of disease attributable to selected major risk factors, vol. 1. WHO, Geneva, Switzerland, pp. 163–209

17. Llewellyn-Jones D (1965) Severe anaemia in pregnancy. Aust N Z J Obstet Gynaecol 5:191–197

18. Tasker PWG (1958) Anaemia in pregnancy. A five-year appraisal. Med J Malaysia 8:3–8

19. Fullerton WT, Turner AG (1962) Exchange transfusion in treatment of severe anaemia in pregnancy. Lancet 1:75–78

20. Harrison KA (1975) Maternal mortality in anaemia in pregnancy. West Afr Med J 92 (Suppl 5):27–31

21. Harrison KA, Rossiter CE (1985) Maternal mortality. Br J Obstet Gynaecol 92(suppl 5) 5:100–115

22. Konar M, Sikdar K, Basak S, Lahiri D (1980) Maternal mortality. J Indian Med Assoc 75:45–51

23. Sarin AR (1995) Severe anaemia of pregnancy, recent experience. Int J Gynaecol Obstet 50:S45–S49

24. Johnson JWC, Ojo OA (1967) Amniotic fluid oxygen tensions in severe maternal anaemia. Am J Obstet Gynecol 97:499–506

25. Harrison KA (1982) Anaemia, malaria, and sickle cell disease. Clin Obstet Gynaecol 9:445–477

26. Sloan NL, Jordan E, Winikoff B (2002) Effects of iron supplementation on maternal hematologic status in pregnancy. Am J Public Health 92:288–293

27. Geelhoed D, Agadzi F, Visser L et al (2006) Maternal and fetal outcome after severe anemia in pregnancy in rural Ghana. Acta Obstet Gynecol Scand 85:49–55

28. Malhotra M, Sharma JB, Batra S, Sharma S, Murthy NS, Arora R (2002) Maternal and perinatal outcome in varying degrees of anemia. Int J Gynaecol Obstet 79:93–100

29. O'Brien ME, Kupka R, Msamanga GI, Saathoff E, Hunter DJ, Fawzi WW (2005) Anemia is an independent predictor of mortality and immunologic progression of disease among women with HIV in Tanzania. J Acquir Immune Defic Syndr 40:219–225

30. Chhabra S, Sirohi R (2004) Trends in maternal mortality due to haemorrhage: two decades of Indian rural observations. J Obstet Gynaecol 24:40–43

31. Stafford JL (1961) Iron deficiency in man and animals. Proc R Soc Med 54:1000–1004

32. Harris C (1957) The vicious circle of anaemia and menorrhagia. Can Med Assoc J 77:98–100

33. Faisel H, Pittrof R (2000) Vitamin A and causes of maternal mortality: association and biological plausibility. Public Health Nutr 3:321–327

34. Dewar MJ (1969) Antenatal anaemia and postpartum haemorrhage. Aust N Z J Obstet Gynaecol 9:18–20

35. Tsu VD (1993) Postpartum haemorrhage in Zimbabwe: a risk factor analysis. Br J Obstet Gynaecol 100:327–333

36. Selo-Ojeme DO, Okonofua FE (1997) Risk factors for primary postpartum haemorrhage: a case control study. Arch Gynecol Obstet 259:179–187

37. Combs CA, Murphy EL, Laros RK Jr (1991) Factors associated with postpartum hemorrhage with vaginal birth. Obstet Gynecol 77:69–76

38. Feerasta SH, Motiei A, Motiwala S, Zuberi NF (2000) Uterine atony at a tertiary care hospital in Pakistan: a risk factor analysis. J Pak Med Assoc 50:132–136

39. Stoltzfus RJ, Dreyfuss ML (1998) Guidelines for the use of iron supplements to prevent and treat iron deficiency anemia. International Life Sciences Institute, Washington, D.C.

40. Mukuria A, Aboulafia C, Themme A (2005) The context of women's health: results from the Demographic and Health Surveys, 1994–2001. Comparative Reports no. 11 2005. ORC Macro, Calverton, Md.

41. Galloway R, Dusch E, Elder L et al (2002) Women's perceptions of iron deficiency and anemia prevention and control in eight developing countries. Soc Sci Med 55:529–544

42. Rosato M, Mwansambo CW, Kazembe PN et al (2006) Women's groups' perceptions of maternal health issues in rural Malawi. Lancet 368:1180–1118

43. Maine D (2000) Role of nutrition in the prevention of toxemia. Am J Clin Nutr 72(Suppl):298S–300S

44. Belizan JM, Villar J (1980) The relationship between calcium intake and edema, proteinuria, and hypertension-gestosis: an hypothesis. Am J Clin Nutr 33:2202–2210

45. Hamlin RHJ (1962) Prevention of preeclampsia. Lancet 1:864–865

46. Belizan JM, Villar J, Repke J (1988) The relationship between calcium intake and pregnancy-induced hypertension: up-to-date evidence. Am J Obstet Gynecol 158:898–902

47. Hofmeyr GJ, Atallah AN, Duley L. Calcium supplementation during pregnancy for preventing hypertensive disorders and related problems. Cochrane Database of Systematic Reviews 2006, issue 3. art. no. CD001059. DOI 10.1002/14651858.CD001059.pub2

48. Villar J, Abdel-Aleem H, Merialdi M et al (2006) World Health Organization randomized trial of calcium supplementation among low calcium intake pregnant women. Am J Obstet Gynecol 194:639–649

49. Bucher HC, Guyatt GH, Cook RJ et al (1996) Effect of calcium supplementation on pregnancy-induced hypertension and preeclampsia: a meta-analysis of randomized controlled trials. J Am Med Assoc 275:1113–1117

50. Levine RJ, Hauth JC, Curet LB, Sibai BM, Catalano PM, Morris CD, DerSimonian R, Esterlitz JR, Raymond EG, Bild DE, Clemens JD, Cutler JA (1997) Trial of calcium to prevent preeclampsia. N Engl J Med 337:69–76

51. Sibai BM (2004) Magnesium sulfate prophylaxis in preeclampsia: Lessons learned from recent trials. Am J Obstet Gynecol 190:1520–1526

52. Duley L, Gülmezoglu AM, Henderson-Smart DJ. Magnesium sulphate and other anticonvulsants for women with preeclampsia. Cochrane Database of Systematic Reviews 2003, issue 2. art. no. CD000025. DOI 10.1002/14651858.CD000025

53. Altman D, Carroli G, Duley L et al (2002) Do women with preeclampsia, and their babies, benefit from magnesium sulphate? The Magpie Trial: a randomised placebo-controlled trial. Lancet 359:1877–1890

54. Sibai B, Dekker G, Kupferminc M (2005) Preeclampsia. Lancet 365:785–799

55. Tsu VD (2004) New and underused technologies to reduce maternal mortality. Lancet 363:75–76

56. Roberts JM, Hubel CA (1999) Is oxidative stress the link in the two-stage model of preeclampsia? Lancet 354:788–789

57. Roberts JM (2003) Nutrient involvement in preeclampsia. J Nutr 133(Suppl 2):1684S–1692S

58. Wang Y, Walsh SW (2001) Increased superoxide generation is associated with decreased superoxide dismutase activity and mRNA expression in placental trophoblast cells in preeclampsia. Placenta 22:206–212

59. Chappell LC, Seed PT, Briley AL et al (1999) Effect of antioxidants on the occurrence of preeclampsia in women at increased risk: a randomised trial. Lancet 354:810–816

60. Poston L, Briley AL, Seed PT, Kelly FJ, Shennan AH, Vitamins in Preeclampsia (VIP) Trial Consortium (2006) Vitamin C and vitamin E in pregnant women at risk for preeclampsia (VIP trial): randomised placebo-controlled trial. Lancet 367:1145–1154

61. Beazley D, Ahokas R, Livingston J, Griggs M, Sibai BM (2005) Vitamin C and E supplementation in women at high risk for preeclampsia: a double-blind, placebo-controlled trial. Am J Obstet Gynecol 192:520–521

62. Rumbold AR, Crowther CA, Haslam RR, Dekker GA, Robinson JS; ACTS Study Group (2006) Vitamins C and E and the risks of preeclampsia and perinatal complications. N Engl J Med 354:1796–1806

63. Rice AL, West KP Jr, Black RE (2004) Chap. 4: Vitamin A deficiency. In: Ezzati M et al (eds) Comparative quantification of health risks: global and regional burden of disease attributable to selected major risk factors, vol. 1. WHO, Geneva, Switzerland, pp. 211–256

64. West KP Jr (2002) Extent of vitamin A deficiency among preschool children and women of reproductive age. J Nutr 132(Suppl):2857S–2866S

65. West KP Jr, Katz J, Khatry SK, LeClerq SC, Pradhan EK, Shrestha SR, Connor PB, Dali SM, Christian P, Pokhrel RP, Sommer A and the NNIPS study group (1999) Low dose vitamin A or β-carotene supplementation reduces pregnancy-related mortality: A double-masked, cluster randomized prevention trial in Nepal. Br Med J 318:570–575

66. Christian P, West KP Jr, Khatry SK et al (2000) Night blindness during pregnancy and subsequent mortality among women in Nepal: effects of vitamin A and beta-carotene supplementation. Am J Epidemiol 152:542–547

67. McGanity WJ, Cannon RO, Bridgforth EB (1945) The Vanderbilt cooperative study of maternal and infant nutrition. IV. Relationship of obstetric performance to nutrition. Am J Obstet Gynecol 67:501–527

68. Green HN, Pindar D, Davis G, Mellanby E (1931) Diet as a prophylactic agent against puerperal sepsis. BMJ ii: 595–598

69. Hakimi M, Dibley MJ, Suryono A et al (1999) Impact of vitamin A and zinc supplements on maternal postpartum infections in rural central Java, Indonesia. International Vitamin A Consultative Group Meeting, 8–11 March 1999, Durban, South Africa

70. Christian P, West KP Jr, Khatry SK et al (2000) Vitamin A or β-carotene supplementation reduces symptoms of illness in pregnant and lactating Nepali women. J Nutr 130:2675–2682

71. Caulfield LE, Zavaleta N, Shankar AH, Merialdi M (1998) Potential contribution of maternal zinc supplementation during pregnancy to maternal and child survival. Am J Clin Nutr 68(Suppl): 499S–508S

72. Christian P (2003) Micronutrients and reproductive health issues: an international perspective. J Nutr 133:1969S–1973S

73. Osendarp SJ, West CE, Black RE, Maternal Zinc Supplementation Study Group (2003) The need for maternal zinc supplementation in developing countries: an unresolved issue. J Nutr 133:817S–827S

74. Dujardin B, Van Cutsem R, Lambrechts T (1996) The value of maternal height as a risk factor for dystocia: a meta-analysis. Trop Med Intern Health 4:510–521

75. World Health Organisation (1995) Maternal anthropometry and pregnancy outcomes. A WHO collaborative study. Bull WHO 73:1S–98S

76. Harrison KA (1985) Child-bearing, health and social priorities: a survey of 22,774 consecutive hospital births in Zaria, Northern Nigeria. Br J Obstet Gynaecol 92(Suppl 5):1–119

77. Garner P, Kramer M, Chalmers I (1992) Might efforts to increase birth weight in undernourished women do more harm than good? Lancet 340:1021–1023

78. Ceesay SM, Prentice AM, Cole TJ et al (1997) Effects on birth weight and perinatal mortality of maternal dietary supplements in rural Gambia: 5 year randomised controlled trial. BMJ 315:786–790

79. Christian P, Khatry SK, Katz J et al (2003) Effects of alternative maternal micronutrient supplements on low birth weight in rural Nepal: double blind randomised community trial. BMJ 326:571

80. Osrin D, Vaidya A, Shrestha Y et al (2005) Effects of antenatal multiple micronutrient supplementation on birthweight and gestational duration in Nepal: double-blind, randomised controlled trial. Lancet 365:955–962

81. Christian P, Osrin D, Manandhar DS, Khatry SK, de L Costello AM, West KP Jr (2005) Antenatal micronutrient supplements in Nepal. Lancet 366:711–712

82. Katz J, Christian P, Dominici F, Zeger SL (2006) Treatment effects of maternal micronutrient supplementation vary by percentiles of the birth weight distribution in rural Nepal. J Nutr 136:1389–1394

83. Christian P, West KP, Khatry SK et al (2003) Effects of maternal micronutrient supplementation on fetal loss and infant mortality: a cluster-randomized trial in Nepal. Am J Clin Nutr 78:1194–1202

84. Schroeder DG (2001) Chap. 16: Malnutrition. In: Semba RD, Bloem MW (eds) Nutrition and health in developing countries. Humana, Totowa, N.J. pp. 393–426

85. The Epic Study (2005) Age at menarche in relation to adult height. Am J Epidemiol 162:623–632

86. Tomkins A (2001) Nutrition and maternal morbidity and mortality. Br J Nutr 85(Suppl 2):S93–S99

87. Christian P, Katz J, Wu L, Pradhan EK, LeClerq SC, Khatry SK, West KP Jr. Risk factors for pregnancy-related mortality: A prospective study in rural Nepak. Public Health 2007, doi:10.1016/s.puhe.2007.06.003

88. ACC/SCN (2000) Fourth Report on the World Nutrition Situation. Geneva: ACC/SCN in collaboration with IFPRI
89. The ESHRE Capri Workshop Group (2006) Nutrition and reproduction in women. Hum Reprod Update 12:193–207
90. Ramakrishnan U, Gonzalez-Cossio T, Neufeld LM, Rivera J, Martorell R (2005) Effect of prenatal multiple micronutrient supplements on maternal weight and skinfold changes: A randomized double-blind clinical trial in Mexico. Food Nutr Bull 26:273–279

22 Anemia and Iron Deficiency in Developing Countries

Usha Ramakrishnan and Beth Imhoff-Kunsch

Summary Iron deficiency and anemia are major public health concerns throughout the world and are of special concern in many developing countries where the incidence and severity of anemia in certain populations is very high. Pregnant women, women of childbearing age, and young children are especially vulnerable to iron deficiency and iron-deficiency anemia (IDA) because of increased iron needs during growth and pregnancy, and iron losses during menstruation and childbirth. The most commonly used indicator of IDA in many resource-poor settings is hemoglobin; however, since anemia can be caused by factors other than iron deficiency, it is recommended that measurement of hemoglobin be combined with a more specific measure of iron status such as ferritin to determine whether the anemia is due to iron deficiency. IDA, especially severe anemia (hemoglobin $< 7\,g/dl$), can influence outcomes in both the mother and her child, including maternal mortality, birth weight, cognition in both the mother and her child, and infant development, among other outcomes. Strategies to improve iron status and reduce the burden of anemia include iron supplementation, staple food fortification, dietary diversification and modification, and public health measures, such as control and prevention of parasitic diseases, which can cause anemia. Although some anemia prevention strategies have proven successful, iron deficiency and anemia continue to impose a considerable public health burden on vulnerable groups such as pregnant women, women of childbearing age, and children. Further research and commitment by all stakeholders involved in public health are necessary to better understand how to prevent and control iron deficiency and anemia, especially in settings where anemia is widespread and very severe.

Keywords: Iron deficiency, Anemia, Pregnancy, Women, Developing countries

22.1 MAGNITUDE AND NATURE OF THE PROBLEM

Iron deficiency is one of the most common nutrient deficiencies worldwide and affects young children and women of reproductive age in both developed and developing countries. It is a dynamic process that begins with depletion of iron stores, leading finally to anemia (Fig. 22.1), which is characterized by low hemoglobin levels and is associated

From: *Nutrition and Health: Handbook of Nutrition and Pregnancy*
Edited by: C.J. Lammi-Keefe, S.C. Couch, E.H. Philipson © Humana Press, Totowa, NJ

- **Iron depletion**
  Reduction of iron stores
  - ↓ Serum ferritin
  - ↑ Total iron binding capacity (TIBC)

- **Iron-deficient erythropoiesis**
  Exhaustion of iron stores
  - ↓ Serum iron
  - ↓ Transferrin saturation
  - ↑ Free erythrocyte protoporphyrin (FEP)
  - ↑ Serum transferrin receptor concentration

- **Iron-deficiency anemia**
  Exhaustion of iron stores and microcytic, hypochromic erythrocytes
  - ↓ Hemoglobin
  - ↓ Hematocrit

Fig. 22.1. Stages of iron deficiency

with several functional consequences such as low birth weight, impaired cognition, and reduced work performance [1, 2].

The first stage of iron deficiency, known as *iron depletion*, occurs when iron stores are low and serum ferritin concentrations drop. The second stage, *iron-deficient erythropoiesis*, occurs when iron stores are depleted and the body does not absorb iron efficiently. Iron-deficient erythropoiesis is characterized by a decrease in transferrin saturation and increases in transferrin receptor expression and free erythrocyte protoporphyrin (FEP) concentration. *Iron-deficiency anemia* (IDA) is the third and most severe stage of iron deficiency and is characterized by low hemoglobin and hematocrit values. Erythrocytes are hypochromic and microcytic during IDA and hemoglobin concentration falls below −2 standard deviations of the age- and sex-specific normal reference. Anemia is the most widely used indicator of iron deficiency in most settings. The World Health Organization (WHO) reference values for anemia are hemoglobin < 11 g/dl for pregnant women and children under 5, < 12 g/dl for nonpregnant women, and < 13 g/dl for men [3].

Anemia is a highly prevalent public health problem that affects more than 2 billion people worldwide, and an estimated 50% of anemia is caused by iron deficiency [4, 5]. Pregnant women are particularly vulnerable to developing anemia and an estimated 18 and 56% of pregnant women are anemic in industrialized and developing countries, respectively (Table 22.1). WHO classifies the public health significance of anemia based on national anemia prevalence estimates and, as evidenced by the prevalence rates in Table 22.1, anemia in pregnant women and children is a severe public health problem (anemia prevalence ≥ 40%) in regions such as Africa, the Eastern Mediterranean, and South-East Asia [3].

Although iron deficiency is the most common cause of anemia, there are other nutritional and non-nutritional causes of anemia [6]. As illustrated in Fig. 22.2, not all anemia is caused by iron deficiency, and not all iron deficiency results in anemia. For example, inadequate intakes of folate and vitamin B_{12} can also cause anemia. Infections and genetic abnormalities such as thalassemia may also contribute to anemia in some populations. One of the major limitations of understanding how much anemia can be attributed to iron deficiency is the lack of data on the causes of anemia in many developing countries.

Table 22.1
Prevalence of Anemia in Developing and Industrialized Countries and in World Health Organization (WHO)-Classified Regions [4, 5]

	Pregnant women (%)	Nonpregnant women (%)	School-age children (%)
Industrialized countries	18	12	9
Developing countries	56	44	53
WHO regions			
Africa	51		52[a]
Americas	35		23[a]
South-East Asia	75		63[a]
Europe	25		22[a]
Eastern Mediterranean	55		45[a]
Western Pacific	43		21[a]

[a]Five- to 14-year-olds

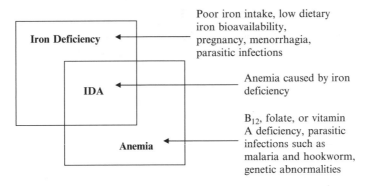

Fig. 22.2. Etiology of anemia

The assumption that 50% of anemia seen in many developing countries is due to iron deficiency is often based on small, nonrepresentative samples, and recent studies indicate that the contribution of iron deficiency to anemia may be overestimated in certain parts of the world. Understanding the etiology of anemia is very important in the development of appropriate public health strategies. Despite considerable efforts since the 1960s, anemia continues to be a problem in many parts of the world and progress in the reduction of iron deficiency and anemia has been less than satisfactory [7]. Interventions aimed at improving iron status, such as supplementation, may not necessarily reduce all forms of anemia and its related consequences in some populations. Similarly, relying only on hemoglobin as an indicator of response for programs that aim to prevent and control iron deficiency may be inadequate.

22.2 ASSESSMENT OF IRON DEFICIENCY AND ANEMIA

Numerous indicators for assessing anemia and iron status are available. These include serum ferritin, transferrin concentration and saturation, transferrin receptor, erythrocyte protoporphyrin, hemoglobin, hematocrit, and erythrocyte morphology and color (Table 22.2).

Table 22.2
Assessment of Iron Deficiency and Iron-Deficiency Anemia [9, 11]

Indicator	Measure	Cutoffs	Indication	Commonly used methods	Special considerations in developing countries
Serum Ferritin	Total body iron stores	<12 mcg/l[a]	Depleted iron stores	Venous or capillary blood, dried blood spots (DBS) ELISA method	Infection and inflammation may cause inflated ferritin values. Use of DBS convenient for field work
Serum transferrin concentration (TIBC)	Concentration of iron-transport protein	360 mcg/dl 390 mcg/dl 410 mcg/dl	Depleted iron stores Iron-deficient erythropoiesis Iron-deficiency anemia	Assay in which transferrin is saturated with excess iron; chromogenic methods	Influenced by infection and inflammation. More complicated laboratory procedure that requires quality-control sera
Serum transferrin saturation	Iron transport protein	<15%	Iron-deficient erythropoiesis	Calculated from TIBC and serum iron values	Influenced by infection and inflammation. Diurnal variation
Soluble serum transferrin receptor (STfR)	Expression of STfR, which bind ferritin for uptake in cells	≥8.5 mg/l 14 mg/l	Iron-deficient erythropoiesis Iron-deficiency anemia	Venous blood ELISA method	Possible to quantify STfR using a dried blood spot. Generally not significantly influenced by infection and inflammation. Can be influenced by other nutritional deficiencies such as B_{12} and folate deficiency and specifically by acute malaria infection

(continued)

Free erythrocyte protoporphyrin (FEP)	Serves as an intermediate in heme biosynthesis	>70 mmol FEP/ mol heme	Iron-deficient erythropoiesis	Whole blood (drop) Hemotofluorometry	Influenced by infection and inflammation Portable hemafluorometer available
Hemoglobin	Blood hemoglobin concentration	<11 g/dl	Anemia in pregnant women	Venous or capillary blood	Influenced by certain parasitic infections and other micronutrient deficiencies
		<12 g/dl	Anemia in nonpregnant women >15 years	Dried blood spot HemoCue or cyanmethemoglobin method	It is necessary to make adjustments to cutoff values for persons living in high altitudes Less expensive, field friendly equipment available
Hematocrit	Packed red blood cell volume	36%	Anemia in nonpregnant women >15y	Whole blood Centrifugation method	Less expensive, but methods can be difficult to standardize in a field setting
		33%	Anemia in pregnant women		
Erythrocyte	Color and shape or erythrocyte	Microcytic or hypochromic	Anemia	Whole blood Microscopy	

aWHO recommends using a serum ferritin cutoff of <15 mcg/l in areas where infections such as malaria are prevalent

As outlined below, certain indicators are more appropriate for use in field settings in developing countries with limited resources and laboratory capacity and high rates of parasitic diseases. In very resource-poor settings where access to a laboratory or laboratory equipment is limited or not possible, clinical examination to detect iron-deficiency anemia might be the only option [3].

Iron is stored in the body primarily in the protein ferritin, which is an indicator used to determine iron stores [8]. Serum ferritin concentrations are commonly determined in venous or capillary blood or dried blood spots using enzyme-linked immunosorbent assays (ELISA) or two-site immunoradiometric assays [9, 10]. Infection and inflammation can falsely elevate serum ferritin concentration, and this is a concern in developing countries where parasitic diseases are common [11]. Iron is transported through the body bound to the transport plasma protein transferrin, which can be measured in venous blood. Both transferrin saturation and transferrin concentration (total iron binding capacity [TIBC]) can serve as indicators for iron deficiency. Transferrin can be measured using chromogenic methods. Ferritin uptake into cells is regulated by transferrin receptors, which are expressed on cell surfaces and can be measured. Elevated expression of serum transferrin receptors, measured using ELISA techniques, can indicate iron-deficient erythropoiesis [12, 13].

Free erythrocyte protoporphyrin (FEP) serves as an intermediate in heme biosynthesis. Elevated FEP concentrations can indicate an interruption in heme synthesis due to iron deficiency and a subsequent build-up of the FEP precursor. FEP can be measured in whole blood, using hematofluorometry [11].

Seventy percent of iron in the body is contained in hemoglobin, an erythrocyte protein that transports oxygen from the lungs to tissues in the body. Hemoglobin concentration is commonly used to diagnose anemia in developing countries because the determination is relatively inexpensive and generally does not require complicated laboratory procedures. Hemoglobin can be measured using a portable photometer such as a HemoCue™, which is battery operated and can be used in a field setting. Determination of hemoglobin concentration using a HemoCue™ requires a capillary blood sample, obtained from either a finger, ear, or heel prick, or a small amount of blood from a whole blood sample (10 µl) (www.hemocue.com). Hemoglobin concentration can also be determined using the cyanmethemoglobin method, which requires the dilution of venous blood and analysis by a spectrophotometer. Hematocrit, the packed red cell volume in whole blood, can be determined by centrifugation using venous or capillary blood. Although this method is relatively simple, factors, including measurement error, can influence the precision, specificity, and sensitivity of the test [9]. Both hemoglobin and hematocrit concentrations can be influenced by other factors that might influence erythrocyte production and cause anemia. These include parasitic infections and other nutritional deficiencies (i.e. B_{12}, folate, vitamin A), which are of special concern in developing countries.

Ideally, these indicators should be used in combination to determine iron deficiency and iron-deficiency anemia. For example, serum ferritin and transferrin receptor measures can be used in conjunction with hemoglobin measures to determine whether anemia is caused by iron deficiency. The most recent recommendations by WHO for monitoring programs that aim to prevent and control iron deficiency and anemia have been to include hemoglobin and serum ferritin [14].

22.3 IRON REQUIREMENTS DURING PREGNANCY

Iron (Fe) requirements increase dramatically during pregnancy due to the rapid expansion of blood volume, tissue accretion, and potential for blood loss during delivery (Table 22.3). Although some of the increased blood volume is available to the mother after delivery, iron requirements still increase severalfold during pregnancy. For a normal pregnancy, it has been estimated that women need at least 6 mg of Fe/day compared with only 1.3 mg of Fe/day when they are not pregnant. This sixfold increase is very difficult to meet from diet alone, especially in settings where diets are poor in quantity and quality. In many developing countries, women enter pregnancy with depleted iron stores and/or iron-deficiency anemia, and this increases both their risk of becoming anemic during pregnancy and the adverse consequences related to iron deficiency and anemia. It should be noted however that hemoglobin drops during mid-pregnancy even among well-nourished women with adequate iron stores as a result of plasma volume expansion (Fig. 22.3). Although WHO recommends a single indicator, i.e., hemoglobin < 11 g/dl, trimester specific cutoff values have been used in developed countries such as the United States. The cut-off value for anemia during mid pregnancy is hemoglobin < 10.5 g/dl in the United States [15–17].

22.4 CONSEQUENCES OF IRON DEFICIENCY AND ANEMIA DURING PREGNANCY

Considerable work has been done to understand the functional consequences of iron deficiency and anemia during pregnancy for both maternal and infant outcomes [18–20]. The conceptual framework that shows the potential pathways by which iron deficiency during pregnancy may influence subsequent outcomes is shown in Fig. 22.4. However, most of the research has been conducted in developed country settings, where the problem is less severe and therefore less likely to detect any effects. Another hurdle in conducting well-designed studies that examine the consequences of iron deficiency during pregnancy is the inability to include a control group that does not receive prenatal iron supplements given the current WHO recommendations for universal supplementation [3]. A brief review of current knowledge on this topic is provided in this section.

22.4.1 Anemia and Maternal Mortality

The relationship between iron deficiency, anemia, and maternal mortality is complex and controversial. Anemia that results from iron deficiency is a physiological condition

Table 22.3
Iron Requirements during Pregnancy: Compartments.

Gross iron loss during pregnancy is 1.2 g for a healthy woman	
Fetus, umbilical cord and placenta	360 mg
Maternal blood loss	150 mg
Basal losses	230 mg
Red cell mass expansion	450 mg
Total	1,190 mg

Net iron losses are 580 mg due to recovery of increased red cell mass (450 mg) at delivery and lack of menstruation (160 mg).

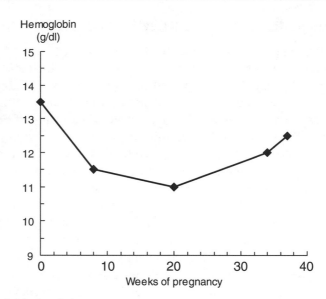

Fig. 22.3. Hemoglobin levels in pregnancy

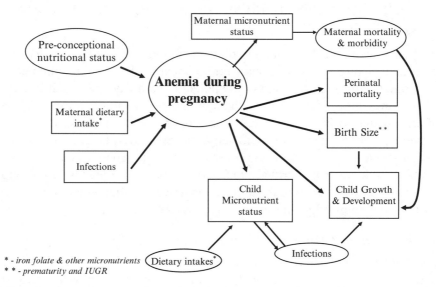

* - iron folate & other micronutrients
* * - prematurity and IUGR

Fig. 22.4. Functional consequences of anemia during pregnancy and early childhood. (From [18])

that is associated with limited function in that aerobic output may be impaired. The result is that women feel more tired and, therefore, quality of life is affected [1]. It is also well recognized that severe anemia (hemoglobin < 7 g/dl) is associated with an increased risk of dying, especially in settings where access to safe delivery practices is limited, as in many developing countries. Severely anemic women tend to be at increased risk of blood loss and cardiac failure, which can result in death. The relationship between iron deficiency per se, as well as mild–moderate anemia, is, however, less clear.

In a meta-analysis of several observational and intervention trials, Ross and Thomas [21] concluded that approximately 20% of the maternal mortality seen in sub-Saharan Africa and South Asia is attributable to anemia that is primarily the result of iron deficiency. However, the causal association between iron deficiency and maternal mortality has been questioned and is limited by the dearth of data from controlled trials [19, 20, 22]. Another concern is evidence suggesting that iron supplementation may increase the risk of infections [23]. A recent study showed an increased risk of dying among young children who received iron supplements in Zanzibar, a region where malaria is endemic [24]. These findings have renewed concerns about the safety of routine iron supplementation during pregnancy in these settings. This complicates current efforts to address iron deficiency and protect women and young children from the potential adverse effects. Screening for anemia and iron deficiency combined with targeted interventions may be required.

22.4.2   Birth Outcomes

Considerable work has been focused on the relationship between anemia, iron deficiency, and birth outcomes such as prematurity and intrauterine growth retardation. Severe anemia has been associated with an increased risk of stillbirth and infant mortality [25, 26]. Based on several observational studies there is an increased risk of delivering a preterm and/or low-birth-weight infant for women who are anemic compared to those who are not. Interestingly, a U-shaped relationship has been observed between hemoglobin levels and birth weight in that the risk of delivering a low-birth-weight infant is increased at both the lower and upper end of the hemoglobin distribution. It is important to note, however, that the mechanisms may differ. Specifically, the increased risk at the upper end may not be due to excess iron but rather it may represent inadequate plasma volume expansion [17].

As is the case for maternal mortality, there are very few well-designed controlled trials in which the efficacy of iron supplementation during pregnancy on improving birth outcomes, such as birth weight, has been evaluated [19, 27]. Pena-Rosas and Viteri concluded in a recent review that currently, strong evidence of iron supplementation during pregnancy and improved birth and pregnancy outcomes is lacking, and that further studies are necessary [27]. Nevertheless, a few recent trials provide findings that support routine iron supplementation during pregnancy. In a large cluster, randomized controlled trial that was conducted in Nepal, where the prevalence of anemia and low birth weight are high, the efficacy of different combinations of micronutrient supplements during pregnancy was assessed [28]. Specifically, the prevalence of low birth weight was reduced significantly from 43 to 34% among women who received iron–folate supplements along with vitamin A during pregnancy compared with those in the control group who received only vitamin A. Interestingly, the prevalence of low birth weight was slightly higher among those who received zinc along with iron–folate and vitamin A and similar to those who received a multivitamin–mineral supplement. There were no differences in prematurity. Two recent studies were also conducted in the United States, where the rates of low birth weight are much lower. Iron supplementation is standard practice for all women who are diagnosed as either anemic and/or iron deficient. Thus, in these studies the benefits of prenatal iron supplementation for women who were iron sufficient were evaluated. Both Cogswell et al. [29] and Siega-Riz et al. [30] found that iron supplementation of iron-replete women during pregnancy significantly reduced the prevalence of low birth weight and prematurity by almost half. It should be noted, however, that there were no significant differences in the prevalence of intrauterine growth retardation, suggesting that most of the effect was mediated through the effect on gestational age. These findings clearly support the current practice of universal iron supplementation, but the effect on other pregnancy outcomes is not known.

22.4.3 Early Childhood Growth and Development

Young children are also at increased risk of iron deficiency and anemia, which have been associated with adverse consequences, such as impaired learning and cognitive development [31, 32]. Several recent nationally representative surveys have shown that more than 50% of infants and preschool children are anemic, and this has raised considerable interest in identifying strategies that prevent this problem as soon as possible [4]. Recent studies have demonstrated that iron stores in infants at birth depend on maternal iron status and that clinical practices, such as delayed clamping of the umbilical cord, could help boost iron stores safely [33]. Healthy term babies typically have adequate iron stores during the first 6 months of life, but this may not be the case for infants who are born to mothers who are iron deficient during pregnancy, as well as premature and low-birth-weight babies who typically have low or no iron stores [1].

Recently, data have emerged to support the notion that women continue to be at risk of developing anemia and/or iron deficiency during the postpartum period, which may be associated with adverse consequences for both mothers and their infants. Women who are iron deficient may be at increased risk of depression and impaired cognitive function and this, in turn, would affect their ability to take care of their child and may indirectly influence child growth and development [34, 35]. Beard et al. [35] demonstrated that iron supplementation of women during the postpartum period improved cognitive functioning.

22.5 STRATEGIES TO COMBAT IRON DEFICIENCY AND IRON-DEFICIENCY ANEMIA

Because the etiology of iron deficiency and anemia is complex in developing countries, a variety of strategies have been pursued, and often more than one approach is necessary. The most common strategy to address this public health problem in many developing countries is supplementation, followed more recently by iron fortification of staple foods such as wheat flour. The lack of progress in reducing the burden of anemia and iron deficiency in many developing countries has, however, heightened the need to pursue a range of strategies that not only address the adequacy of iron intakes in these settings, but also the causes of iron loss, including parasitic infections such as hookworm, malaria, etc., and increased burden of reproduction due to frequent closely spaced pregnancies. A summary of the key strategies recommended for the prevention and control of iron deficiency and anemia is shown in Table 22.4.

Forging both food based strategies that address both diet quality and quantity with public health interventions such as improving hygiene and sanitation, routine deworming, increased access to health services, and so forth are clearly needed in many of these settings [36]. Some of the progress that has been made in these various areas is described below. Another important concern is the need to adopt a life-cycle approach, as in many of

Table 22.4
Strategies to Combat Iron Deficiency and Anemia in Developing Countries

Strategy	Target groups	Intervention
Supplementation	Pregnant women, women of child-bearing age, and children at risk of iron deficiency	Routine daily or weekly oral iron supplementation (tablets or liquid) in conjunction with education to improve compliance Effective health care delivery system necessary
Food fortification	Populations at risk of low iron intake (entire population should benefit from staple food fortification)	Fortification (with appropriate fortificant) of widely consumed staple foods such as wheat flour or rice Ensure quality control and monitoring and evaluation of fortification program
Dietary modification and diversification	Groups at risk of low iron intake, or who consume diets low in bioavailable iron (vulnerable groups)	Education about methods to reduce phytates in plant-based foods (i.e., germination), home gardening of micronutrient-rich foods, consumption of iron enhancers, and avoidance of iron inhibitors during the same meal
Parasitic infection control	Groups at risk of parasitic infections such as malaria or helminth infection (geographical areas with endemic rates of parasitic infections)	Treatment and prevention of parasitic infections (bed nets, deworming, health education, improved sanitation)

these settings iron deficiency begins early in life and continues through the reproductive years for most women.

22.5.1   Supplementation

Routine supplementation with iron and folic acid is recommended by the WHO for all pregnant women, especially in developing country settings where the prevalence of anemia and iron deficiency is high. Specifically, the current recommendation is a 6-month regimen of a daily supplement containing 60 mg of elemental iron along with 400 mcg of folic acid. In settings where the prevalence of anemia is ≥40%, supplementation is recommended for an additional 3 months during the postpartum period [3]. Also, in cases of individuals with severe anemia that is detected by clinical signs (pallor) and/or very low hemoglobin (<70 g/l) or hematocrit (<33%), a higher dose of 120 mg iron/day along with 400 mcg folic acid is recommended for 3 months.

Many women in developing countries are often already iron deficient and/or anemic before they become pregnant, and thus supplementation has been considered a preventive strategy for all women of reproductive age, especially adolescents who are at greatest risk. Weekly supplementation with the same dose recommended during pregnancy for a 3-month period has been suggested by the WHO. Although there is considerable evidence that treatment of iron deficiency with supplements is efficacious in improving hemoglobin levels during pregnancy, intervention programs have not been very successful. Some of the major problems with the distribution of iron supplements during pregnancy occur both at the level of the provider and consumer [4, 37, 38]. For example, quality of health care services, which include issues related to access, adequate and timely supply of the supplements, problems in the delivery system, poor quality of supplements (packaging, taste, appearance), etc., are some examples of hurdles that occur at the provider level. Lack of knowledge about the benefits of consuming the supplements, lack of motivation and side effects are some of the issues at the level of the consumer. Considerable efforts have been made in addressing many of these obstacles, and recent attempts that addressed many of these issues have shown that it is possible to improve the coverage of prenatal iron supplementation in many resource poor settings. For example, training of traditional birth attendants to distribute supplements and motivate women to consume them, social marketing strategies to make supplements more widely available through outlets besides health care, improved packaging and appearance of the supplement, etc., are some of the innovative strategies that have been pursued. In spite of these efforts, however, there have been concerns about the adequacy of providing only iron supplements during pregnancy in settings where several nutrient deficiencies coexist even before a woman is pregnant. The potential for providing multiple micronutrient supplements during pregnancy and early childhood has been considered by many international agencies [39], and issues related to the ideal dosage, frequency (weekly vs. daily), and age groups for targeting supplements need to be evaluated. A brief summary of the current knowledge regarding multivitamin–mineral supplements and timing of supplementation is provided in the following sections.

22.5.1.1 ROLE OF MULTIVITAMIN–MINERAL SUPPLEMENTS

Recent studies have documented the coexistence of several micronutrient deficiencies in many developing countries as a result of poor diets, both in terms of quantity and quality combined with the increased requirements during pregnancy and early childhood [40]. However, the evidence to date on the benefits of providing prenatal multivitamin–mineral (MVTM) supplements compared with the standard iron–folate or iron only supplements remains mixed [41]. Although some studies have shown improvements in birth weight [42–44], two studies from Mexico and Nepal failed to show improvements in the prevalence of anemia or iron status [45, 46]. Of greater concern are the pooled results of two recent trials conducted in Nepal suggesting that multiple micronutrient supplements during pregnancy may be associated with an increased perinatal and neonatal death [47]. These findings need to be considered cautiously while we await the results of other large ongoing trials to better understand the potential mechanisms and determine if they are specific to the South Asian population [48]. In contrast, studies conducted among HIV-positive women indicate that multiple micronutrients may delay the progression of HIV/AIDS in women with more advanced disease [42] (See Chap. 23, "Micronutrient status and pregnancy Outcomes in HIV-Infected Women".) The role of micronutrients with the implementation of assisted reproductive technology, however, remains to be evaluated. Similarly, other potential benefits, such as improved child growth and development and improved micronutrient status, also need to be considered before making any changes to the current recommendations. It is in this context that the current concerns regarding the appropriateness of universal supplementation with iron and folate during pregnancy is highly relevant and merits discussion (see below).

22.5.1.2 TIMING OF SUPPLEMENTATION

Current recommendations for universal supplementation with iron during pregnancy have been challenged recently, especially in light of the adverse effects of oxidative stress that can be induced by iron [49]. Recent studies point to excess iron, as indicated by high serum ferritin, as being associated with an increased risk of gestational diabetes mellitus. This is especially a concern when women who are not iron deficient but may be anemic receive iron supplements. Beaton [50] challenged the public health community and researchers to rethink the need for iron supplementation during pregnancy and made the case that using improvements in hemoglobin, in the absence of improvements in functional outcomes such as birth weight, constitutes a disservice. At the same time, however, we know that many women in many developing countries are iron deficient and/or anemic even before they are pregnant, and that iron supplementation during pregnancy is not adequate. It is in this context that a shift to recommending weekly supplementation for women of reproductive age is highly appealing. Viteri and Berger [51] recently demonstrated how weekly supplementation of women of reproductive age both before and during pregnancy improved iron reserves effectively and safely. Ekstrom et al. [52] also found that weekly supplementation was effective during pregnancy. Clearly, this approach has the potential of being more feasible and less confounded by the typical problems of lack of availability of supplements, poor utilization of antenatal care services, poor compliance due to side effects, etc., that plague prenatal iron supplementation programs in many developing countries [37, 38, 53].

22.5.2 Fortification

Fortification of staple foods serves as a cost effective means to increase iron intake in a population. Important factors that influence the success of a fortification program include availability of fortification technology, proper quality control at factories and/or mills, choosing a staple food that is widely and regularly consumed by all socioeconomic and geographical groups (including the most vulnerable groups), determining the most appropriate fortificant and the level of fortification necessary, assessing the bioavailability of the fortificant in food, and monitoring and evaluation [54]. Several countries including Venezuela, Chile, and the United States have successfully implemented food fortification programs. However, monitoring and evaluation of the effect of iron fortification programs on anemia prevalence is scarce. An analysis of anemia in school-age children suggested that fortification of wheat and corn flour in Venezuela in 1993 reduced anemia; however, this program's effectiveness needs further evaluation [55]. Several efficacy trials of staple food fortification have shown a reduction in iron deficiency and/or anemia, such as fortification of fish sauce in Vietnam, curry powder in India, soy sauce in China, and salt in Morocco [55–58]. Chile's compulsory fortification program, in which wheat flour is fortified with ferrous sulfate, folic acid, and other B vitamins, has successfully reduced the incidence of neural tube defects and folate deficiency in the population and has likely reduced the incidence of IDA, which is very low compared with other South American countries [59, 60]. A recent analysis of the consumption of fortified wheat flour in Guatemala illustrated that households bought an average of 84 g/day of fortified wheat flour equivalents (wheat flour plus breads made with fortified wheat flour) per adult equivalent unit, and this varied by socioeconomic status, ethnicity, and area [61]. Extremely poor, poor, and nonpoor households purchased 14, 48, and 124 g/day of wheat flour equivalents, respectively. Overall, purchases were lowest in extremely poor, indigenous, and rural households. This analysis illustrates that fortification of a staple food such as wheat flour can potentially benefit a population in which iron intake is low; however, the staple food must be consumed by all segments of the population in order to benefit the most vulnerable groups, such as the extremely poor. Fortification of staple foods, if implemented and monitored correctly, provides a cost-effective additional source of dietary iron for entire populations.

22.5.3   Dietary Diversification and Modification

Dietary diversification and modification, which can include home gardening, food processing techniques, reducing consumption of foods that inhibit non–heme iron absorption, and increasing consumption of foods that enhance non–heme iron absorption, serve as methods to increase dietary intake and bioavailability of iron (Table 22.5) [62, 63]. Nutrition education about home gardening of micronutrient-rich foods, and drying of meats and fish, for example, can potentially improve micronutrient content of the diet. Increasing iron-rich flesh food consumption serves as an ideal dietary solution to improving iron intake; however, flesh foods are expensive and certain cultural or religious beliefs might preclude the intake of these foods [63]. Iron in food exists in two forms: non–heme iron and heme iron. Plant foods and dairy products contain non–heme iron and flesh foods, such as meat and fish, contain heme iron, which is much more bioavailable than non–heme

Table 22.5
Inhibitors and Enhancers of Non–Heme Iron Absorption

Inhibitors of iron absorption
  Phytates
    Legumes such as black beans, lentils and soybeans, peanuts, and foods containing grains such as whole wheat bread and corn tortillas
  Polyphenols
    Tea, coffee, certain fruits and vegetables
  Fiber and bran
    Whole grain foods such as whole wheat flour
  Calcium
Enhancers of iron absorption
  Ascorbic acid
  Citrus fruits such as oranges and lemons
  Meat
    Meat, fish, or Poultry

iron. Efficiency of heme iron absorption is approximately two to three times greater than that of non–heme iron [63]. Many plant-based foods such as legumes and cereals contain high levels of phytates, which can inhibit dietary non–heme iron absorption. Certain food processing techniques such as fermentation and germination can reduce the phytate concentration in foods, and thereby improve the bioavailability of non-heme dietary iron. A variety of foods enhance or decrease the bioavailability of dietary non–heme iron; for example, vitamin C–rich foods or drinks and meat increase the bioavailability of non–heme iron, whereas phytate-rich foods and tannin-containing foods or drinks decrease bioavailability of non–heme iron (Table 22.5). Education about foods that enhance and inhibit iron absorption could potentially improve the bioavailability of dietary iron.

22.5.4 Infection Control

Parasitic infections such as malaria, hookworm, whipworm, and schistosomiasis can cause or exacerbate anemia, especially when the infection is moderate to heavy, and when women are coinfected with multiple parasites [64]. Helminthes attach to the intestines and/or bladder and feed on blood, causing regular host blood loss due to blood loss at the site of helminth attachment, and the blood consumed by the parasite. Parasitic infections can lead to, among other symptoms, anorexia, malabsorption of nutrients, nutrient loss through fecal or urinary blood loss, nausea, diarrhea, and vomiting, which can result in depletion of iron stores and iron-deficiency anemia [65, 66]. Efforts to control and prevent parasitic infections such as the use of bed nets, routine deworming using chemotherapeutic agents, malaria prevention and control, and improved sanitation can help combat anemia (Table 22.6). Specific to pregnant women, it is likely safe to provide deworming therapy after the first trimester of pregnancy, and one study showed that pregnant women infected with hookworm who were given antihelminthic chemotherapy at the end of their first trimester had higher hemoglobin concentrations than did the controls [62, 66].

22.6 CONCLUSION

Iron deficiency and anemia continue to be major public health challenges in many parts of the world. Considerable progress has been made in our understanding of the potential consequences and a range of strategies is available to address this complex

Table 22.6
Prevention and Control of Parasitic Infections

Bed nets (impregnated with insecticide)
Vector control (i.e., use of insecticides)
Routine antihelminthic treatment using chemotherapeutic drugs
Prevention and treatment of malaria
Improved sanitation (i.e., proper latrines)
Access to clean water
Improved hygiene (i.e., hand washing, bathing in clean water)
Health education
Specifically for children:
Breastfeeding
Appropriate introduction of solid foods (timing, sanitation)

problem. While several research questions remain to be answered, there is clearly a need for mobilizing populations and policy makers to address this problem in many developing countries. The successful experiences with programs, such as fortification of staples, need to be translated and adapted to different settings. Resources that will not only improve infrastructure, but also the capacity to monitor ongoing efforts in the prevention and control of anemia and iron deficiency, are needed. This will require the efforts of not only the public health community but also the commitment of other stakeholders, such as industry, policy makers, and civic society.

REFERENCES

1. Yip R (2001) Iron. In: Bowman B, Russell R (eds) Present knowledge in nutrition, 8 edn. ILSI, Washington, D.C.
2. Allen LH (2000) Anemia and iron deficiency: effects on pregnancy outcome. Am J Clin Nutr 71:1280S–1284S
3. WHO (2001) Iron deficiency anemia: assessment, prevention and control: a guide for programme managers. UNICEF, United Nations University, WHO, Geneva, Switzerland
4. Galloway R (2003) Anemia prevention and control: what works. USAID, The World Bank, UNICEF, PAHO, FAO, The Micronutrient Initiative
5. SCN (2000) The fourth report on the world nutrition situation: nutrition throughout the life cycle. Standing Committee on Nutrition, United Nations System
6. Allen L, Casterline-Sabel J (2001) Prevalence and causes of nutritional anemias. In: Ramakrishnan U (ed) Nutritional anemias. CRC Press, Boca Raton, Fla, pp. 7–21
7. Mason JB, Lotfi M, Dalmiya N, Sethuraman K, and Deitchler M, with Geibel S, Gillenwater K, Gilman A, Mason K, and Mock N (2001) Progress in controlling micronutrient deficiencies. MI/Tulane University/UNICEF. The Micronutrient Initiative
8. Brody T (1994) Nutritional biochemistry. Academic, San Diego, Calif.
9. Gibson R (2005) Principles of nutritional assessment, 2nd edn. Oxford University Press, Oxford, UK
10. Ahluwalia N, Lonnerdal B, Lorenz SG, Allen LH (1998) Spot ferritin assay for serum samples dried on filter paper. Am J Clin Nutr 67:88–92
11. Lynch S, Green R (2001) Assessment of Nutritional Anemias. In: Ramakrishnan U (ed) Nutritional anemias. CRC Press, Boca Raton, Fla.
12. Cook JD, Flowers CH, Skikne BS (2003) The quantitative assessment of body iron. Blood 101:3359–3364
13. Baynes R, Stipanuk M (2001) Iron. In: Stipanuk M (ed) Biochemical and physiological aspects of human nutrition. Saunders, Philadelphia, Pa., pp 711–740
14. CDC/WHO recommendations for monitoring iron status (unpublished)
15. Scholl TO (2005) Iron status during pregnancy: setting the stage for mother and infant. Am J Clin Nutr 81:1218S–1222S
16. Institute of Medicine (2001) Dietary Reference Intakes. Vitamin A, vitamin K, arsenic, boron, chromium, copper, iodine, iron, manganese, molybdenum, nickel, silicon, vanadium, zinc. National Academy of Sciences, Washington, D.C.
17. Yip R (2000) Significance of an abnormally low or high hemoglobin concentration during pregnancy: special consideration of iron nutrition. Am J Clin Nutr 72:272S–279S
18. Ramakrishnan U (2001) Functional consequences of nutritional anemia during pregnancy and early childhood. Prevalence and causes of nutritional anemias. In: Ramakrishnan U (ed) Nutritional anemias. CRC Press, Boca Raton, Fla., pp 43–68
19. Rasmussen K (2001) Is there a causal relationship between iron deficiency or iron-deficiency anemia and weight at birth, length of gestation and perinatal mortality? J Nutr 131:590S–601S; discussion, 601S–603S
20. Brabin B, Hakimi M, Pelletier D (2001) An analysis of anemia and pregnancy-related maternal mortality. J Nutr 131:604S–614S; discussion, 614S–615S
21. Ross JS, Thomas EL (1996) Iron deficiency anemia and maternal mortality. PROFILES 3 working notes series no. 3. Academy for Education Development, Washington D.C.

22. Rush D (2000) Nutrition and maternal mortality in the developing world. Am J Clin Nutr 72:212S–240S
23. Oppenheimer SJ (2001) Iron and its relation to immunity and infectious disease. J Nutr 131:616S–633S; discussion, 633S–635S
24. Sazawal S, Black RE, Ramsan M, Chwaya HM, Stoltzfus RJ, Dutta A, Dhingra U, Kabole I, Deb S, Othman MK, Kabole FM (2006) Effects of routine prophylactic supplementation with iron and folic acid on admission to hospital and mortality in preschool children in a high malaria transmission setting: community-based, randomised, placebo-controlled trial. Lancet 367:133–143
25. Lone FW, Qureshi RN, Emanuel F (2004) Maternal anaemia and its impact on perinatal outcome. Tropi Med Int Health 9:486–490
26. Brabin BJ, Premji Z, Pelletier D (2001) An analysis of anemia and child mortality. J Nutr 131:2S
27. Pena-Rosas JP, Viteri FE (2006) Effects of routine oral iron supplementation with or without folic acid for women during pregnancy. Cochrane Database of Systematic Reviews 2006, issue 3. art. no. CD004736. DOI: 10.1002/14651858.CD004736.pub2
28. Christian P, Khatry SK, Katz J, Pradhan EK, LeClerq SC, Shrestha SR, Adhikari RK, Sommer A, West KP Jr. (2003) Effects of alternative maternal micronutrient supplements on low birth weight in rural Nepal: double blind randomised community trial. Br Med J 326:571– 576
29. Cogswell ME, Parvanta I, Ickes L, Yip R, Brittenham GM (2003) Iron supplementation during pregnancy, anemia, and birth weight: a randomized controlled trial [see comment]. Am J Clin Nutr 78:773–781
30. Siega-Riz AM, Hartzema AG, Turnbull C, Thorp J, McDonald T, Cogswell ME (2006) The effects of prophylactic iron given in prenatal supplements on iron status and birth outcomes: a randomized controlled trial. Am J Obstet Gynecol 194:512–519
31. Lozoff B, Wachs TD (2001) Functional correlates of nutritional anemias in infancy and early childhood – child development and behavior. In: Ramakrishnan U (ed) Nutritional anemias. CRC Press, Boca Raton, Fla., pp 69–88
32. Grantham-McGregor S, Ani C (2001) A review of studies on the effect of iron deficiency on cognitive development in children. J Nutr 131:649S–666S; discussion, 666S–668S
33. Chaparro CM, Neufeld LM, Tena G, Liz RE, Dewey KG (2006) Effect of timing of umbilical cord clamping iron states in Mexican infants: a randomised controlled trial. Lancet, 367(9527): 1997–2004.
34. Makrides M, Crowther CA, Gibson RA, Gibson RS, Skeaff CM (2003) Efficacy and tolerability of low-dose iron supplements during pregnancy: a randomized controlled trial. Am J Clin Nutr 78:145–153
35. Beard JL, Hendricks MK, Perez EM, Murray-Kolb LE, Berg A, Vernon-Feagans L, Irlam J, Isaacs W, Sive A, Tomlinson M. (2005) Maternal iron deficiency anemia affects postpartum emotions and cognition. J Nutr 135:267–272
36. Ramakrishnan U (2001) Conclusions. In: Ramakrishnan U (ed) Nutritional anemias. CRC Press, Boca Raton, Fla., pp 241–245
37. Ekstrom E-C (2001) Supplementation for nutritional anemias. In: Ramakrishnan U (ed) Nutritional anemias. CRC Press, Boca Raton, Fla., pp 129–152
38. Galloway R, Dusch E, Elder L, Achadi E, Grajeda R, Hurtado E, Favin M, Kanani S, Marsaban J, Meda N, Moore KM, Morison L, Raina N, Rajaratnam J, Rodriquez J, Stephen C (2002) Women's perceptions of iron deficiency and anemia prevention and control in eight developing countries. Soc Sci Med 55:529–544
39. UNICEF/WHO/UNU Study Team (2002) Multiple micronutrient supplementation during pregnancy (MMSDP): efficacy trials. London, Centre for International Child Health, Institute of Child Health, University College London, 4–8 March 2002
40. Ramakrishnan U, Huffman S (2001) Multiple micronutrient malnutrition—what can be done? In: Semba RD, Bloem M (ed) Nutrition and health in developing countries. Humana, Totowa, N.J., pp 365–392
41. Sommer A (2005) Innocenti micronutrient research report no 1. ILSI, Washington, D.C.
42. Fawzi WW, Msamanga GI, Spiegelman D, Urassa EJ, McGrath N, Mwakagile D, Antelman G, Mbise R, Herrera G, Kapiga S, Willett W, Hunter DJ (1998) Randomised trial of effects of vitamin supplements on pregnancy outcomes and T cell counts in HIV-1-infected women in Tanzania. Lancet 351:1477–1482
43. Osrin D, Vaidya A, Shrestha Y, Baniya RB, Manandhar DS, Adhikari RK, Filteau S, Tomkins A, Costello AM (2005) Effects of antenatal multiple micronutrient supplementation on birthweight and gestational duration in Nepal: double-blind, randomised controlled trial. Lancet 365:955–962

44. Kæstel P, Michaelsen KF, Aaby P, Friis H (2005) Effects of prenatal multimicronutrient supplements on birth weight and perinatal mortality: a randomised, controlled trial in Guinea-Bissau. Eur J Clin Nutr 59:1081–1089

45. Ramakrishnan U, Neufeld LM, González-Cossío T, Villalpando S, García-Guerra A, Rivera J, Martorell R (2004) Multiple micronutrient supplements during pregnancy do not reduce or improve iron status compared to iron-only supplements in semirural Mexico. J Nutr 134:898–903

46. Christian P, Shrestha J, LeClerq SC, Khatry SK, Jiang T, Wagner T, Katz J, West KP Jr. (2003). Supplementation with micronutrients in addition to iron and folic acid does not further improve hematologic status of pregnant women in rural Nepal. J Nutr 133:3492–3498

47. Christian P, Osrin D, Manandhar DS, Khatry SK, de L Costello AM, West KP Jr. (2005) Antenatal micronutrient supplements in Nepal. Lancet 366:711–712

48. Shrimpton R, Dalmiya N, Darnton-Hill I, Gross R (2005) Micronutrient supplementation in pregnancy: Lancet 366:2001–2002

49. Casanueva E, Viteri FE (2003) Iron and oxidative stress in pregnancy. J Nutr 133:1700S–1708S

50. Beaton GH (2000) Iron needs during pregnancy: do we need to rethink our targets? Am J Clin Nutr 72:265S–271S

51. Viteri FE, Berger J (2005) Importance of pre-pregnancy and pregnancy iron status: can long-term weekly preventive iron and folic acid supplementation achieve desirable and safe status? Nutr Rev 63: S65–S76

52. Ekström EC, Hyder SM, Chowdhury AM, Chowdhury SA, Lönnerdal B, Habicht JP, Persson LA (2002) Efficacy and trial effectiveness of weekly and daily iron supplementation among pregnant women in rural Bangladesh: disentangling the issues. Am J Clin Nutr 76:1392–1400

53. Cavalli-Sforza T, Berger J, Smitasiri S, Viteri F (2005) Weekly iron-folic acid supplementation of women of reproductive age: impact overview, lessons learned, expansion plans, and contributions toward achievement of the millennium development goals. Nutr Rev 63:S152–S158

54. Allen L (2006) New approaches for designing and evaluating food fortification programs. J Nutr 136:1055–1058

55. Mannar V, Gallego EB (2002) Iron fortification: country level experiences and lessons learned. J Nutr 132:856S–858S

56. Thuy PV, Berger J, Davidsson L (2003) Regular consumption of NaFeEDTA-fortified fish sauce improves iron status and reduces the prevalence of anemia in anemic Vietnamese women. Am J Clin Nutr 78:284–290

57. Ballot DE, MacPhail AP, Bothwell TH, Gillooly M, Mayet FG (1989) Fortification of curry powder with NaFe(111)EDTA in an iron-deficient population: report of a controlled iron-fortification trial. Am J Clin Nutr 49:16–19

58. Zimmermann MB, Wegmueller R, Zeder C (2004) Triple fortification of salt with microcapsules of iodine, iron, and vitamin A. Am J Clin Nutr 80:1283–1290

59. Hertrampf E, Cortes F (2004) Folic acid fortification of wheat flour: Chile. Nutr Rev 62:S44–S48; discussion, S49

60. Walter T, Olivares M, Pizarro F, Hertrampf E (2001) Fortification. In: Ramakrishnan U (ed) Nutritional anemias. CRC Press, Boca Raton, Fla., pp 153–184

61. Imhoff-Kunsch BC, Flores R, Dary O, Martorell R (2007) Wheat flour fortification is unlikely to benefit the neediest in Guatemala. J Nutr 137:1017–1022

62. Yip R (2001) Iron deficiency and anemia. In: Semba R, Bloem M (eds) Nutrition and health in developing countries. Humana, Totowa, N.J. pp. 327–342

63. Gibson RS, Hotz C (2001) Dietary diversification/modification strategies to enhance micronutrient content and bioavailability of diets in developing countries. Br J Nutr 85(Suppl 2):S159–S166

64. Larocque R, Casapia M, Gozz E, Gyorkos TW (2005) Relationship between intensity of soil-transmitted helminth infections and anemia during pregnancy. Am J Trop Med Hyg 73:783–789

65. Stephenson LS, Latham MC, Ottesen EA (2000) Malnutrition and parasitic helminth infections. Parasitology 121(Suppl):S23–S38

66. Hall A (2001) Public health measures to control helminth infections. In: Ramakrishnan U (ed) Nutritional anemias. CRC Press, Boca Raton, Fla., pp 215–239

23 Micronutrient Status and Pregnancy Outcomes in HIV-Infected Women

Saurabh Mehta, Julia L. Finkelstein, and Wafaie W. Fawzi

Summary Micronutrient deficiencies are widely prevalent in developing regions, especially in vulnerable groups such as HIV-infected pregnant women. Micronutrient supplementation is considered one of the most cost-effective strategies to reduce malnutrition and improve health outcomes.

Maternal multivitamin supplementation (vitamins B-complex, C, and E) has been shown to reduce the incidence of adverse pregnancy outcomes in HIV-infected pregnant women, such as fetal loss, infant low birth weight, small for gestational age, prematurity, and mother-to-child transmission (MTCT) of HIV. On the other hand, supplementation with vitamin A has not demonstrated such benefits; in fact, some trials have found that vitamin A supplementation increases the risk of MTCT of HIV.

The role of other micronutrients such as zinc and selenium has not been well established, and warrants further investigation; indiscriminate supplementation with these nutrients is to be avoided until this becomes clear. Routine iron and folate supplementation for HIV-infected pregnant women is to be continued, as recommended by the World Health Organization; however, there are some concerns regarding the safety of iron supplementation in HIV-infected pregnant women, and further research is urgently needed.

Multivitamin supplementation of HIV-infected pregnant women is strongly recommended. Overall, there is no evidence to support the use of vitamin A supplementation in HIV-positive pregnant women, due to concerns regarding vertical HIV transmission. Additional research is needed to elucidate the effect of other micronutrients including iron, zinc, and selenium on pregnancy outcomes in HIV-infected pregnant women.

Keywords: HIV, AIDS, Micronutrients, Multivitamins, MTCT, Pregnancy outcomes

23.1 INTRODUCTION

The Copenhagen Consensus is a project that seeks to establish priorities for advancing global welfare and economic development. In its 2004 report, this panel of economic experts identified HIV/AIDS and malnutrition as the two leading obstacles to human

From: *Nutrition and Health: Handbook of Nutrition and Pregnancy*
Edited by: C.J. Lammi-Keefe, S.C. Couch, E.H. Philipson © Humana Press, Totowa, NJ

betterment. HIV/AIDS control strategies and provision of micronutrient supplementation were identified as the most cost-effective methods to improve global welfare [1].

Unfortunately, micronutrient deficiencies and HIV/AIDS occur in tandem, predominantly in impoverished settings, where investments in control strategies have traditionally been difficult. Of the estimated 33 million people living with HIV/AIDS worldwide, more than 95% reside in developing nations [2]. Sub-Saharan Africa is the most heavily affected region, which comprises 10% of the world's population but accounts for 60% of the total number of people living with HIV/AIDS globally [2]. Malnutrition and food insecurity are prevalent in HIV-infected populations, and most people living with HIV/AIDS experience some degree of micronutrient deficiency [3].

In this chapter, we focus on the interplay of micronutrient deficiencies and HIV/AIDS among pregnant women. We begin with a brief review of the current evidence for the use of various micronutrients in HIV-negative pregnant women. The relationship between micronutrient status and pregnancy outcomes among HIV-infected women is then explored in detail, by reviewing the available evidence from observational studies and randomized trials. Finally, we conclude with a discussion of the implications of these findings on clinical and public health practice.

23.2 MICRONUTRIENT STATUS AND PREGNANCY OUTCOMES IN HIV-UNINFECTED WOMEN

23.2.1 Iron and Folate

Iron and folate supplementation is the standard of prenatal care in most developing countries; current World Health Organization guidelines recommend prenatal daily iron–folate supplements (400 mcg folate and 60 mg iron) for a duration of six months to prevent anemia, and iron supplementation two times daily for the treatment of severe anemia [4]. Iron supplementation is recommended based on its demonstrated benefit in preventing maternal anemia and its related complications such as premature birth, low birth weight, postpartum hemorrhage, and mortality [5–8]. Folate supplementation is also well-established as an effective strategy for preventing adverse birth outcomes, particularly neural tube defects (Relative risk [RR]: 0.28; 95% confidence interval [CI]: 0.13, 0.58) [9]. To date, there has been no evidence to suggest that folate supplementation should differ in the context of HIV/AIDS. However, recently, there have been some concerns regarding the benefit of iron supplementation in HIV-infected pregnant women; Ramakrishnan and Imhoff-Kunsch discuss these in greater detail in Chapter 22, ("Anemia and Iron Deficiency in Developing Countries") in this volume [10].

23.2.2 Other Micronutrients

There is relatively limited evidence for the use of other individual micronutrients during pregnancy. In a randomized trial by West et al. in Nepal, vitamin A and beta-carotene supplementation resulted in a 44% overall reduction in maternal mortality [11]. These findings are being confirmed in ongoing trials in Bangladesh and Ghana. Vitamin A deficiency during pregnancy can also lead to increased occurrence of fetal loss; however, given teratogenic concerns, it is recommended that maternal vitamin A intake should not exceed 10,000 IU per day (3,000 mcg/d) [12, 13].

Several trials have failed to identify a consistent benefit of maternal zinc supplementation on pregnancy outcomes such as preterm labor, premature rupture of membranes, postpartum hemorrhage, perinatal mortality, and fetal growth [14, 15]. However, some studies have suggested a beneficial effect of zinc on infant neurobehavioral development and immune function, as well as some role in the prevention of congenital malformations, such as cleft lip or palate [14–17].

23.2.3 Multiple Micronutrient Supplementation

The evidence for benefits of multiple micronutrient supplementation in pregnancy from randomized trials has been equivocal to date. In a trial in semirural Mexico, Ramakrishnan et al. randomized pregnant women to receive daily iron supplementation (60 mg) either alone or in combination with multiple micronutrients; these supplements contained several vitamins and minerals (vitamins A, B-complex, C, D, E, and folic acid; iron, zinc, and magnesium) at doses of one to 1.5 times the Recommended Dietary Allowance (RDA) levels. Multiple micronutrients did not confer any additional benefit on maternal weight gain during pregnancy [18], maternal hematological status [19], or infant birth weight or length [20], compared with iron-only supplementation.

Similar results were observed in a double-blind, cluster-randomized controlled trial in Nepal, in which four combinations of micronutrients (folic acid; folic acid and iron; folic acid, iron, and zinc; and a multiple micronutrient supplement) plus vitamin A, or vitamin A alone, were administered to women daily during pregnancy. The multiple micronutrient supplement contained folic acid, iron, zinc, and 11 other nutrients (vitamins B-complex, C, D, E, K; copper and magnesium). Multiple micronutrient supplementation did not improve maternal hematological status [21], or reduce the risk of low birth weight [22], compared with folic acid and iron supplementation. None of the supplements reduced the occurrence of fetal loss [23].

On the other hand, micronutrient supplementation was found to have beneficial effects in randomized trials in Tanzania and Nepal. In the randomized trial in Tanzania, the administration of a micronutrient-fortified beverage (vitamins A, B-complex, C, and E; iron, iodine, zinc, folate) resulted in a 4.16-g/l mean increase in maternal hemoglobin concentrations, and reduced the risks of anemia and iron deficiency anemia by 51 and 56%, respectively [24]. In the trial in Nepal, multiple micronutrient (vitamins A, B-complex, C, D, and E; iron, zinc, copper, selenium and iodine) supplementation of pregnant women resulted in an increase in birth weight and a lower proportion of low birth weight infants [25].

In another trial in Tanzania, 8,468 HIV-uninfected pregnant women were randomized to receive daily multivitamins (vitamins B-complex, C, and E) or placebo from enrollment to six weeks after delivery. The results demonstrated that multivitamin supplements significantly reduced the risks of infant low birth weight and small size for gestational age by 18 and 23%, respectively [26].

23.3 MICRONUTRIENT STATUS AND PREGNANCY OUTCOMES IN HIV-INFECTED WOMEN

Micronutrient deficiencies may increase the risk of adverse pregnancy outcomes among HIV-infected women *via* a number of biological mechanisms. For example, sub-optimal micronutrient status may increase the risk of mother-to-child transmission

(MTCT) of HIV by impairing systemic immune function and by affecting the epithelial integrity of the maternal lower genital tract [27, 28]. Deficiencies of various micronutrients may amplify the risk of postpartum HIV transmission by increasing the risk of clinical or subclinical mastitis and subsequent viral shedding, and by impairing the epithelial integrity of the infant gastrointestinal tract [27, 29]. Micronutrient deficiencies may also accelerate clinical, immunologic, and virologic HIV disease progression, and consequently increase maternal morbidity and risk of HIV transmission [3, 27]. Further, HIV infection itself may affect nutrient absorption and contribute to the development of micronutrient deficiencies and wasting, thus perpetuating a vicious cycle [30].

23.3.1 Observational Studies

Observational studies of the relationship between micronutrient status and pregnancy outcomes in HIV-positive populations have been predominantly focused on the role of vitamin A. In studies in sub-Saharan Africa, low maternal vitamin A status was associated with an increased risk of adverse pregnancy outcomes. In studies in Malawi, Semba and colleagues demonstrated that low maternal plasma vitamin A levels were associated with an increased risk of low birth weight (<2,500 grams), infant mortality, and MTCT of HIV [31, 32]. Low plasma concentrations of vitamin A were also associated with a greater risk of MTCT of HIV in an observational study in Rwanda [33].

Findings regarding vitamin A status and adverse pregnancy outcomes in the United States have been divergent. In a study conducted by Burns et al., maternal plasma vitamin A concentrations were associated with a significant decrease in the risk of low birth weight, and a nonsignificant reduction in the rate of MTCT of HIV [34]. Greenberg and colleagues also identified an increased risk of vertical HIV transmission in the presence of low vitamin A status among HIV-infected pregnant women in two urban areas in the United States [35]. However, in a study by Burger et al., maternal plasma concentrations of vitamin A and beta-carotene were not associated with the risk of MTCT of HIV [36].

The association between other micronutrients and pregnancy outcomes in HIV-infected women were examined in other observational studies in resource-limited settings. In a cross-sectional study of pregnant women in Zimbabwe, HIV infection was associated with reduced concentrations of serum retinol, beta-carotene, folate, ferritin, and hemoglobin [37, 38]. The observed reductions in hemoglobin were particularly profound among HIV-positive women with low serum retinol and nondepleted iron stores (12.9 g/l lower hemoglobin level; 95% CI: 8.9, 16.8); the decrease was less pronounced in women with other combinations of serum ferritin and retinol (7–8 g/l lower hemoglobin levels; P for interaction = 0.038). In an observational study in Haiti, lower serum zinc concentrations were associated with an increased risk of vertical HIV transmission, although this finding was not statistically significant [39].

In an observational study of HIV-infected pregnant women in Tanzania, lower baseline plasma selenium concentrations were associated with a significant increase in risk of mortality over a median follow-up period of 5.7 years [40]. Low maternal selenium status was also associated with a greater risk of fetal death, child mortality, and HIV transmission *via* the intrapartum route. Although low maternal selenium concentrations were associated with a reduced risk of small size for gestational age, no significant relationships were noted between maternal selenium status and infant low birth weight or preterm birth [41].

23.3.1.1 COMMENT

The aforementioned observational studies have a few limitations, which warrant caution while interpreting the results. HIV infection may lead to decreased nutrient absorption and increased excretion, resulting in lower serum concentrations of the micronutrient and an apparent deficiency. Reductions in concentrations of micronutrients such as vitamin A, zinc, and selenium may also be attributable to the acute phase response to infection, rather than being a marker of actual micronutrient status. The observed associations between micronutrient deficiencies and adverse pregnancy outcomes may therefore be due to reverse causation. Further, although the observational studies mentioned above adjusted for some confounders in multivariate analyses, most did not adjust for important potential confounders such as micronutrient supplement use, intake of other dietary nutrients, and presence of lower genital infections. Therefore, residual confounding of the relationship between micronutrient status and pregnancy outcomes by these covariates may lead to biased results.

23.3.2 Randomized Trials

The results and limitations of the observational studies referred to above prompted the undertaking of a number of randomized controlled trials to further investigate the relationship between micronutrient supplementation and pregnancy outcomes in HIV-infected populations. Randomized trials have also chiefly focused on the role of vitamin A status in pregnancy outcomes among HIV-infected women. All of these studies were conducted on antiretroviral-naïve pregnant women.

In a trial in Malawi, 697 HIV-infected pregnant women were randomized to receive daily iron-folate supplementation either alone or in combination with vitamin A (10,000 IU), from 18 to 28 weeks of gestation until delivery. Maternal vitamin A supplementation significantly reduced the occurrence of low birth weight (14 vs. 21.1%; $P = 0.03$) and neonatal anemia (23.4 vs. 40.6%; $P < 0.001$) at six weeks postpartum [42]. However, supplementation with vitamin A did not decrease the risk of other adverse pregnancy outcomes, including vertical HIV transmission, prematurity, fetal death, and infant mortality.

In a trial in South Africa, Coutsoudis et al. examined the impact of vitamin A and beta-carotene supplementation on adverse pregnancy outcomes in HIV-positive women. The investigators randomized 728 HIV-infected pregnant women to receive either vitamin A or placebo. The vitamin A treatment consisted of a daily dose of 5,000 IU vitamin A and 30 mg beta-carotene during the third trimester, and a 200,000-IU dose of vitamin A at delivery. Vitamin A supplementation significantly reduced the occurrence of preterm delivery (11.4 vs. 17.4%; $P = 0.03$); however, vitamin A had no effect on the risks of vertical HIV transmission, low birth weight, small size for gestational age, or fetal death [43].

In a trial in Zimbabwe, Humphrey et al. evaluated the efficacy of vitamin A supplementation in the prevention of adverse pregnancy outcomes in HIV-positive women. Vitamin A treatment consisted of a single postpartum dose of vitamin A administered to women (400,000 IU) and/or infants (50,000 IU) at birth. Administration of the vitamin A regimen to either the mother or infant significantly increased the risks of vertical HIV transmission and infant mortality. However, vitamin A administered to both the mother and infant did not increase the risk of these outcomes, compared to the placebo. Additionally, all three vitamin A regimens resulted in a two-fold increase in the risk of mortality among infants who were HIV-negative at six weeks postpartum ($P \leq 0.05$) [44].

The Trial of Vitamins (TOV) study was conducted in Tanzania in order to investigate the role of micronutrient status in perinatal health outcomes among HIV-infected women and their children. Investigators enrolled 1,078 HIV-infected pregnant women in a 2×2 factorial study design, and randomized participants to receive vitamin A alone, vitamin A and multivitamins, multivitamins alone, or placebo. Vitamin A treatment consisted of 30 mg beta-carotene and 5,000 IU of preformed vitamin A; multivitamin supplementation included vitamins B-complex, C, and E in doses that were six to 10 times the RDA levels. Multivitamin supplements significantly decreased the risks of severe preterm birth (<34 weeks gestation) by 39%, low birth weight (<2,500 g) by 44%, small size for gestational age by 43%, and fetal death by 39% [45]. Although vitamin A supplementation had no effect on these pregnancy outcomes, it significantly increased the risk of vertical HIV transmission by 38% [46].

Multivitamin supplementation demonstrated a protective effect against vertical HIV transmission through breastfeeding in a subgroup of women who were nutritionally and/or immunologically compromised at baseline [46]. Multivitamin supplementation also significantly increased prenatal weight gain, maternal CD4 and CD8 cell counts, hemoglobin levels [47–49], and placental weight [47, 48], whereas vitamin A had no effect on these outcomes [47, 48].

Prenatal multivitamin supplementation (including vitamins B-complex, C, and E) also significantly reduced the risk of developing hypertension during pregnancy (RR: 0.62; 95% CI: 0.40, 0.94); vitamin A supplementation had no effect on the risk of hypertension [50]. Further, among infants born to HIV-infected mothers in Tanzania, maternal multivitamin supplementation resulted in a significant improvement in psychomotor child development and a decrease in the risk of developmental delay, whereas vitamin A supplementation had no effect on these outcomes [51].

The effect of zinc supplementation on pregnancy outcomes was recently examined among HIV-positive women in a trial in Tanzania. Investigators enrolled 400 HIV-infected pregnant women, and randomized participants to receive either daily zinc supplementation (25 mg) or placebo until six weeks postpartum. Intervention groups did not significantly differ on levels of viral load, or CD3, CD4, or CD8 cell counts; and zinc supplementation had no effect on the risks of low birth weight, MTCT of HIV, preterm delivery, fetal death, or neonatal mortality. However, maternal zinc supplementation resulted in a significant three-fold increase in the risk of wasting (mid–upper arm circumference [MUAC] < 22 cm) during an average 22-week follow-up period (RR: 2.7; 95% CI: 1.1, 6.4; $P = 0.03$), with an average 4-mm mean reduction in MUAC during the second trimester ($P = 0.02$). Although hemoglobin concentrations increased in both groups, this effect was blunted in the zinc-supplemented group. This finding is likely attributable to zinc's interference with iron absorption [52, 53].

The supplementation trials referred to above have been summarized in Table 23.1. The role of other micronutrients in pregnancy outcomes has not yet been examined in randomized trials in HIV-infected women; however, a trial examining the impact of selenium supplementation on maternal and child health outcomes in HIV-infected women is currently underway in Tanzania.

Table 23.1
Randomized Trials Examining the Role of Micronutrient Supplements in Pregnancy Outcomes in HIV-Infected Women

Study site	N	Supplementation regimes	Results
Malawi [42]	697 women	Iron and folate alone or with vitamin A daily	• Reduced risk of low birth weight and neonatal anemia • No effects on MTCT of HIV, prematurity, fetal death, or infant mortality
South Africa [43]	728 women	Vitamin A and beta-carotene or placebo	• Lower risk of preterm delivery • No effect on risks of MTCT of HIV, low birth weight, small size for gestational age, or fetal death
Zimbabwe [44]	4,495 infants born to HIV-infected women	Vitamin A to both mother and infant; Vitamin A to mother, placebo to infact; placebo to mother, vitamin A to infant; or placebo to both mother and infant	• Higher risks of MTCT of HIV and infant mortality when vitamin A given to mother alone or to infant alone • No differences in risk of perinatal outcomes were observed when both mother and infant were supplemented • Among infants who were HIV-negative at six weeks of age, all three vitamin A regimes resulted in a two-fold increase in mortality
Tanzania [45–51]	1,078 women	Vitamin A plus multivitamins (vitamins B-complex, C, and E), vitamin A alone, multivitamins alone, or placebo	• 38% increase in risk of MTCT of HIV with vitamin A supplementation • Lower risk of MTCT of HIV with multivitamins only, among women who were nutritionally and/or immunologically compromised at baseline • Lower risk of hypertension during pregnancy, severe preterm birth, infant low birth weight, small size for gestational age, and fetal death with multivitamins • Increased prenatal weight gain, maternal CD4 and CD8 counts, hemoglobin levels, and placental weight with multivitamins • Lower risk of developmental delay and improved psychomotor development in children born to women supplemented with multivitamins
Tanzania [52, 53]	400 women	Zinc supplementation or placebo daily	• Lower increase in hemoglobin in the zinc group • No effect of zinc supplementation on risks of low birth weight, MTCT of HIV, preterm delivery, fetal death, or neonatal mortality • Increased risk of wasting and blunted increase in hemoglobin levels, in women supplemented with zinc

361

23.3.2.1 COMMENT

Findings regarding the impact of vitamin A supplementation on the risk of vertical HIV transmission have been unexpected and contradictory to the original hypothesis. In the trials in Tanzania and Zimbabwe, vitamin A supplementation significantly increased the risk of MTCT of HIV; however, this effect was not observed in studies in Malawi and South Africa. This discrepancy may be attributable to differences in trial study designs, including supplementation dosages, and duration and schedule of administration (e.g., continued micronutrient use during breastfeeding, versus restriction to the antenatal period). Further, since the vitamin A treatment regimens in the Tanzania trial also included beta-carotene, it is possible that beta-carotene supplementation alone may result in an increased risk of transmission. Researchers have also postulated that vitamin A may lead to increased density of CCR5 receptors, *via* increased multiplication and differentiation of lymphoid and myeloid cells; CCR5 receptors are critical for attachment and replication of HIV [54]. Vitamin A may also modulate HIV replication, as the virus genome has been found to contain a retinoic acid receptor element [55].

23.4 CONCLUSIONS AND IMPLICATIONS FOR PRACTICE

Micronutrient deficiencies increase the risk of adverse pregnancy outcomes in HIV-infected women. Multivitamin supplementation (including B-complex, C, and E) has demonstrated a consistent benefit on pregnancy outcomes among HIV-infected women, including a reduced risk of prematurity, low birth weight, HIV transmission *via* breast-feeding, and fetal death. Current epidemiological evidence supports the use of multivitamin supplements to reduce the risk of adverse pregnancy outcomes in HIV-infected women. However, the aforementioned studies were conducted on antiretroviral-naïve pregnant women; it is not evident if the observed effect of multivitamin supplementation on pregnancy outcomes is generalizable to HIV-infected women taking antiretroviral therapy. There is also insufficient evidence regarding the relative benefit of administering single versus multiple RDA levels of micronutrients in prenatal supplements for HIV-infected women. Ongoing randomized trials in Tanzania may provide evidence regarding the generalizability of these findings, inform multivitamin supplementation dosage and administration, and elucidate the role of micronutrients among HIV-infected individuals receiving antiretroviral therapy.

Vitamin A supplementation has not demonstrated a consistent benefit on the risk of adverse pregnancy outcomes in HIV-infected women. The increased risk of vertical HIV transmission following vitamin A use observed in trials in Tanzania and Zimbabwe is particularly disconcerting. Vitamin A supplementation is therefore not recommended for HIV-infected pregnant women, and should be avoided. There is currently no strong epidemiological evidence to support the use of other micronutrient supplements, such as zinc and selenium, to prevent adverse pregnancy outcomes in HIV-positive women. However, a selenium supplementation trial currently underway in Tanzania may elucidate the role of selenium in perinatal health outcomes among HIV-infected women and their children.

Micronutrient supplementation, however, is unlikely to be a stand-alone mantra for success in preventing adverse pregnancy outcomes in HIV-infected women. The importance of ensuring access to appropriate antiretroviral therapy cannot be overemphasized. The role of nutrition in HIV-infected women taking HAART has not been well-established and warrants further investigation. Further, vulnerable groups such as HIV-infected women, particularly in developing countries, are likely to have multiple micronutrient deficiencies. This is additionally complicated by the fact that micronutrients can have either synergistic or antagonistic interactions with regard to biological effects. For example, iron supplements can interfere with zinc absorption [56, 57], and zinc in high doses may reduce the absorption of iron or copper [58, 59]. Further attention should also be focused on complementary dietary approaches such as food fortification and dietary diversification as potential beneficial and sustainable adjuncts to micronutrient supplementation.

REFERENCES

1. Copenhagen Consensus: the results. 2004. Available via http://www.copenhagenconsensus.com/Admin/Public/DWSDownload.aspx?File=Files%2fFiler%2fCC%2fPress%2fUK%2fcopenhagen_consensus_result_FINAL.pdf
2. UNAIDS. AIDS Epidemic Update: 2007. In. Geneva: World Health Organization; 2007.
3. Fawzi W (2003) Micronutrients and human immunodeficiency virus type 1 disease progression among adults and children. Clin Infect Dis 37(Suppl)2:S112–S116
4. Stoltzfus RJ, Dreyfuss ML (1998) Guidelines for the use of iron supplements to prevent and treat iron deficiency anemia. International Nutritional Anemia Consultative Group (INACG)/WHO/UNICEF. World Health Organization, Geneva, Switzerland
5. Mungen E (2003) Iron supplementation in pregnancy. J Perinat Med 31:420–426
6. Cavalli-Sforza T, Berger J, Smitasiri S, Viteri F (2005) Weekly iron-folic acid supplementation of women of reproductive age: impact overview, lessons learned, expansion plans, and contributions toward achievement of the millennium development goals. Nutr Rev 63:S152–S158
7. Viteri FE, Berger J (2005) Importance of pre-pregnancy and pregnancy iron status: can long-term weekly preventive iron and folic acid supplementation achieve desirable and safe status? Nutr Rev 6:S65–S76
8. Allen LH (1997) Pregnancy and iron deficiency: unresolved issues. Nutr Rev 55:91–101
9. Lumley J, Watson L, Watson M, Bower C (2001) Periconceptional supplementation with folate and/or multivitamins for preventing neural tube defects. Cochrane Database Syst Rev 2001:CD001056
10. Ramakrishnan U, Imhoff-Kunsch B (2007) Anemia and iron deficiency in developing countries. In: Lammi-Keefe CJ, Couch SS (eds) Handbook of nutrition and pregnancy. Humana, Totowa, N.J. 337–354
11. West KP, Jr., Katz J, Khatry SK, LeClerq SC, Pradhan EK, Shrestha SR, Connor PB, Dali SM, Christian P, Pokhrel RP, Sommer A. Double blind, cluster randomised trial of low dose supplementation with vitamin A or beta carotene on mortality related to pregnancy in Nepal. The NNIPS-2 Study Group. Bmj 1999;318:570–5.
12. Azais-Braesco V, Pascal G (2000) Vitamin A in pregnancy: requirements and safety limits. Am J Clin Nutr 71(Suppl):1325S–33S
13. Food and Nutrition Board (FNB) and Institute of Medicine (IOM) (2000) Dietary Reference Intakes for vitamin A, vitamin K, arsenic, boron, chromium, copper, iodine, iron, manganese, molybdenum, nickel, silicon, vanadium, and zinc. National Academies Press, Washington, D.C.
14. Shah D, Sachdev HP (2001) Effect of gestational zinc deficiency on pregnancy outcomes: summary of observation studies and zinc supplementation trials. Br J Nutr 85 Suppl 2:S101–S108
15. Shah D, Sachdev HP (2006) Zinc deficiency in pregnancy and fetal outcome. Nutr Rev 64:15–30
16. Osendarp SJ, West CE, Black RE (2003) The need for maternal zinc supplementation in developing countries: an unresolved issue. J Nutr 133:817S–827S
17. Osendarp SJ, van Raaij JM, Darmstadt GL, Baqui AH, Hautvast JG, Fuchs GJ (2001) Zinc supplementation during pregnancy and effects on growth and morbidity in low birthweight infants: a randomised placebo controlled trial. Lancet 357:1080–1085
18. Ramakrishnan U, Gonzalez-Cossio T, Neufeld LM, Rivera J, Martorell R (2005) Effect of prenatal multiple micronutrient supplements on maternal weight and skinfold changes: a randomized double-blind clinical trial in Mexico. Food Nutr Bull 26:273–280

19. Ramakrishnan U, Neufeld LM, Gonzalez-Cossio T, Villalpando S, Garcia-Guerra A, Rivera J, Martorell R. Multiple micronutrient supplements during pregnancy do not reduce anemia or improve iron status compared to iron-only supplements in Semirural Mexico. J Nutr 2004;134:898–903

20. Ramakrishnan U, Gonzalez-Cossio T, Neufeld LM, Rivera J, Martorell R (2003) Multiple micronutrient supplementation during pregnancy does not lead to greater infant birth size than does iron-only supplementation: a randomized controlled trial in a semirural community in Mexico. Am J Clin Nutr 77:720–725

21. Christian P, Shrestha J, LeClerq SC, Khatry SK, Jiang T, Wagner T, Katz J, West KP, Jr. Supplementation with micronutrients in addition to iron and folic acid does not further improve the hematologic status of pregnant women in rural Nepal. J Nutr 2003;133:3492–8

22. Christian P, Khatry SK, Katz J, Pradhan EK, LeClerq SC, Shrestha SR, Adhikari RK, Sommer A, West KP, Jr. Effects of alternative maternal micronutrient supplements on low birth weight in rural Nepal: double blind randomised community trial. Bmj 2003;326:571

23. Christian P, West KP, Khatry SK, Leclerq SC, Pradhan EK, Katz J, Shrestha SR, Sommer A. Effects of maternal micronutrient supplementation on fetal loss and infant mortality: a cluster-randomized trial in Nepal. Am J Clin Nutr 2003;78:1194–202

24. Makola D, Ash DM, Tatala SR, Latham MC, Ndossi G, Mehansho H (2003) A micronutrient-fortified beverage prevents iron deficiency, reduces anemia and improves the hemoglobin concentration of pregnant Tanzanian women. J Nutr 133:1339–1346

25. Osrin D, Vaidya A, Shrestha Y, Baniya RB, Manandhar DS, Adhikari RK, Filteau S, Tomkins A, Costello AM. Effects of antenatal multiple micronutrient supplementation on birthweight and gestational duration in Nepal: double-blind, randomised controlled trial. Lancet 2005;365:955–62

26. Fawzi W, Msamanga G, Urassa W, Hertzmark E, Petraro P, Willett W, Spiegelman D. Vitamins and perinatal outcomes among HIV-negative women in Tanzania. N Engl J Med 2007;356:1423–31

27. Dreyfuss ML, Fawzi WW (2002) Micronutrients and vertical transmission of HIV-1. Am J Clin Nutr 75:959–970

28. Kupka R, Fawzi W (2002) Zinc nutrition and HIV infection. Nutr Rev 60:69–79

29. Willumsen JF, Filteau SM, Coutsoudis A, Uebel KE, Newell ML, Tomkins AM (2000) Subclinical mastitis as a risk factor for mother-infant HIV transmission. Adv Exp Med Biol 478:211–223

30. Keusch GT, Farthing MJ (1990) Nutritional aspects of AIDS. Annu Rev Nutr 10:475–501

31. Semba RD, Miotti PG, Chiphangwi JD, Liomba G, Yang LP, Saah AJ, Dallabetta GA, Hoover DR. Infant mortality and maternal vitamin A deficiency during human immunodeficiency virus infection. Clin Infect Dis 1995;21:966–72

32. Semba RD, Miotti PG, Chiphangwi JD, Saah AJ, Canner JK, Dallabetta GA, Hoover DR. Maternal vitamin A deficiency and mother-to-child transmission of HIV-1. Lancet 1994;343:1593–7

33. Graham N, Bulterys M, Chao A, Humphrey J, Clement L, Dushimmana A, Kurawige J, Flynn C, Saah A. Effect of maternal vitamin A deficiency on infant mortality and perinatal HIV transmission. In: National Conference on Human Retroviruses and Related Infections; 1993; Baltimore: Johns Hopkins University; 1993

34. Burns DN, FitzGerald G, Semba R, Hershow R, Zorrilla C, Pitt J, Hammill H, Cooper ER, Fowler MG, Landesman S. Vitamin A deficiency and other nutritional indices during pregnancy in human immunodeficiency virus infection: prevalence, clinical correlates, and outcome. Women and Infants Transmission Study Group. Clin Infect Dis 1999;29:328–34

35. Greenberg BL, Semba RD, Vink PE, Farley JJ, Sivapalasingam M, Steketee RW, Thea DM, Schoenbaum EE. Vitamin A deficiency and maternal-infant transmissions of HIV in two metropolitan areas in the United States. Aids 1997;11:325–32

36. Burger H, Kovacs A, Weiser B, Grimson R, Nachman S, Tropper P, van Bennekum AM, Elie MC, Blaner WS. Maternal serum vitamin A levels are not associated with mother-to-child transmission of HIV-1 in the United States. J Acquir Immune Defic Syndr Hum Retrovirol 1997;14:321–6

37. Friis H, Gomo E, Koestel P, Ndhlovu P, Nyazema N, Krarup H, Michaelsen KF. HIV and other predictors of serum beta-carotene and retinol in pregnancy: a cross-sectional study in Zimbabwe. Am J Clin Nutr 2001;73:1058–65

38. Friis H, Gomo E, Koestel P, Ndhlovu P, Nyazema N, Krarup H, Michaelsen KF. HIV and other predictors of serum folate, serum ferritin, and hemoglobin in pregnancy: a cross-sectional study in Zimbabwe. Am J Clin Nutr 2001;73:1066–73

39. Ruff A. Zinc deficiency and transmission and progression of HIV infection. In: Kelley L, Black RE, eds. Zinc for child health. Page 10. Proceedings of the conference, 1996, November 18-20. Baltimore, MD: Child Health Research Project (USAID), 1997

40. Kupka R, Msamanga GI, Spiegelman D, Morris S, Mugusi F, Hunter DJ, Fawzi WW. Selenium status is associated with accelerated HIV disease progression among HIV-1-infected pregnant women in Tanzania. J Nutr 2004;134:2556–60

41. Kupka R, Garland M, Msamanga G, Spiegelman D, Hunter D, Fawzi W (2005) Selenium status, pregnancy outcomes, and mother-to-child transmission of HIV-1. J Acquir Immune Defic Syndr 39:203–210

42. Kumwenda N, Miotti PG, Taha TE, Broadhead R, Biggar RJ, Jackson JB, Melikian G, Semba RD. Antenatal vitamin A supplementation increases birth weight and decreases anemia among infants born to human immunodeficiency virus-infected women in Malawi. Clin Infect Dis 2002;35:618–24

43. Coutsoudis A, Pillay K, Spooner E, Kuhn L, Coovadia HM (1999) Randomized trial testing the effect of vitamin A supplementation on pregnancy outcomes and early mother-to-child HIV-1 transmission in Durban, South Africa. South African Vitamin A Study Group. Aids 13:1517–1524

44. Humphrey JH, Iliff PJ, Marinda ET, Mutasa K, Moulton LH, Chidawanyika H, Ward BJ, Nathoo KJ, Malaba LC, Zijenah LS, Zvandasara P, Ntozini R, Mzengeza F, Mahomva AI, Ruff AJ, Mbizvo MT, Zunguza CD. Effects of a Single Large Dose of Vitamin A, Given during the Postpartum Period to HIV-Positive Women and Their Infants, on Child HIV Infection, HIV-Free Survival, and Mortality. J Infect Dis 2006;193:860–71

45. Fawzi WW, Msamanga GI, Spiegelman D, Urassa EJ, McGrath N, Mwakagile D, Antelman G, Mbise R, Herrera G, Kapiga S, Willett W, Hunter DJ. Randomised trial of effects of vitamin supplements on pregnancy outcomes and T cell counts in HIV-1-infected women in Tanzania. Lancet 1998;351:1477–82

46. Fawzi WW, Msamanga GI, Hunter D, Renjifo B, Antelman G, Bang H, Manji K, Kapiga S, Mwakagile D, Essex M, Spiegelman D. Randomized trial of vitamin supplements in relation to transmission of HIV-1 through breastfeeding and early child mortality. Aids 2002;16:1935–44

47. Villamor E, Msamanga G, Spiegelman D, Antelman G, Peterson KE, Hunter DJ, Fawzi WW. Effect of multivitamin and vitamin A supplements on weight gain during pregnancy among HIV-1-infected women. Am J Clin Nutr 2002;76:1082–90

48. Fawzi WW, Msamanga GI, Spiegelman D, Wei R, Kapiga S, Villamor E, Mwakagile D, Mugusi F, Hertzmark E, Essex M, Hunter DJ. A randomized trial of multivitamin supplements and HIV disease progression and mortality. N Engl J Med 2004;351:23–32

49. Fawzi WW, Msamanga GI, Kupka R, Spiegelman D, Villamor E, Mugusi F, Wei R, Hunter D. Multivitamin supplementation improves hematologic status in HIV-infected women and their children in Tanzania. Am J Clin Nutr. 2007 May;85(5):1335–43

50. Merchant AT, Msamanga G, Villamor E, Saathoff E, O'Brien M, Hertzmark E, Hunter DJ, Fawzi WW. Multivitamin supplementation of HIV-positive women during pregnancy reduces hypertension. J Nutr 2005;135:1776–81

51. McGrath N, Bellinger D, Robins J, Msamanga GI, Tronick E, Fawzi WW (2006) Effect of maternal multivitamin supplementation on the mental and psychomotor development of children who are born to HIV-1-infected mothers in Tanzania. Pediatrics 117:e216–e225

52. Fawzi WW, Villamor E, Msamanga GI, Antelman G, Aboud S, Urassa W, Hunter D. Trial of zinc supplements in relation to pregnancy outcomes, hematologic indicators, and T cell counts among HIV-1-infected women in Tanzania. Am J Clin Nutr 2005;81:161–7

53. Villamor E, Aboud S, Koulinska IN, Kupka R, Urassa W, Chaplin B, Msamanga G, Fawzi WW. Zinc supplementation to HIV-1-infected pregnant women: Effects on maternal anthropometry, viral load, and early mother-to-child transmission. Eur J Clin Nutr 2006

54. MacDonald KS, Malonza I, Chen DK, Nagelkerke NJ, Nasio JM, Ndinya-Achola J, Bwayo JJ, Sitar DS, Aoki FY, Plummer FA. Vitamin A and risk of HIV-1 seroconversion among Kenyan men with genital ulcers. Aids 2001;15:635–9

55. Semba RD, Lyles CM, Margolick JB, Caiaffa WT, Farzadegan H, Cohn S, Vlahov D. Vitamin A supplementation and human immunodeficiency virus load in injection drug users. J Infect Dis 1998;177:611–6.

56. Hambidge KM, Krebs NF, Sibley L, English J (1987) Acute effects of iron therapy on zinc status during pregnancy. Obstet Gynecol 70:593–596

57. Solomons NW (1986) Competitive interaction of iron and zinc in the diet: consequences for human nutrition. J Nutr 116:927–935

58. Festa MD, Anderson HL, Dowdy RP, Ellersieck MR (1985) Effect of zinc intake on copper excretion and retention in men. Am J Clin Nutr 41:285–292

59. Porter KG, McMaster D, Elmes ME, Love AH (1977) Anaemia and low serum-copper during zinc therapy. Lancet 2:774

Index

About the Editors

Dr. Carol J. Lammi-Keefe is Alma Beth Clark Professor of Nutrition and Division Head, Human Nutrition and Food, School of Human Ecology at Louisiana State University; Adjunct Professor at Pennington Biomedical Research Center, Baton Rouge; and Professor Emeritus at the University of Connecticut. She has devoted over two decades to research in maternal and fetal nutrition, with an emphasis on lipids, especially *n*-3 fatty acids. She has been recognized by the American Dietetic Association Foundation, with the Ross Award in Women's Health and the Award for Excellence in Dietetic Research. Her research findings have been published in journals such as the *American Journal of Clinical Nutrition*, the *Journal of the American Dietetic Association*, *Lipids*, and the *Journal of Pediatric Gastroenterology and Nutrition*. Recent research findings include the benefit of docosahexaenoic acid during pregnancy on infant functional outcomes, including visual acuity and problem solving. As principal or co-investigator, research support has come from the U.S. Department of Agriculture, the National Institutes of Health and various other foundations, institutes, or industry.

Dr. Sarah C. Couch is an associate professor in the Department of Nutritional Sciences, College of Allied Health Sciences at the University of Cincinnati. Dr. Couch received her master's degree and Ph.D. from the University of Connecticut and was a research associate at Columbia University in the Department of Pediatrics prior to her appointment at the University of Cincinnati. Dr. Couch's research focuses on lipid alterations and nutrition-related risk factors in the prevention and treatment of cardiovascular disease. She has numerous publications in nationally recognized journals including the *American Journal of Clinical Nutrition*, the *Journal of the American Dietetic Association*, *Lipids*, and the *Journal of Pediatrics*. She has been principal and co-investigator on external grants received to support her research, including grants from the National Institutes of Health, the American Heart Association, and the American Dietetic Association. Dr. Couch's most recent funded research study focuses on examining dietary patterns that alter blood pressure in children and teenagers with hypertension. Dr. Couch serves on the Board of Editors for the *Journal of the American Dietetic Association* and the *Journal of Hunger* and *Environmental Nutrition*.

Dr. Elliot H. Philipson is Vice-Chairman of the Department of Obstetrics and Gynecology and Section Head of Maternal–Fetal Medicine at the Cleveland Clinic, Cleveland, Ohio. He is also a clinical professor of Obstetrics and Gynecology at the Cleveland Clinic Lerner College of Medicine, and the site director for the Obstetrics and Gynecology residency program at the Cleveland Clinic. After medical school in Rome, Italy, he did his obstetrical and gynecology residency at the Albany Medical Center, Albany, New York, and a maternal–fetal medicine fellowship at Metrohealth Medical Center in Cleveland, Ohio. His research interests have been in gestational diabetes and most recently, in several aspects of maternal/neonatal infection, ultrasound, and multiple pregnancies. He has more than 50 publications and has always been interested in perinatal nutrition and the importance of maternal diet and its caloric components.

About the Series Editor

Dr. Adrianne Bendich is Clinical Director of Calcium Research at GlaxoSmithKline Consumer Healthcare, where she is responsible for leading the innovation and medical programs in support of several leading consumer brands including TUMS and Os-Cal. Dr. Bendich has primary responsibility for the coordination of GSK's support for the Women's Health Initiative (WHI) intervention study. Prior to joining GlaxoSmithKline, Dr. Bendich was at Roche Vitamins Inc., and was involved with the groundbreaking clinical studies proving that folic acid-containing multivitamins significantly reduce major classes of birth defects. Dr. Bendich has co-authored more than 100 major clinical research studies in the area of preventive nutrition. Dr. Bendich is recognized as a leading authority on antioxidants, nutrition and bone health, immunity, and pregnancy outcomes, vitamin safety, and the cost-effectiveness of vitamin/mineral supplementation.

In addition to serving as Series Editor for Humana Press and initiating the development of the 20 currently published books in the *Nutrition and Health™* series, Dr. Bendich is the editor of 11 books, including *Preventive Nutrition: The Comprehensive Guide for Health Professionals.* She also serves as Associate Editor for *Nutrition: The International Journal of Applied and Basic Nutritional Sciences,* and Dr. Bendich is on the Editorial Board of the *Journal of Women's Health and Gender-Based Medicine,* as well as a past member of the Board of Directors of the American College of Nutrition. Dr. Bendich also serves on the Program Advisory Committee for Helen Keller International.

Dr. Bendich was the recipient of the Roche Research Award, was a Tribute to Women and Industry Awardee, and a recipient of the Burroughs Wellcome Visiting Professorship in Basic Medical Sciences, 2000–2001. Dr. Bendich holds academic appointments as Adjunct Professor in the Department of Preventive Medicine and Community Health at UMDNJ, Institute of Nutrition, Columbia University P&S, and Adjunct Research Professor, Rutgers University, Newark Campus. She is listed in *Who's Who in American Women.*

Printed in the United States of America

Portland Community College